AF849767

Industrial Decarbonization

Industrial Decarbonization: Materials, Methods, and Developments offers a timely and in-depth exploration of the technologies and strategies reshaping high-emission industries in the global push toward carbon neutrality. As sectors such as chemical, steel, cement, concrete, asphalt, and construction face mounting pressure to reduce their environmental impact, this book provides a comprehensive, multidisciplinary perspective on the innovations driving industrial transformation.

Drawing from cutting-edge research and real-world implementation across academia, national labs, and industry, the book highlights both the challenges and scalable solutions in industrial decarbonization. Key features include:

- Innovative Technologies: Investigates emerging thermo-, electro-, and photocatalytic processes, microwave-assisted techniques, and energy-efficient manufacturing systems aimed at reducing industrial emissions.
- Analytical Frameworks: Introduces life cycle assessment (LCA) and techno-economic analysis (TEA) tools to assess the viability and environmental impact of new decarbonization approaches.
- Carbon Management: Explores advanced materials and integrated system designs for carbon capture, utilization, and storage (CCUS), including novel separation and process integration technologies.
- Sector-Specific Solutions: Showcases targeted innovations in cement, concrete, steel, and chemical production, alongside cross-cutting strategies such as circular economy models and thermal energy storage.
- Digital Tools: Discusses the role of computational modeling and data-driven methods in optimizing decarbonization pathways.

With a rich collection of case studies and practical examples, this book is an essential resource for researchers, engineers, and policymakers working at the intersection of materials science, industrial engineering, and sustainable development.

Qingxu (Bill) Jin is an assistant professor in the Department of Civil and Environmental Engineering at Michigan State University (MSU) and the principal investigator of the Resilient, Intelligent, Sustainable and Energy-efficient (RISE) Infrastructure Materials Lab at MSU. His research focuses on the decarbonization of construction materials and the development of advanced infrastructure materials with enhanced durability, adaptability, and multifunctionality, particularly for extreme environments. Dr. Jin earned his Ph.D. in Civil and Environmental Engineering from the Georgia Institute of Technology, with a minor in Materials Science. He also holds dual M.Sc. degrees in Civil Engineering and Natural Resources & Environment from the University of Michigan, Ann Arbor. He previously served as a guest researcher at the National Institute of Standards and Technology (NIST). A passionate advocate for STEM education, Dr. Jin is an American Society of Civil Engineers (ASCE) Excellence in Engineering Education (ExCEEd) Fellow, a STEM Ambassador for MSU, and a National Science Foundation (NSF) CMMI Game Changer Academies (C-GCA) Panel Fellow. He is a licensed Professional Engineer (PE) in Michigan and Maryland, and a member of the honor societies, Sigma Xi and Chi Epsilon.

Xue (Ida) Chen is a Global R&D Fellow in the Dow Chemical Company, specializing in energy transition and sustainability research. Dr. Chen has successfully commercialized over 20 innovative and sustainable products aimed at decarbonization across various markets. She currently serves on the industrial technical advisory board for several Department of Energy (DOE) and Department of Transportation (DOT)-funded programs, focusing on decarbonization in the chemical industry and advanced electricity heating technologies. Dr. Chen currently serves as Chair of the Sustainability Engineering Forum (SEF) within the American Institute of Chemical Engineers (AIChE), a member of the Committee of Science of the American Chemical Society (ACS), and a leadership team member of the American Association for the Advancement of Science (AAAS) Steering Committee. Dr. Chen has published more than 30 peer-reviewed journal papers and has filed over 700 global patent applications, with approximately 190 granted patents. She has delivered numerous technical presentations and invited talks to both industry and academia. Dr. Chen has received more than ten prestigious national and international individual awards from organizations such as the American Chemical Society, American Institute of Chemical Engineers, Society of Asian Scientists and Engineers, Society of Global Chinese Chemical Engineers, and Society of Women Engineers. Her patented decarbonization technologies have won more than ten global technology awards, including the R&D 100 Award, Edison Award, and Business Intelligence Group Award. Dr. Chen earned her Ph.D. in Chemical Engineering from the University of Michigan, USA., and her Bachelor's degree in Chemical Engineering from Tianjin University, China.

Fan Shi is a Senior Scientist at the National Energy Technology Laboratory (NETL) under the U.S. Department of Energy in Pittsburgh, USA. Prior to joining NETL, he served as a Technical Fellow at Leidos. Dr. Shi has edited two books and authored or co-authored over 40 peer-reviewed articles focused on the recovery of critical metals, energy conversion, and decarbonization technologies. He is also an inventor of ten patents. His research projects earned him two R&D 100 Awards in 2020 and 2021, and a Bronze Edison Award in 2022. Dr. Shi has established academic and industrial connections globally in the field of industrial decarbonization, facilitating his work on related publications. He holds a Ph.D. in chemical engineering from the University of Pittsburgh and is a member of the American Chemical Society (ACS), AIChE, and the North American Catalysis Society.

Dushyant Shekhawat is the team supervisor for the Reaction Engineering team at the NETL, U.S. Department of Energy, Morgantown, USA. His research focuses on fuel processing, reaction engineering, plasma and microwave-assisted catalysis, and energy systems. He co-developed a pyrochlore-based reforming catalyst, licensed to Pyrochem Catalyst Company, and won the Federal Laboratory Consortium 2011 Award for Excellence in Technology Transfer. He has over 200 publications, seven book chapters, three books, and 15 patents. A registered Professional Engineer in West Virginia, Dr. Shekhawat holds a B.S. from the University of Minnesota and a Ph.D. from Michigan State University.

Green Chemistry and Chemical Engineering

Series Editor
Dominic C.Y. Foo
University of Nottingham, Malaysia

Managing Biogas Plants: A Practical Guide
Mario Alejandro Rosato

The Water-Food-Energy Nexus: Processes, Technologies, and Challenges
I. M. Mujtaba, R. Srinivasan, and N. O. Elbashir

Hemicelluloses and Lignin in Biorefineries
Jean-Luc Wertz, Magali Deleu, Séverine Coppée, and Aurore Richel

Materials in Biology and Medicine
Sunggyu Lee and David Henthorn

Resource Recovery to Approach Zero Municipal Waste
Mohammad J. Taherzadeh and Tobias Richards

Hydrogen Safety
Fotis Rigas and Paul Amyotte

Nuclear Hydrogen Production Handbook
Xing L. Yan and Ryutaro Hino

Water Management: Social and Technological Perspectives
Iqbal Mohammed Mujtaba, Thokozani Majozi and Mutiu Kolade Amosa

Efficiency and Sustainability in the Energy and Chemical Industries: Scientific Principles and Case Studies, Third Edition
Krishnan Sankaranarayanan

Process Integration for Resource Conservation, Second Edition
Dominic C.Y. Foo

Industrial Decarbonization: Materials, Methods, and Developments
Qingxu (Bill) Jin, Xue (Ida) Chen, Fan Shi, and Dushyant Shekhawat

For more information about this series, please visit:
https://www.routledge.com/Green-Chemistry-and-Chemical-Engineering/book-series/CRCGRECHECHE

Industrial Decarbonization

Materials, Methods, and Developments

Edited by
Qingxu (Bill) Jin
Xue (Ida) Chen
Fan Shi
Dushyant Shekhawat

CRC Press
Taylor & Francis Group
Boca Raton London New York

CRC Press is an imprint of the
Taylor & Francis Group, an **informa** business

Designed cover image: © 2026 Shutterstock

First edition published 2026
by CRC Press
2385 NW Executive Center Drive, Suite 320, Boca Raton FL 33431

and by CRC Press
4 Park Square, Milton Park, Abingdon, Oxon, OX14 4RN

CRC Press is an imprint of Taylor & Francis Group, LLC

© 2026 selection and editorial matter, Qingxu (Bill) Jin, Xue (Ida) Chen, Fan Shi, and Dushyant Shekhawat; individual chapters, the contributors

Reasonable efforts have been made to publish reliable data and information, but the author and publisher cannot assume responsibility for the validity of all materials or the consequences of their use. The authors and publishers have attempted to trace the copyright holders of all material reproduced in this publication and apologize to copyright holders if permission to publish in this form has not been obtained. If any copyright material has not been acknowledged please write and let us know so we may rectify in any future reprint.

Except as permitted under U.S. Copyright Law, no part of this book may be reprinted, reproduced, transmitted, or utilized in any form by any electronic, mechanical, or other means, now known or hereafter invented, including photocopying, microfilming, and recording, or in any information storage or retrieval system, without written permission from the publishers.

For permission to photocopy or use material electronically from this work, access www.copyright.com or contact the Copyright

Clearance Center, Inc. (CCC), 222 Rosewood Drive, Danvers, MA 01923, 978-750-8400. For works that are not available on CCC please contact mpkbookspermissions@tandf.co.uk

Trademark notice: Product or corporate names may be trademarks or registered trademarks and are used only for identification and explanation without intent to infringe.

ISBN: 978-1-032-82179-5 (hbk)
ISBN: 978-1-041-10513-8 (pbk)
ISBN: 978-1-003-65544-2 (ebk)

DOI: 10.1201/9781003655442

Typeset in Times
by SPi Technologies India Pvt Ltd (Straive)

Contents

PART III Decarbonization of the Iron and Steel Industry

PART IV Strategies for Decarbonizing Construction Materials

PART V Strategies and Innovations for a Decarbonized Future

Preface

As industries worldwide confront the urgent need for sustainability, decarbonization has become a critical priority. *Industrial Decarbonization: Materials, Methods, and Developments* explores the technologies, strategies, and innovations transforming key industrial sectors toward a low-carbon future. This book offers both theoretical and practical approaches to the decarbonization process, addressing the challenges and opportunities industries encounter as they strive to reduce their environmental impact.

Focusing on several energy- and emissions-intensive industries – including chemical, iron and steel, cement and concrete, and other construction-related sectors – this book highlights advancements, case studies, and real-world applications that exemplify the future of industrial decarbonization. These sectors are among the largest contributors to global industrial emissions, yet their products are essential to daily life and serve as critical inputs for numerous industries. Decarbonizing these sectors will not only transform manufacturing processes but also have far-reaching effects on global supply chains, influencing economies worldwide.

The book begins with *Chapter 1: Industrial Decarbonization: An Overview*, which introduces the importance of reducing industrial emissions and presents various decarbonization strategies. *Chapter 2: Toward a Greener Industry: Decision-Making and Optimization Strategies* then examines the decision-making tools that guide industrial transformation, utilizing multi-criteria decision analysis and real options analysis to balance economic, environmental, and social goals.

Subsequent chapters explore decarbonization within specific industries. The chemical industry is examined through emerging technologies such as electrothermal and electrochemical processes (*Chapters 3 and 4*), microwave heating (*Chapter 5*), and industrial-scale energy conversion systems (*Chapter 6*). The iron and steel industry is explored through hydrogen-based processes (*Chapter 7*) and digital technologies for emissions reduction and production optimization (*Chapter 8*). The cement and concrete sectors feature discussions on methods of emissions reduction and material alternatives (*Chapters 9 and 10*). Beyond these industries, the book addresses emissions reduction in asphalt pavements (*Chapter 11*), thermal energy storage for building energy conservation (*Chapter 12*), and the role of the circular economy in the construction sector (*Chapter 13*). Finally, *Chapter 14* highlights various separation technologies that are critical for advancing emissions reduction across industries.

Through research, technological innovations, and practical case studies, this book provides a comprehensive understanding of the evolving landscape of industrial decarbonization. Researchers, industry professionals, and policymakers will find that *Industrial Decarbonization: Materials, Methods, and Developments* offers valuable insights to help accelerate the transition toward lower-emission industrial systems. We hope the insights presented in this book inspire innovation, foster collaboration, and drive meaningful action toward a future with reduced industrial emissions, ultimately contributing to a more resilient global economy.

–Editors: Qingxu (Bill) Jin, Xue (Ida) Chen, Fan Shi, and Dushyant Shekhawat

Contributors

Michael J. Adams
Georgia Institute of Technology
Atlanta, Georgia, USA

Mahdi Ahmed
The University of Sheffield
Sheffield, UK

Saket S. Bhargava
The Dow Chemical Company
Freeport, Texas, USA

Hanna Breunig
Lawrence Berkeley National Laboratory
Berkeley, California, USA

Brandon S. Byers
ETH Zurich
Zurich, Switzerland

Kangyi Cai
Missouri University of Science and Technology
Rolla, Missouri, USA

Xue (Ida) Chen
The Dow Chemical Company
Lake Jackson, Texas, USA

Yuanjun Chen
Northwestern University
Evanston, Illinois, USA

Joan Cordiner
The University of Sheffield
Sheffield, UK

Gao Deng
Missouri University of Science and Technology
Rolla, Missouri, USA

Qi Dong
Purdue University
West Lafayette, Indiana, USA

Jianan (Erick) Huang
Northwestern University
Evanston, Illinois, USA

Yuyan Huang
Missouri University of Science and Technology
Rolla, Missouri, USA

Muhammad Haris Javed
Missouri University of Science and Technology
Rolla, Missouri, USA

Tuo Ji
Nanjing Tech University
Nanjing, China

Qingxu (Bill) Jin
Michigan State University
East Lansing, Michigan, USA

Zain Karsan
ETH Zurich
Zurich, Switzerland

Wenyu Liao
Missouri University of Science and Technology
Rolla, Missouri, USA

Junqiang Liu
The Dow Chemical Company
Freeport, Texas, USA

Luke Lombardi
Buro Happold
Los Angeles, California, USA

Xiaohua Lu
Nanjing Tech University
Nanjing, China

Hongyan Ma
Missouri University of Science and Technology
Rolla, Missouri, USA

Heidelberg Materials
Irving, Texas, USA

Aditya Menon
Northwestern University
Evanston, Illinois, USA

Akanksha K. Menon
Georgia Institute of Technology
Atlanta, Georgia, USA

Sepehr Mohammadi
Michigan Technological University
Houghton, Michigan, USA

Priyam Mondal
Purdue University
West Lafayette, Indiana, USA

Tyamo Okosun
Purdue University Northwest
Hammond, Indiana, USA

Xinkang Peng
Nanjing Tech University
Nanjing, China

Joseph B. Powell
University of Houston
Houston, Texas, USA

Fabian Rosner
University of California Los Angeles
Los Angeles, California, USA

Abhirup Sen
Purdue University
West Lafayette, Indiana, USA

Armin Silaen
Purdue University Northwest
Hammond, Indiana, USA

Cong Tian
Northwestern University
Evanston, Illinois, USA

Orlando J. Ugarte
Purdue University Northwest
Hammond, Indiana, USA

Jason Woods
National Renewable Energy Laboratory
Golden, Colorado, USA

Rong Xia
Northwestern University
Evanston, Illinois, USA

Zhanping You
Michigan Technological University
Houghton, Michigan, USA

Nadia Zaleski
Georgia Institute of Technology
Atlanta, Georgia, USA

Nannan Zhang
Missouri University of Science and Technology
Rolla, Missouri, USA

Chenn Q. Zhou
Purdue University Northwest
Hammond, Indiana, USA

Part I

Global Industrial Decarbonization

1 Industrial Decarbonization
An Overview

Joseph B. Powell

1.1 INTRODUCTION

The concept of "Industry" emerged in the 16th century to describe systematic labor to produce goods and services of value.[1] The "industrial" revolution of the early 1800s was based on the discovery of coal (Figure 1.1), which transformed human life from a subsistence lifestyle devoted 24/7 to obtaining food and shelter, to our modern economy. Humans throughout their existence had previously used biomass as the only known primary energy source. Coal, the first widely used fossil fuel, provided the energy needed to heat and power the machines required to free humankind from manual labor.

Coal was subsequently augmented by petroleum (oil) and natural gas, which are fossil fuels produced over geologic time from sunlight and photosynthetic plants and organisms to provide highly concentrated, storable energy sources. Today, more than 80% of global energy is supplied by fossil fuels.[3] Release of fossil or prehistoric carbon back into the atmosphere as CO_2 is implicated in climate change or global warming, and hence we seek to either capture and sequester the released fossil CO_2, or move to alternative clean (renewable) energy sources. "Decarbonization" is used to describe this goal, which specifically entails a reduction in fossil or anthropogenic greenhouse gas emissions to the atmosphere. Here, we describe the challenge of continuing to enjoy the conveniences of modern life provided by "industry," while reducing prehistoric fossil carbon emissions.

1.2 ENERGY SOURCES AND THE DECARBONIZATION CHALLENGE

Coal provided the substantial increase in energy density needed to drive the industrial revolution, relative to previous reliance upon biomass (Figure 1.2). As a concentrated form of prehistoric energy, coal's "power density," or the area of land needed for supply, was less than the power and energy demand for the cities it served,[3,10] such that today less than 1% of the population in the developed world is needed for food production.

As shown in Figure 1.1, the initial use of coal was augmented by petroleum, which provided more efficient liquid fuels for the transportation sector, where steam

DOI: 10.1201/9781003655442-2

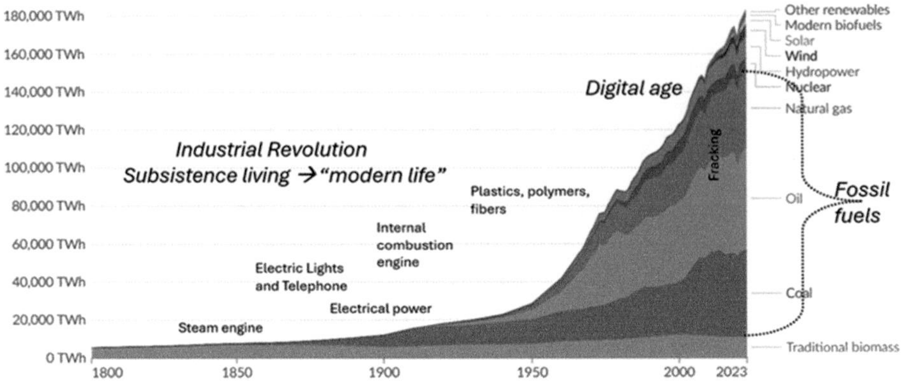

FIGURE 1.1 Global primary energy consumption by source (terawatt-hours).

(Adapted from Ritchie and Rosado, Our World in Data.[2])

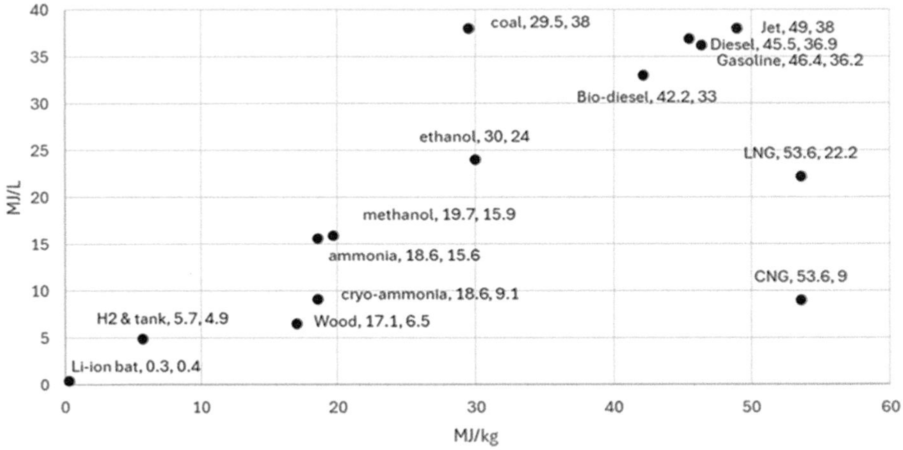

FIGURE 1.2 Energy density of fuels.

(Sources: Compiled from DOE, EIA, others.[4–9])

engines were replaced by internal combustion engines using gasoline, diesel, or jet fuel. Natural gas was then developed as an optimal resource for heating and power applications. Each added fossil fuel provided enhanced efficiencies and convenience in use at a lower cost to consumers for the energy service provided. Growth in demand occurred because of economic advantage. A key challenge for "decarbonization" is how to mitigate the net emission of fossil carbon and other greenhouse gases into the atmosphere, including optimal use of alternatives such as renewable wind and solar power that lack the energy density and "storability" advantages of the energy supply they seek to replace.

1.3 INDUSTRIAL SECTORS

The "Energy Industry" considers the broad aspects of energy supply across the economy, as shown in Sankey diagrams for U.S. Energy, in energy flows from resources through producers to specific end-use sectors.[11] After carving out end-use sectors in transportation, commercial, and residential, the industry end-use sector comprises the remainder. Electrical power generation is allocated to the end-use sectors. With this definition, "industry" was responsible for 30% or 1,360 million metric tons of the total 4,563 MM tons of energy-related CO_2 emissions in 2020, according to the U.S. EIA (Figure 1.3).[13] Energy-related CO_2 emissions include those from the combustion of fossil fuels and process emissions associated with byproducts from petroleum feedstocks. Other greenhouse gases were considered elsewhere, and emissions from flaring of natural gas were not included. The U.S. Industrial Decarbonization Roadmap further broke down these emissions via industry subsectors.[14] By 2022, emissions grew to 4,854 MM tons with transportation increasing to 37%.[15]

Similar analyses are available for global energy,[16,17] though sector definitions differ across many of the reports. The International Panel on Climate Change (IPCC)[18] lists Agriculture (22%), Electricity and Heat Production (23%), Transportation (15%), Buildings (6%), and other (10%) along with Industry (24%) as the end-use sector categories. The U.S. EPA also designates Agriculture as a separate sector at 10% of total, subtracting from residential and commercial, with industry at 30% of U.S. totals.[19] Ritchie (Our World in Data)[20] reports an Energy Sector responsible for 73% of emissions as a broad umbrella including energy generation by industry (24.2%) and CO_2 emissions from industrial processing including cement and chemicals (5.2%), giving a 29.4% share for "industry" of global CO_2 emissions.

"Decarbonization" refers to decreasing or eliminating release of stored fossil carbon to the atmosphere, from the combustion of fossil fuels to provide energy, or as a byproduct of industrial processes using subsurface or fossil carbon as feedstock.

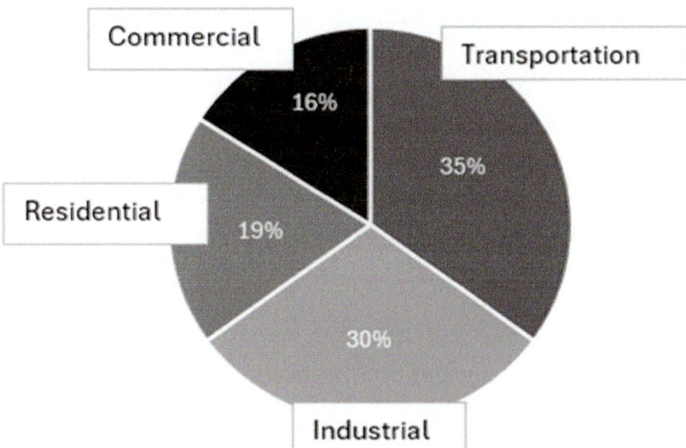

FIGURE 1.3 U.S. CO_2 emissions for end-use sectors 2020.

(Data source: U.S. EIA Annual Energy Outlook.[12])

"Bio-based" resources capture CO_2 from the air and convert it to storage energy via photosynthesis by green plants or microorganisms:

$$CO_2 + H_2O + \text{solar energy} \rightarrow \text{organic hydrocarbons}\left(\text{photosynthesis}\right) \quad (1.1)$$

This capture occurs via annual or perennial plant species, which re-release CO_2 upon decomposition.[21] Biomass is made primarily of fibrous carbohydrate polymers, or units of -C(H_2O)-, which can be processed into edible sugars, as well as inedible lignin comprising repeat units with an approximate formula of or $C_1H_1O_{1/3}$.[22,23] Oxygen content in biomass range from about 30 to 45% by weight, depending on origin and structure.

Biodiesel is made from edible portions of green plants and has oxygen content of an average of 11%.[24] Biodiesel has substantially higher energy density versus bio-ethanol (35% oxygen) but comprises only 12% of the total biofuel supply today, with prospects for growth limited by competition with food supply and utilization of land and water resources.[25] Overall, bio-energy can at best provide 100 exa-joules (EJ) or 20% of the 550 EJ required in 2050 for the International Energy Agency's NetZero scenario, or only about 15% of total energy for projected increase to 750 EJ by 2050 with current global policies for decarbonization.[26] The presence of oxygen is responsible for reduced energy value: oxygenated hydrocarbons can be considered as "partially combusted" fuel relative to the energy content of fossil fuels (Figure 1.2).[1]

One can also synthetically produce hydrocarbon fuels and products synthetically from CO_2 captured from the air via "Direct Air Capture" (DAC), or indirectly from seawater, and use of hydrogen formed from splitting of water using renewable energy using human-made systems[27–30]:

$$H_2O \rightarrow H_2 + \tfrac{1}{2}O_2 \quad dH_R = +237.2\,\text{kJ/mol electricity} + 48.6\,\text{kJ / mol heat} \quad (1.2)$$

followed by a synthetic reaction similar to photosynthesis:

$$CO_2 + H_2 + \text{energy} \rightarrow \text{synthetic natural gas, gasoline, diesel, or jet / marine fuel} \quad (1.3)$$

The resulting "E-fuels" reflect the use of renewable energy for conversion either directly or via generation of hydrogen and require less land use than biomass[31] but are projected to cost more due to the additional energy requirements to reduce CO_2 (72% oxygen) to high-energy-density hydrocarbons. Lower costs may be realized via use of bio-derived feedstocks, which have already been partially reduced by photosynthesis and biomolecular conversion to 11–45% oxygen.

Fossil fuels were formed over hundreds of millions of years starting also with photosynthesis, then via geologic pressure and temperatures acting on biomass derived from prehistoric animals and plants, including carbonization to coal proceeding via dehydration, decarboxylation, and de-methanation reactions.[32,33] These reactions served to concentrate energy and deoxygenate the subsurface hydrocarbons. Crude oil ranges from 0.06 to 0.4 weight percentage (wt%) oxygen.[34–36]

Cost of production of final fuels increases with increasing oxygen content of feedstock, given that energy is required to eliminate chemically bound oxygen. Oil is directly produced and refined with low (less than 10% energy consumption) as a stable, high-energy fuel (38 MJ/L) and can be transformed with minimal energy inputs into final fuel products gasoline, diesel, or jet (aviation) fuel. Early production required less than one barrel of oil energy to process 20 barrels of oil, which today continues at average production ratios of more than 6 barrels of oil per barrel of energy provided.[37] First-generation biodiesel (11% oxygen) is produced at a marginally higher cost but is limited in availability. Conversion of biomass (30–45% oxygen) to advanced biofuel requires more energy and is projected to remain higher cost than today's production from petroleum or natural gas. Biomass has the potential to supply substantially more energy than biodiesel, but sustainable use would limit total production to less than 20% of global energy needs. Production of hydrocarbon E-fuel is not limited by available supply: atmospheric CO_2 and renewable energy are ubiquitous in availability. But E-fuel pathways require higher energy inputs versus biomass or fossil fuel options and are therefore expected to continue to bear higher production costs despite future advancements in technology.[38–41]

1.3.1 Renewable and Other Clean Forms of Energy

As stated above, the Industrial Revolution was powered by fossil fuels. Electricity to date has been produced primarily from the combustion of fossil fuels, such that direct combustion of fuel for heat and process energy, where plausible, was more efficient than the production and use of electricity. Energy generation via fuel combustion to form heat is subject to the inefficiency of the second law of thermodynamics, which for the Carnot (most efficient) cycle heat engine is:

$$\eta = 1 - T_{cold} / T_{hot}$$

Internal combustion engines for vehicles have efficiencies on the order of 30%,[42] while "single cycle" gas turbines for dispatchable power generation average 26% efficiency and, combined cycle with secondary power generation, provides an average of 58% efficiency,[43] with newest models generating efficiencies exceeding 60%, or potentially 80% if considering utilization of water heat.[44]

With the advent of low-cost renewable wind and solar power (Figure 1.4), electrical power can be made directly from the energy source, and not subject to the heat engine inefficiencies related to fuel combustion. Electrochemical storage in batteries also bypasses the second law and can obtain efficiencies greater than 90%. Battery storage is however an order of magnitude (10×) less dense than the energy in hydrocarbon fuels, and hence 24/7 GW-scale industrial applications are challenged in costs.

1.4 INDUSTRIAL SUBSECTORS

To quantify the industrial decarbonization challenge, the U.S. DOE published a U.S. Industrial Decarbonization Roadmap[45] using energy data from 2020, followed by a Liftoff report using data from 2021.[46] The roadmap broke down more segments of the industry, as shown in Figure 1.5, for a total industry footprint of 1,360 million tons of

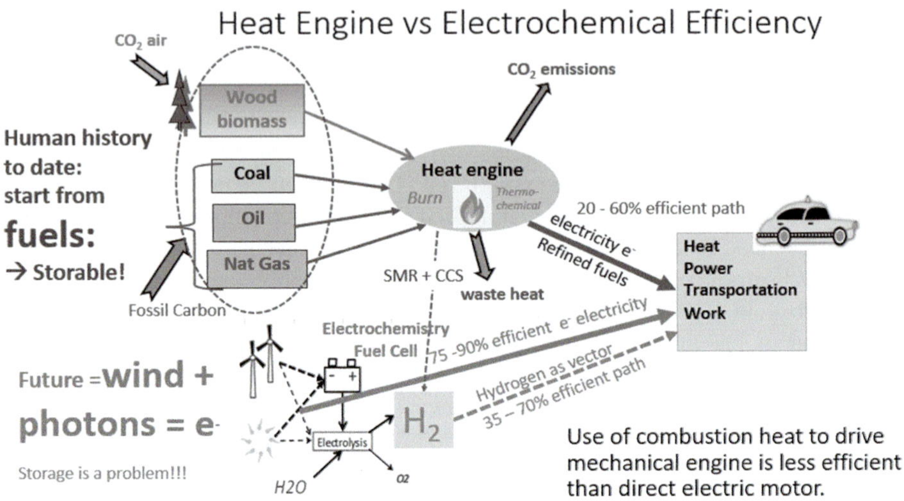

FIGURE 1.4 Direct generation of power from wind and solar.

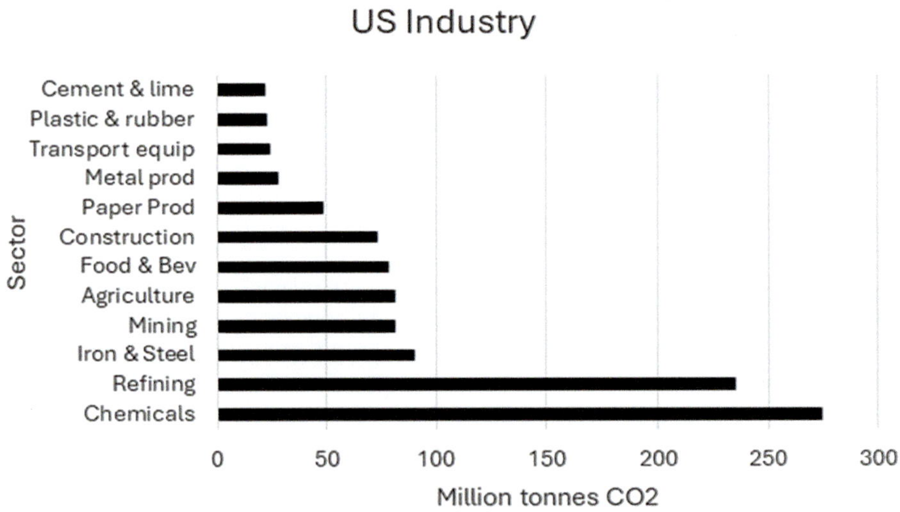

FIGURE 1.5 Million (MM) metric tons CO_2 emissions from U.S. Industry subsectors. **(Data source: US Industrial Decarbonization Roadmap.[47])**

CO_2 emissions. The Liftoff report focused on a smaller subset of eight subsectors of the industry, whose emissions totaled only 14% of the U.S. Economy or 876 million tons of CO_2.

The U.S. Department of Energy roadmap and most corporate commitments focus on "scope 1" and "scope 2" emissions. "Scope 1" refers to direct emissions from fossil fuel purchased and combusted, or that portion of fossil feedstocks lost to CO_2 as an undesired byproduct. "Scope 2" refers to the embedded emissions in the electrical power purchased for these sectors, to run their operations and services. "Decarbonizing" scope 1 and scope 2 emissions in a given sector reduces greenhouse gas emissions from current energy systems and infrastructure in that sector but does not solve the integrated decarbonization problem across the overall economy, which requires consideration of scope 3.

For the Industrial Decarbonization Roadmap, the combination of chemicals and refining contributes more than 37% of U.S. CO_2 scope 1 and 2 emissions (Figure 1.5). Currently, 97% of the scope 3 emissions of the fuels manufactured in the refining sector pass through to provide scope 1 emissions for the transportation sector, which relies primarily on the combustion of fossil fuels.

Typical temperatures required for industrial processing versus subsectors are provided in Figure 1.6, and Table 1.1. Required temperatures vary significantly across industry subsectors, which impact heating methods and fuel requirements. Biomass with its lower energy density cannot reach target temperatures for all industrial applications. Allowed temperature and heat transfer rates can also be problematic for some proposed electrical heating applications, though very high temperatures can be achieved using resistance heating or plasmas for steel production.

Much of the fuel in refinery and petrochemical operations is from byproduct "off-gas," which if substituted by renewable electricity, would result in flaring and no CO_2 savings. Off-gas is currently used as fuel for process heat and power generation, and

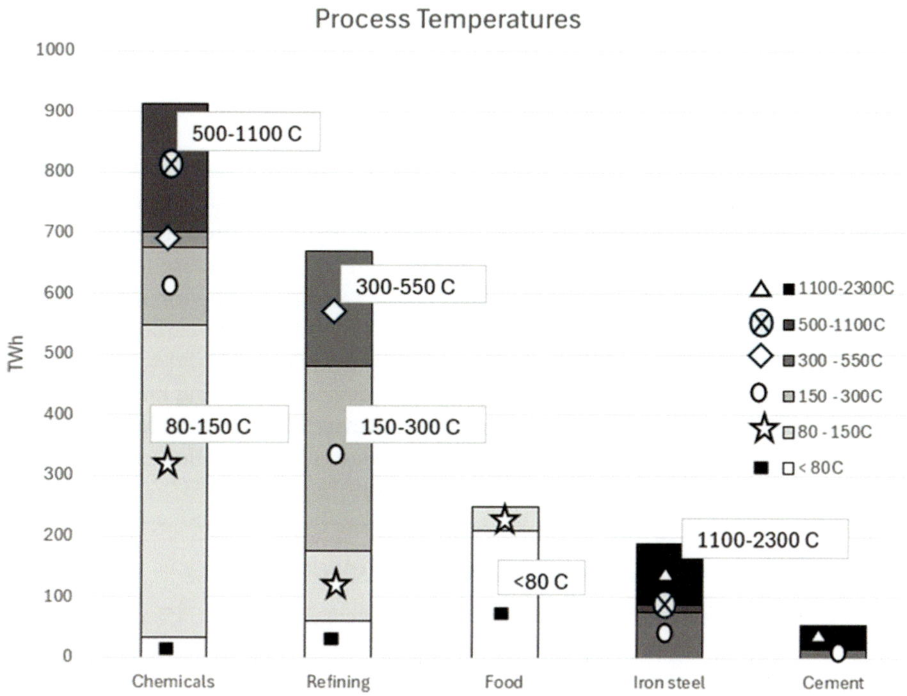

FIGURE 1.6 Temperatures required versus industry subsector.

(Data Source: U.S. Industrial Decarbonization Roadmap.[29])

TABLE 1.1
Heat Sources versus Industrial Application

Industrial Application	°C
Steel blast or electric arc furnace	1,650–1,760
Cement kiln	1,350–1,800 (indirect); 2,300 rotary
Glass	1,600
Ethylene steam cracker	700–950
H2 via steam methane reforming	700–850
Steam turbine (power)	600
Ammonia synthesis	350
Methanol synthesis	250
Pulp and paper	200
Food drying	80–100
Heat Source	**Max Temperature (°C)**
Electric resistance, arc	1,750–2,000; 2,300 kiln
Oil	2,200
Hydrogen	2,100
Coal	2,000
Wood	1,100
Nuclear (advanced)	850

Source: Friedman et al.[48]; Kutscher, Logan, and Coburn.[49]

can be combined with post-combustion CO_2 capture and storage (CCS) as a means for decarbonization. Alternately, the hydrocarbon off-gas fuel can be reformed to H_2 and CO_2 with pre-combustion capture and storage of CO_2, followed by the use of decarbonized H_2 for combustion to provide heat and power. The presence of byproduct fuel gas streams is one reason that CCS is often a preferred option for decarbonizing refining and chemicals compared to electrification with low-carbon power.

Across all subsectors, McKinsey & Company estimates that 50% of industrial heat can be replaced via electrification with efficiency savings, using known technology.[50] Research and development is needed however for electrical heating above 800–1,000 °C, to consider differences in heat transfer characteristics as well as unit design for large, high-voltage industrial applications.[51]

1.4.1 Efficiency of Fuel versus Electrical Power

U.S. energy generation for industry is shown in Table 1.2. Overall, 34% of all primary energy is "applied" to process for end-use, while 66% is waste heat.[53] This reflects the inefficiency of fuel combustion, constrained by the second law of thermodynamics. Direct electrification via renewable or clean power, including generation via high-efficiency combined cycle gas turbines can provide pathways for improved efficiency.[54]

TABLE 1.2
Industrial Energy Efficiency[52,53]

End-Use (TWh)	Feed	Gen Loss	Usage Loss	Amt Used	Eff (η) (%)
Electricity	2,824	1,872	456	496	18
Steam	1,863	765	448	651	35
Direct fuel	1,751	0	694	1,057	60
Total	6,438				

η = Efficiency of fuel usage.

1.4.2 Scope 3 Emissions

An industry's Scope 3 emissions are those of customers using products manufactured by the industry, or "embedded" emissions or those emissions resulting from the supply chains that the industry uses for its feedstock, materials, or purchased fuels. A refinery producing gasoline, diesel, or aviation fuel would have scope 3 emissions from its customers combusting these fuels to provide energy for transportation. Similarly, energy used to produce or transport purchased fuels, chemicals, or materials (steel) is an "embedded" emission for the purchasing industry, which adds to the carbon footprint of its products.

Industrial subsectors vary extensively with the extent of indirect scope 3 emissions, versus scope 1 and 2 emissions, which are directly dependent upon fuel and power choices (Figure 1.7).

Steel and cement are inorganic products that are not combusted at the end of life. Scope 3 emissions are low for its products, relative to the scope 1 and 2 emissions associated with fuel and electricity used in processing. Chemicals and refining

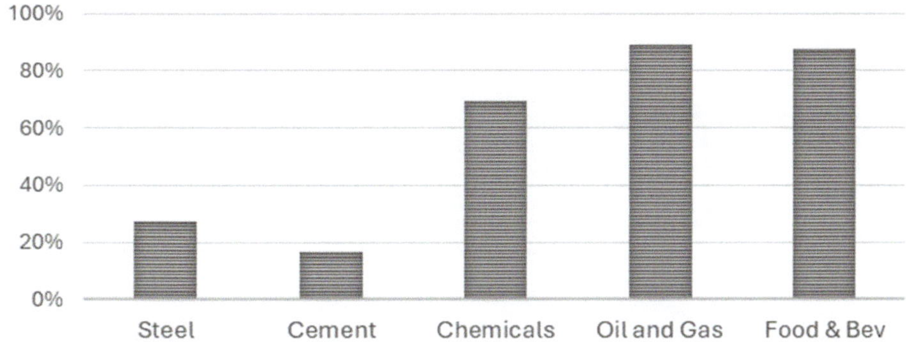

FIGURE 1.7 Scope 3 emissions for industrial subsectors.

(Data source: Ipieca.[55])

subsectors, as well as food and beverage, have substantial scope 3 emissions, shown in the "feedstock" energy versus primary energy used to convert and process feedstock. Refinery products are combusted as fuel, and combustion or degradation of chemical products (plastics) is also significant for end of life in many economies.

Where scope 3 emissions represent a smaller portion of total emissions, an industry subsector can focus on decarbonization of its own scope 1 and 2 emissions, with reduced concern over decarbonization needs for the remainder of the economy. Where an industry subsector has high scope 3 emissions for its products or feedstocks, it must be concerned that the processes and technologies it must decarbonize may require substantial change. As an example, a refinery producing aviation fuel from petroleum can reduce its scope 1 and 2 emissions by substituting low-carbon fuels (hydrogen) and electricity, for current process steps including distillation and hydrocracking. The resulting hydrocarbon jet fuel emits fossil CO_2 when combusted by its scope 3 customers as the largest source of emissions across the supply chain. Hence, alternative processes such as the use of bio-based feedstocks or synthetic manufacture of jet fuel from captured CO_2 and H_2 produced by renewable electrolysis may be the preferred investment.

1.4.3 Circularity

Coupled with the "decarbonization" challenge to address and mitigate climate change, there is an environmental waste problem to address in plastics, and increasingly electronics and batteries. Today, approximately 8% of chemicals and plastics are recycled.[56] In the future the target is 75%, but recycling of polymer and plastic streams is energy intensive and can require high-temperature process heat, 24/7 over and above that required currently with petroleum refining. To decarbonize a future circular economy, industry must consider not the linear economy of today, but a future economy with substantial recycling, as the process technology and capital investment will differ (Figure 1.8).

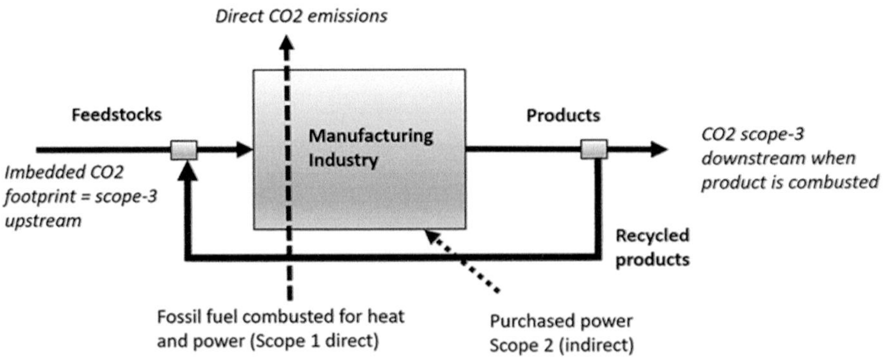

FIGURE 1.8 System boundaries for scope 1, 2, and 3 emissions with material recycle (circularity).

TABLE 1.3
Net Exported CO_2 Emissions from Outsourcing U.S. Manufacturing[57]

MM ton/yr CO_2	Import (% of tot)		Export	Net	% U.S. Ind
China	226	55	19	207	15
Russia	21	5	0	21	2
ASEAN	28	7	0	28	2
Japan & Korea	37	9	0	37	3
India	56	14	0	56	4
EU	42	10	23	19	1
Total	410	100	42	368	27

Note: U.S. million (MM) tons per year CO_2 emissions imported vs. exported via manufactured goods, and percentage of U.S. 1,360 million tons of CO_2 emissions from U.S. DOE's definition of industrial sector scope 1 and scope 2 emissions for operations within the United States.

1.5 GLOBAL SUPPLY CHAINS

The United States has outsourced a substantial portion (27%) of its manufacturing to China and the Far East (ASEAN, Japan, Korea, India) and other lower-cost manufacturing regions of the world, which is partially responsible for the apparent flattening of scope 1 and 2 CO_2 emissions per unit of economic activity (GDP) in the United States (Table 1.3). Outsourced emissions are evident if scope 3 emissions are accounted and must be considered in future balances in goals to achieve total emissions reductions to meet NetZero targets across the economy. The net outsourced emissions are more than seven times the amount of CO_2 captured globally by carbon capture and storage projects completed to date.

1.6 OPTIONS FOR INDUSTRY DECARBONIZATION

Methods and challenges for industry decarbonization are described in the U.S. DOE's Industrial Decarbonization Roadmap, the DOE's Liftoff report, and International Energy Agency's (IEA's) global report on NetZero by 2050, and other third-party publications and reports.[51,58–60] To "decarbonize" or reduce fossil CO_2 emissions from the subset of the energy economy that has been defined as "industry," consider:

- Electrification: use low-carbon electricity to replace fossil fuels for process heat and electrical power.
- Low-carbon fuel feedstocks and energy sources (LCFFES): replace fossil fuels with low-carbon fuels where the fossil carbon has been captured or is not present (renewable, nuclear, or geothermal energy).

- Carbon capture and storage (CCS): capture CO_2 from heating, power generation, or process byproducts, and store it permanently subsurface.
- Energy efficiency: reduce the amount of energy consumption via improvement in efficiency and control, including the use of waste heat.

"Electrification" is the method to consider first for decarbonization, given the new ability to produce low-cost instantaneous power from renewable wind and solar power, or decarbonize a centralized fuel-fired power plant via carbon capture and storage with fossil fuels for ease of energy distribution, or use of biomass gasification or renewable natural gas for power generation. Direct use of renewable power from wind and solar avoids efficiency losses from fuel combustion (2nd law of thermodynamics), while battery storage can provide greater than 90% cycle efficiency.

"Low-carbon fuels" can include hydrogen as a vector, which provides zero carbon emissions during use, and also can avoid second-law inefficiencies if reacted electrochemically via a fuel cell to release energy. Energy is consumed in converting low-carbon electricity to hydrogen via electrolysis of water ("Green Hydrogen), or if hydrocarbons such as natural gas or fuel gases are reformed into H_2 with capture and storage of CO_2 ("Blue Hydrogen"). Hence, hydrogen is a second choice to use where the benefits of energy storage and transport at higher energy density as a molecular fuel are warranted. Hydrogen can be further reacted with biomass or captured CO_2 to form biofuels or synthetic fuels, which again consume more energy for formation but can result in a "drop in" of liquid fuels that can be transported and stored at ambient pressures and temperatures as replacements for today's fossil gasoline, diesel, jet, or marine fuels. Natural gas can also be synthesized via anaerobic fermentation ("renewable") or via reaction with captured CO_2 (synthetic or "E-fuel), for direct replacement of fossil natural gas.

Hydrogen can serve as a low-carbon energy vector or fuel, derived from the electrolysis of water or reforming of natural gas with carbon capture and storage. Efficiency of energy use is about half that of direct use of electricity, but longer-term and higher-capacity storage is plausible relative to the limitations of battery storage. Hydrogen is thus viewed as a low-carbon vector for replacement of natural gas in areas serviced via pipelines, and as a replacement for diesel when used as a compressed gas or liquid for transport applications in fuel cells or combustion engines. Flame speed and combustion characteristics require burner re-design and retrofit, while coating materials are required for the retrofit of existing natural gas pipeline systems to avoid embrittlement or permeation. Longer duration (beyond 4 hours) storage for grid and commercial services can be best achieved via storage in salt domes, technology proven on the U.S. Gulf Coast for more than 50 years. Geologically suitable salt formations are not universally available, however.

Hydrogen can be reacted with N_2 to form ammonia to serve as a moderate-density energy carrier and storage media. Ammonia is currently handled at more than 100 ports via more than 100 dedicated ships for international trade of fertilizer, and locally via farmers. However, it is considered too toxic for use by the general public, such that hydrogen when used as a means for transport and storage of clean energy, is cracked back to H_2 at location for distribution as H2, for most scenarios now considered.

Both electrification and low-carbon fuels can be produced in large, centralized facilities with distribution. Electrification can also use dedicated renewable power (wind and solar) or small modular nuclear reactors. Hydrogen can be produced on-site or locally via electrolysis, especially via the use of low-carbon grid power. Biomass as a feedstock is limited in supply, capable of supplying ca. 20% of global energy needs, is a lower density feedstock compared to petroleum, and has a lower energy density and heating value due to the incorporated water and oxygen ("carbohydrates") and hence is a more expensive fuel than petroleum. Advanced biofuels or synthetic or E-fuels made from CO_2 and H_2 (derived from water via electrolysis) re-create the energy density of petroleum, but at a cost of large energy inputs to overcome feedstock collection costs and conversion to eliminate oxygen. CO_2 is available everywhere via direct air or ocean capture, and together renewable wind and solar power are unlimited in supply, but costs of producing synthetic fuels or chemicals and hydrocarbon materials are higher than refining from petroleum, or even the use of biomass.

Energy efficiency has been a well-worked problem for decades, and many efficiency gains such as cogeneration of heat and power, and LED lighting have already been or are being implemented. New opportunities include increased use of heat pumps to utilize waste or low-valued heat, and use of artificial intelligence (AI) and machine learning to drive a new era of increased process optimization, combined with electrification of process heat to enable more adaptative and targeted application of heat and process energy.

Carbon capture and storage can be retrofitted to the extensive array of existing industrial assets. Given the widely variable temperature, operating conditions, and compositions of industrial process streams, carbon capture and storage (CCS) is considered to be the lowest-cost option for many aspects of industrial decarbonization, especially chemicals and refining.[26]

1.7 CHEMICALS AND REFINING

As described in Figure 1.5, chemicals and refining have the largest scope 1 and 2 emissions of any industrial subsector. Carbon capture and storage is particularly attractive for this sector to enable use of existing capital investments. This includes capturing CO_2 from reforming off-gas or fuel streams to generate hydrogen, which can be used in process furnaces and boilers as a low-carbon fuel. The resulting products have high scope 3 emissions at the end of life, however, and societal concerns are also driving a desire to greatly increase recycling. Enhanced recycling and reduction in scope 3 emissions both lead to very substantial changes in process technology and feedstocks. A key decision for chemicals and refining is therefore whether to decarbonize today's manufacturing infrastructure to reduce scope 1 and 2 emissions, or to change feedstocks and technologies to address scope 3 emissions and recycle.

A move to more sustainable feedstocks (biomass, or CO_2 in the atmosphere at 450 ppm) requires large energy inputs for upgraded performance of fuels and products.[59] Approximately 0.17 barrels of oil are consumed as process energy or equivalent CO_2 emissions (methane) across the complete supply chain for production of refined

TABLE 1.4
Energy Profit Ratio[62] Fossil versus E-Fuel Synthetic Liquid

Energy Profit Ratio	Out/In	Relative
Oil production (mean)	5.6	6.0
E-fuel liquids (max)	0.93	1.0

Source: IEA[63] and NIST[64] thermochemical tables.
Note: Energy addition requirements for power-to-X fuels and chemicals vs. use of petroleum or natural gas.

hydrocarbon fuels and chemical feedstock, such that a mean energy yield of approximately 5.6 is obtained for CO_2 emissions versus energy content for the production of oil and gas. This compares with a maximum net energy output ratio of 0.93 for the formation of a synthetic C6 range gasoline molecule ("E-fuel") from captured CO_2 and hydrogen produced by electrolysis of water. CO_2 and H_2O are fully thermodynamically degraded and require full energy addition to create energy out of these feedstocks for the production of "energy carriers."[61]

Table 1.4 shows details of the energy balance comparison. Oil production and refining efficiency is based upon mean 2020 analysis of all global production by IEA[63] including methane emissions from associated gas, and all energy requirements for production, transport, and refining. E-fuels assume 100% efficiency for H_2 production by water electrolysis (Solid Oxide Electrolyzer with waste heat integration[65]) and CO_2 capture at zero energy from biogenic CO_2 fermentation off-gas, with no costs for transport. Hence, 0.93 is the theoretical maximum thermodynamic yield of C6 energy out versus the energy potentials of CO_2 and H_2O feedstocks.

Rather than supply the "Energy Economy" through syngas via coal, stranded natural gas, or biomass as deployed or proposed during various global energy crises from 1900 to today, one can use syngas routes to make and replace all fossil fuel and chemical products from captured CO_2 and hydrogen produced from renewable energy and water. All steps are fully commercial except (high technology readiness levels TRL[66]) the "reverse water gas shift reaction" (TRL 7+), where the forward reaction is currently practiced (Figure 1.9).[68]

One can thus make other carriers such as H_2 itself, ammonia, or methanol, which is a precursor to synthetic gasoline or jet fuel via further reaction.[38,69,70] Per the energy density of Figure 1.2, these provide lower volumetric and mass energy densities but require less H_2 and hence energy addition. Hydrogen can be used in fuel cells at higher efficiency than open-cycle combustion.

The available scale for renewable energy can be a challenge relative to current industrial complexes. Today's largest solar parks include 2.5 GW installed or nameplate capacity (Bhadla, India, 14,000 acres) to 5 GW (Xinjiang, China, 494,000 acres).[71] This compares with an estimated 18 GW of power[72] with 9 GW backup power required for Texas petrochemical and refining industry centered in the Houston Gulf Coast area. For the use of variable renewable energy (VRE), energy storage or overcapacity build is required to accommodate the need for 24/7 power for industry.

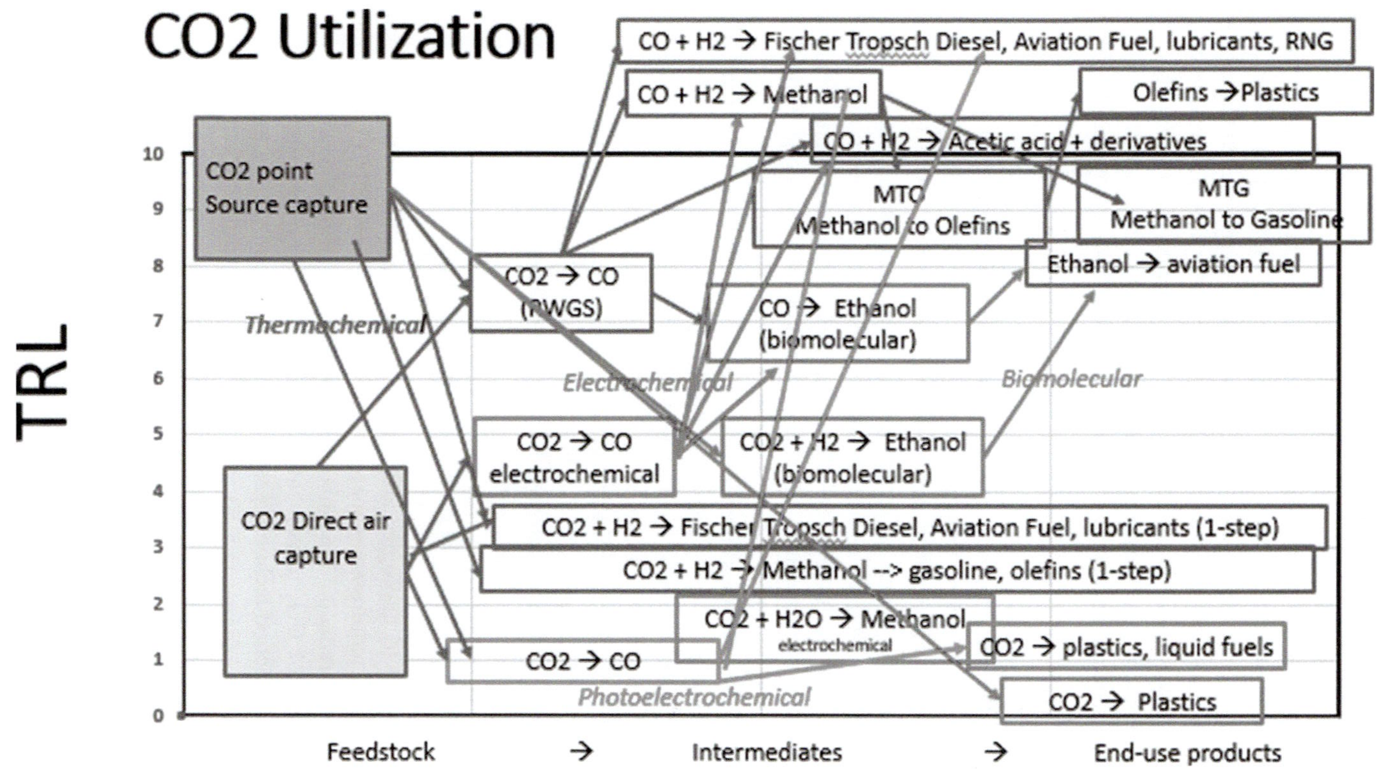

FIGURE 1.9 Rebuilding the energy economy via renewable energy from captured CO_2 via Syngas.

(Data source: Concawe,[67] Shell, others. TRL = Technology Readiness Level (1 = low, 9 = high). RNG = renewable or synthetic natural gas.)

As discussed above, the energy required to reverse CO_2 into usable hydrocarbons via reaction with hydrogen derived from the energy-intensive splitting of water is high. However, technology exists to rebuild any aspects of a hydrocarbon fuels and chemicals economy that are desired, with synthetic aviation fuels representing an early target. Bio-based feedstocks allow some chemical intermediates to be made selectively without having to go through gasification to a C_1 building block (CO) and hence can be more efficient since they already incorporate some of the energy needed via photosynthesis (sunlight). Thus, bio-based feedstocks can potentially provide the lowest costs. Synthesis of "E-Fuels" and "E-chemicals" from the reaction of CO_2 and H_2 derived from the electrolysis of water can meet larger global demand, given limited availability of bio-resources, and the land and water use required for their deployment.

1.8 STEEL

Steel starts with the reduction of iron ore (Fe_2O_3) via carbon from coal or petroleum solids (coke) to form carbon steels and then higher alloys. Circularity is active for the global steel industry, as the United States has led the way in the deployment of electric arc furnaces for the use of scrap recycled steel, where scrap is available. The United States is thus a boutique steel producer using recycled metal as feedstock, processed by electric arc furnace. Substitution of low-carbon electricity (renewable or nuclear) is therefore a means for decarbonization of the U.S. scope 1 and 2 footprints, using proven technology. Scrap metal is of limited supply, however, and as shown in Figure 1.3, one must make up for losses via technologies to reduce iron ore mined and reduced at some places on the globe. The United States has largely offshored iron-ore reduction to the Far East or Europe, and hence offshored CO_2 footprints needed to sustain a global supply, hidden as scope 3 emissions.

Technology to decarbonize iron ore reduction for the required make-up supply is in development. Direct reduction of iron ore with natural gas and now low-carbon hydrogen (Hybrit[73]) provides pathways for an order-of-magnitude reduction in CO_2 footprints for global steel manufacturing.

$$Fe_2O_3 + 3H_2 = 2Fe + 3H_2O \quad (\text{iron ore reduction})$$

Low-carbon steel technologies are thus in development and at the pilot stage, subsidized by policy requirements for automakers to reduce overall CO_2 footprints of vehicle transport by reducing the embodied CO_2 footprint of purchased steel.[74]

Levelized costs of CO_2 abatement using CO_2 capture, low-carbon fuels, and hydrogen direct reduction technologies have been reviewed by Friedman et al.[59]

1.9 CEMENT AND CONCRETE

Cement and concrete are the largest building materials in quantity for supplying transportation infrastructure and materials for residential and commercial buildings. Cement represents 8–10% of global CO_2 emissions, and 1–2% of U.S. emissions (scope 1 and 2).[75] Cement is made from limestone (calcium carbonate) and necessarily eliminates CO_2 in kilns used to produce reduced clinker, which can react and set when activated by water:

$$Ca(CO_3) \rightarrow CaO + CO_2 \quad dH_R = +178\,kJ/mol$$

The European Union considers CO_2 released from this reaction to be inevitable; it represents half the CO_2 footprint from cement manufacture. The other half is the process energy fuel (diesel or natural gas) used in cement kilns for manufacture. Electrification of cement kilns is thus considered as a means for decarbonization,[76] using rotating kiln designs that allow high-temperature heating up to 2,300 °C (resistive or otherwise) to be used with direct electrification.[77] VTT (Finland) has developed a rotary kiln.[78] Though actually fossil in origin (calcium carbonate), the CO_2 off-gas can have high purity and may be treated as unavoidable CO_2 emission to allow its capture and use as a low-carbon synthetic fuel, via some of the policy and regulations under consideration. Integrated electrochemical splitting of calcium carbonate and water is considered in early-stage research.[79]

Carbon capture and storage is thus considered to be an optimal solution for cement decarbonization, given the inherent CO_2 emissions from the process itself, and from use of natural gas to supply the large endothermic heat of reaction.[80,81] Capture costs can range as low as \$60/ton CO_2.[82] "Oxy-combustion," where O_2 is first separated from the air and then used for combustion to produce a high-concentration CO_2 off-gas for ease of capture, is one option for the capture of process-side CO_2 co-product.[83] Use of low-carbon hydrogen, renewable natural gas, or biofuel (biodiesel) can be used to decarbonize the utility heat required, noting that biomass itself cannot be used because required temperatures (to 2,300 °C) cannot be attained (Table 1.1). Some new technologies examine the use of carbon-free (non-carbonate) rock to form cement.[80] However, changing chemistry to form untested materials is a challenge for the risk-averse building industry.

Cement plants are distributed throughout the United States.[84] While local production of cement and clinker used to make concrete is most economical due to shipping costs, international shipping is feasible to enable high plant capacity utilization, resulting in being manufactured centrally in areas with low energy costs, and shipped globally, with more than 100 million tons of international trade, nearly equal to domestic shipping.[85] Manufacturing of cement and clinker is thus one industry that can relocate to areas of the globe rich in renewable energy.[86]

1.10 FOOD AND BEVERAGE

Food and beverage entail low-temperature heat (Figure 1.3). Less than 10% of the scope 1 and 2 carbon footprint for this supply chain is from the F&B industry involved in processing – most of the footprint of the processing industry are scope 3 emissions from agriculture and goods transport.

Decarbonization options include electrification with low-carbon power, energy efficiency, and electrification of heating equipment, including boilers, ovens, fryers, and cookers.[87] Much of the processing is distributed across low- to medium-scale production units of lower capex, relative to other industrial processes. This provides additional options for integration of variable renewable energy with less need for storage, such that the industry may serve as variable demand response loads for grid stabilization, which can lead to reduced power costs.

Steam is often used for process heat, such that electrification of utility boilers can improve energy efficiency and reduce carbon footprint. Direct electrification of heating can also be considered but may impact product quality given the sensitivity of food chemistry to changes in rates of heat transfer and ultimately reaction.

1.11 CROSSCUTTING APPLICATIONS

Drying is a ubiquitous operation for many generic "industry" applications, including food and beverage, and pharmaceuticals. Use of fuel-fired steam boilers has been long used to provide heat for low-temperature drying. Substitute of resistance heating can be done with high efficiency (90%) for steam generation. Alternately, direct heating via electromagnetic microwaves and induction heating can be substituted and, in some cases, improve control. Electromagnetic or electrical heating can also be used in chemical processing to directly heat catalysts used in transformation versus bulk heating of entire fluid streams, potentially reducing energy needs and improving yields by targeting only the desired catalyst itself versus the entire stream.

Electric steam boilers can be highly efficient at moderate scale,[88] though steam as a heat transfer medium is limited to 500–600 °C in superheat mode. Voltage and connection requirements for large-scale furnaces can be a challenge for energy-intensive process requirements such as ethane cracking to ethylene. Large 100 MW chemical process heaters are challenged relative to demonstrated electrical heating options.[89]

1.12 OPPORTUNITIES AND CHALLENGES

The recent U.S. Liftoff report estimates the cost of industrial manufacturing CO_2 mitigation, including subsidies and incentives from the U.S. Inflation Reduction Act (IRA).[90] Despite credits and subsidies from recent IRA legislation, including 45Q for sequestration of CO_2, only a few options such as capture of nearly pure CO_2 from ethylene oxide manufacture, or by-product of ethanol production via fermentation, are economically viable with the subsidies (Figure 1.10).

With renewable energy, land use is a concern, given the move away from higher-energy density liquid, liquefied, or solid fuels (Figure 1.2).[31] Biofuels greatly increase (10× or more) the land use for energy relative to the previous fossil-based economy. Sixty percent of energy land use in the United States results from bioethanol production, which contributes 2% of total energy via blending into gasoline. Biodiesel as a similar first-generation fuel requires similar land use, while second-generation (and beyond) technology to convert biomass to liquid fuels is currently uneconomic without subsidy, but if plausible would have a 10-fold reduction in land area required per unit of energy generated versus first-generation biofuels.

One option for direct utilization of low-cost variable renewable energy is for industrial manufacturing processes to act as a demand sink. The capital intensity of industrial sectors such as chemicals and refining renders this uneconomic for many scenarios. These sectors consider energy sources with high-capacity factor of "up time," such as nuclear power, geothermal, or natural gas with CCS as low-carbon solutions versus variable renewable energy, given the high cost and low density of battery storage (Table 1.5).

Given cost challenges in energy storage for renewable energy, many lowest-cost scenarios consider overbuilding 3× installed capacity relative to 1× demand, where

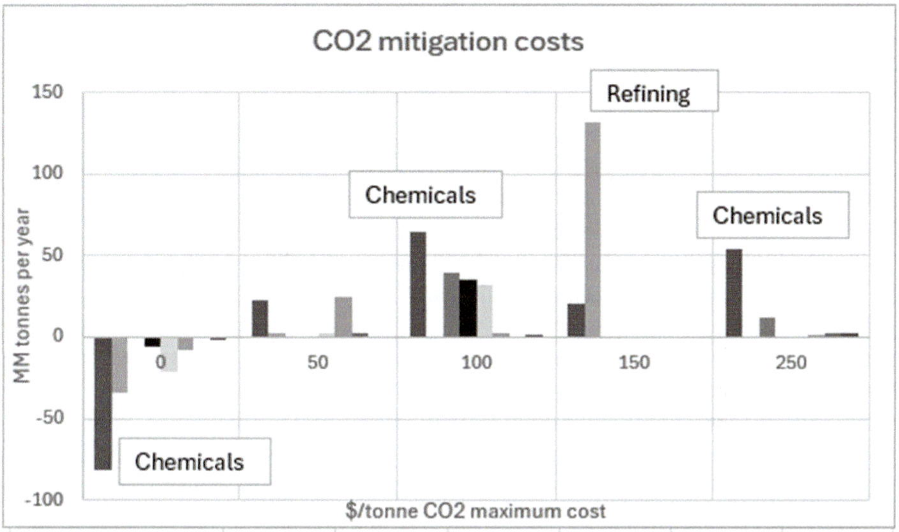

Sector	<$0	$1-50	$51-100	$101-150	$151-250	Net > 0
Chemicals	-81	22	64	20	54	79
Refining	-34	2		132	0	100
Iron and steel			39	0	12	51
Food & Bev	-6		35			29
Cement	-22	3	32			13
Pulp and Paper	-8	25	2		1	20
Aluminum	-1	2			2	3
Glass	-2	0	1		2	1
Units: millions of metic tonnes of CO2 per year					Total:	296

FIGURE 1.10 Costs for decarbonization versus subsector.

TABLE 1.5
Battery Storage Costs

$/kWh	10-Day Bat	3-Day Bat	4h Bat	3d CAES	10d NG
Capital	$2.61	$0.78	$0.04	$0.21	$0.02
Fuel	0.021	0.021	$0.02	0.021	$0.0345
Total	$2.63	$0.80	$0.06	$0.23	$0.05

Source: NREL.[91]
Note: CAES = Compressed air energy storage. NG = Natural gas.

overcapacity is cheaper than storage and demand can be partially scheduled to offset some of the need for additional.[92] As a best-case scenario for integration of land use for energy generation and agriculture or other use, required land area for the use of solar in achieving 100% clean electrical power for the United States in 2035 may require less than the land used previously for coal production, infrastructure, and storage.[92]

Industry can also relocate to regions of low-cost, low-carbon energy. Solar and wind intensity vary more than 10-fold globally,[86] including combined wind and solar capacity factors or "up time" as high as 75%. While DC power transmission is developing, electrical power transmission is challenged at distances longer than about 1,000 km.

Distributed production is a cost challenge for capital-intensive industries. CO_2 utilization including distributed production from biogenic sources, cement plants, or direct capture from air is one option considered for future low-carbon fuels. Given the low energy density of feedstock, these options result in smaller scale facilities than current petroleum refineries, and the multistep nature of converting CO_2 to fuels and products is highly capital intensive.[93] CO_2-derived products are thus high cost and energy intensive relative to petroleum- or natural gas-derived products today, where costs scale to the 0.6-power of throughput rate or annual capacity.

McKinsey estimates an additional $4 trillion a year is needed for re-capitalization of the global energy economy through 2050, to achieve "Net Zero" fossil carbon emissions.[88] That is not a large number relative to global GDP; however, only 20% of the world (OECD) had the economic prosperity to pay for a just energy transition; the remainder (80%) of the world must expand energy supply to rise above "energy poverty," given the strong correlation between per-capital energy consumption and human well-being.[94] Willingness to pay for higher-cost products is thus a challenge for decarbonization. Most of the cost lies in the large energy inputs over and above fossil fuel costs required for 24/7, high-energy density and power services required by our current economy. This is a thermodynamic reality for starting with CO_2 or in part biomass as the molecular carriers and products for these services.

1.13 CONCLUSIONS

Industrial processes represent 25–30% of U.S. and global greenhouse gas emissions, or more if one counts the "energy industry" and its Scope 3 emissions in supplying energy for other sectors of the economy. "Decarbonization," meaning reduction in fossil CO_2 emissions, is relevant for those sectors that do not entail carbon-based products and can be achieved by electrification of energy services via clean power supplied from renewable or other non-fossil resources, or fossil fuels with carbon capture and storage. Alternately, one can supply a "clean molecular fuel" in the form of hydrogen or one of its derivatives, or a hydrocarbon fuel where the carbon is obtained from biogenic sources, or via capture of CO_2 from the atmosphere either directly or from the ocean. Reduction in energy demand including improvements in energy efficiency can further reduce greenhouse gas emissions, with new opportunities emerging in smart systems and application of AI and machine learning to energy systems management. CO_2 can also be captured from industrial and stored or sequestered underground as a means for decarbonizing existing industrial applications, which is often the lowest-cost approach, given the high capital investment in the current industry.

Temperatures and operating conditions vary widely across industrial subsectors and processes. This can present challenges in heat transfer and operating temperature for switching from fuel-fired heating to low-carbon electrical heat. Capital costs to convert fuel-fired furnaces to electric heating can be extensive. Use of hydrogen as a low-carbon vector for the retrofit of fuel-fired heaters may require less extensive capital modifications. On the other hand, electrical heating can offer benefits of improved heat control, heating of only desired components or catalysts, and ability to manage dynamic loads and conditions. Where energy density and heating performance require a carbon-based fuel, drop-in hydrocarbon fuels can be produced from biomass or synthetically by reaction of captured CO_2 with hydrogen, for use in transportation or industrial applications. Given the large inputs of energy required to make these synthetic fuels, costs are likely to remain higher than those of current fossil fuels, and hence policy or premiums are needed for large-scale adoption.

Most industries entail globally competitive supply chains, with manufacturing over a wide range of operating conditions favored by low energy prices. 24/7 operation is typically required for competitive use of capital. Energy storage is likely required where low-cost renewable energy is deployed, which drives up costs. While most roadmaps and industry commitment concern scope 1 and 2 emissions from use of fuels and electricity, industry must ultimately also consider the "scope 3" emissions of the products it sells, or feedstocks and materials it buys, as well as other sustainability drivers for a future economy where increased recycle of materials is desired. Feedstocks and processes for future industry may require the use of waste streams, feedstocks derived from bio-resources, or direct air capture of CO_2, resulting in a dilemma of decarbonizing existing manufacturing assets, versus moving to future systems using low-carbon energy and feedstock as inputs.

A multi-pronged approach is thus needed for industry decarbonization, given the breadth of operating conditions and service needs encountered. Use of systems analysis to assess lowest-cost optimal pathways including intersections of other sectors of the economy will be essential for success.

NOTE

1 "Bio-based" refers to both edible sugars or carbohydrates, which serve as feedstocks for biofuels ("first generation"), including corn and sugar cane (fermented to ethanol fuel), or edible oils (soybean), which can be upgraded to biodiesel or aviation fuel. It also includes the much larger quantity of "biomass," which can be converted to "second generation and beyond" biofuels, which is comprised of polymeric celluloses and lignin and must be chemically converted to form liquid or gaseous biofuels.

REFERENCES

1. "Industry", *Merriam-Webster* https://www.merriam-webster.com/dictionary/industrial
2. Hannah Ritchie and Pablo Rosado, Energy Mix. 2020. Published online at OurWorldinData.org. Retrieved from: https://ourworldindata.org/energy-mix [Online Resource].
3. Vaclav Smil, *Energy Transitions: History, Requirements, Prospects*. Praeger, 2010.
4. R. Heinberg and D. Fridley, *Our Renewable Future: Laying the Path to One Hundred Percent Clean Energy*.
5. DNV: Comparison of Alternative Marine Fuels SEA\LNG Ltd Report No.: 2019-0567, Rev. 3 Document No.: 11C8I1KZ-1; Date: July 05, 2019.

6. EIA, *Today in Energy*, 2013. https://www.eia.gov/todayinenergy/detail.php?id=9991. Visualizing Energy (2024) https://visualizingenergy.org/fuel-energy-density-what-is-it-and-why-is-it-important/
7. Hydrogen Storage, 2024. https://www.energy.gov/eere/fuelcells/hydrogen-storage
8. V. Smil, *Waiting for Super Batteries: They are Still a Long Way from Matching the Energy Density of Liquid Fuel*. https://spectrum.ieee.org/ev-battery-2658649740
9. Wikipedia, *Energy Density*, 2024. https://en.wikipedia.org/wiki/Energy_density
10. Vaclav Smil, *Energy and Civilization: A History*. MIT Press, 2018.
11. Otherlab, *US Energy Flow Super Sankey*, 2018. https://www.otherlab.com/blog-posts/us-energy-flow-super-sankey
12. U.S. EIA Annual Energy Outlook, 2021 https://www.eia.gov/outlooks/archive/aeo21/pdf/00AEO2021ChartLibrary.pdf
13. Annual Energy Outlook 2021 with Projections to 2050, U.S. Energy Information Administration, February 3, 2021, https://www.eia.gov/outlooks/archive/aeo21/. See Table 18. Energy-Related Carbon Dioxide Emissions by Sector and Source.
14. U.S. DOE, *Industrial Decarbonization Roadmap*, DOE/EE 2635. September 2022. https://www.energy.gov/eere/doe-industrial-decarbonization-roadmap
15. U.S. Energy Information Administration, Annual Energy Outlook 2023, *Table 18. Energy-Related Carbon Dioxide Emissions by Sector and Source*. https://www.eia.gov/outlooks/aeo/data/browser/
16. Rhodia Group, 2024. https://rhg.com/research/global-greenhouse-gas-emissions-2022/
17. S. Gupta, From grey and brown to green and blue: Low-carbon hydrogen in refining, *Wood Mackenzie*, June 2022. https://www.woodmac.com/news/opinion/from-grey-and-brown-to-green-and-blue-low-carbon-hydrogen-in-refining/
18. IPCC, Climate Change 2014: Mitigation of Climate Change. In Edenhofer, O., Pichs-Madruga, R., Sokona, Y., Farahani, E., Kadner, S., Seyboth, K., Adler, A., Baum, I., Brunner, S., Eickemeier, P., Kriemann, B., Savolainen, J., Schlömer, S., von Stechow, C., Zwickel, T. and Minx, J. C. (Eds.), *Contribution of Working Group III to the Fifth Assessment Report of the Intergovernmental Panel on Climate Change*. Cambridge University Press, 2019.
19. US EPA, *Inventory of US Greenhouse Gas Emissions and Sinks*, 2022. https://www.epa.gov/ghgemissions/inventory-us-greenhouse-gas-emissions-and-sinks
20. Hannah Ritchie, *Sector by Sector: Where do Global Greenhouse Gas Emissions Come from?*, 2020. Published online at OurWorldinData.org. Retrieved from: https://ourworldindata.org/ghg-emissions-by-sector [Online Resource].
21. National Energy Technology Laboratory (NETL), *Carbon Dioxide 101*, 2024. https://netl.doe.gov/carbon-management/carbon-storage/faqs/carbon-dioxide-101
22. J. Huang, S. Fu and L. Gan, *Lignin Chemistry and Applications*, Chapter 2: Structure and Characteristics of Lignin, Elsevier. 2019.
23. PubChem, NIH, *Organosolv Lignin*. https://pubchem.ncbi.nlm.nih.gov/#query=C81H92O28
24. U.S DOE Alternative Fuels Data Canter. 2024. https://afdc.energy.gov/fuels/biodiesel-basics
25. A. Muscat, E. M. De Olde, I. J. M. De Boer and R. Ripoll-Bosch, The battle for biomass: A systematic review of food-feed-fuel competition, *Global Food Security*, 2020, 25, 100330.
26. IEA, *Net Zero by 2050*. IEA, 2021. https://www.iea.org/reports/net-zero-by-2050
27. National Academies of Sciences, Engineering, and Medicine, *Carbon Dioxide Utilization Markets and Infrastructure: Status and Opportunities: A First Report*. The National Academies Press, 2023. https://doi.org/10.17226/26703

28. National Academies of Sciences, Engineering, and Medicine, *Carbon Utilization Infrastructure, Markets, and Research and Development: A Final Report*. The National Academies Press, 2024. https://doi.org/10.17226/27732
29. A Soler, V. Gordillo, W. Lilley, P. Schmidt, W. Werner, T. Houghton and S. Dell'Orco. *E-Fuels: A Techno Economic Assessment of European Domestic Production and Imports Towards 2050*, Report 17/22, Concawe, 2022. www.concawe.eu/wp-content/uploads/Rpt_22-17.pdf
30. Z. Han, S. Yuan, D. Liu, Q. Zheng, Y. A. Huang, S. Yan and Z. Zou, Physical basis of multi-energy coupling-driven water oxidation, *Frontiers in Chemistry*, 2022, 10, 902814.
31. P. Gabrielli, M. Gazzani and M. Mazzotti, The role of carbon capture and utilization, carbon capture and storage, and biomass to enable a net-zero-CO_2 emissions chemical industry, *Industrial & Engineering Chemistry Research*, 2020, 59, 7033–7045. https://doi.org/10.1021/acs.iecr.9b06579
32. P. G. Hatcher, J. L. Faulon, K. A. Wenzel and G. D. Cody, A structural model for lignin-derived vitrinite from high-volatile bituminous coal (coalified wood), *Energy Fuels*, 1992, 6, 813–820.
33. IEA, *Bioenergy Review 2023: How Bioenergy Contributes to a Sustainable Future*, 2023. https://www.ieabioenergyreview.org/
34. C. C. Seah, C. H. Tan, N. A. Arifin, R. S. R. M. Hafriz, A. Salmiaton, S. Nomanbhay and A. H. Shamsuddin, Co-pyrolysis of biomass and plastic: Circularity of wastes and comprehensive review of synergistic mechanism, *Results in Engineering*, 2023, 17, 100989.
35. C. L. Williams, R. M. Emerson and J. S. Tumuluru, *Biomass Volume Estimation and Valorization for Energy*, ed. Tumuluru J. S., InTech, 2017. https://www.intechopen.com/chapters/52751
36. L. R. Radovic, *Energy and Fuels in Society: Analysis of Bills and Media Reports*, McGraw-Hill Custom Publishing, 1997. https://personal.ems.psu.edu/~radovic/matsc101.html
37. S. Ong, C. Campbell, P. Denholm, R. Margolis and G. Heath, *Land-Use Requirements for Solar Power Plants in the United States*, NREL/TP-6A20-56290, 2013. https://www.nrel.gov/docs/fy13osti/56290.pdf
38. IEA, Putting CO2 to use, 2019. https://www.iea.org/reports/putting-co2-to-use
39. Z. Huang, R. G. Grim, J. A. Schaidle and L. Tao, The economic outlook for converting CO2 and electrons to molecules, *Energy & Environmental Science*, 2021, 14, 3664–3678.
40. J. Perner, M. Unteutsch, A. Lövenich (Frontier Economics/Agora) *The Future Cost of Electricity-Based Synthetic Fuels, Study*, 2018 https://static.agora-energiewende.de/fileadmin/Projekte/2017/SynKost_2050/Agora_SynKost_Study_EN_WEB.pdf
41. J.-P. Lange, Towards circular carbo-chemicals – The metamorphosis of petrochemicals, *Energy & Environmental Science*, 2021, 14, 4358–4376.
42. U.S. EPA *Fuel Economy*. https://www.epa.gov/fueleconomy
43. U.S. EIA, *Use of Natural Gas-Fired Generation Differs in the United States by Technology and Region*, February 22, 2024. https://www.eia.gov/todayinenergy/detail.php?id=61444#:~:text=CCGT%20plants%20that%20started%20operating,1980s%20and%20up%20until%201998
44. U.S. DOE *How Gas Turbine Power Plants Work*, 2024. https://www.energy.gov/fecm/how-gas-turbine-power-plants-work
45. Doe industrial decarbonization roadmap. *Energy.gov*. Accessed May 20, 2023. https://www.energy.gov/eere/doe-industrial-decarbonization-roadmap
46. U.S. DOE, Pathways to commercial liftoff, industrial decarbonization, 2024. https://liftoff.energy.gov/wp-content/uploads/2024/02/LIFTOFF_DOE_Industrial-Decarbonization_REV022724.pdf

47. DOE Industrial Decarbonization Roadmap, *Energy.gov*, 2024. Accessed 20 May 2023. https://www.energy.gov/eere/doe-industrial-decarbonization-roadmap
48. S. J. Friedman, Z. Fan and K. Tang, *Low-Carbon Heat Solutions for Heavy Industry: Sources, Options, and Costs Today*, Columbia/SIPA Center on Global Energy Policy, October 2019.
49. Charles F. Kutscher, Jeffrey S. Logan and Timothy C. Coburn, Renewable and Sustainable Energy Institute, University of Colorado Boulder Accelerating the US Clean Energy Transformation Challenges and Solutions by Sector. https://missionzero.io/accelerating-the-us-clean-energy-transformation-challenges-and-solutions-by-sector/
50. Occo Roelofsen, Ken Somers, Eveline Speelman and Maaike Witteveen, *Plugging In: What Electrification Can Do for Industry*, McKinsey & Co, Energy Insights Practice, May 2020, https://www.mckinsey.com/industries/electric-power-and-natural-gas/our-insights/plugging-in-what-electrification-can-do-for-industry#/
51. Arnout de Pee, Dickon Pinner, Occo Roelofsen, Ken Somers, Eveline Speelman and Maaike Wittevee, *Decarbonization of Industrial Sectors: The Next Frontier*. McKinsey & Co., June 2018. https://www.mckinsey.com/industries/oil-and-gas/our-insights/decarbonization-of-industrial-sectors-the-next-frontier#/
52. S. Brueske, R. Sabouni, C. Zach and H. Andres, *Energetics Inc, U. S. Manufacturing Energy Use and Greenhouse Gas Emission Analysis*, ORNL/TM-2012/504, November 2012. https://info.ornl.gov/sites/publications/files/Pub39685.pdf
53. S. Brueske and R. Sabouni, *Energy Use in U.S. Manufacturing: Analysis of Energy Supply, End Use, Loss, and Greenhouse Gas Emissions, Energetics*, 2013 https://www.aceee.org/files/proceedings/2013/data/papers/2_219.pdf.
54. U.S. DOE (2024): How Gas Turbine Power Plants Work. https://www.energy.gov/fecm/how-gas-turbine-power-plants-work
55. Ipieca, *Estimating Petroleum Industry Value Chain (Scope 3) Greenhouse Gas Emissions. Overview of Methodologies*, 2023. https://www.ipieca.org/resources/estimating-petroleum-industry-value-chain-scope-3-greenhouse-gas-emissions-overview-of-methodologies
56. Ellen MacArthur Foundation, *Plastics and the Circular Economy–Deep Dive*, 2019. https://www.ellenmacarthurfoundation.org/plastics-and-the-circular-economy-deep-dive
57. Z. Liu, J. Meng, Z. Deng et al. Embodied carbon emissions in China-US trade, *Science China Earth Sciences*, 2020, 63, 1577–1586. https://doi.org/10.1007/s11430-019-9635-x
58. David Gardiner and Associates, *Electrifying U.S. Industry: A Technology- and Process-Based Approach to Decarbonization*, January 2021. https://www.dgardiner.com/electrifying-u-s-industry/
59. D. S. Mallapragada, Y. Dvorkin, M. A. Modestino, D. V. Esposito, W. A. Smith, B.-M. Hodge, M. P. Harold, V. M. Donnelly, A. Nuz, C. Bloomquist, K. Baker, L. C. Grabow, Y. Yan, N. N. Rajput, R. L. Hartman, E. J. Biddinger, E. S. Aydil and A. D. Taylor, Decarbonization of the chemical industry through electrification: Barriers and opportunities, *Joule*, 2023, 7, 23–41.
60. A. Gailani, S. Cooper, S. Allen, A. Pimm, P. Taylor and R. Gross, Assessing the potential of decarbonization options for industrial sectors, *Joule*, 2024, 8, 576–603.
61. T. A. Atsbha, T. Yoon, B.-H. Yoo and C.-J. Lee, Techno-economic and environmental analysis for direct catalytic conversion of CO_2 to methanol and liquid/high-calorie-SNG fuels, *Catalysts*, 2021, 11, 687.
62. Hidenori Nakamura, Revisiting the operational principle of sustainability: Physical and economic aspects, *Journal of Sustainable Development*, 2012, 5. https://doi.org/10.5539/jsd.v5n9p98

63. IEA, Emissions from Oil and Gas Operations in Net Zero Transitions: A World Energy Outlook Special Report on the Oil and Gas Industry and COP28, 2020. https://www.iea.org/reports/the-oil-and-gas-industry-in-net-zero-transitions
64. NIST *Chemistry WebBook*. https://webbook.nist.gov/
65. A. Hauch, R. Küngas, P. Blennow, A. B. Hansen, J. B. Hansen, B. V. Mathiesen and M. B. Mogensen, Recent advances in solid oxide cell technology for electrolysis, *Science*, 2020, 370, eaba6118. https://science.sciencemag.org/content/370/6513/eaba6118.editor-summary
66. C. G. Manning, NASA, Technology Readiness Levels, 2023. https://www.nasa.gov/directorates/somd/space-communications-navigation-program/technology-readiness-levels/
67. A. Soler, *Role of e-Fuels in the European transport System: Literature Review*, Concawe Report 14/19, 2019. https://www.concawe.eu/wp-content/uploads/Rpt_19-14.pdf.
68. L. E. Sandoval-Diaz, R. Schlögl and T. Lunkenbein, Quo vadis dry reforming of methane?—A review on Its chemical, *Environmental, and Industrial Prospects, Catalysts*, 2022, 12, 465. https://doi.org/10.3390/catal12050465, https://www.mdpi.com/2073-4344/12/5/465
69. L. Magalhaes, A. Ferreira, M. Rocha and A. Silva, Techno-economic Analysis of Sustainable Aviation Fuels: e-fuels, Aerospace Europe Conference 2023 – 10TH EUCASS – 9TH CEA021-164560-article/2023, 2023. https://doi.org/10.13009/EUCASS2023-952
70. IATA, *SAF Handbook*, 2024. https://www.iata.org/contentassets/d13875e9ed784f75bac90f000760e998/saf-handbook.pdf
71. T. Gill, *The World's Biggest Solar Farms*, March 2024. https://www.theecoexperts.co.uk/solar-panels/biggest-solar-farms#:~:text=1all%20others%20on%20this%20list.
72. U. S Energy Information Agency (EIA), *Texas State Profile and Energy Estimates*, 2024. https://www.eia.gov/state/analysis.php?sid=TX
73. Hybrit, *Fossil-free Steel, a Shared Opportunity*. https://www.hybritdevelopment.se/
74. M. Pei, M. Petäjäniemi, A. Regnell and O. Wijk, Toward a fossil free future with HYBRIT: Development of Iron and steelmaking technology in Sweden and Finland, *Metals*, 2020, 10, 972. https://www.mdpi.com/2075-4701/10/7/972
75. U.S. DOE, *Pathways to Commercial Liftoff–Low-Carbon Cement*, September 2023. https://liftoff.energy.gov/wp-content/uploads/2023/12/20230921-Pathways-to-Commercial-Liftoff-Cement.pdf
76. S. Quevedo Parra and M. C. Romano, Decarbonization of cement production by electrification, *Journal of Cleaner Production*, 2023, 425, 138913
77. J. Friedmann, Z. Fan, Z. Byrum, E. Ochu, A. Bhardwaj and H. Sheerazi, *Levelized Cost of Carbon Abatement: An Improved Cost-Assessment Methodology for a Net-Zero Emissions World*. Columbia U., 2020. https://www.energypolicy.columbia.edu/publications/levelized-cost-carbon-abatement-improved-cost-assessment-methodology-net-zero-emissions-world/
78. VTT Electric Rotary Kiln, 2022. https://cementproducts.com/2022/12/09/vtt-electric-rotary-kiln-ready-to-be-scaled/
79. L. D. Ellis, A. F. Badel, M. L. Chiang, R. J.-Y. Park and Y.-M. Chiang, Toward electrochemical synthesis of cement—An electrolyzer-based process for decarbonating $CaCO_3$ while producing useful gas streams, *Proceedings of the National Academy of Sciences of the United States of America*, 2020, 117, 12584–12591
80. W. Carlsen, A. Gangotra and K. Kennedy: *Cutting-Edge Projects Aim to Decarbonize US Cement Emissions*, World Resources Institute, July 17, 2024. https://www.wri.org/insights/low-carbon-cement-technology

81. IEA, World Energy Outlook, 2024. https://iea.blob.core.windows.net/assets/c036b390-ba9c-4132-870b-ffb455148b63/WorldEnergyOutlook2024.pdf
82. IEA, Energy Technology Perspectives 2020 - Special Report on Carbon Capture Utilisation and Storage: CCUS in clean energy transitions, OECD Publishing, 2020. https://www.oecd-ilibrary.org/energy/energy-technology-perspectives-2020-special-report-on-carbon-capture-utilisation-and-storage_208b66f4-en
83. D. Bizley, *An Overview of Oxy-combustion*, World Cement, Friday, 27 October 2023. https://www.worldcement.com/special-reports/27102023/an-overview-of-oxy-combustion/
84. USGS, Location of cement plants and counties producing construction sand and gravel and crushed stone, 2024. https://www.usgs.gov/media/images/location-cement-plants-and-counties-producing-construction-sand-and
85. Ad Ligthart, *Seaborne Trade and Distribution of Cement and Clinker* (2016). https://cementdistribution.com/wp-content/uploads/2016/10/seaborne_trade_and_distribution_of_cement_and_clinker.pdf
86. O. Inderwildi, C. Zhang, X. Wang and M. Kraft. The impact of intelligent cyber-physical systems on the decarbonization of energy, *Energy & Environmental Science*, 2020, 13(3), 744–771.
87. U.S. DOE, *Pathways to Commercial Liftoff – Industrial Decarbonization*, September 2023. https://liftoff.energy.gov/industrial-decarbonization/
88. D. Long and R. Jultano, Watlow Electrification and medium voltage process heater systems, Heat Exchanger World, May 13, 2022 https://heat-exchanger-world.com/electrification-and-medium-voltage-process-heater-systems/
89. Gaumer Process: Medium Voltage Heater and Panel. https://www.gaumer.com/products/medium-voltage-4-kv-heater--panel.html
90. U.S. Dept. of Treasury, *Inflation Reduction Act*, 2022. https://home.treasury.gov/policy-issues/inflation-reduction-act
91. Wesley Cole and Akash Karmakar. 2023. Cost Projections for Utility-Scale Battery Storage: 2023 Update. Golden, CO: National Renewable Energy Laboratory. NREL/TP-6A40-85332. https://www.nrel.gov/docs/fy23osti/85332.pdf
92. Paul Denholm, Patrick Brown, Wesley Cole, et al. 2022. Examining Supply-Side Options to Achieve 100% Clean Electricity by 2035. Golden, CO: National Renewable Energy Laboratory. NREL/TP6A40-81644. https://www.nrel.gov/docs/fy22osti/81644.pdf
93. IEA, *Energy Technology Perspectives 2023*. IEA, 2023. https://www.iea.org/reports/energy-technology-perspectives-2023
94. Max Roser, The world's energy problem, 2020. Published online at *OurWorldinData.org*. Retrieved from (2024). https://ourworldindata.org/worlds-energy-problem

2 Toward a Greener Industry

Decision-Making and Optimization Strategies

Mahdi Ahmed and Joan Cordiner

2.1 INTRODUCTION

Globally, nations are advancing toward net-zero emissions within the coming decades as part of the Paris Agreement, with a specific focus on industrial sectors. These sectors are responsible for approximately 25% of global CO_2 emissions and 40% of final energy consumption, making them critical targets for decarbonization efforts [1]. Despite their strategic importance in supplying essential commodities and resources, the financial burden of decarbonization remains significant, threatening to escalate costs for consumers, and stall progress [2].

Industries play a crucial role in driving economic growth and innovation, producing high-value products essential for sectors such as defense, electronics, and public health. This dichotomy highlights the complexity of aligning climate action with economic sustainability in a competitive, global marketplace. Currently, the transition toward decarbonization is predominantly policy-driven, reliant on mechanisms such as emissions trading schemes (ETS), tax exemptions, and carbon border adjustments. This dependence introduces uncertainty for investments in decarbonization technologies, especially for sectors operating with narrow profit margins.

The diversity of available decarbonization technologies, including hydrogen, electrification, carbon capture, and process intensification, adds another layer of complexity. Industries must navigate this uncertainty against the evolving policy landscape and market dynamics, emphasizing the importance of selecting pathways that are both economically viable and environmentally sustainable. The geopolitical climate, particularly the implications of energy dependency highlighted by recent global conflicts, accentuates the urgency for industries to adopt decarbonization strategies that ensure resilience and self-sufficiency. Thus, the industrial sector's pursuit of decarbonization is not only about adhering to regulatory standards but also about securing a sustainable future in an increasingly uncertain energy market.

DOI: 10.1201/9781003655442-3

2.1.1 Scope and Structure

The urgency of decarbonizing industrial sectors stems from their significant contribution to global CO_2 emissions and their critical role in the global economy. This chapter addresses the question: What are the best ways for industries, particularly in the manufacturing sectors, to achieve full decarbonization? By analyzing technological options, economic considerations, and policy frameworks, this chapter aims to equip stakeholders with the knowledge needed to make informed decisions that balance environmental goals with economic viability. Additionally, effective decision-making is emphasized for optimizing decarbonization efforts, exploring various frameworks and tools such as multi-criteria decision analysis (MCDA) to support decision-making.

2.2 BACKGROUND AND CONTEXT

2.2.1 An Overview of UK Industry

The term "industry" encompasses a wide range of sectors, including power generation, manufacturing, finance, aerospace, and defense. This chapter focuses on the manufacturing industries, which are viewed as "hard to abate" sectors due to the immense cost of transitioning and the highly specific process operations involved [3].

Steel, chemicals, and cement make up over 50% of all UK industrial emissions, with the majority of this energy use being fossil fuel-derived, such as coal and natural gas. In many other countries across the entire world, the situation is very similar. These fuels are essential for various applications, including fertilizers, plastics, process heating, and feedstocks [4]. Existing infrastructure and supply chains built on fossil fuel use have enabled affordable manufacturing and consumption. However, this reliance now poses a hurdle in shifting to sustainable alternatives. A large-scale overhaul of industrial processes is likely unviable due to the vast capital already invested in fossil fuel systems that would need replacing. Relocating or shutting down these systems due to decarbonization costs will pose supply-chain risks and dependencies on geopolitically unstable or hostile states. Yet, meeting climate goals is vital, raising the question of the most optimal decarbonization route.

2.2.2 Decarbonization Policies and Strategies

The method by which a site may pursue decarbonization is largely dependent on the country's specific decarbonization strategy and policy framework. Insight into policy is important because it is the only way to make any sort of decarbonization initiative economically viable, given the nascent market and limited consumer demand for green products. This section therefore explores what (current) policies enable economic industrial decarbonization, but also discusses the uncertainty about being reliant on such policies.

2.2.2.1 Emissions Trading Scheme

ETS provide an inherent value to decarbonization by creating a market for carbon [5]. It is a cap-and-trade system, where there is a cap on the total level of greenhouse gas (GHG) emissions for specific industry sectors. Companies must hold enough allowances to cover their emissions; if they exceed their allowance, they must either purchase additional allowances or acquire carbon offsets. This creates an economic incentive for companies to reduce their emissions, as it can be more cost-effective to do so than purchasing additional allowances.

Many industrialized countries, including the EU, US (certain states), UK, China, Canada, and Japan, have implemented or are developing an ETS, highlighting the global commitment to decarbonization [6]. The participation of these countries in the Paris Agreement also creates a framework for international carbon allowance trading (linking mechanism), which helps to achieve reduction targets, encourages global cooperation, and promotes reliable accounting practices.

As the cap for CO_2 emissions is gradually reduced through policy, the number of allowances available for companies will also decrease, making it more expensive to emit GHGs. This will incentivize industries to invest in emission-reduction technologies and practices to avoid the need to purchase additional allowances. The price of carbon credits should also increase, providing potential avenues for profit as companies can hold or sell their allowances to other firms.

However, if these policies are not consistently implemented or enforced, this could lead to a risk of reduced viability for non-carbon-trading dependent states. The lack of a stable and predictable carbon market could undermine the financial incentives necessary for sustained decarbonization efforts.

2.2.2.2 A UK Perspective

The UK has implemented numerous policies to enable decarbonization, particularly through the development of hydrogen and CO_2 pipelines tailored for industrial clusters. These clusters are strategically positioned to optimize the use of shared infrastructure, enhancing efficiency and reducing costs. For instance, the UK Industrial Decarbonization Strategy emphasizes the creation of four net-zero industrial clusters by 2030, leveraging joint collaborations between companies and the government to fund these projects, build business cases, and exploit economies of scale [7, 8]. Such collaborative efforts ensure that plants within or near these clusters are effectively provided with the necessary solutions for their decarbonization initiatives. The UK's North Sea is recognized as one of the most promising areas for CO_2 storage globally, leveraging its existing infrastructure and expertise from the oil and gas industry [9, 10]. This strategic advantage aligns with broader decarbonization strategies, as seen in other nations adopting similar approaches to leverage local geographic and geological benefits for CO_2 storage [11].

Despite such policies, financial uncertainty is still a concern, driven by the high capital costs of deploying new technologies such as carbon capture and storage (CCS) and hydrogen infrastructure, posing significant risks for investors without guaranteed returns or stable long-term policy support [12, 13]. The potential for policy shifts exacerbates these financial risks, as changes in subsidies or carbon pricing can alter the economic landscape [14]. Technical challenges further complicate decarbonization efforts, such as the safe integration of hydrogen into existing gas networks and the establishment of CO_2 pipelines and storage facilities, which are complex and resource-intensive [15].

Policy-related issues further contribute to the uncertainty. The dynamic and often unpredictable nature of policy frameworks creates an unstable environment for long-term investments. Frequent regulatory updates can deter investment in large-scale projects that depend on predictable conditions [16]. Additionally, the lack of cohesive policies across sectors and regions can lead to fragmented efforts and inefficiencies. This decarbonization strategy primarily targets industrial clusters, which account for approximately 50% of the country's industrial emissions. These clusters benefit from shared infrastructure and coordinated efforts for decarbonization. However, the remaining emissions arise from dispersed industrial sites that are not located within these clusters. These sites face significant challenges due to their isolation from major infrastructure developments [17]. Including this, procuring low-carbon fuels and integrating them into existing operations at these remote sites presents technical and logistical difficulties [18].

There are currently no plans for dispersed sites, and so this chapter aims to explore how to develop business cases for these sites specifically, where little policy support is available.

2.2.2.3 A European Perspective

The European Union (EU) has introduced a significant policy measure known as the Carbon Border Adjustment Mechanism (CBAM), which was ratified in 2023. This policy aims to address the issue of embedded carbon emissions in imported goods (focused on cement, electricity, fertilizers, iron and steel, aluminum, and hydrogen), leveling the playing field for EU industries, and maintaining the competitiveness of EU industries [19].

The primary goal of CBAM is to prevent industries from relocating to countries with less stringent carbon regulations, a phenomenon known as "carbon leakage." This mechanism supports local industries by ensuring that foreign producers also bear the cost of carbon emissions, encouraging global adherence to carbon reduction practices [20]. Consequently, EU industries are not forced to decarbonize while losing competitive advantage to foreign producers, who do not face similar carbon costs.

Moreover, CBAM serves as a model for global carbon pricing policies, potentially influencing other regions to adopt similar mechanisms. By incorporating the cost of carbon into the price of imported goods, CBAM promotes fair competition and incentivizes global industries to reduce their carbon footprint [21]. This approach not only supports the EU's internal climate goals but also drives international efforts toward sustainable industrial practices.

There are, however, some moral and political issues with policies like these, often being labeled as "protectionist," among other criticisms [22]. This will, in fact, generate significant economic advantages for the host nation and obviously benefit the manufacturers themselves. Given we are headed toward a more protectionist market globally, the intricate balance between protectionism and free trade becomes increasingly apparent [23]. For the purposes of this chapter, geopolitical uncertainty risks are out of scope; here, we simply focus on policies relevant for industries looking to decarbonize and the economic incentives.

2.2.2.4 Supporting Industries

Through ETS and CBAM, industries need to understand how to exploit these policies to maximize economic benefit.

For manufacturers, there are several key considerations:

1. Manufacturers Outside a Joint CBAM but within an ETS: The UK serves as a prime example. Despite not being an EU member, the UK has implemented its own ETS and is in the process of developing a CBAM policy. UK and EU manufacturers operate within separate carbon markets, but a UK manufacturer exporting to the EU would pay the difference between the UK and EU carbon prices. For instance, if the UK carbon price is \$50/tCO_2 and the EU ETS is \$70/tCO_2, the manufacturer would only pay the \$20 difference [24, 25]. Without such agreements, there is a risk of double taxation.
2. Manufacturers Outside a CBAM Completely: Manufacturers in countries like the US, which currently do not have a CBAM, would incur costs based on the ETS of the importing region. For example, US manufacturers exporting to the EU would face the full cost of the EU ETS without any offsets from a local carbon pricing mechanism [26].
3. Business as Usual for Countries with No Carbon Pricing Mechanisms: Manufacturers in countries without carbon pricing mechanisms exporting to other nations without such systems continue business as usual and have an economic advantage. However, they risk future policy changes and may find themselves at a competitive disadvantage if global carbon pricing becomes more prevalent [27].

There is also the added risk of being reliant on a nascent carbon market, where prices may be unpredictable. Financial derivatives could be useful tools to lock in or hedge against certain costs of carbon. Utilizing instruments such as carbon futures and options can help companies manage the volatility in carbon prices by providing a way to secure prices for future carbon credits. However, this is a complex task that requires expertise in financial markets and is beyond the scope of this chapter.

The key message for the reader is the importance of identifying relevant variables and factors. Manufacturers must plan ahead and consider different outcomes to reduce uncertainty and increase resilience, by staying aware of ongoing international policies.

2.2.2.5 Energy Markets and Procurement

Equally important as a carbon market is the energy market, particularly the changes in the cost of current fossil fuels and future low-carbon fuels. This is especially relevant for manufacturers considering the use of hydrogen or electrifying their processes. While a detailed analysis of energy market dynamics, future predictions, and the integration of markets with policies and external factors is beyond the scope of this chapter, it is up to the reader to understand the relevant factors and make informed decisions based on them.

When considering hydrogen markets, or perhaps fuel procurement in general, given that it is a nascent market, purchasing agreements are hugely important. Purchase agreement contracts can be used to mitigate financial uncertainty. These could include a single long-term contract to cover base demand and multiple short-term contracts to exploit market volatility, which is important for a nascent market

[28, 29]. Industry needs to examine how these contracts can be arranged to mitigate financial uncertainty.

Yet, being reliant on an emerging energy market requires the development of sufficient infrastructure, production, and demand. An electricity market already exists, but the key issue here is the need for green electricity. This relies on countries developing their green electricity infrastructure, which varies from country to country. For instance, countries with established green electricity sources like France (nuclear) and Norway (hydropower) have lower CO_2 emissions for electrified processes [30, 31].

Increasing electricity demand can also strain the electrical grid, but it presents opportunities to participate in ancillary markets. If the industrial process in question is flexible enough, load shedding can be economically feasible [32].

2.2.3 Industrial Heat

Industrial processes require significant amounts of heat, known as process heat, which accounts for, on average, 50% of all energy use in a typical process, making it an attractive target for decarbonization efforts [33]. Depending on the heating requirements, industry can be classified as low-, medium-, or high-temperature processes (Table 2.1) [34].

The amount of heat needed can differ greatly depending on the specific process in question. Applications of process heat include fluid heating, distillation, drying, smelting, and many others. The overall heating requirements play an important role in the feasibility of making a low-carbon transition.

This "heat" can manifest in different forms. Direct heat implies that the material to be heated is in direct contact with the flame or heat source, common in high-temperature industries such as metallurgy and cement production. Indirect heating uses an intermediate source, such as steam or thermal fluids, due to their excellent thermal and transport properties [35].

2.2.4 Current State of Industrial Energy

Industrial process heat can be generated through various methods. In the UK alone, there are over 4,300 pieces of equipment, each consuming more than 1 MW of energy [36]. Understanding the existing infrastructure and equipment is important, as it influences the decisions necessary for a transition, including practical considerations and their significance.

TABLE 2.1
Classifying Process Heating Requirements for Industry

Classification	Temperature (°C)
Low	Below 100
Medium	Between 100 and 500
High	Over 500

2.2.4.1 Furnaces

Industrial furnaces are essential for generating high-temperature heat and are typically used in situations requiring direct heat application. They can run on various fuel types like natural gas, propane, coal, and diesel.

The long economic lifespans of these large-scale assets – typically lasting between 15 and 35+ years depending on their usage – inhibit decarbonization progress. They can't easily be replaced due to the large capital and expertise requirements [37]. However, electric arc furnaces, powered by electricity, are gradually being adopted by the metallurgical industries [38]. These recent innovations offer superior control over the heating process, ensuring more uniform heating and improved energy efficiency.

2.2.4.2 Boilers and Thermal Fluid Systems

Boilers, typically fueled by natural gas, provide indirect heating to produce process steam, constituting nearly 40% of all gas-consuming equipment with a lifespan of 25 to 40+ years [39]. Many boilers are maintained beyond this period, and so industries are debating their timing for decarbonization if investments have to be made regardless of their assets' end of life.

Thermal fluid systems, using transfer fluids as heating mediums, are often preferred over steam-based systems when high temperatures are needed. Achieving similar temperatures with steam would demand higher operating pressures [40]. For example, mineral oils display excellent thermal stability and pumpability at low- to medium-temperature chemical plants. At 250 °C, steam would reach 50 bar pressure, while a common thermal fluid would only reach 6 bar.

2.2.4.3 Combined Heat and Power

Combined Heat and Power (CHP) involves the simultaneous generation of electrical and thermal energy and is generally used for medium-temperature processes [41]. Fuel is combusted to produce steam by heating water. This steam is then directed through a turbine to produce electrical power, while the remaining thermal energy is utilized for indirect process heating.

CHP systems, a relatively recent development adopted for industrial process heat in the last 10–20 years, are considered "greener" than alternatives due to their efficiency, exempting them from regulations like the UK's Climate Change Levy. However, given their primary consumption of natural gas, there are still considerable emissions.

2.2.4.4 Dryers and Microwaves

Drying processes, often found in industries like food, chemicals, and pharmaceuticals, span the low-temperature process range and include fluidized beds and spray dryers. Electricity often powers these processes to ensure control and consistent heating for superior product quality. Where control isn't necessary, conventional fossil fuels or waste heat can be used.

Microwaves (electrically powered) are typically used instead of dryers when rapid heating in small amounts is required [42]. This, in fact, gives a "simple" decarbonization opportunity, as already electricity-driven processes can simply be powered by green electricity already available at a higher cost.

2.3 DECARBONIZATION TECHNOLOGIES AND STRATEGIES

This section gives a brief overview of the various technologies, both current and emerging, that can help industrial plants achieve decarbonization. We'll look at a range of options that should be factored into any sizing, optimization, and decision-making processes. By understanding these technologies, including their practical limitations, stakeholders can make more informed choices.

2.3.1 Carbon Capture Technologies

Carbon capture utilization and storage involves capturing CO_2 directly from a source and finding methods to transport and utilize or store it. It's a key decarbonization option for hard-to-abate sectors, allowing minimal process changes to continue "business as usual." However, its current high cost makes it unattractive unless it becomes more affordable or policy measures mandate a shift.

2.3.1.1 Amine Capture

Amine systems are employed for post-combustion carbon capture, directly capturing CO_2 from flue gas streams. The process involves an absorption column where CO_2 reacts with a solvent, such as monoethanolamine (MEA), followed by a stripping column where CO_2 is recovered through high-energy thermal reclamation and the solvent is recycled [43].

The cost of amine-based CO_2 capture varies widely depending on flue gas composition, CO_2 concentration, and desired purity [44]. For example, from $29 to $250 per ton of CO_2 captured for oil refineries, and $17 to $164 per ton of CO_2 captured for cement production [45].

2.3.1.2 Operational Challenges

Factors like amine concentration, CO_2 loading, flue gas temperature, and solvent corrosion characteristics influence safety and cost-efficiency [46]. Understanding these factors is critical for successful implementation.

- Flue gas concentration – Higher concentrations of CO_2 consume less energy.
- Solvent corrosion – Harms equipment and lowers efficiency. Selecting solvents for their corrosion traits and incorporating inhibitors can mitigate this.
- Solvent concentration – Higher amine concentration improves CO_2 absorption up to a point, beyond which effects diminish.
- Solvent degradation – Prolonged exposure to compounds like CO_2, O_2, SO_x, and fly ash in flue gases leads to degradation. Regulating these factors is vital for efficiency.

2.3.1.3 Membrane capture

The application of membrane separation for carbon capture is new, but promising, as it demonstrates low energy consumption and operational cost in comparison to amine technologies [47]. Membrane systems are also more easily retrofitted as installation and operation are far less complex [48]. It is more effective to use an amine system at lower concentrations of CO_2 but membrane systems at higher concentrations. Yet, its novelty limits commercial usage, and a significant gap remains in practical industry applications due to uncertainty about their long-term performance under real-world conditions [49]. Membrane systems are, however, a perfect example of a promising technology expected to be highly relevant in the future. Therefore, they should be considered in investment timing models, with similar examples discussed later in the chapter.

2.3.2 Hydrogen

Hydrogen is a fuel-switching technique where it is directly substituted into existing systems, such as natural gas-fired CHP plants, to minimize the need for significant plant-wide infrastructure changes.

2.3.2.1 Hydrogen Production and Emissions

Hydrogen production methods include gray hydrogen (steam methane reforming), blue hydrogen (with CO_2 capture), and green hydrogen (renewable energy-driven electrolysis) [50]. Of course, CO_2 isn't the only greenhouse gas, and the combustion of green hydrogen emits three times more NO_x than natural gas at temperatures above 1,300 °C [51]. Methods such as flue gas recirculation (FGR) or steam injection can mitigate high NO_x levels [52].

2.3.2.2 Industrial Applications of Hydrogen

Hydrogen can be blended with other fuels for use in conventional industrial energy systems such as CHP, boilers, and furnaces. A mixture of up to 25% hydrogen and 75% methane can be used without system modifications [53]. Recent developments enable the use of 100% hydrogen as a fuel source through retrofitting, facilitating complete decarbonization [54].

2.3.2.3 Operational Considerations

Compared to natural gas, hydrogen has a faster flame speed, higher flame temperature, and lower stoichiometric air requirement [55]. These characteristics require specialized burner systems and safety measures to handle hydrogen's properties. Hydrogen's high flammability and potential for embrittlement necessitate comprehensive monitoring and safety protocols.

2.3.3 Biomass

Biomass, a renewable resource, and another fuel-switching technique, refers to organic material from plants or animals, such as wood, crops, and waste [56]. It can be used for process heating through direct combustion or gasification.

2.3.3.1 Biomass Emissions

Biomass generates higher CO_2 emissions than coal and natural gas when combusted, offset by CO_2 absorption during growth [57]. Factors such as origin, growth duration, and harvest time affect CO_2 sequestration. BECCS (Bioenergy with CCS) can achieve negative emissions by capturing CO_2 from biomass combustion [58].

2.3.3.2 Biomass Combustion in Industrial Applications

Direct combustion of solid biomass is used for large-scale CHP power generation. For instance, the Drax power station in the UK utilizes compressed wood pellets to generate 2.6 GW of electricity, accounting for 6% of the country's total electricity demand [59]. Fluidized bed systems are also employed for biomass combustion due to their ability to handle diverse fuels and high combustion efficiencies [60].

Bio-coal, resulting from solid biomass pyrolysis, serves as a direct replacement for coal, yielding less ash, sulfur, and NO_x emissions [61].

2.3.3.3 Operational Considerations

Biomass can meet any process heating requirements, though its form depends on the specific application. Biomass feedstocks exhibit significant differences in properties, requiring specific methods for collection, storage, and transportation [62]. Key characteristics include moisture content, energy content, ash content, and particle size.

The direct combustion of biomass poses challenges such as tar and ash formation, causing fouling and slagging issues [63]. Measures like soot blowing and monitoring temperature variations are necessary to maintain burner system performance [64].

Biomass is already widely used in industries like paper and pulp (where paper is considered as the biomass source), and energy consumption is regarded as low-carbon, leaving fewer opportunities for decarbonization through fuel switching. Enhancing efficiency and process intensification are alternative options for such industries [65].

2.3.4 Electrification

Electrification involves using low-carbon electricity to process heat. Outlined here are the current technologies that exist that enable an industrial plant to electrify itself.

2.3.4.1 Electric Resistance Heating

Electric resistance heating converts electricity to heat through electrical resistors and is commonly used in industries like glass and ceramics for precise control and higher-quality end products [66]. Electric boilers can replace existing steam generation methods, providing zero-emission operation and ease of maintenance [39]. Electric arc furnaces are also increasingly used in steel-making, offering an efficient and cleaner alternative for melting scrap steel and producing new steel products. Both natural gas furnaces and electrical systems need water treatments like softening or ion exchange. But electrical systems can degrade faster due to resistors' sensitivity to feed-water impurities [67].

2.3.4.2 Induction Heating

Induction heating directly heats fluids like hydrogen, natural gas, and air through a magnetic field in an electrically conductive material, inducing eddy currents. It is used in metalworking, automotive, aerospace, and other manufacturing sectors for precise control and consistent heating [68].

2.3.4.3 Heat Pumps

Heat pumps operate by using electrical energy to transfer heat from a low-temperature source to a higher temperature, effectively upgrading the temperature of a stream for process heat. They work on a thermodynamic refrigeration cycle, which involves the compression and expansion of a refrigerant. This process absorbs heat from the surroundings (low-temperature source) and releases it at a higher temperature, making heat pumps highly efficient within a low-temperature range. This makes them particularly suitable for industries such as chemicals and paper [69].

2.3.4.4 Operational Considerations

Electrification enables flexible use of renewable energy but requires stable electricity supply and infrastructure. The economic feasibility of electrification depends on the cost and carbon intensity of electricity, which varies by location [70].

Traditional fossil fuel sources generally have more stable infrastructure, but as global electricity grids aim to decarbonize, the increased proportion of volatile renewable energy sources could lead to greater grid instability [71]. Electrified industries need to ensure resilience to frequent grid outages by implementing sufficient backup storage and other contingency measures.

2.4 DESIGNING OPTIMAL DECARBONIZATION STRATEGIES

We've discussed the existing technologies and policies necessary for decarbonization. The question at hand is: How do we use this to decarbonize our industrial plant? The answer is highly specific to the industrial process in question. To address this, we can use optimization techniques, which help answer sizing, economic, and emission-related questions and provide insight on how to manage and mitigate risk.

2.4.1 Investment Timing and Technological Evolution

Understanding what to invest in is essential, but timing these investments is perhaps even more important in such a dynamic market. It can be influenced by market conditions (e.g., avoiding investments when the emissions cap is low) and the impact of technological lock-in. If your current system is relatively new/operates efficiently, it will deter immediate investment, to await further system degradation or technological advancements.

Example:

Timing Investments Using Mixed Integer Linear Programming (MILP)

Consider the decision to invest in new low-carbon technology. The binary variable y_t represents the investment decision at time t:

$$y_t = \begin{cases} 1 & \text{if investment is made at time } t, \\ 0 & \text{otherwise.} \end{cases}$$

Using the Big-M method, we can express the investment constraints by introducing a large constant M to linearize binary (decision logic) constraints, enabling the model to switch between investment and non-investment states as explained by the logic below:

$$x_t \leq M \cdot y_t \forall t \in \mathcal{T} \tag{2.1}$$

$$x_t \geq y_t \forall t \in \mathcal{T} \tag{2.2}$$

$$y_t \in \{0,1\} \forall t \in \mathcal{T} \tag{2.3}$$

These constraints ensure that:

- When $y_t = 0$, x_t must be zero.
- When $y_t = 1$, x_t can take any value up to M.

In this context, x_t is a continuous variable representing the level of action taken at time t. When an investment is made at time t ($y_t = 1$), x_t can assume a non-zero value, subject to an upper bound M. Conversely, if no investment is made ($y_t = 0$), x_t must also be zero.

Below is an example of incorporating this into an optimization framework, using heat pumps as a case study for technological lock-in. This can be adapted for various systems, such as fuel switching.

In the given formulation, the indices i and p are defined as follows:

- $i \in \mathcal{I}$ represents the discrete time periods within the planning horizon.
- $p \in \mathcal{P}$ denotes heat pumps that can be purchased.

Investment decisions:

$$\forall (i,p) \in \mathcal{I} \times \mathcal{P}, \quad A_{\text{pump}}(i,p) = \begin{cases} P_{\text{purchase}}(i) & \text{if } i = 0, \\ A_{\text{pump}}(i-1,p) + P_{\text{purchase}}(i,p) & \text{otherwise,} \end{cases} \tag{2.4}$$

$$\forall p \in \mathcal{P}, \quad I_{\text{pump}}(p) = \sum_{i \in \mathcal{I}} P_{\text{purchase}}(i,p), \tag{2.5}$$

$$\forall (i,p) \in \mathcal{I} \times \mathcal{P}, \quad O_{\text{pump}}(i,p) \leq C_{\text{installed}}(p) \cdot A_{\text{pump}}(i,p), \tag{2.6}$$

$$N_{\text{pumps}} = \sum_{i \in \mathcal{I}} \sum_{p \in \mathcal{P}} P_{\text{purchase}}(i,p), \tag{2.7}$$

$$\sum_{p \in \mathcal{P}} C_{\text{installed}}(p) \leq W_{\text{maxwaste}}. \tag{2.8}$$

Equation (2.4) defines $A_{pump}(i, p)$, indicating heat pump availability as a binary outcome influenced by the decision variable $P_{purchase}(i, p)$. This binary outcome reflects activation status, dependent on prior and current purchase decisions. Equation (2.5) aggregates these decisions across all time periods, providing a cumulative investment timing metric. Equation (2.6) limits operational output, tying it to the product of availability and installed capacity. Equation (2.8) imposes a constraint on total capacity, ensuring that the summed installed capacities do not exceed the maximum waste heat resources available.

2.4.1.1 Objective Function

The objective function (Equation 2.9) aims to minimize the total costs (simplified for clarity). The installation costs are a function of capacity, working temperature, and the number of pumps. Electricity cost is estimated based on the heat pump's coefficient of performance (COP), which is calculated considering the extent of temperature lift, waste heat composition, volume, and other factors. The cost of heat is also included, as the purpose of a heat pump is to minimize overall process heating requirements, and so it is assumed that a heat exchanger will be used to offset existing process heat, reducing overall costs if used optimally.

$$\min : C_{installation} + C_{electricity} + C_{heat} \tag{2.9}$$

Figure 2.1 illustrates the outcomes of these constraints. In Scenario 1, an investment in heat pumps is made at time interval 4 since it's more valuable to utilize waste heat earlier to exploit the time value of money. This contrasts with Scenario 2, where

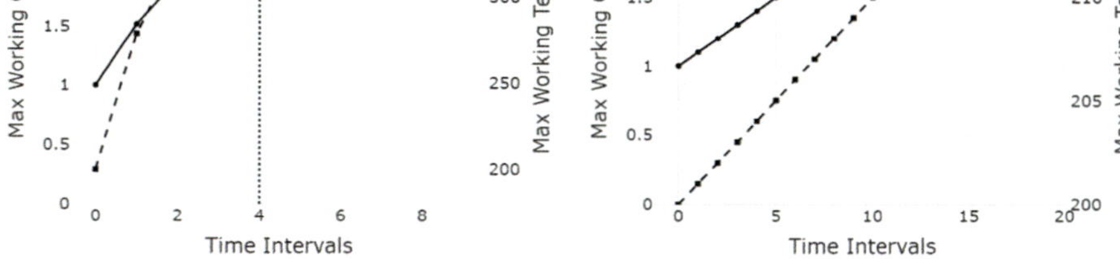

(a) Scenario 1: The performance characteristics of the heat pump are expected to increase rapidly in the near future but then slow down, prompting an investment at time interval 4.

(b) Scenario 2: The performance characteristics of the heat pump are expected to linearly increase steadily, yet the improvement rate is too slow to justify a worthy investment.

FIGURE 2.1 Comparison of two hypothetical scenarios based on the expected performance improvements of a heat pump, to determine the optimal timing for investment. In Scenario 1, rapid initial performance gains justify an investment at time interval 4. In Scenario 2, steady linear improvements are insufficient for an investment within the observed period.

such an investment is not justified. This demonstrates a basic framework for making temporal investment decisions, which can be expanded to include the impact of an ETS, integration with other systems, weather conditions, and more.

It is important to note that the data used is forecasted, meaning that as a decision-maker, you can only infer how a technology might evolve in the future, introducing uncertainty. This uncertainty can be managed through sensitivity analysis, as outlined in the next section.

2.4.2 Uncertainty

Uncertainty is inherent in the process of decarbonization due to the dynamic nature of technological advancements, market conditions, and policy changes. Effectively managing these uncertainties is crucial for making informed decisions.

2.4.2.1 Sensitivity Analysis

Sensitivity analysis evaluates how changes in key variables impact outcomes, helping to identify critical factors that influence decision-making. For instance, a company might examine how variations in carbon prices, technology costs, and regulatory requirements affect the viability of carbon capture investments.

2.4.2.2 Scenario Planning

Scenario planning involves developing multiple plausible futures to explore the impacts of different assumptions and uncertainties. Companies can create scenarios such as aggressive climate policies, technological breakthroughs, economic downturns, or energy crises to assess how their decarbonization strategies perform under each condition.

2.4.2.3 Generalized Real Options Analysis

Real options analysis provides a framework for evaluating the flexibility of different investment decisions, considering the uncertainties in technological advancements and market conditions. By treating investments as options that can be exercised in the future, companies can make more adaptive and resilient decisions. Scenario analysis allows us to handle uncertainty by optimizing investment decisions under various future conditions. Each scenario represents a different set of assumptions about technological advancements, market prices, and policy changes.

Step 1: Define Decision Variables

- $y_{t,k}$: Binary variable indicating whether to invest in technology k at time t.
- $x_{t,k}$: Continuous variable representing the level of investment in technology k at time t.

Step 2: Objective Function

The objective function (Equation 2.10) in this case aims to maximize the net present value (NPV) of the investment, considering the costs and benefits of each technology across different scenarios.

$$\max \sum_{s \in \mathcal{S}} p_s U \left(\sum_{t \in \mathcal{T}} \sum_{k \in \mathcal{K}} \left(B_{t,k,s} \cdot x_{t,k,s} - C_{t,k,s} \cdot y_{t,k,s} \right) \cdot e^{-rt} \right) \tag{2.10}$$

where:

- $\mathcal{S}$ is the set of scenarios.
- p_s is the probability of scenario s.
- $B_{t,k,s}$ is the expected benefit from investing in technology k at time t under scenario s.
- $C_{t,k,s}$ is the cost of investing in technology k at time t under scenario s.
- $U(\cdot)$ is the utility function representing the decision-maker's risk preferences.
- r is the discount rate.

The term e^{-rt} represents the discount factor, which adjusts future cash flows to their present value, reflecting the time value of money. Note that this can be pre-solved.

Step 3: Constraints

Investment Constraints:

$$x_{t,k,s} \leq M \cdot y_{t,k,s} \forall t \in \mathcal{T}, \forall k \in \mathcal{K}, \forall s \in \mathcal{S} \tag{2.11}$$

This constraint ensures that the investment level $x_{t,k,s}$ in technology k at time t under scenario s can only be positive if the binary decision variable $y_{t,k,s}$ is 1, indicating that an investment is made.

$$x_{t,k,s} \geq 0 \quad \forall t \in \mathcal{T}, \forall k \in \mathcal{K}, \forall s \in \mathcal{S} \tag{2.12}$$

This ensures that the investment level $x_{t,k,s}$ is non-negative, preventing the model from assigning negative investment levels, which would be unrealistic in practical scenarios.

$$y_{t,k,s} \in \{0,1\} \quad \forall t \in \mathcal{T}, \forall k \in \mathcal{K}, \forall s \in \mathcal{S} \tag{2.13}$$

Budget Constraints:

$$\sum_{k \in \mathcal{K}} C_{t,k,s} \cdot y_{t,k,s} \leq B_t \quad \forall t \in \mathcal{T}, \forall s \in \mathcal{S} \tag{2.14}$$

where B_t is the budget available at time t.

This constraint ensures that the total cost of investments in all technologies k at time t under scenario s does not exceed an available budget B_t.

Technology Evolution Constraints:

$$x_{t,k,s} \leq E_{t,k,s} \cdot y_{t,k,s} \forall t \in \mathcal{T}, \forall k \in \mathcal{K}, \forall s \in \mathcal{S} \tag{2.15}$$

This constraint ensures that the level of investment $x_{t,k,s}$ does not exceed the efficiency or capacity $E_{t,k,s}$ of technology k at time t under scenario s, where $E_{t,k,s}$ represents the efficiency or capacity of technology k at time t under scenario s.

This framework enables a more generalized comparison of systems and scenarios, to even consider a decision-maker's risk preferences, discount rates, and probabilities of scenarios.

2.4.3 Emission Reduction Targets and Constraints

The emissions reduction target is specific to corporate objectives or the local, national, and international regulations governing the site. Below is a basic example framework for optimizing carbon trading.

2.4.3.1 Carbon Credits and Emissions

$$\forall i \in \mathcal{I}, \quad C_{\text{credits}}(i) + C_{\text{offset}}(i) \geq E_{\text{total}}(i) - E_{\max}(i), \tag{2.16}$$

$$\forall i \in \mathcal{I} \setminus \{0\}, \quad C_{\text{earned}}(i) + \left(E_{\max}(i) - E_{\text{total}}(i) + C_{\text{offset}}(i)\right) \cdot \left(1 - \beta(i)\right), \tag{2.17}$$

$$\forall i \in \mathcal{I}, \quad C_{\text{credits}}(i) + C_{\text{offset}}(i) \leq M \cdot \beta(i), \tag{2.18}$$

$$\forall i \in \mathcal{I} \setminus \{0\}, \quad C_{\text{sold}}(i) \leq C_{\text{bank}}(i-1) + C_{\text{earned}}(i) - C_{\text{offset}}(i), \tag{2.19}$$

$$\forall i \in \mathcal{I}, \quad C_{\text{bank}}(i) \leq C_{\text{bank}}(i-1) + C_{\text{earned}}(i) - C_{\text{offset}}(i) - C_{\text{offset}}(i), \tag{2.20}$$

$$\forall i \in \mathcal{I}, \quad E_{\text{total}}(i) - E_{\max}(i) \leq M \cdot \beta(i). \tag{2.21}$$

These equations form a framework for managing carbon emissions and credits within a carbon trading system. Equation (2.16) ensures that the sum of carbon credits and offsets for each interval i compensates for any excess emissions beyond the maximum allowed, $E_{\max}(i)$. Equation (2.17) calculates carbon credits earned based on emissions savings relative to $E_{\max}(i)$, modulated by $\beta(i)$, a binary variable indicating compliance with emission caps or not using the Big-M method. Equation (2.18) sets a cap on the combined carbon credits and offsets, controlled by policy or market constraints. Equations (2.19) and (2.20) govern the dynamics of carbon credit transactions, balancing sold and banked credits over time, and factoring in newly earned credits used to offset future obligations. Lastly, Equation (2.21) ensures total emissions do not exceed the cap by more than the mitigated amount through purchased credits, hence enforcing compliance.

It is important to note that this framework relies on inherently uncertain future data, such as the price of carbon. It assumes known future prices, but in reality, these prices are always unknown. While the model allows for banking carbon credits (holding onto them) to offset future emissions, it could theoretically enable the purchase of unlimited credits if future prices were known, ensuring a guaranteed return. Therefore, carbon credits should only be purchased when the plant is operating below its emissions cap.

Banking carbon credits also has a risk component, akin to hedging. This can be beneficial if managed with the correct expertise and appropriate constraints are applied considering the risk appetite of the investor.

It is also important to remember that carbon is typically not traded on a spot market. Instead, there is an end-of-period settlement process, where all carbon-related trades, transactions, and accounts are finalized at designated intervals, such as on an annual basis.

This framework is flexible; by multiplying the number of credits needed by their cost, one can determine the cost of emitting. This cost varies by region, and when exporting, the difference between markets encapsulates the entire trading effect. E_{max} captures the effect of a variable cap. If more stringent corporate emissions policies are in place, this variable can reflect that. In an optimization framework, it's also simple to add additional constraints to reflect other factors, such as tax incentives.

2.4.4 Application in CHP System Optimization

2.4.4.1 Model Formulation

The objective function (Equation 2.22) is designed to minimize the total costs over a planning horizon, accounting for capital and operating expenditure (CAPEX and OPEX) of both the new hydrogen system and the existing natural gas system, as well as the cost of carbon credits for emissions. In this case, a manufacturer is considering either retrofitting or purchasing an entirely new CHP burner.

Objective Function:

$$\min \sum_{i \in \mathcal{I}} \left(C_{\text{capex}} \cdot y_t + C_{\text{opex,hydrogen}}(i) \cdot x_t + C_{\text{opex,naturalgas}}(i) \cdot (1 - x_t) \right) + \sum_{i \in \mathcal{I}} C_{\text{carbon}}(i) \cdot P_{\text{carbon}}(i) \tag{2.22}$$

where:

- C_{capex} is the capital expenditure of the new/retrofitted hydrogen system.
- $C_{\text{opex,hydrogen}}$ is the operating cost of the new/retrofitted hydrogen system.
- $C_{\text{opex,natural gas}}$ is the operating cost of the existing natural gas system.
- C_{carbon} is the number of carbon credits purchased at interval *i*.
- y_t is the binary variable indicating investment in the new system at time *t*.
- x_t is the continuous variable representing the operation of the new system at time *i*.

After optimizing under different scenarios, sensitivity analysis can be performed to understand the impact of changes in key parameters (e.g., market prices, discount rate, volatility) on investment decisions. This involves varying one parameter at a time while keeping others constant to observe how sensitive the optimal solution is to changes in that parameter.

For the purposes of this chapter, we only focus on fundamental examples that address investment timing decisions. This simplified model serves as an introductory framework and can be expanded to incorporate additional complexities, such as operational downtime resulting from new investments, variations in system efficiency, and other relevant factors.

2.4.4.2 Model Implementation

The above example model can be implemented in various ways, incorporating integer variables and linear terms. However, some functions are inherently non-linear and include quadratic or bilinear terms, such as capital cost terms or degradation, which would otherwise be misrepresentative.

Therefore, it is important to use a solver capable of handling integer values and bilinear terms to ensure the solution is as close to a global minimum as possible. Numerous open-source and commercial solvers are available for this purpose, including Gurobi [72], Couenne [73], and Baron [74]. These solvers can be implemented in various modeling languages such as GAMS [75], Pyomo [76], AMPL [77], and more. In this case, the model was solved using Gurobi within Pyomo.

First, we assume the presence of an emissions cap that gradually reduces over time. We also assume the existence of hydrogen infrastructure, ensuring the safe and reliable procurement of hydrogen to fuel the system. This infrastructure includes storage, transportation, and distribution networks necessary for the integration of hydrogen as a primary energy source. For the specific scenario considered, an existing CHP system is retrofitted to be compatible with hydrogen combustion. This retrofitting includes modifications to burners and control systems to handle hydrogen. Once an investment is made, it is assumed the system itself is fully operational instantaneously with no downtime.

In this case, a CHP model has already been built, which considers mechanical constraints such as ramping rates, energy ratios, and efficiencies based on hourly loads. Energy produced from the CHP system is used either to cover the plant's energy requirements or sold to the grid as excess profit. If an investment in hydrogen is made, the electricity produced is also assumed to be low-carbon. This specific model is run for 8,440 hours (representing one year of operation), and the results are then extrapolated to cover the entire investment horizon of 30 years. The model's outputs for one year are assumed to be representative of each subsequent year, considering factors such as degradation, maintenance, and efficiency improvements. This gives the operational characteristics used to model capital and operating costs, as well as emissions based on hourly demand data.

2.4.4.3 Model Output

According to the model, hydrogen should start being used after three years to maximize economic viability. This recommendation is primarily driven by the

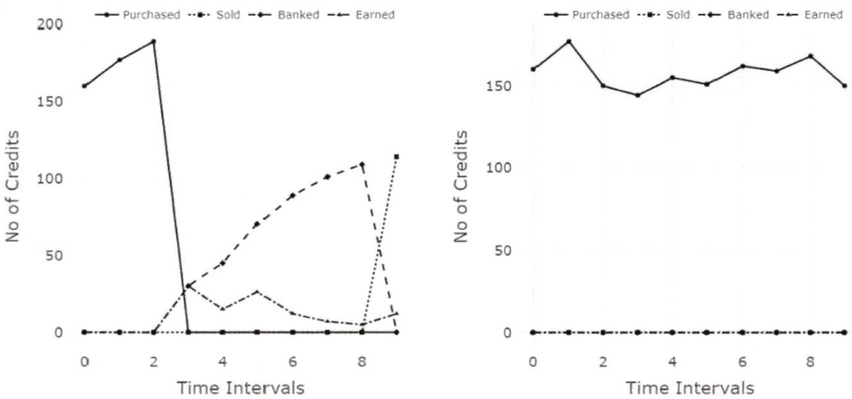

(a) Scenario 1: The cost of emitting CO_2 increases significantly over time, prompting an investment into H_2 technology to reduce emissions

(b) Scenario 2: The cost of emitting CO_2 remains stable, and so no capital investment is made as this would work out more expensive over the investment horizon.

FIGURE 2.2 Comparison of two hypothetical scenarios based on the cost dynamics of carbon credits and their impact on investment decisions in a CHP system integrated with an ETS. The figures illustrate the first 10 years of a 30-year investment horizon, showing the number of carbon credits purchased, sold, banked, and earned over time.

anticipated increase in the cost of emissions over time. While the model's output aligns with our expectations, it raises a significant question: Is this information genuinely useful?

The model relies on hypothetical data regarding the future behavior of the carbon market – essentially, how we anticipate it will evolve. However, a decision-maker can envision numerous potential futures, each with varying degrees of likelihood and impact. Therefore, the true value of this framework lies in its ability to illustrate how the optimal investment timeline might shift under different market scenarios and technological breakthroughs (Figure 2.2).

To maximize its utility, the framework should be applied to evaluate all available technologies, presenting all results to the decision-maker. This allows for a comprehensive comparison of solutions, considering optimized economic costs under various scenarios. By doing so, decision-makers can identify not only the most cost-effective solution but also one that remains resilient across different market conditions, has stable investment horizons, and adapts to technological advancements.

2.5 DECISION-MAKING FRAMEWORKS

2.5.1 Decision-Making Theory

Modeling has allowed us to size and design specific decarbonization technologies, providing insights into their operation in an uncertain world. However, we must consider more than just economic factors; qualitative aspects are equally important. When faced with a range of technologies, how does a decision-maker make the final choice? What is the optimal method for selecting the best option when all necessary information is available?

This section explores several key theories that guide the selection of pathways and technologies to decarbonize, while not solely focusing on economic factors.

2.5.1.1 The Decision-Making Process

This involves a systematic process of defining the problem, identifying alternatives, evaluating these alternatives, and selecting the best option based on specific criteria. It assumes that decision-makers have access to all necessary information and can process it without cognitive limitations. However, Herbert Simon's theory of bounded rationality suggests that decision-makers always operate under constraints such as limited information and cognitive limitations [78, 79]. This theory acknowledges that while decision-makers strive to make rational choices, their ability to do so is limited by the complexity of the environment and their cognitive capacity.

Supporting decision-makers in this context involves acknowledging and addressing these constraints. This complexity includes factors such as technological feasibility, economic viability, regulatory requirements, and stakeholder interests. For instance, when deciding between different carbon capture technologies or low-carbon energy sources, decision-makers must consider not only the cost and efficiency of each option but also the long-term sustainability and potential policy incentives, many of which are still unknown and exogenous, as we've explored in the previous section.

2.5.1.2 Behavioral Economics

Incorporating behavioral economics insights, such as cognitive biases and heuristics, further refines the decision-making process. Cognitive biases (overconfidence and anchoring) and heuristics (availability and representativeness) illustrate that decisions are not always purely rational [80, 81]. For example, a decision-maker might overestimate the benefits of a well-known technology (availability heuristic) or rely too heavily on initial cost estimates (anchoring bias), leading to sub-optimal choices.

To mitigate these biases, decision-making frameworks can integrate behavioral insights in several ways:

- Diverse Perspectives: Ensuring that a wide range of stakeholders are involved in the decision-making process can reduce individual biases and provide a more comprehensive evaluation of options.
- Structured Decision-Making Tools: Tools such as decision trees, MCDA, and scenario planning can help structure the decision-making process, making it more transparent and less susceptible to cognitive biases.
- Behavioral Nudges: Implementing simple interventions, such as framing effects and default options, can guide decision-makers toward more rational choices without restricting their freedom.

By integrating these behavioral insights into the evaluation process, we can design better decision-making frameworks that account for human behavior, ultimately leading to more effective and economic decarbonization strategies.

2.5.2 Multi-Criteria Decision Analysis

Theoretically, any framework that provides additional information about alternatives within a defined problem can be considered a decision-making model, though some structures are more effective than others. Not only that, but this chapter thus far has primarily addressed the economic factors. Any rational individual knows that decisions cannot be made based solely on economics, as all decision-making variables are interrelated, often in complex and indirect ways. Presenting an investor with only the economic impact of a low-carbon technology might seem convincing, but an industrial plant requires considering social and safety aspects, which are not as easily quantifiable. For instance, the adoption of a new hydrogen-based technology may offer substantial economic benefits, but it also introduces safety concerns due to hydrogen's flammability and storage challenges. Also, the social impact on the local community, such as job creation or displacement, must be considered. These factors, while critical, do not lend themselves to straightforward economic valuation and require a broader qualitative assessment. Therefore, MCDA models are a wonderful tool to integrate both quantitative and qualitative factors. Key MCDA models include:

- Analytic Hierarchy Process (AHP): A structured technique for organizing and analyzing complex decisions.
- Technique for Order of Preference by Similarity to Ideal Solution (TOPSIS): A method based on the concept that the chosen alternative should have the shortest distance from the ideal solution.
- Elimination and Choice Expressing Reality (ELECTRE): Methods used to choose, rank, and sort alternatives by comparing them pairwise.

Although the specific intricacies of each MCDA technique can differ, they all serve the same fundamental purpose: to evaluate and rank alternatives based on predefined criteria and, as such, help complex decision-making.

2.5.3 Fuzzy AHP Methodology

The following section focuses on the application of a Fuzzy Analytical Hierarchical Process (FAHP) to an industrial decarbonization problem. This example illustrates how a classical MCDA tool can be applied to a classical industrial decarbonization problem. In this case, FAHP is chosen for its ease and flexibility of application and its ability to handle imprecision in decision-making, as explained in the following section.

FAHP is a technique that employs fuzzy logic to deal with vague and imprecise information [82, 83]. Fuzzy logic is an approach that allows for varying degrees of truth rather than a strict true/false binary, effectively handling uncertainty and imprecision. The application of fuzzy logic enables the examination of both quantitative and qualitative elements and offers a method of weighing each component. This is significant because not everything can be converted directly into numerical values, yet these factors are still critical for decision-making processes.

The first step is formulating the problem. A hierarchy is established (example shown in Figure 2.3) to represent the structure of the problem, typically consisting of three levels:

1. Goal: The primary aim of the decision-making process, such as choosing low-carbon energy systems for a particular case study.
2. Criteria: The elements to be considered during the decision-making process, such as the scoring mechanisms.
3. Alternatives: The different options available, such as alternate technologies.

Decision-makers then carry out pairwise comparisons among elements within each level of the hierarchy to quantify their relative significance. Linguistic evaluations can also be utilized, managing uncertainty by allowing decision-makers to articulate their judgments using fuzzy linguistic terms like “low” or “high” [84, 85].

For n criteria, we have an $n \times n$ matrix M:

$$M = \begin{pmatrix} (l_{11}, m_{11}, u_{11}) & \cdots & (l_{1n}, m_{1n}, u_{1n}) \\ \vdots & \ddots & \vdots \\ (l_{n1}, m_{n1}, u_{n1}) & \cdots & (l_{nn}, m_{nn}, u_{nn}) \end{pmatrix} \tag{2.23}$$

where (l_{ij}, m_{ij}, u_{ij}) is a triangular fuzzy number (TFN) representing the importance of criterion i over criterion j. The diagonal elements satisfy $(l_{ii}, m_{ii}, u_{ii}) = (1, 1, 1)$ since a criterion compared with itself is always equally important [82].

Each of these terms is linked with a membership function, symbolizing its affinity to that function. These membership functions can be represented using triangular, trapezoidal, or Gaussian forms.

A TFN is represented as $A = (l, m, u)$, where l, m, and u stand for the lower, middle, and upper values, respectively. The membership function $\mu_A(x)$ for A is defined as:

$$\mu_A(x) = \begin{cases} \dfrac{x - l}{m - l} & \text{for} \quad l \leq x \leq m \\ \dfrac{u - x}{u - m} & \text{for} \quad m \leq x \leq u \\ 0 & \text{otherwise} \end{cases} \tag{2.24}$$

If given multiple judgments, the responses are aggregated.

In the process of “defuzzification,” fuzzy numbers are transformed into quantifiable values. These values represent weightings or the relative significance of corresponding criteria and alternatives, thus incorporating uncertainty. Typically, the centroid method is used:

$$\text{Centroid} = \frac{l + m + u}{3} \tag{2.25}$$

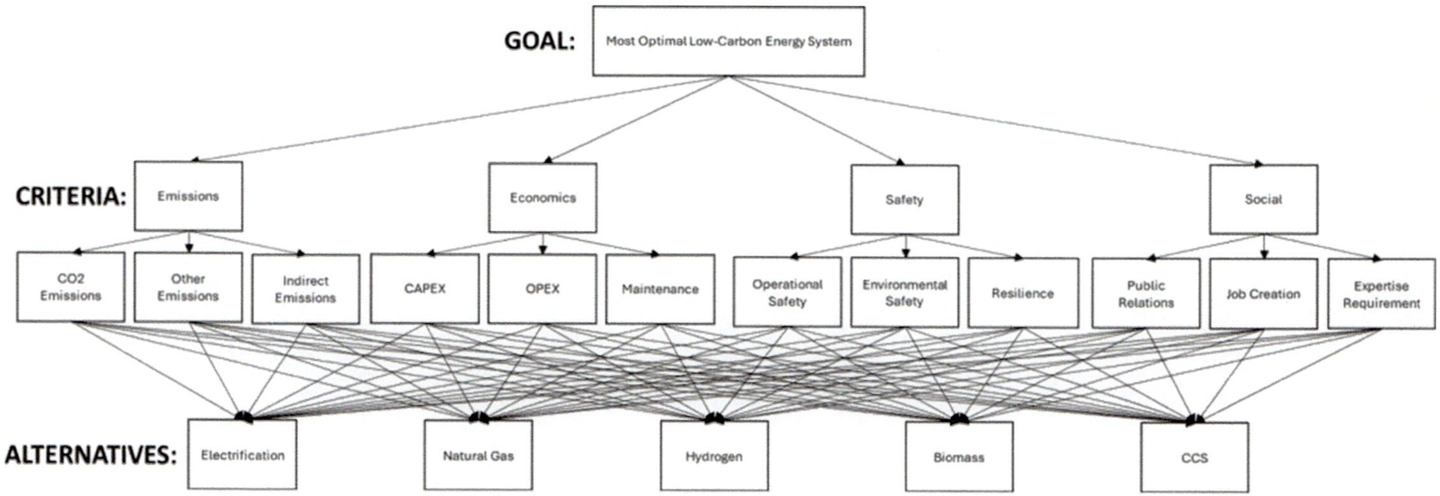

FIGURE 2.3 Example structure for carrying out a fuzzy analytical hierarchy process (FAHP) to choose the most optimal low-carbon energy system for a large energy-consuming specialty chemicals plant.

2.5.3.1 Fuzzy Synthetic Extent and Degree of Possibility

Chang's method of Fuzzy Synthetic Extent Analysis is used to handle the fuzzy comparison matrix. This involves calculating the degree of possibility of one fuzzy number being greater than another [82].

For two triangular fuzzy numbers $A = (l_1, m_1, u_1)$ and $B = (l_2, m_2, u_2)$, the degree of possibility $V(A \geq B)$ is defined as:

$$V(A \geq B) = \min\left(\mu_A(x), \mu_B(y)\right) \tag{2.26}$$

This is further simplified into:

$$V(A \geq B) = \begin{cases} 1 & \text{if} \quad m_1 \geq m_2 \\ 0 & \text{if} \quad l_1 \geq u_2 \\ \dfrac{l_1 - u_2}{(m_2 - u_2) - (m_1 - l_1)} & \text{otherwise} \end{cases} \tag{2.27}$$

From this, the synthetic extent value S for each criterion can be calculated. The overall weight of each criterion is then determined by normalizing these synthetic extent values [82].

$$S_i = \sum_{j=1}^{n} R_j \times A_{ij} \tag{2.28}$$

where S_i is the fuzzy synthetic extent of the ith object, and R_j is the TFN representing the importance of the jth criterion.

To calculate the degree of possibility:

$$V(S_i \geq S_j) = \frac{1}{2} + \frac{1}{2} \times \frac{a_{2i} - a_{2j}}{a_{3i} - a_{1j} + a_{2i} - a_{2j}} \tag{2.29}$$

V is the degree of possibility of S_i being greater than S_j. The normalized weights W_i rank each object:

$$W_i = \frac{\sum_{j=1}^{n} V(S_i \geq S_j)}{n} \tag{2.30}$$

2.5.3.2 Consistency Check Using Fuzzy Consistency Index

To ensure the consistency of the pairwise comparisons, a Fuzzy Consistency Index (FCI) can be calculated. This involves finding the maximum eigenvalue of the fuzzy pairwise comparison matrix and then computing the FCI [83].

1. Calculate and normalize the geometric mean for each row i:

$$W_i = \frac{\left(\prod_{j=1}^{n} a_{ij}\right)^{\frac{1}{n}}}{\sum_{k=1}^{n} x_k} \tag{2.31}$$

2. Compute the Fuzzy Consistency Index (FCI):

$$FCI = \frac{n \times \sum_{i=1}^{n} W_i \times \left(\sum_{j=1}^{n} a_{ij}\right) - n}{n-1} \tag{2.32}$$

3. Calculate the Generalized Consistency Index (GCI):

$$GCI = \frac{FCI}{\lambda_{\max} - n} \tag{2.33}$$

   where $\lambda_{\max}$ is the maximum eigenvalue of the pairwise comparison matrix. We assume $\lambda_{\max} = n$ is perfectly consistent.
   Given:

$$x > y \wedge y > z \Rightarrow x > z \tag{2.34}$$

   A perfectly consistent matrix should maintain this transitive property.

Judgments are deemed "acceptably" consistent if GCI < 0.10. Otherwise, judgments should be reconsidered. These final weightings offer insights into the importance of each evaluated factor. For instance, operating a hydrogen system's cost-effectiveness and safety risks can be weighed appropriately based on their significance. Given quantifiable numbers for each variable through modeling, a decision can therefore be made. Yet, this approach leans heavily on decision-makers' expertise and is thus prone to potential biases (an expert may have an inherent bias against hydrogen technologies, for example). To mitigate this, multiple decision-makers should be used, though it may be time-consuming. Given the novelty of many of these technologies, reliance on experienced individuals can also be limited given little "actual" experience.

2.5.4 Implementing FAHP

FAHPs are most effective when applied to a specific problem. Consider the following example: A specialty chemicals plant aims to select a new low-carbon energy system to replace its current CHP plant. Various aspects of each system, such as economics and emissions, are already known from prior modeling.

First, we define the criteria for decision-making. Using domain knowledge, an FAHP structure is created to select relevant criteria and sub-criteria for making an overall decision. While emissions and economics are already quantified, other aspects like social and safety factors, which are not as easily modeled, also must be considered. The essential criteria and sub-criteria are outlined in Figure 2.3.

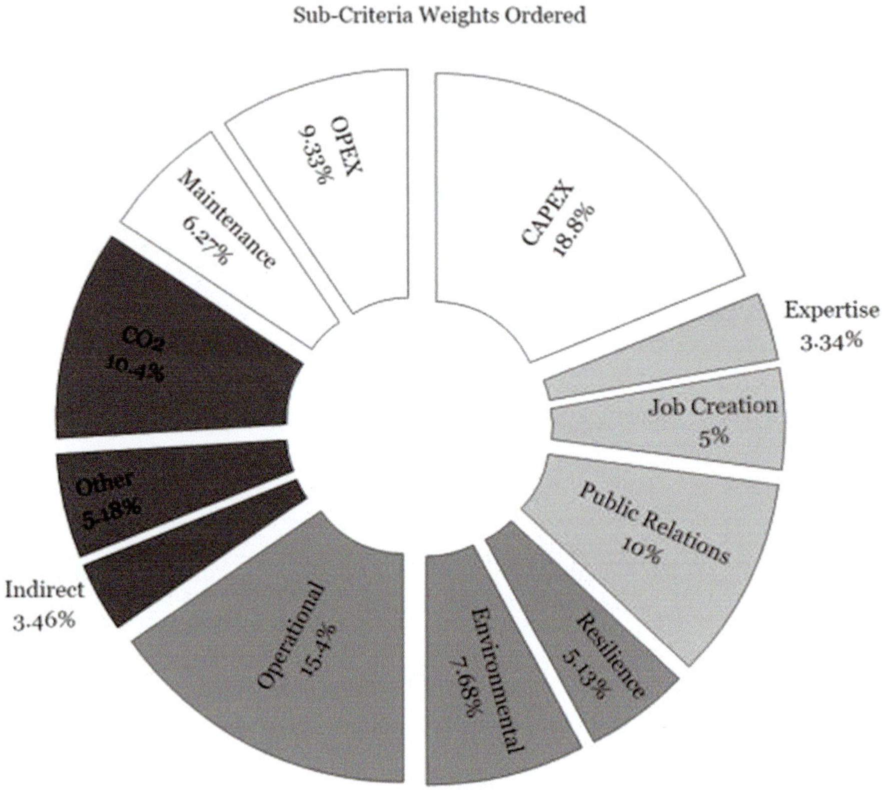

FIGURE 2.4 Weight distribution of sub-criteria for selecting the optimal low-carbon energy system in a large power plant, reflecting their relative importance in the decision-making process.

Once a qualified decision-maker is selected (in this case, a leading chemical engineer with over 30 years of experience), they perform pairwise comparisons between each criterion as described in the previous section. This process results in each criterion having its own weighting. The resulting weights are illustrated in Figure 2.4.

CAPEX, weighted the highest at 18.8%, reflects the significant financial commitment required for these long-term investments. Operational safety, at 15.4%, shows the necessity of ensuring technical feasibility and regulatory compliance. Public relations, weighted at 10%, emphasize the value of maintaining a positive public image and consumer trust for a competitive advantage in a market increasingly driven by sustainability. The significant weighting of CO2 emissions at 10.4% indicates the priority placed on environmental responsibility and adherence to emissions regulations, aligning with corporate (and political) sustainability goals (Figure 2.5).

After determining the weights, economic and emission values are incorporated into the FAHP framework. Qualitative aspects such as safety and social impact are derived using linguistic statements, allowing the decision-maker to influence these

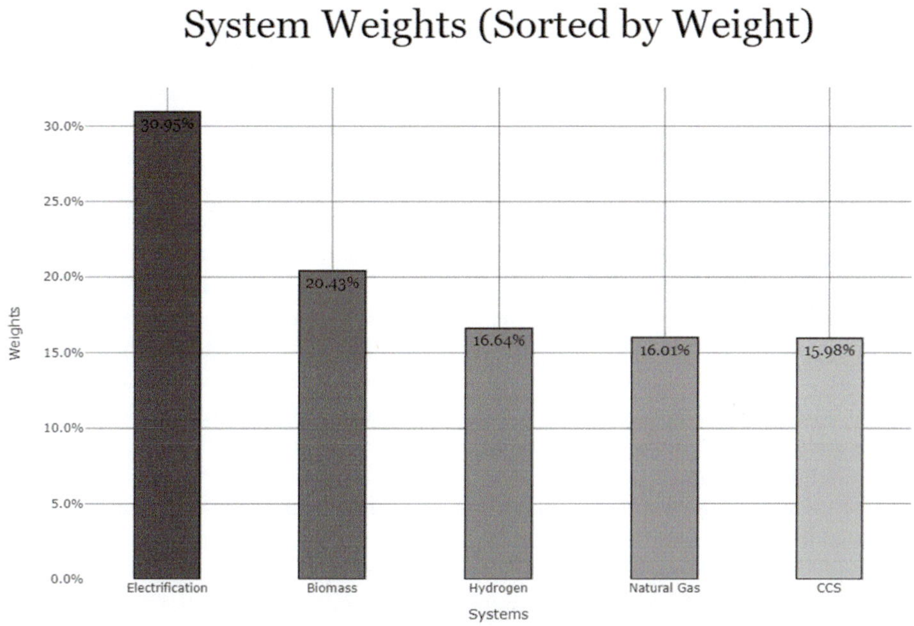

FIGURE 2.5 Final system weights from the FAHP analysis. The graph shows the relative importance of each low-carbon energy system. Electrification is the most favorable option with a weight of 30.95%, followed by biomass (20.43%), hydrogen (16.64%), natural gas (16.01%), and CCS (15.98%). Electrification is approximately twice as good as CCS in this scenario.

factors based on their perceptions – such as the perceived safety (or danger) of hydrogen.

However, this methodology has limitations. First, the FAHP process must be repeated whenever new information is presented or for different sites, making it difficult to generalize. Second, the trustworthiness of these weightings is debatable. As discussed earlier in decision-making theory, decision-makers operate under cognitive biases and limited information. This method does not eliminate these issues. In this case, the bias of a single expert opinion affects the weightings, introducing potential subjectivity.

While models can provide objective economic and emissions data, they cannot determine weightings, especially for qualitative factors like safety and social impact. The problem itself, though well-defined, is inherently complex and subjective. A typical decision-making process would involve multiple experts, whose opinions might differ, leading to varying weightings and potential disagreements. This highlights the importance of consulting multiple experts to reduce bias and create a sensitivity analysis within the "disagreed" range to observe how the final system choice might vary.

To a non-expert, these weights may appear arbitrary without understanding the underlying justification and you need to assess the weightings and consider whether they make sense in the broader context of the decision-making process.

2.5.5 Conclusion

This chapter has explored various aspects of industrial decarbonization, including technological options, economic considerations, policy frameworks, and decision-making tools. We have highlighted the significant contributions of industrial sectors to global CO_2 emissions and the critical role they play in the global economy.

By analyzing technological options such as CCS, hydrogen, biomass, and electrification, we have provided insights into their operational challenges and potential benefits. The discussion on policy frameworks, including ETS and carbon border adjustment mechanisms, highlights the importance of supportive policies in driving decarbonization efforts. Decision-making algorithms like FAHP and real options analysis offer robust methodologies for stakeholders to evaluate and select optimal decarbonization pathways. The practical examples and case studies illustrate the application of these tools and technologies in real-world scenarios, demonstrating their effectiveness and challenges.

In conclusion, integrating technological, economic, and policy-driven strategies is essential for achieving effective and resilient industrial decarbonization. Stakeholders must navigate the complexities of market conditions, technological advancements, and policy changes to develop comprehensive and adaptive decarbonization strategies. This chapter provides a roadmap for achieving these goals, contributing to a sustainable and low-carbon future.

REFERENCES

[1] D. Hodgson, T. Vass, P. Levi, and P. Hugues, "Sectoral overview: Industry," [Accessed June 5, 2023]. [Online]. Available: https://www.iea.org/reports/industry

[2] McKinsey & Company, "Decarbonizing industry will take time and money," [Accessed May 26, 2023]. [Online]. Available: https://www.mckinsey.com/capabilities/sustainability/our-insights/sustainability-blog/decarbonizing-industry-will-take-time-and-money-but-heres-how-to-get-a-head-start

[3] S. Paltsev, J. Morris, H. Kheshgi, and H. Herzog, "Hard-to-abate sectors: The role of industrial carbon capture and storage (ccs) in emission mitigation," *Applied Energy*, vol. 300, Oct. 2021, issn: 03062619. doi: 10.1016/J.APENERGY.2021.117322

[4] M. Balat, "Status of fossil energy resources: A global perspective," *Energy Sources, Part B: Economics, Planning, and Policy*, vol. 2, pp. 31–47, 1 Jan. 2007, issn: 15567249. doi: 10.1080/15567240500400895

[5] European Commission, "EU emissions trading system (EU ETS)," [Accessed June 14, 2023]. [Online]. Available: https://climate.ec.europa.eu/eu-action/eu-emissions-trading-system-eu-ets_en

[6] International Carbon Action Partnership, "ICAP ETS map," [Accessed June 2, 2023]. [Online]. Available: https://icapcarbonaction.com/en/ets

[7] G. P. Hammond, "The UK industrial decarbonisation strategy revisited," *Proceedings of the Institution of Civil Engineers - Energy*, vol. 174, no. 3, pp. 93–101, 2021. doi: 10.1680/jener.21.00056

[8] S. Ren, F. Gozalpour, and B. Tohidi, "CO_2 injection for IOR and storage: Opportunities and challenges for the north sea," *First Break*, vol. 23, no. 4, pp. 31–36, 2005. doi: 10.3997/2214-4609-PDB.12.D14

[9] S. Agada, C. Kolster, G. Williams, and S. Krevor, "Modelling basin-scale CO_2 storage in the bunter sandstone of the UK southern north sea," *Energy Procedia*, vol. 114, pp. 4300– 4311, 2016. doi: 10.3997/2214-4609.201601501
[10] S. Ghanbari, E. Mackay, N. Heinemann, J. Alcalde, A. James, and M. Allen, "Impact of CO_2 mixing with trapped hydrocarbons on CO_2 storage capacity and security: A case study from the captain aquifer (north sea)," *Applied Energy*, vol. 278, p. 115634, 2020. doi: 10.1016/j.apenergy.2020.115634
[11] N. Strachan, R. Hoefnagels, A. Ramírez, et al., "CCS in the north sea region: A comparison on the cost-effectiveness of storing Co_2 in the Utsira formation at regional and national scales," *International Journal of Greenhouse Gas Control*, vol. 5, no. 4, pp. 1000–1009, 2011. doi: 10.1016/J.IJGGC.2011.08.009
[12] R. Haszeldine, "Carbon capture and storage: How green can black be?" *Science*, vol. 325, no. 5948, pp. 1647–1652, 2009. doi: 10.1126/science.1172246
[13] H. Yang, "Is carbon capture and storage really needed?" *Environmental Science & Technology*, vol. 44, no. 16, pp. 6021–6022, 2010. doi: 10.1021/es903626u
[14] T. C. Zhang, R. Y. Surampalli, C. M. Kao, and W. Huang, "Carbon capture and storage: Major issues, challenges and the path forward," *Environmental Engineering and Management Journal*, vol. 14, no. 1, pp. 81–92, 2015. doi: 10.1061/9780784413678.CH18
[15] D. Lewis, "The global state of carbon capture and storage in 2009," *Engineering & Technology*, vol. 4, no. 10, pp. 82–85, 2009. doi: 10.1049/ET.2009.1210
[16] A. I. Osman, M. Hefny, M. I. A. Abdel Maksoud, A. M. Elgarahy, and D. W. Rooney, "Recent advances in carbon capture storage and utilisation technologies: A review," *Environmental Chemistry Letters*, vol. 19, pp. 797–849, 2020. doi: 10.1007/s10311-020-01133-3
[17] Element Energy Ltd, "Industrial decarbonisation strategy: A comparative analysis," Element Energy Ltd Report, vol. 1, pp. 1–50, 2021. [Online]. Available: https://www.element-energy.co.uk/element-energy/
[18] E. Lewis, R. Edwards, and J. Howe, "Delivering the industrial decarbonisation challenge: Geographical considerations for decarbonisation," *Geographical Journal*, vol. 189, no. 2, pp. 125–141, 2023. doi: 10.1080/00167487.2023.2217631
[19] T. Beaufils, H. Ward, M. Jakob, and L. Wenz, "Assessing different European carbon border adjustment mechanism implementations and their impact on trade partners," *Communications Earth & Environment*, vol. 4, no. 1, pp. 1–8, 2023. doi: 10.1038/s43247-023-00788-4
[20] A. L. Siy, A. Wang, T. Zheng, and X. Hu, "Research on the impact of the EU's carbon border adjustment mechanism: Based on the GTAP model," *Sustainability*, vol. 15, no. 6, p. 4761, 2023. doi: 10.3390/su15064761
[21] Z. Yan and Z. Yuan, "Discussion on the impact of EU carbon border adjustment mechanism (CBAM) for China-EU trade," *Environmental Research Communications*, vol. 5, no. 4, pp. 10053–10065, 2023. doi: 10.1088/2515-7620/ad04f6
[22] I. Overland and R. Sabyrbekov, "Know your opponent: Which countries might fight the European carbon border adjustment mechanism?" *Energy Policy*, vol. 164, p. 113175, 2022. doi: 10.1016/j.enpol.2022.113175
[23] T. R. Peter Holmes and J. Rollo, "Border carbon adjustments and the potential for protectionism," *Climate Policy*, vol. 11, no. 2, pp. 883–900, 2011. doi: 10.3763/cpol.2009.0071
[24] Centre for European Reform, "The EU's carbon border adjustment mechanism," Centre for European Reform Policy Brief, 2021. [Online]. Available: https://www.cer.eu/sites/default/files/pbrief_cbam_sl_21.4.21.pdf

[25] Boston University, "The global impact of a carbon border adjustment mechanism," 2022. [Online]. Available: https://www.bu.edu/gdp/files/2022/03/TF-WP-001-FIN.pdf
[26] Carnegie Endowment for International Peace, "A political economy perspective on the EU's carbon border tax," 2023. [Online]. Available: https://carnegieendowment.org/2023/05/09/political-economy-perspective-on-eu-s-carbon-border-tax-pub-89706
[27] Hydro, "Greenwashing via CBAM: Loopholes threaten European green products market," *Hydro News*, 2023. [Online]. Available: https://www.hydro.com/en-SE/about-hydro/stories-by-hydro/greenwashing-via-cbam-loopholes-threaten-european-green-products-market/
[28] National Renewable Energy Laboratory, "Economic assessment of hydrogen technologies participating in electricity markets," NREL Report, 2016. [Online]. Available: https://www.nrel.gov/docs/fy16osti/65856.pdf
[29] Royal Society of Chemistry, *Future costs of hydrogen: A quantitative review*, RSC Publishing, 2024. [Online]. Available: https://pubs.rsc.org/en/content/articlehtml/2024/se/d4se00137k
[30] Baker McKenzie, "Shaping tomorrow's global hydrogen market," Baker McKenzie Report, 2020. [Online]. Available: https://www.bakermckenzie.com/-/media/files/insight/publications/2020/01/hydrogen_report.pdf
[31] PricewaterhouseCoopers, "Green hydrogen economy - predicted development of tomorrow," PwC Report, 2024. [Online]. Available: https://www.pwc.com/gx/en/industries/energy-utilities-resources/future-energy/green-hydrogen-cost.html
[32] Resources for the Future, "Global energy outlook 2024: Peaks or plateaus?" RFF Report, 2024. [Online]. Available: https://www.rff.org/publications/reports/global-energy-outlook-2024/
[33] G. P. Thiel and A. Stark, "To decarbonize industry, we must decarbonize heat," *Joule*, vol. 5, pp. 531–550, 3 Mar. 2021, issn: 2542-4351. doi: 10.1016/J.JOULE.2020.12.007
[34] D. Vine, "Clean industrial heat: A technology inclusive framework," Center for Climate and Energy Solutions, Tech. Rep., 2021.
[35] M. H. Bade and S. Bandyopadhyay, "Targeting minimum heat transfer fluid flow for multiple heat demands," in *11th International Symposium on Process Systems Engineering, ser. Computer Aided Chemical Engineering*, I. A. Karimi and R. Srinivasan, Eds., vol. 31, Elsevier, 2012, pp. 675–679. doi: 10.1016/B978-0-444-59507-2.50127-X. [Online]. Available: https://www.sciencedirect.com/science/article/pii/B978044459507250127X
[36] Department for Business, Energy Industrial Strategy, "Industrial decarbonisation strategy," 2021. [Online]. Available: https://www.gov.uk/government/publications/industrial-decarbonisation-strategy
[37] V. Vogl, O. Olsson, and B. Nykvist, "Phasing out the blast furnace to meet global climate targets," *Joule*, vol. 5, no. 10, pp. 2646–2662, 2021, issn: 2542-4351. doi: 10.1016/j.joule.2021.09.007. [Online]. Available: https://www.sciencedirect.com/science/article/pii/S2542435121004359
[38] B. Lee and I. Sohn, "Review of innovative energy savings technology for the electric arc furnace," *JOM*, vol. 66, pp. 1581–1594, 9 Sep. 2014, issn: 15431851. doi: 10.1007/s11837-014-1092-y. [Online]. Available: https://link.springer.com/article/10.1007/s11837-014-1092-y
[39] C. Schoeneberger, J. Zhang, C. McMillan, J. B. Dunn, and E. Masanet, "Electrification potential of U.S. industrial boilers and assessment of the GHG emissions impact," *Advances in Applied Energy*, vol. 5, p. 100089, 2022. doi: 10.1016/j.adapen.2022.100089
[40] T. Ennis, "Safety in design of thermal fluid heat transfer systems," in *Symposium series*, vol. 155, 2009, pp. 162–169.

[41] Department for Business, Energy & Industrial Strategy, "Combined heat and power-technologies," 2021.

[42] R. Nagahata and K. Takeuchi, "Encouragements for the use of microwaves in industrial chemistry," *The Chemical Record*, vol. 19, no. 1, pp. 51–64, 2019. doi: 10.1002/tcr.201800064

[43] T. Gelles, S. Lawson, A. A. Rownaghi, and F. Rezaei, "Recent advances in development of amine functionalized adsorbents for CO_2 capture," *Adsorption*, vol. 26, pp. 5–50, 123. doi: 10.1007/s10450-019-00151-0

[44] P. Panja, B. McPherson, and M. Deo, "Techno-economic analysis of amine-based CO_2 capture technology: Hunter plant case study," *Carbon Capture Science & Technology*, vol. 3, p. 100041, Mar. 2022. doi: 10.1016/j.ccst.2022.100041

[45] A. Olabi, T. Wilberforce, K. Elsaid, E. T. Sayed, H. M. Maghrabie, and M. A. Abdelkareem, "Large scale application of carbon capture to process industries – A review," *Journal of Cleaner Production*, vol. 362, p. 132300, 2022, issn: 0959-6526. doi: 10.1016/j.jclepro.2022.132300

[46] Z. Liang, W. Rongwong, H. Liu, et al., "Recent progress and new developments in post-combustion carbon-capture technology with amine based solvents," *International Journal of Greenhouse Gas Control*, vol. 40, pp. 26–54, 2015. doi: 10.1016/J.IJGGC.2015.06.017

[47] G. Ji and M. Zhao, "Membrane separation technology in carbon capture," in *Recent advances in carbon capture and storage*, Y. Yun, Ed., Rijeka: IntechOpen, 2017, ch. 3. doi: 10.5772/65723

[48] O. Akeeb, L. Wang, W. Xie, R. Davis, M. Alkasrawi, and S. Toan, "Post-combustion CO_2 capture via a variety of temperature ranges and material adsorption process: A review," *Journal of Environmental Management*, vol. 313, p. 115026, Jul. 2022, issn: 0301-4797. doi: 10.1016/J.JENVMAN.2022.115026

[49] P. Luis, T. V. Gerven, and B. V. D. Bruggen, "Recent developments in membrane-based technologies for CO_2 capture," *Progress in Energy and Combustion Science*, vol. 38, pp. 419–448, 3 Jun. 2012, issn: 03601285. doi: 10.1016/J.PECS.2012.01.004

[50] A. M. Oliveira, R. R. Beswick, and Y. Yan, "A green hydrogen economy for a renewable energy society," *Current Opinion in Chemical Engineering*, vol. 33, p. 100701, 2021, issn: 2211-3398. doi: 10.1016/j.coche.2021.100701. [Online]. Available: https://www.sciencedirect.com/science/article/pii/S2211339821000332

[51] "*No_x emissions from hydrogen-methane fuel blends*," Georgia Institute of Technology, Tech. Rep., 2023. [Online]. Available: https://research.gatech.edu/sites/default/files/inline-files/gt_epri_nox_emission_h2_short_paper.pdf

[52] K. Kikuchi, T. Hori, and F. Akamatsu, "Fundamental study on hydrogen low-No_x combustion using exhaust gas self-recirculation," *Processes*, vol. 10, 1 Jan. 2022, issn: 22279717. doi: 10.3390/pr10010130

[53] M. Götz, J. Lefebvre, F. Mörs, et al., "Renewable power-to-gas: A technological and economic review," *Renewable Energy*, vol. 85, pp. 1371–1390, 2016, issn: 0960-1481. doi: 10.1016/j.renene.2015.07.066. [Online]. Available: https://www.sciencedirect.com/science/article/pii/S0960148115301610

[54] Centrica, "Hydrogen chp units – Centrica Business Solutions," [Accessed May 11, 2023]. [Online]. Available: https://www.centricabusinesssolutions.com/energy-solutions/combined-heat-and-power-chp/hydrogen-chp-units

[55] J. Guarco, B. Langstine, and M. Turner, *Practical considerations for firing hydrogen versus natural gas*, 2023.

[56] U.S. Energy Information Administration, Biomass explained, [Accessed June 1, 2023]. [Online]. Available: https://www.eia.gov/energyexplained/biomass/

[57] Partnership for Policy Integrity, "Carbon emissions from burning biomass for energy," Partnership for Policy Integrity, Tech. Rep., 2020. [Online]. Available: https://www.pfpi.net/carbon-emissions-from-burning-biomass-for-energy
[58] International Energy Agency Bioenergy, *Deployment of BECCS/U value chains*, 2020.
[59] F. Affan and A. Maaroof, "A critical evaluation of the UK Drax power station," *Journal of Ecosystem Ecography*, vol. 6, Jan. 2016. doi: 10.4172/2157-7625.1000206
[60] K. Y. Kwong and E. J. Marek, "Combustion of biomass in fluidized beds: A review of key phenomena and future perspectives," *Energy & Fuels*, vol. 35, no. 20, pp. 16303–16334, 2021. doi: 10.1021/acs.energyfuels.1c01947
[61] B. H. Cheng, B. C. Huang, R. Zhang, et al., "Bio-coal: A renewable and massively producible fuel from lignocellulosic biomass," *Science Advances*, vol. 6, 1 Jan. 2020, issn: 23752548. doi: 10.1126/SCIADV.AAY0748. [Online]. Available: https://www.science.org/doi/10.1126/sciadv.aay0748
[62] US Environmental Protection Agency, "Biomass combined heat and power catalog of technologies," Tech. Rep., 2007. [Online]. Available: www.epa.gov/chp
[63] J. Lachman, M. Baláš, M. Lisý, H. Lisá, P. Milčák, and P. Elbl, "An overview of slagging and fouling indicators and their applicability to biomass fuels," *Fuel Processing Technology*, vol. 217, Jun. 2021, issn: 03783820. doi: 10.1016/J.FUPROC.2021.106804
[64] Heat Management, "Challenges for power boilers," 2016. [Online]. Available: https://heatmanage.com/portfolio-items/challenges-for-power-boilers/
[65] V. Lenz, N. Szarka, M. Jordan, and D. Thrän, "Status and perspectives of biomass use for industrial process heat for industrialized countries," *Chemical Engineering & Technology*, vol. 43, pp. 1469–1484, 8 Aug. 2020, issn: 1521-4125. doi: 10.1002/CEAT.202000077
[66] A. Schmitz, J. Kamiński, B. Maria Scalet, and A. Soria, "Energy consumption and CO_2 emissions of the European glass industry," *Energy Policy*, vol. 39, no. 1, pp. 142–155, 2011, issn: 0301-4215. doi: 10.1016/j.enpol.2010.09.022
[67] C. Ranaraja, J. Devasurendra, M. Maduwantha, G. Madhuwantha, and R. Hansa, "Optimization of an industrial boiler operation," *Journal of Research Technology and Engineering*, vol. 1, pp. 126–134, 2020, issn: 2714-1837.
[68] C. Kuhwald, S. Türkhan, and A. Kirschning, "Inductive heating and flow chemistry – A perfect synergy of emerging enabling technologies," *Beilstein Journal of Organic Chemistry*, vol. 18, p. 688, 2022, issn: 18605397. doi: 10.3762/BJOC.18.70
[69] G. Ondrey, "High-temperature heat pump promises significant energy savings," [Accessed June 2, 2023]. [Online]. Available: https://www.chemengonline.com/heat-pump/
[70] R. Liu, L. Chen, Y. Zhao, M. Liu, W. S. Djati, and N. Siswanto, "Optimization operation strategy of electric boilers by mixed integer programming method in nickel production process," *IOP Conference Series: Materials Science and Engineering*, vol. 1072, p. 012044, 1 Feb. 2021, issn: 1757-899X. doi: 10.1088/1757-899X/1072/1/012044
[71] D. Paolo, G. Thomas, and T. Humayun, "How to increase grid resilience through targeted investments," 2021.
[72] L. Gurobi Optimization, Gurobi optimizer reference manual, https://www.gurobi.com/documentation/current/refman/index.html, 2024.
[73] P. Belotti, P. Bonami, M. Fischetti, A. Lodi, J. Smith, and A. Taws, "Branching and bounds tightening techniques for non-convex MINLP," *Optimization Methods and Software*, vol. 24, no. 4-5, pp. 597–634, 2009.
[74] N. V. Sahinidis, "Baron: A general purpose global optimization software package," *Journal of Global Optimization*, vol. 8, no. 2, pp. 201–205, 1996.
[75] G. D. Corporation, Gams: The solver manuals, https://www.gams.com/latest/docs/S_MAIN.html, 2024.

[76] W. E. Hart, C. D. Laird, J.-P. Watson, et al., *Pyomo – Optimization modeling in Python*. Springer, 2017, vol. 67.

[77] R. Fourer, D. M. Gay, and B. W. Kernighan, *AMPL: A modeling language for mathematical programming*. Duxbury Press, 2002.

[78] H. A. Simon, "A behavioral model of rational choice," *Quarterly Journal of Economics*, vol. 69, no. 1, pp. 99–118, 1955. doi: 10.2307/1884852

[79] H. A. Simon, "Rational decision making in business organizations," *American Economic Review*, vol. 69, no. 4, pp. 493–513, 1979. [Online]. Available: https://www.jstor.org/stable/1808698

[80] A. Tversky and D. Kahneman, "Judgment under uncertainty: Heuristics and biases," *Science*, vol. 185, no. 4157, pp. 1124–1131, 1974. doi: 10.1126/science.185.4157.1124

[81] D. Kahneman, "Maps of bounded rationality: Psychology for behavioral economics," *American Economic Review*, vol. 93, no. 5, pp. 1449–1475, 2003. [Online]. Available: https://www.jstor.org/stable/3132137

[82] D. Y. Chang, "Applications of the extent analysis method on fuzzy AHP," *European Journal of Operational Research*, vol. 95, no. 3, pp. 649–655, 1996. doi: 10.1016/0377-2217(95)00300-2

[83] J. J. Buckley, "Fuzzy hierarchical analysis," *Fuzzy Sets and Systems*, vol. 17, no. 3, pp. 233–247, 1985. doi: 10.1016/0165-0114(85)90090-9

[84] C. Kahraman, U. Cebeci, and D. Ruan, "Multi-attribute comparison of catering service companies using fuzzy AHP: The case of Turkey," *International Journal of Production Economics*, vol. 87, no. 2, pp. 171–184, 2004. doi: 10.1016/S0925-5273(03)00099-9

[85] P. J. M. van Laarhoven and W. Pedrycz, "A fuzzy extension of Saaty's priority theory," *Fuzzy Sets and Systems*, vol. 11, no. 1-3, pp. 229–241, 1983. doi: 10.1016/S0165-0114(83)80082-7

Part II

Chemical Industry Decarbonization

3 Materials and Processes for Chemical Industry Decarbonization

On the Electrothermal Chemical Synthesis and Manufacturing

Priyam Mondal, Abhirup Sen, and Qi Dong

3.1 INTRODUCTION

The chemical industry is undergoing a transformative shift driven by the urgent need for decarbonization and the increasing availability of renewable electricity. As traditional fossil fuel-based processes face growing scrutiny due to their environmental impact, electrification has emerged as a promising pathway to achieve more sustainable and efficient chemical synthesis and manufacturing. Among various electrification strategies, electrothermal chemical synthesis has gained substantial attention for its industrial scalability and potential to revolutionize conventional processes. Unlike electrochemical approaches, which often rely on low or medium temperatures, electrothermal synthesis utilizes direct heating methods to reach relatively high operation temperatures. These methods include resistive heating (Joule heating), microwave heating, induction heating, and plasma-assisted reactions. Compared to conventional thermochemical reactions powered by combustion heating, these techniques often offer precise process control, reduced environmental impact, and high energy efficiency, enabling more sustainable conversion of hydrocarbons, carbon dioxide, biomass, and other precursors into value-added chemicals. In particular, recent advancements in electrothermal technologies have demonstrated promising improvements in product selectivity, energy efficiency, and feedstock versatility, positioning them as strong contenders for large-scale industrial adoption. This chapter explores the latest fundamental and applied advances in electrothermal chemical synthesis, with a focus on hydrocarbon conversion, carbon dioxide utilization, biomass processing, and ammonia synthesis. By comprehensively reviewing existing literature, we aim to highlight key breakthroughs, ongoing challenges, and future research directions. Ultimately, this discussion seeks to foster interdisciplinary collaboration

DOI: 10.1201/9781003655442-5

between academia and industry, paving the way for a more sustainable future for chemical manufacturing.

3.2 ELECTROTHERMAL HYDROCARBON CONVERSION

3.2.1 Steam Reforming of Hydrocarbons

Steam reforming of hydrocarbons is a critical industrial process for producing hydrogen (H_2) and syngas. Due to the abundant natural gas and other methane-rich resources on earth, steam methane reforming (SMR) has gained the greatest attention and industrial progress. As such, this section will use SMR as a model reaction to discuss recent advances made for decarbonizing hydrocarbon steam reforming reactions. At present, SMR contributes to more than half of the global H_2 production.[1] SMR typically occurs at high temperatures (e.g., 1023–1223 K) and mild pressures (14–20 atm) following the reaction below:

$$CH_4 + H_2O \leftrightarrow CO + 3H_2 \quad \Delta H^{\circ}_{298} = +206\,\text{kJ/mol}$$

The SMR reaction was first discovered by Neumann and Jacob in 1924.[2] By 1930, SMR had its first industrial application. Over the past century, researchers have been striving to improve the efficiency of the process.[3] In traditional SMR reactors, the desired temperature is raised and then maintained by combustion heating via gas burners. As heat is supplied from the outside across the reactor wall to the inside, thermal insulation and heat circulation are needed to achieve desirable energy efficiency while supporting the energy required for Reaction 1. In a typical industrial setting, up to 35% of the methane (CH_4) feedstock can be used as fuel, while the remaining serves as the reactant. Due to the large energy consumption because of the temperature requirement, SMR is often associated with heavy carbon dioxide (CO_2) emission and variable CH_4 leakage.[4] These challenges have driven ongoing efforts to enhance reactor design and improve reaction performance. In particular, a range of electrified approaches have been developed over the past few decades, aiming to decarbonize the process heat while improving efficiency.[5–7]

3.2.1.1 Electrothermal SMR Using Resistive Heating

In 1992, Spagnolo and colleagues pioneered electrified SMR at the laboratory scale, where SMR catalyst is heated resistively (i.e., Joule heating) inside of the reactor instead of being heated by an external furnace (mimicking industrial combustion heating). Specifically, a metal screen coated with nickel-dispersed alumina was placed inside a large ceramic-lined pressure vessel connected to an AC power supply. This setup provided several benefits as follows: (1) by generating heat directly at the catalytic surface, heat transfer efficiency is greatly improved; (2) by removing the burners, the reactor becomes more compact, gases are easier to mix, and arrangement of the catalyst bed is improved for better reaction performance. This process for the first time hints at the potential of leveraging the increasingly available renewable energy sources for chemical synthesis and manufacturing. This innovative method requires only 5–12 kJ/L_{H2}, comparable to the conventional reformers

(5–10 kJ/L_{H2}). The authors pointed out that further enhancement of catalyst support, enriched understanding of diffusion mechanisms, and optimization of reactor design are needed. This work opened the door to electrothermal SMR operations, encouraging a number of later studies for the decarbonization of H_2 and syngas production processes.[8]

A series of studies were later carried out by Kameyama and colleagues on Ni-based catalysts for electrified SMR. The team first demonstrated a 2D dynamic model to simulate the performance of a CH_4 steam reformer using an electrically heated Ni/Al_2O_3/alloy catalyst. The simulation showed that the internal heating improved heat transfer coefficient by ten-fold compared to the external heating. Meanwhile, as resistive heating induced different temperature distributions, the catalyst placement was found to significantly impact the reaction performance. Experimentally, an optimal staggering coefficient of 0.45 under internal heating increased CH_4 conversion from 62% to 83%.[9] Following this work, the team further synthesized a 17.9 wt% Ni/$NiAl_2O_4$/γ-Al_2O_3/alloy catalyst with an interfacial $NiAl_2O_4$ layer, which demonstrated excellent SMR reactivity at 700 °C. The interfacial $NiAl_2O_4$ layer anchored the metallic nickel particles, effectively preventing catalyst sintering. This catalyst achieved a high CH_4 conversion rate of 97.3% with 20 hours of stationary stability and maintained stable operation for 130 hours.[10] Furthermore, the study confirmed that the resistively heated pattern did not exhibit any negative influence on the SMR reactivity of anodic nickel catalysts. Rather, it markedly reduced the reformer's start-up time from one to two hours to just a few minutes.[11] The catalyst structure was further modified by the same team, featuring a plate-type Al/Fe–Cr alloy/Al clad material anchored into a metal-monolithic anodic alumina support. The anodic alumina-supported Ni catalyst demonstrated excellent durability and hydrothermal stability, performing well for 3000 hours in a stationary SMR test and 500 cycles in a daily start-up and shut-down test.[12]

A major breakthrough was recently reported by Wismann and coworkers on a highly compact approach to achieve green SMR powered via Joule heating.[1] The laboratory scale reactor was constructed using FeCrAl alloy, leveraging the material's low-temperature dependence on its electrical resistance. The laboratory scale reactor's internal surface features a 50 cm FeCrAl alloy tube and a 130 µm nickel-impregnated ZrO_2-based washcoat with copper sockets at both ends. This setup provides near-uniform heat distribution and temperature gradient (e.g., < 2 °C difference across the reactor's internal coating, minimal gradient across the reactor wall) via passing it through an alternating current (Figure 3.1(A)). The ZrO_2-modified FeCrAl alloy tube addresses the thermal conductivity issues of conventional reformers; meanwhile, the close contact between the FeCrAl alloy and the reaction sites quickly drives the reaction close to the thermal and chemical equilibrium. As a result, product selectivity and yield are improved, carbon accumulation is minimized, and catalyst lifetime is extended. Notably, the electrothermal reactor has a significantly smaller footprint (~5 m^3) compared to a conventional reactor (1100 m^3) for the same H_2 yield (2230 kmol/h).[13]

The FeCrAl alloy was later used in other forms of resistive heaters, as demonstrated by Ledesma and others. Using Ni/ZrO_2-coated FeCrAl wires, Ledesma and coworkers achieved a high fractional conversion rate of 0.87 at 860 °C for SMR.

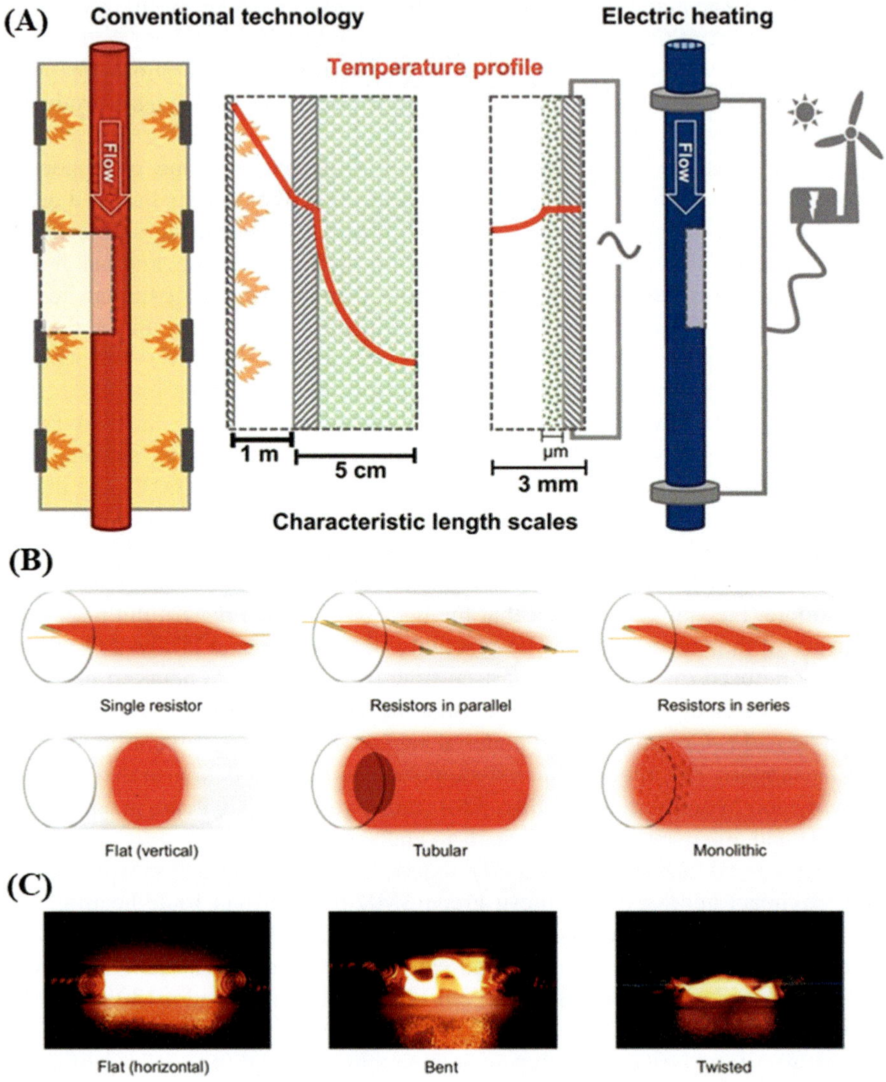

FIGURE 3.1 (A) Temperature profile comparison between conventionally heated reactors and electrically heated reactors. Adapted from Ref. 1, DOI: 10.1126/science.aaw8775. (B) Schematic showing different configurations of resistive heaters. (C) Images of flexible carbon fiber paper heaters. Adapted from Ref. 21, DOI: 10.1038/s44286-024-00134-1.

Notably, CH_4 conversion was found to increase dramatically with input power reaching above 60 W, corresponding to a 750 °C wire temperature. In comparison, conventional setting lacks such a sharp transition in CH_4 conversion, indicating that the applied current in resistively heated reactors directly activates the catalyst. The authors attribute this observation to a redox process, where "thermal electrons" reduce NiO crystallites, enhancing its reforming activity.[14]

In addition to metal-based resistive heaters, ceramic heating elements have been explored to address metal's lack of chemical stability for electrothermal applications. In particular, silicon carbide (SiC)-based ones have become the most popular choices owing to their superior chemical stability and relatively good conductivity. Renda and colleagues synthesized a structured catalyst by washcoating a SiC heating element with silica-mullite slurry and nickel precursor, then tested it in SMR conditions. When heated to 800 °C, the SiC-based structured catalyst achieved over 85% CH_4 conversion, with energy consumption of 4.8 kWh/Nm^3 H_2, which is comparable to modern electrolyzers.[15] In parallel, Tronconi and coworkers reported a suite of electrified SMR platforms based on SiSiC. The SiSiC's bulk resistivity enabled direct resistive heating; meanwhile, its structure can be made into porous media to help reduce heat and mass transfer limitations, resulting in an efficient low-carbon H_2 production system via SMR. Using a Rh/Al_2O_3 wash-coated SiSiC foam, the authors demonstrated nearly full CH_4 conversion above 700 °C and near-equilibrium conversions under various conditions. At 650 °C, the system demonstrated an energy efficiency of 61%, with a catalyst density of 46 g/L and a specific power consumption of only 2.0 kWh/Nm^3 H_2.[16] Further improvements reached an 81% energy efficiency with an intensified catalyst density of 86.3 g/L and a specific energy demand of 1.24 kWh/Nm^3 H_2, surpassing modern electrolyzers. Scale-up calculations indicated that small-scale electrified SMR units with H_2 productivities up to 200 Nm^3/h were feasible and compatible with industrial electricity supplies.[17] Later on, Tronconi and colleagues investigated design protocols for electrified SMR reactors, finding that flow rate and mass transfer strongly impact thermal efficiency. The team compared open-cell foam and honeycomb structures, discovering that while honeycomb has five times lower resistivity, higher currents, and reduced power losses, it suffers from poorer bulk mass transfer. Consequently, the authors recommended open-cell foam for electrified SMR with scrub-coated structured catalysts.[18]

Indeed, the optimization and implementation of heater structures are equally, if not more, impactful than the heater material itself on the performance of electrified SMR. Balakotaiah and colleagues systematically evaluated six model resistive heater configurations for electrified SMR and several endothermic reactions. These model configurations include parallel wire, parallel plate, perforated thin plate, metal mesh, monolithic, and parallel tube reactors, which can have numerous related structural derivations. This study revealed the critical role played by the reactor and heater structure on the reaction performance considering mass and heat transfer, catalyst-reactant interaction, temperature control, and timescale, among others. As pointed out by the authors, additional multi-scale modeling and analysis as well as experimental validation at relevant scales are needed when it comes to specific reaction conditions and catalyst choices. For hydrocarbon conversions in particular, coking, catalyst deactivation, impact of magnetic field on the reaction, and integration with downstream processes also deserve thorough consideration.[19] Fortunately, recent advances in the field of 3D printing enable geared structural designs for electrothermal chemical synthesis. For electrified SMR, Peloquin and coworkers have pioneered an integrated microreactor with radiative electrical heaters. The entire microreactor was 3D-printed using Inconel and equipped with a custom catalyst. The optimized structure led to nearly complete CH_4 conversion above 900 °C and a gas hourly space velocity (GHSV) of 71,000 h^{-1}. Its low thermal mass enabled rapid start-up and cool-down to accommodate the intermittent renewable electricity supply. Notably, the microreactor operated for 170 hours with minimal performance decay and no structural degradation after 400 hours of operation.[20]

Recently in a perspective on commodity chemical synthesis, the authors overviewed the versatile heater designs of resistively heated reactors, which can be used to enhance heat transfer and enable precise control for reaction modulation. Simple configurations like wires and plates are commonly used for fundamental studies, as they ensure reproducibility, facilitate temperature and intermediate measurements. Advanced designs, such as zone-heating arrays or tubular structures, improve spatial and temporal temperature control for complex reactions and scalability (Figure 3.1(B)). Flexible heaters, like bendable carbon paper, enhance gas-solid interactions, improving energy efficiency, and can even be dynamically adjusted during operation to influence reaction outcomes (Figure 3.1(C)).[21]

Along with the promising experimental and computational progress at smaller scales, several system-level analyses were reported recently, supporting the high potential of electrified SMR for practical implementation. For example, Song and coworkers integrated electrified SMR into power-to-gas technology based on the principles of electrothermal reaction and energy cascade utilization. The authors highlighted a high thermal efficiency (97.27%) of the proposed system compared to current industrial SMR operations, as well as superior electricity-to-H_2 efficiency (88.68%) over commercial electrolyzers. The process also features a low CO_2 emission (5.5 kg/kg H_2) and potentially reduces H_2 cost to \$2.47/kg.[22] Despite these promising predictions, there still exists a daunting gap between laboratory-scale experimental results and industrially relevant operation outcomes. Current limitations on the scalability are primarily due to the high capital equipment and material cost, as well as the expensive industrial electricity. While the latter may be complemented by the increasingly available renewable electricity, the former will likely require more efficient material and reactor designs that can be scaled as well as supported by a thorough technoeconomic analysis (TEA).

3.2.1.2 Electrothermal SMR Using Induction Heating

Vinum and coworkers first reported using induction heating as an alternative electrothermal approach for SMR with a focus on magnetic catalyst development. Two model CoNi-based nanocatalysts were synthesized and compared for their SMR activity.[23] In another study, Scarfiello and colleagues reported ferromagnetic $Ni_{60}Co_{40}$ nanoparticles supported on commercial γ-Al_2O_3 pellets for inductively powered SMR using a radiofrequency alternating magnetic field. These studies demonstrated the dual role of nanoparticles as both heat-dissipating agents and catalysts for promoting inductively heated processes.[24] Dotsenko and colleagues further explored the impact of loading and Co content of CoNi catalysts for inductively heated SMR. Higher catalyst loading resulted in higher temperatures (735 °C vs. 700 °C by control sample) and conversion values (80% vs. 75%) due to increased number of energy dissipation sites. Meanwhile, higher Co content was found to lead to even higher temperatures (780 °C) and conversion (90%) due to Co's larger magnetic anisotropy thus stronger hysteretic dissipation. This study offered several useful catalyst design guidelines for electrified SMR powered by induction heating.[25]

In addition to catalyst developments to couple with inductive heating, several studies focused on process optimization and reactor design. For example, Almind and coworkers revealed the impact of structural settings and process parameters on the reactor's performance. For example, increasing the frequency from 68 kHz to 189 kHz was found to show improved energy efficiency. Additionally, switching from a short, wide coil (height-to-radius ratio of 3.8) to a long, narrow coil (height-to-radius ratio of 10.8) further enhanced the reaction performance. Overall, energy efficiency increased from 11% in the initial bench-scale reactor setup to 23% using the optimized version. Their theoretically modeled framework indicated that when extrapolated to large-scale H_2 production, the system's efficiency could increase substantially, potentially reaching close to 90%.[26] Ma and coworkers introduced a high-temperature heat pipe reactor with induction heating, demonstrating a start-up time of just 250 seconds and a temperature uniformity of 96.7%. At 620 °C, the CH_4 conversion reached 80% and H_2 mole fraction reached 70.5%. Both metrics were found to increase with temperature; meanwhile, the temperature gradient decreased with increased bed porosity.[27]

Ma and team introduced a novel high-temperature heat pipe tubular reactor that utilizes both concentrated solar power and photovoltaic electricity for highly effective heat transfer. This innovative design offers another viable option in addition to electrified approaches to address industrial decarbonization challenges and to meet clean energy and low-emission H_2 production demands. Initial performance tests using a concentrated solar simulator and a high-frequency induction device demonstrated a start-up time of only 9 minutes. Notably, the high-temperature heat pipe tubular reactor achieved a high CH_4 conversion rate of 90% and an H_2 purity of 74%, which are close to the thermodynamic equilibrium values of SMR. Continuous operation of this reactor with the hybrid energy supply (via both induction and concentrated solar power) maintained CH_4 conversion at 91% and H_2 purity at 72% over 24 hours, showing great promise for further exploration toward scale-up and more practical settings.[28]

3.2.1.3 Electrothermal SMR Using Microwave

Meloni and coworkers reported electrified SMR powered by microwave heating. The authors investigated the impact of reactor configuration on SMR efficiency using a Ni-based catalyst (7 wt% with respect to the washcoat) on SiC monoliths. Both Ni-based catalyst and SiC are microwave susceptors, making them effectively heated during the microwave operation. The reaction at 750 °C and with a GHSV of 5000 h^{-1} reached thermodynamic equilibrium for CH_4 conversion. Comparing two reactor setups, the optimized configuration, which was configured to intensify the electric field in the catalytic zone by restricting the placement of the structured catalyst, was found to enhance the microwave efficiency compared to the classical design. The optimized setup demonstrated higher energy efficiency, reaching 73%, compared to 50% by the classical setup. The optimized reactor also achieved an energy consumption of 2.5 kWh/Nm^3H_2 at 750 °C, closely matching the efficiency of the best resistive SMR systems (~2 kWh/Nm^3 H_2). This development underscores the potential of microwave-assisted electrothermal catalytic reactors in achieving competitive energy efficiencies for low-emission and highly efficient H_2 production.[29]

3.2.2 Non-Oxidative Methane Conversion

The non-oxidative decomposition of CH_4 yields a variety of compounds such as ethane (C_2H_6), ethylene (C_2H_4), acetylene (C_2H_2), benzene/toluene/xylene (BTX), heavier aromatics, H_2, wax, soot, and coke. The product composition depends on factors including the reaction temperature, pressure, catalyst, reaction process parameters, and many more. Typically, non-oxidative CH_4 conversion follows a chain reaction mechanism[30,31]:

$$2CH_4 \rightarrow C_2H_6 + H_2 \rightarrow C_2H_4 + 2H_2 \rightarrow C_2H_2 + 3H_2 \rightarrow 2C + 4H_2$$

One major application of this chain process is for H_2 production with a coproduct of solid carbon. The H_2 generated through CH_4 decomposition (often referred to as CH_4 pyrolysis) is called the "turquoise H_2," in contrast with "green H_2" produced via electrolysis, "gray H_2" via SMR and "blue H_2" via SMR combined with carbon capture. The byproduct, solid carbon, is easily storable and applicable in industries such as tire and rubber manufacturing, improving the product value of this process. On the other hand, halting the chain reaction in the middle can lead to other useful product streams such as C_2H_x and aromatic compounds (often referred to as CH_4 dehydrodimerization (MDD) and dehydroaromatization (MDA), respectively), which usually require unique catalyst choices, process parameters, or both. Compared to SMR, non-oxidative CH_4 conversions to H_2, hydrocarbons, aromatics, and others typically necessitate higher temperatures due to the endothermic nature of these reactions, which is accompanied by larger energy consumption and lower energy efficiency. These issues have made electrified processes particularly appealing for non-oxidative CH_4 conversions.

3.2.2.1 Electrothermal Non-Oxidative Methane Conversion Using Resistive Heating

Porsin and coworkers investigated a high-temperature process for the pyrolysis of CH_4 on an electrically heated tungsten wire, achieving improved CH_4 conversion and C_2H_2 selectivity while reducing carbon black production. The authors discovered that diluting CH_4 to 10–15% with inert gas and maintaining the temperature between 1800 and 2000 °C resulted in 80% CH_4 conversion and 80% C_2H_2 selectivity using short contact times with the heating element/zone. Below 1500 °C, C_2H_4 was found to be the main product with 30% selectivity. As the temperature increases, the rate of C_2H_4 decomposition increases, leading to higher C_2H_2 yields.[32]

To understand the reaction mechanism, Sigaeva and coworkers studied CH_4 pyrolysis at 750–1100 °C using a Joule-heated FeCrAl wire, both in the absence of oxygen (O_2) and at CH_4:O_2 ratios of 15:1 and 9:1. The process proceeded in two distinct temperature ranges with different product selectivity. In the absence of O_2, CH_4 decomposition primarily leads to C_2H_6 and C_2H_4 below 950 °C, above which C_2H_2 formation starts to increase accompanied by intensive carbon deposition on the

FeCrAl wire surface. When O_2 was present, the temperature ranges shifted to higher values. For CH_4:O_2 ratios of 15:1 and 9:1, the transition temperature shifted to 1050 °C and 1100 °C, respectively, where the most extensive carbon and condensable product formation were observed.[33]

Studies conducted by Shekunova and coworkers demonstrated that the cracking of C_1–C_4 light hydrocarbons mediated by electrically heated W, Mo, or NiCr alloy wires leads to hydrocarbon decomposition, forming mostly alkenes. In the temperature range of 300–600 °C, W was found to provide the highest C_2H_4 and C_3H_6 yields and the least coke formation. Conducted in a tubular quartz reactor with an electric current (25–40 V, 7.0–9.5 A, and 200–350 W) flowing through the coils, the process was found to exhibit reduced specific energy consumption for olefin production, making it potentially useful for industrially relevant settings.[34]

It is important to note that conventional thermochemical approaches to conduct non-oxidative CH_4 decomposition face challenges to further improve the synthesis rate, selectivity, and yield beyond what is dictated by the chemical equilibrium. To break such constraints, Dong and coworkers developed a non-equilibrium continuous CH_4 pyrolysis technique using pulsed Joule heating (Figure 3.2(A) and (B)). This method allows rapid temperature swing by over several hundreds of Kelvin within milliseconds. The fast ramping of temperature and the high peak temperature during pulsed heating enable high conversion and reaction rate; meanwhile, the rapid quenching after a transient duration of heating allows the fine control of product distribution in a kinetically controlled fashion. Using porous carbon as a model Joule heater, the authors demonstrated fast heating and cooling rates of approximately 10^4 K s^{-1}, leading to high selectivity for value-added C_2 products (~80%) and suppressed coke formation from CH_4 pyrolysis (Figure 3.2(C)). Scaling up the pulsed heating and quenching process requires enlarged Joule heaters, which should remain porous for facile gas diffusion but also possess low heat capacity to ensure high heating and cooling rates.[35]

3.2.2.2 Electrothermal Non-Oxidative Methane Conversion Using Microwave

CH_4 can be converted into valuable products such as C_2H_4, BTX, and H_2 through processes such as MDD and MDA. Due to the high stability of CH_4, both processes require relatively high temperatures to achieve decent conversion efficiency. For example, MDA requires temperatures of at least 700 °C in the presence of metal-zeolite catalysts, yet the conversion is often limited to ~15% dictated by the chemical equilibrium. Additionally, the high temperatures lead to inevitable coke formation, causing rapid catalyst deactivation.[36]

Recently, microwave-assisted non-oxidative CH_4 conversion received notable attention due to its selective heating nature, which can potentially improve energy efficiency. Deng and coworkers investigated microwave-assisted direct conversion of CH_4, with a focus on the design and evaluation of microwave-sensitive H-(Fe) -ZSM-5 catalysts for hotspot formation. Isomorphous substitution of Al and Fe in the zeolite lattice was found to increase microwave sensitivity due to improved dielectric properties, enhancing microwave heating of the catalyst bed. The performance of these catalysts in conventional thermally heated fixed-bed and microwave reactors was compared to

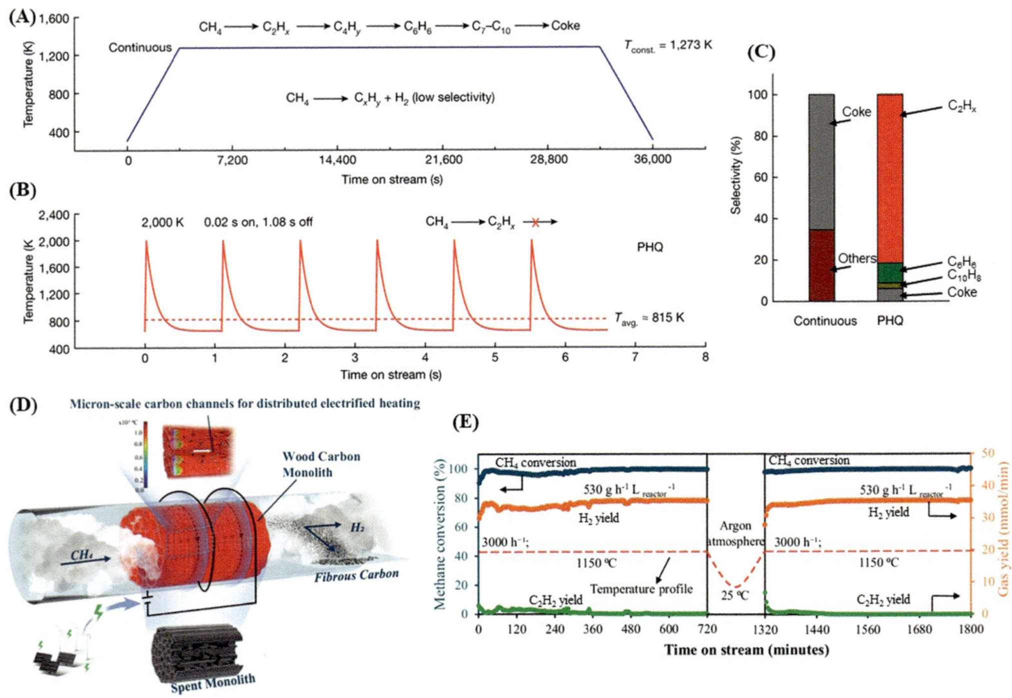

FIGURE 3.2 (A, B) Comparison of temperature profiles between continuous heating and pulsed Joule heating using CH_4 pyrolysis as a model reaction. Continuous heating yields diverse products due to limited control over reaction pathways (A), while pulsed Joule heating (also known as PHQ) enables selective C_2 product formation via transient heating durations (B). (C) Comparison of CH_4 pyrolysis performance: continuous heating produces coke as the major product, while the pulsed Joule heating achieves > 75% selectivity for value-added C_2 products at similar CH_4 conversion (~13%). Adapted from Ref. 35, DOI: 10.1038/s41586-022-04568-6. (D) Schematic of CH_4 pyrolysis using an induction-heated wood carbon monolith, which efficiently produces pure H_2 and carbon nanotubes/fibers. (E) Diagram showing extended durability of pyrolysis reaction and CH_4 conversion using the monolith reactor. Adapted from Ref. 40, DOI: 10.1038/s41467-024-47534-8.

assess the effect of microwave irradiation on the catalyst activity. CH_4 conversion was found to substantially increase under microwave conditions (3% by thermal heating, 40% by microwave), showing accelerated CH_4 activation over the metal site and the evolution of hot metal sites even at lower bulk temperatures.[37]

In the following study, Bai and colleagues studied the MDA process under microwave irradiation using both experimentation and finite-element simulation. The authors discovered that in the microwave reactor, a CH_4 conversion of 18% was achieved at 550 °C, whereas the conventional reactor required temperatures above 800 °C to obtain a comparable CH_4 conversion. Using finite-element simulations, the authors ascribed this enhanced conversion at lower bulk temperatures to the evolution of hot spots as well as an electric field of ~10^4 V/m intensity on the catalyst bed. These effects led to molecular polarization and deformation, which enhances CH_4 activation.[38]

To understand the scalability of microwave-assisted MDA from laboratory to larger configurations, Julian and colleagues proposed a selective microwave heating strategy combined with structured catalytic materials, which effectively created a temperature difference that suppressed non-selective gas-phase reactions and catalyst deactivation by coking. The larger-scale setup achieved a 150-fold increase in feeding rate, a 125-fold increase in CH_4 conversion, and a 450-fold increase in benzene production compared to the laboratory scale tests. Interestingly, only a 6-fold increase of microwave input power is required to achieve the aforementioned enhancement. The larger-scale setup exhibited stable operation for up to 295 hours under unmanned conditions, with catalytic performance maintained over seven consecutive cycles before a slight decay. This work revealed a non-linear scaling relationship of reactor performance against energy consumption under electrothermal conditions.[39]

3.2.2.3 Electrothermal Non-Oxidative Methane Conversion Using Induction Heating

Recently, Yang and coworkers introduced an induction heating approach using wood carbon monolith, whose unique structure enables rapid and uniform heating of gaseous reactants in numerous micron-scale channels (Figure 3.2(D)). The accurate reaction temperature control with minimal gradients across the micro-scale channels drives the reaction toward the equilibrium conversion and eliminates the need for elongated reactor tubes and external furnaces. Demonstrated using catalyst-free CH_4 pyrolysis, the inductively heated reactor achieved a stable H_2 production rate of 530 g h^{-1} L^{-1} along with carbon nanotubes/fibers as coproduct. 100% CH_4 conversion was achieved at 1150 °C, surpassing those achieved by various metal-based catalysts (Figure 3.2(E)).[40]

3.2.2.4 Electrothermal Non-Oxidative Methane Conversion Using Plasma

Microsecond and nanosecond pulsed spark discharge plasma for H_2 production via CH_4 pyrolysis has been systematically investigated by Gao and colleagues. In particular, the authors examined the effects of discharge power, gap length, and gas flow rate on CH_4 conversion and gas product composition. Results showed that H_2 and C_2H_2 were the primary products. The highest CH_4 conversion (91.2%) and H_2 yield (38.4%) were achieved with a microsecond pulsed spark discharge at a 6 mm gap length and a 50 mL/min gas flow rate, translating to an energy efficiency of 44.3%.[41]

Fulcheri and colleagues developed a thermal plasma process to achieve economically viable commercial-scale production of H_2 and carbon black. Unlike non-thermal plasmas, thermal plasmas offer high ionization rates, making them a versatile heat source that is highly suitable for endothermic processes with reduced or even zero carbon emission. In this study, the authors demonstrated CH_4 conversions of over 99% and H_2 selectivity of over 95% at the laboratory scale. Meanwhile, the solid product recovery rate remained high (>90%). The authors also reported an energy intensity of approximately 25 kWh per kg of H_2 measured at the first commercial plant, which is substantially lower than the energy intensity of water electrolysis for H_2 production (~60 kWh per kg of H_2).[42]

Scapinello and coworkers demonstrated that gas-phase plasma-assisted non-oxidative CH_4 coupling could generate C_2H_4 as a major product. Using nanosecond pulsed discharge plasma, the authors achieved a single-pass C_2H_4 yield of ~20% at the laboratory scale. The team revealed the synergy of the H_2 content in the reaction feed and the pressure of the plasma reactor. A proper combination of these factors was found to contribute to excellent C_2H_4 selectivity and yields, featuring a unique product stream compared to previous studies using plasma where C_2H_2 and H_2 are usually the major products.[43]

3.2.3 Non-Oxidative Dehydrogenation of $C_{\geq 2}$ Alkanes

Light alkenes are vital in producing polymers, oxygenates, and other industrial intermediates,[44] with non-oxidative dehydrogenation being a popular method to harvest them from alkanes with a coproduct of H_2.[45,46] This endothermic reaction, however, requires relatively high temperatures for industrially meaningful reaction rate and conversion.[19] Among the dehydrogenation reactions, propane dehydrogenation (PDH) and ethane dehydrogenation (EDH) have been widely implemented in the industry to produce monomers and precursors for many downstream processes toward commodity chemicals. Despite the long history of alkane dehydrogenation reactors, improving product selectivity, yield, and energy efficiency while reducing the carbon footprint remains the top pursuit of the community. Emerging process electrification technologies, such as Joule heating, plasma, and microwaves, offer the potential to decarbonize the process heat while enhancing the reaction outcomes.

3.2.3.1 Electrothermal Dehydrogenation of $C_{\geq 2}$ Alkanes Using Resistive Heating

Liu and coworkers pioneered an electrifiable membrane reactor for non-oxidative alkane dehydrogenation. The reactor employs metal-zeolite catalysts and an H_2-permselective asymmetric carbon molecular sieve (CMS) hollow fiber membrane. The asymmetric CMS hollow fiber membranes, with a thin separation layer, enabled rapid H_2 removal, which can be used to shift the reaction equilibrium toward higher alkane conversions at low temperatures (Figure 3.3(A)). The synergy between the metal-zeolite catalyst and the CMS membrane markedly enhanced alkane conversion, alkene yield, and catalyst stability. Using PDH as a model reaction, the membrane reactor achieved nearly three times the equilibrium conversion of propane

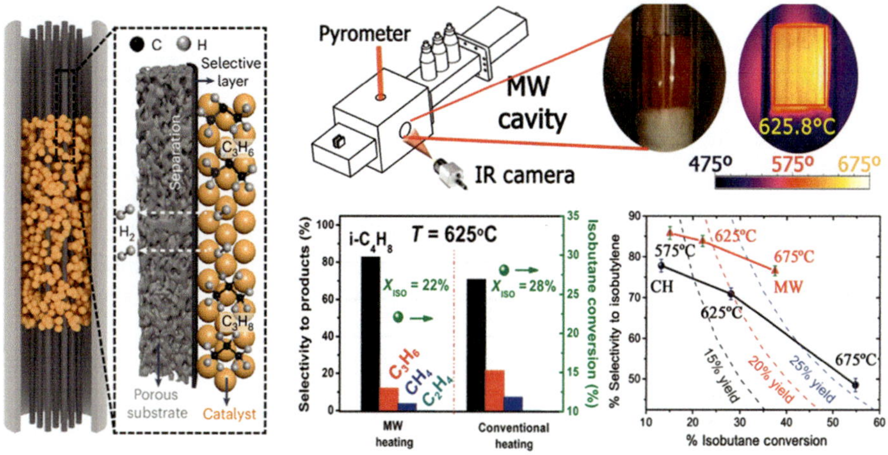

FIGURE 3.3 (A) An electrified hollow carbon fiber reactor enabling selective non-oxidative PDH reaction using a metal-supported zeolite catalyst and with simultaneous H_2 separation from the hydrocarbons. Adapted from Ref. 47, DOI: 10.1016/j.xcrp.2023.101692. (B) Schematic of the microwave reactor setup for oxidative dehydrogenation reaction and in-situ temperature measurement, enlarging nodal placement of the monolith at the electric field maximum and real-time lateral temperature mapping during CODH. (C) Catalytic performance in isobutane CODH at 625 °C: product selectivity and isobutane conversion (solid green circles). (D) Selectivity-conversion curves for isobutane CODH under conventional heating (CH) and microwave heating. At identical temperatures, CH results in higher conversion driven by gas-phase effects, despite that this increase is mainly non-selective in nature. Adapted from Ref. 51, DOI: 10.1126/sciadv.aau9000.

(C_3H_8) and increased propylene (C_3H_6) selectivity while reducing CO_2 emissions by up to 20%. Notably, the membrane reactor operated at temperatures at least 150 °C lower than commercial PDH systems, suppressing coke formation and improving system stability. While the study was conducted using furnace heating, the CMS fibers exhibit suitable electrical conductivity to potentially enable Joule heating, which can be further coupled with renewable electricity.[47]

3.2.3.2 Electrothermal Dehydrogenation of $C_{\geq 2}$ Alkanes Using Microwave

Chen and colleagues developed a packed monolith reactor to address challenges such as thermal energy loss, catalyst sintering, and catalyst screening, among others. This setup, featuring a microwave-absorbing scaffold loaded with catalytic pellets, was systematically evaluated and compared with a fixed-bed reactor using SiC monolith. The authors employed a Ga_2O_3/Al_2O_3 catalyst for EDH, where uniform catalyst utilization, suppressed hot spots, and facile periodic regeneration were achieved. By eliminating hot spots, the packed monolith minimized catalyst sintering. The Ga_2O_3/Al_2O_3 catalyst maintained its activity and selectivity after each regeneration cycle, highlighting the packed monolith's excellent stability under microwave heating. Combined with a high catalyst inventory, this configuration is a promising platform for scaling up the microwave-assisted heterogeneous catalytic processes.[48]

In a recent study, Kwak and coworkers demonstrated PDH using a microwave-heated reactor with $PtSn/SiO_2$ catalyst pellets loaded in a SiC monolith. The authors demonstrated minimal coke formation by suppressing unwanted gas-phase reaction pathways. At 500 °C with higher flow rates and in the absence of H_2 in the feed, the PDH reaction rate was higher under microwave heating compared to that using conventional furnace heating, especially at long time-on-stream values. The micro-fixed-bed showed uniform temperatures and stable microwave operation even above 700 °C without arcing. Although catalyst loading in the micro-fixed-bed reactor was lower than that in the conventional fixed bed, monoliths and foams enhanced heat and mass transfer, leading to higher yields to compensate for lower catalyst loading.[49]

3.2.4 Partial Oxidation of Hydrocarbons

Compared to the aforementioned reaction schemes, the partial oxidation of hydrocarbons, either using O_2 or other oxidants (e.g., CO_2), has not been extensively studied using electrothermal approaches. Nonetheless, resistive heating and microwave-assisted reactors have shown great promise to decarbonize the partial oxidation of hydrocarbons while improving the overall reaction performance metrics.

3.2.4.1 Electrothermal Partial Oxidation of Hydrocarbons Using Resistive Heating

Jung and colleagues pioneered the application of a Joule-heated catalyst, namely the Fe-Cr alloy loaded with Pd, for the partial oxidation of CH_4 for syngas production. The authors achieved a short start-up time of less than 1.5 minutes and a syngas production rate of 95 Nm^3/h using the Joule-heated catalyst and reactor. This approach addressed the deep cracking and coking issues by conventional high-temperature processes. The compact partial oxidation reactor, equipped with 96 cc of such Joule-heated catalyst, allowed the reaction to sustain without external heating. The CH_4 conversion was further enhanced by preheating the reactant mixture or supplying excess O_2 for additional heat release and a higher reaction rate.[50]

3.2.4.2 Electrothermal Partial Oxidation of Hydrocarbons Using Microwave

In addition to resistive heating (Joule-heated catalyst), Ramirez and coworkers demonstrated the utility of microwave heating to overcome upper-temperature barriers in various heterogeneous catalytic processes. Using the oxidative dehydrogenation of isobutane as a model reaction, the authors leveraged the selective heating of the solid phase by microwave to maintain a stable temperature difference of around 50 °C between the solid catalyst and the surrounding gases (Figure 3.3(B)). This temperature gradient led to high conversion rates from isobutane to isobutene on the catalyst while minimizing undesired homogeneous gas-phase reactions. The same team further studied CO_2-assisted dehydrogenation (CODH) of isobutane, where a non-microwave-susceptive V/Al_2O_3 catalyst and a commercial SiC monolith support. Comparisons of isobutylene selectivity at various reaction temperatures revealed the advantage of microwave heating over conventional furnace heating. Specifically, at an approximate 25% yield, selectivity to isobutylene was more than 30% higher by microwave heating (28.7% yield with 76.5% selectivity) compared with that using conventional heating (25.3% yield with 45.8% selectivity) at the same catalyst temperature of 675 °C. The concept of using non-microwave-sensitive catalysts on microwave susceptive SiC can be extended to various systems to suppress unwanted gas-phase chemistries (Figure 3.3(C) and (D)).[51]

In another recent study, Ramirez and coworkers applied microwave heating to enhance C_3H_6 selectivity in the oxidative dehydrogenation of C_3H_8. The selective heating of a SiC monolith allowed for lower gas-phase temperatures, reducing undesired byproducts from homogeneous reaction pathways. Using a V/MgO catalyst, microwave-heated monolithic reactor achieved ~21% C_3H_8 conversion and up to 70% C_3H_6 selectivity. This result outperforms conventional heating in fixed-bed reactors. Under microwave heating, the non-microwave-susceptive V/MgO catalyst in the straight monolith achieved a high C_3H_6 production rate of 3.4 $kg_{C_3H_6}\ kg_{cat}^{-1}\ h^{-1}$.[52]

3.2.5 Dry Reforming of Hydrocarbons

Dry reforming of hydrocarbons (e.g., CH_4, C_2H_6) has received significant attention due to its ability to produce syngas from naturally abundant resources. Among various hydrocarbons, dry methane reforming (DMR) has been most widely studied. Compared to SMR, DMR does not rely on steam production, thereby partially simplifying the process design and eliminating the energy cost for steam generation. More importantly, DMR allows for the production of syngas at a lower H_2-to-CO ratio that is beneficial for many downstream processes such as Fischer-Tropsch synthesis. However, like many dry reforming reactions, DMR is a highly endothermic reaction that typically operates at very high temperatures ranging from 700 to 900 °C, which is associated with high energy cost.[53,54]

$$CH_4 + CO_2 \leftrightarrow 2CO + 2H_2 \quad \Delta H^{\circ}_{298} = +247\,kJ/mol$$

In addition to the high energy requirement, industrial-scale implementation of DMR is hindered by the strong tendency of catalyst deactivation due to coke formation at high temperatures via the CH_4 pyrolysis and dehydrogenation pathways.[55,56] In addition, the large amount of process heat generation is often accompanied by heavy CO_2 emissions due to the combustion of fossil fuels or natural gas. This fact counteracts the benefits of consuming CH_4 and CO_2 as reactants. As such, it makes good sense to use electrified processes to replace combustion for generating the process heat for DMR, or couple electrified processes with combustion to reduce the energy cost and carbon footprint.

3.2.5.1 Electrothermal Dry Reforming of Hydrocarbons Using Resistive Heating

In an early study reported by Labrecque and coworkers, resistive heating was employed for DMR. Steel wool was used as a catalyst as well as a heater in a continuous flow reactor setup. Electrical current was found to substantially improve CH_4 conversion with almost no coking observed.[57] Later, Rieks and colleagues reported an electrically heated reformer with FeCrAl alloy as the Joule heating elements coated with a $LaNi_{0.95}Ru_{0.05}O_3$ catalyst at various washcoat thicknesses. The authors evaluated the effects of different operating parameters on the DMR as well as the reverse water-gas shift (RWGS) reactions (*vide infra*). Results demonstrated a maximum CH_4 conversion of 29.4% with one heating element at 900 °C. An increased CO_2/CH_4 feed ratio at constant CH_4 partial pressure decreased the DMR reaction rate, likely due to suppressed CH_4 adsorption on the catalyst. The catalyst showed good activity and no deactivation at temperatures above 700 °C and with an excess of CO_2.[58]

As discussed above in the non-oxidative CH_4 conversion section (*vide supra*), a similar reactor designed by Yang and coworkers was tested for DMR as another model reaction. Since the reactant mixture contains high concentration of CO_2 that can oxidize the carbonized wood monolith, the authors employed a metallic monolith loaded with unmodified Ni/MgO catalyst washcoat for the DMR application. The reactor demonstrated excellent conversion rates for CH_4 and CO_2 and high syngas production. Specifically, at a weight hourly space velocity (WHSV) of 75.8 Lg_{cat}^{-1} h^{-1}, nearly 100% conversion of CH_4 and CO_2 was achieved at 750 °C, with a consistent H_2-to-CO ratio of close to 1.[40]

In addition to the aforementioned metal heaters, Renda and coworkers performed DMR utilizing commercial SiC heating elements (16 mm diameter, 280 mm length) coated with a thin layer of Ni-based catalysts in a tubular reactor setup. This setup efficiently transferred heat from the resistor to the catalyst, achieving temperatures exceeding 700 °C at the catalytic bed outlet while pushing the DMR reaction toward the thermodynamic equilibrium. At 760 °C, the authors achieved a high CH_4 conversion of 84% and an H_2 yield of 75%.[15] Zheng and coworkers demonstrated the utility of catalytically activated open-cell porous SiSiC foams for DMR, which served as both catalyst supports and heating elements (Figure 3.4(A) and (B)). The reactor, featuring Joule-heated Rh/Al_2O_3-coated foam, exhibited exceptional stability, maintaining both stable catalytic and electrical performance for over 75 hours at temperatures up to 800 °C.[59]

Other than metals and ceramics, Yu and coworkers demonstrated the effectiveness of carbon Joule heaters for the DMR reaction. A bimetallic $PtNi/SiO_2$ catalyst was found to outperform its monometallic counterparts through a dynamic Joule heating process using carbon-based Joule heater. Notably, the authors achieved rapid heating rates up to ~14,000 °C/s, facilitated by the low thermal mass of the Joule heater. The rapid ramping of temperature allows the catalyst to reach temperatures exceeding 1000 °C, largely enhancing syngas productivity and reaction rates. Importantly, the pulsed Joule heating effectively mitigated issues such as coke formation, catalyst sintering, and catalyst phase segregation, leading to improved system stability.[60]

3.2.5.2 Electrothermal Dry Reforming of Hydrocarbons Using Plasma

In an early study by Kameshima and coworkers, non-thermal plasma was found to enable low-temperature DMR reaction, which is promising for efficiency improvement. Specifically, the authors employed a pulsed dielectric barrier discharge (DBD) plasma reactor with a 12 wt.% Ni/Al_2O_3 catalyst with continuously supplied CO_2 and intermittently introduced CH_4 for reducing solid carbon formation (Figure 3.4(C) and (D)). Systematic analysis showed that CH_4 dehydrogenation and the RWGS shift

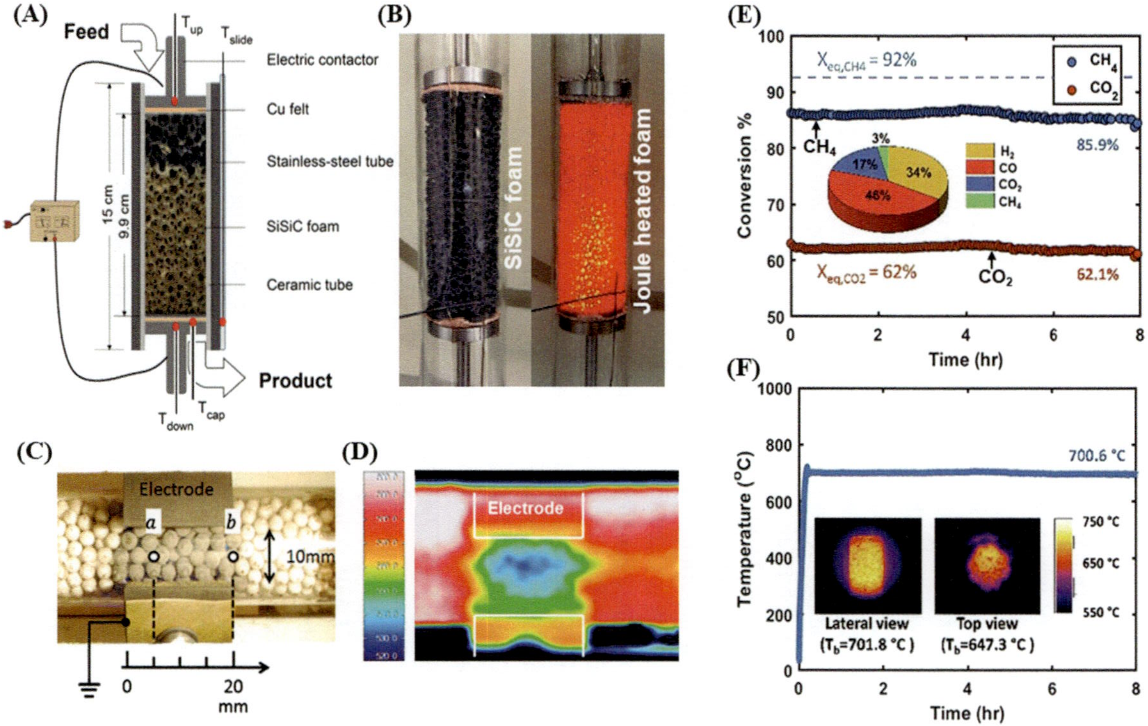

FIGURE 3.4 (A) Illustration of a Joule-heated reactor for DMR reaction using a SiSiC foam as the heater. (B) Images of the reactor setup involving bare SiSiC open-cell foam placed inside a quartz tube at room temperature (left) and uniformly heated to 700 °C using Joule heating under a nitrogen flow rate of 25 Nl/h (right). Adapted from Ref. 59, DOI: 10.1016/j.cej.2023.143154. (C) Image of a catalyst bed setup for DMR reactions using a DBD reactor. (D) Temperature distribution of the DBD reactor showing a thermograph. Adapted from Ref. 61, DOI: 10.1016/j.cattod.2015.05.011. (E) DMR performance in a microwave-assisted heterogeneous catalytic reactor, showing CH_4 and CO_2 conversions over time. Inset shows the gas product composition. (F) The corresponding temperature distribution in (E), with the inset presenting axial and radial infrared temperature profiles captured by side and top IR cameras, respectively. Adapted from Ref. 48, DOI: 10.1016/j.cej.2021.130476.

reactions occurred rapidly, producing syngas with an H_2/CO ratio of 0.8-0.9.[61] The Boudouard reaction,[62] essential for solid carbon removal, was largely enhanced by DBD hybridization, which was driven by radical injection and selective surface heating from catalyst surface charge recombination. This effect promoted carbon diffusion through Ni particles and oxidation by adsorbed CO_2, leading to higher CH_4 and CO_2 conversion without severe coking. Additionally, pulsed CH_4 injection enhanced the non-equilibrium nature of the reactions, initially producing more H_2 and later promoting reactions involving vibrationally excited CO.

While early works on DMR with non-thermal plasmas mainly focused on syngas production,[63,64] Wang and colleagues looked into other valuable product streams such as acetic acid, methanol, ethanol, and formaldehyde from CO_2 and CH_4.[65] At room temperature (~30 °C), under atmospheric pressure and using a novel non-thermal plasma reactor with a water electrode, the total selectivity from CH_4 and CO_2 to oxygenates was approximately 50–60%, with acetic acid being the major component at a selectivity of 40.2%, among the highest in the literature. The unique DBD reactor with a groundwater electrode enabled this thermodynamically unfavorable reaction. In the "plasma only" mode, the total liquid selectivity was 59.1%, with 33.7% for acetic acid, 11.9% each for ethanol and methanol, and 1.6% for acetone. CO selectivity was ~20.0%, with CH_4 and CO_2 conversions of approximately 18.3% and 15.4%, respectively. Trace amounts of formic acid, propanol, and butanol were also detected.

Ahasan and coworkers investigated the effects of catalyst support and morphology on non-equilibrium plasma-assisted DMR using 1 wt% Ru supported on irreducible SiO_2 and reducible CeO_2 with nanorod (NR) and nanocube (NC) morphologies. This study, conducted in a DBD fixed-bed reactor, demonstrated that non-thermal plasma enhances CH_4 and CO_2 conversions and syngas yield compared to thermal catalysis. Reducible CeO_2-supported Ru catalysts outperformed SiO_2 counterparts, with Ru/CeO_2-NR showing superior activity at 450 °C, achieving 51% CH_4 and 37% CO_2 conversion. The enhanced performance is attributed to the unique surface chemistry, active facets, oxygen vacancies, and defect structures of CeO_2-NR. The findings highlight the critical roles of catalyst support reducibility and morphology in optimizing plasma-assisted DMR at low temperatures.[66] The team further investigated the effect of NiO loading on CeO_2 nanorod supports for non-thermal plasma-promoted DMR. Using a wet-impregnation method, the authors synthesized 5 wt% and 15 wt% NiO/CeO_2-NR catalysts and evaluated their performance under non-thermal plasma conditions. The presence of plasma significantly enhanced DMR activity, with the 15 wt% NiO catalyst achieving CH_4 and CO_2 conversions of 66% and 48%, respectively, at 500 °C. These findings demonstrated the synergistic effects of non-thermal plasma and optimized catalyst design, as well as the ability to improve DMR efficiency at low temperatures.[67]

In a more recent study by Li and coworkers, an innovative DMR approach using a 16 W pulsed in-focus laser in combination with a cost-effective Mo_2C catalyst was proposed, enabling reforming reactions under milder conditions. The laser-catalytic DMR achieved high yields of H_2 (14300.8 mmol h^{-1} g^{-1}) and CO (14949.9 mmol h^{-1} g^{-1}) along with good energy efficiency (0.98 mmol k J^{-1}), presenting a promising alternative to high-energy-consuming conventional catalytic systems. Notably, the pulsed laser induced localized high temperatures and generated $^*CH$ species (plasma) on the Mo_2C surface, activating CH_4 and CO_2 to accelerate the DMR reaction. This method bypassed the traditional step-by-step dehydrogenation process and eliminated the rate-limiting step of CH_4 cracking by directly generating $^*CH$ plasma from CH_4 through thermionic emission and cascade ionization. Moreover, this innovative approach maintained a carbon-oxygen balanced environment, preventing the deactivation of the Mo_2C catalyst.[68]

3.2.5.3 Electrothermal Dry Reforming of Hydrocarbons Using Microwave

Many studies using microwave to conduct DMR have focused on the development of catalysts to improve the reaction performance. Gangurde and colleagues used microwave to first synthesize a series of Ru-doped $SrTiO_3$ perovskite catalysts. The 7 wt.% Ru-doped $SrTiO_3$ catalyst was examined for DMR under microwave heating in a custom fixed-bed quartz reactor. The study varied microwave power, CH_4 and CO_2 feed ratio, and GHSV to determine optimal conditions for high reactant conversions and optimal H_2/CO ratio. The 7 wt% Ru-doped $SrTiO_3$ catalyst achieved stable maximum CH_4 and CO_2 conversions of approximately 99.5% and 94%, respectively, with an H_2/CO ratio of around 0.9 at temperatures near 940 °C.[69] In another study, Sharifvaghefi and coworkers found that combining a microwave setup with Ni-MgO on activated carbon catalysts can substantially improve the DMR's syngas selectivity and CH_4 conversion rates. The addition of Mg as a promoter enhanced the catalyst's activity by increasing Ni dispersion, resulting in a higher H_2/CO ratio and reducing the effect of secondary reactions.[70]

Other than catalyst development, Chen and coworkers advanced the field by introducing a novel reactor design. The authors applied microwave-absorbing monolith filled with catalytic pellets, which improved interactions between the electromagnetic field and the susceptor material to improve tunability and efficiency. This design proved to be stable up to 900 °C while maintaining a catalyst packing density comparable to traditional fixed beds. The reactor demonstrated good DMR performance with a commercial Rh/Al_2O_3 catalyst, achieving ~86% CH_4 conversion and a notably higher H_2 throughput (~85 $m^3/kg/h$) than previous laboratory scale systems (Figure 3.4(E)). Electromagnetic simulations revealed that the monolith's electrical conductivity reduced the local power dissipation by directly absorbing heat, unlike conventional fixed beds that tend to develop hot spots easily (Figure 3.4(F)).[48]

3.2.6 Epoxidation of Alkene

Commodity epoxides can be synthesized through the epoxidation reaction of alkenes under thermochemical conditions. For example, ethylene oxide (EO) and propylene oxide (PO) are often made via ethylene and propylene epoxidation (selective oxidation) in the industry, respectively. The reaction scheme below shows the EO formation process:[71,72]

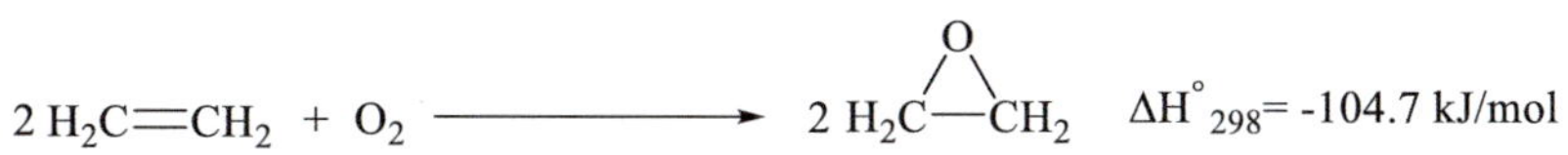

EO and PO are both indispensable to the chemical industry, particularly in polymer manufacturing, forming the basis of polyethylene oxide (PEO), polypropylene oxide (PPO), and polyurethane (PU). In addition, EO and PO are also essential in producing ethylene carbonate (EC) and propylene carbonate (PC), which are key electrolyte components for lithium-ion batteries. The chemical transformation from alkenes to these valuable products involves a selective oxidation process, typically promoted by silver-based heterogeneous catalysts. While these reactions occur at relatively mild temperatures that can be powered up by waste heat from upstream processes where energy consumption is not a major burden to the manufacturers. Enhancing selectivity to the desired epoxide compound against products formed by the more potent over-oxidation (such as CO and CO_2) remains a daunting challenge for further efficiency improvement, as these side reactions require lower activation energy barriers and are more thermodynamically favorable. To this end, coupling electrothermal processes with material innovations, especially electrically responsive catalysts, could potentially improve the overall reaction outcomes such as selectivity, rate, and/or product yield.

3.2.6.1 Electrothermal Epoxidation Using Resistive Heating

The first electrothermal approach to synthesize PO from propylene was reported by Yuan and coworkers using a microreactor with resistive heating. Using Au/TiO_2 as a catalyst (Figure 3.5(A)), the initial rate for epoxidation reaction was low but increased substantially with higher concentrations of H_2, O_2, or C_3H_6, despite that these concentrations were in the explosive limit. The catalyst, under conventional conditions, quickly lost its activity and was highly sensitive to temperature changes. To address these challenges, Yuan and colleagues designed a microchannel reactor coated with the Au/TiO_2 catalyst (Figure 3.5(B)). The reactor features small flow channels to mitigate explosion risks and incorporates internal resistive heating for rapid temperature adjustments, which offers improved process control. Constructed with a FeCrAl alloy, the microreactor can rapidly increase its temperature from 50 °C to 300 °C within just 2 minutes, allowing for efficient catalyst regeneration and full recovery of its activity, largely improving from conventional reactor settings.[73]

3.2.6.2 Electrothermal Epoxidation Using Microwave

Ramirez and coworkers explored the efficacy of microwave-assisted epoxidation of C_2H_4. The team demonstrated the effectiveness of microwave-induced heating for cordierite monolithic reactors, which contained a thin layer of a silver and copper oxide-based catalyst (Figure 3.5(D) and (E)). This catalyst showed a strong and selective heating susceptibility when exposed to microwave irradiation. This method features lower power consumption and higher energy efficiencies compared to conventional heating processes. Importantly, the microwave radiation selectively heated both the catalyst and the monolith walls, while the gas stream remained relatively cooler, which is beneficial for suppressing side reactions such as over-oxidation (Figure 3.5(C)). Specifically, the selective spatial heating created a gas/solid temperature gradient of up to approximately 70 °C at a reaction temperature of 225 °C, high enough for notable selectivity improvement. In addition, the homogeneous deposition of the catalyst was found crucial for microwave-heated catalytic reactors, as the catalyst also acted as a microwave susceptor. Without this even distribution (Figure 3.5(F) and (I)), the local accumulation of catalytic material could lead to hot

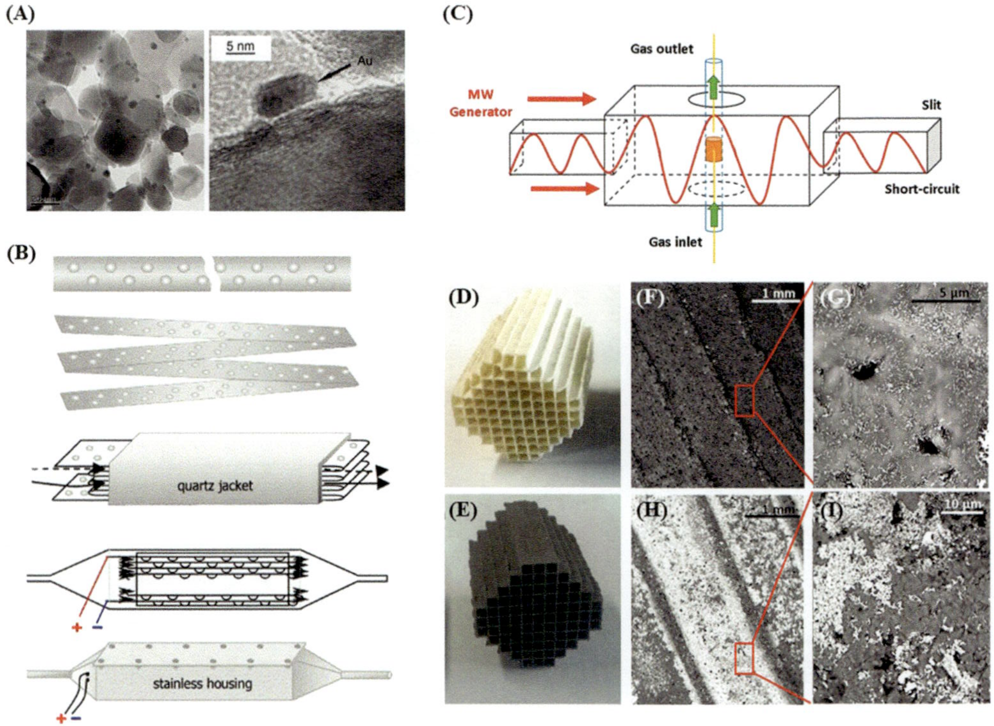

FIGURE 3.5 (A) TEM images of the Au/TiO_2 catalyst used in electrified microreactor for C_3H_6 epoxidation. (B) Schematic of microchannel reactor set up with metal alloy belt featuring dimples, curved configuration of the belt, quartz jacket with flow pathways, internal structure, and external view of the assembly. Adapted from Ref. 73, DOI: 10.1016/j.cattod.2005.06.047. (C) Diagram of a microwave reactor for the C_2H_4 epoxidation reaction, showing single-mode cavity illustrating the direction of microwave propagation while the monolith is positioned at the electric field maximum. (D, E) Digital images of the cordierite monolith before and after catalyst loading. Scanning electron microscopy (SEM) images at varying magnifications demonstrate a uniform coating on the inner channel of a monolith, prepared through vertical immersion in an ethanolic suspension of the microwave-susceptor catalyst (F, G), and with irregular agglomerates and coating inhomogeneities resulting from horizontal positioning during the immersion-sonication-deposition process (H, I). Adapted from Ref. 74, DOI: 10.1016/j.cattod.2016.01.007.

spots, which markedly reduced EO selectivity.[74] To measure gas temperatures within the channels, the team further examined gas-phase reactions of C_2H_4 epoxidation. The authors employed a microwave-sensitive silver-copper oxide hybrid catalyst deposited on a cordierite monolith and applied infrared thermographic techniques to monitor the reaction under microwave heating conditions. The study revealed a gas-solid temperature gradient of more than 50 °C at a temperature of 200 °C at the solid surface, which likely contributed to the improved reaction outcomes via suppressing gas-phase processes.[75]

3.3 ELECTROTHERMAL CARBON DIOXIDE CONVERSION

3.3.1 Carbon Dioxide Hydrogenation to Methanol

In recent years, the growing availability of green H_2 produced from renewable energy sources, coupled with CO_2 captured through advanced carbon capture technologies, has made methanol synthesis via CO_2 hydrogenation a highly promising approach. Particularly, this process offers a dual benefit: (1) effective green H_2 storage and transportation, which is enabled by methanol's liquid state under ambient condition, and (2) sustainable and distributed chemical synthesis and manufacturing, as methanol serves as a key precursor for numerous commodity chemicals, such as formaldehyde, dimethyl ether, and acetic acid, among others. The industrial-scale synthesis of methanol began in 1923, pioneered by BASF using carbon monoxide (CO) and H_2 over a sulfur-resistant ZnO/Cr_2O_3 catalyst.[76,77] In 1966, Imperial Chemical Industries introduced their commercial catalyst named ICI 51-1 based on $Cu/Zn/Al_2O_3$.[78] This catalyst composition has been predominately studied and used for methanol synthesis afterward, despite undergoing continuous refinement through additional element incorporation in order to enhance catalytic activity as well as sintering resistance. In the context of methanol synthesis from CO_2 hydrogenation, Cu-based catalysts remain the most effective candidate, in which metallic copper (Cu^0) serves as the active site while metal oxides such as ZnO serve as the promoter.[79–81] Industrially, CO_2 hydrogenation to methanol operates at moderate temperatures of up to ~573 K, with heat supplied through continuous fossil fuel combustion—an energy-intensive and carbon-heavy approach.[82,83] Interestingly, while CO_2 hydrogenation to methanol is an exothermic reaction where CO_2 conversion favors low temperature according to the Van't Hoff equation, elevated temperature (by heating) or an alternative energy input is still critical to effectively and rapidly activate the reactant species, especially CO_2.

$$CO_2 + 3H_2 \leftrightarrow CH_3OH + H_2O \quad \Delta H^{\circ}_{298} = -49.4\,kJ/mol$$

To this end, it is beneficial to develop electrified methods that can reduce carbon emission by replacing combustion, accommodate intermittent green H_2 supply with instant tunability and rapid start-up/shut-down capability, improve energy efficiency by reducing the reaction temperature, and enhance reaction performance with low-temperature activation. Among various electrothermal techniques, methanol syntheses powered by non-thermal plasma stand out. Non-thermal plasma offers fast activation and the generation of non-equilibrium intermediates to promote chemical reactions at reduced temperatures. These features make non-thermal plasma and its catalysis particularly appealing for exothermic reactions like CO_2 hydrogenation to methanol and ammonia (NH_3) synthesis (*vide infra*). So far, electrothermal CO_2 hydrogenation to methanol has been mostly conducted using plasma-assisted methods, covering many aspects including reactor design, process innovation, catalyst development, substrate effect, and many more.

Wang and colleagues studied the impact of reactor configuration on plasma-assisted CO_2 hydrogenation to methanol.[84] The authors found that the structure of the DBD reactor has a critical effect on the distribution of gas (CO and CH_4) and liquid (methanol and ethanol) products. Specifically, the team explored three different designs, where reactor I features a conventional cylindrical DBD reactor consisting of a stainless-steel rod as the high-voltage electrode and an aluminum foil sheet along the wall as the ground electrode (Figure 3.6(A)). In comparison, reactors II and III replaced the ground electrode with a water-circulating electrode, which also assisted in heat dissipation. For reactor II, the stainless-steel rod was further covered by a quartz tube, which affected the direct contact of catalytic materials with the electrode. While the reactant conversion remained comparable in all these cases (10–14%), reactor III showed 57% selectivity toward methanol, substantially higher than those obtained using reactors I and II where CH_4 and CO were the primary products (Figure 3.6(B) and (C)).

Many researchers studied the effect of electric field, ionic conductivity, and other reactor/process characteristics on the plasma's performance for CO_2 hydrogenation to methanol. For example, Feliz and coworkers studied the Cu loaded on Al_2O_3 and

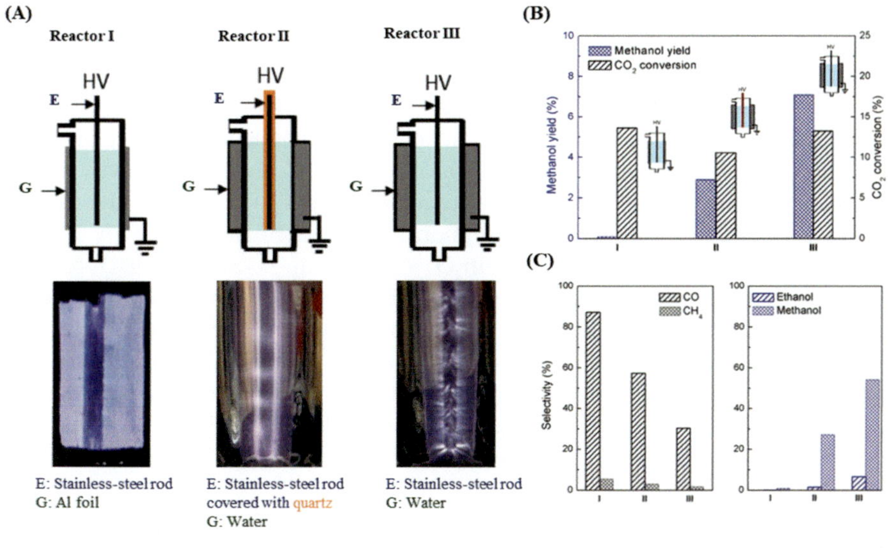

FIGURE 3.6 (A) Schematic representation and images of H_2/CO_2 discharge generated in different DBD reactor configurations (I, II, and III) without catalysts. (B) Methanol yield and CO_2 conversion. (C) Selectivity of gas and oxygenates products (reaction pressure 1 atm, H_2/CO_2 molar ratio 3:1). Adapted from Ref. 84, DOI:10.1021/acscatal.7b02733.

ZSM-5 promoted by Zn in a non-thermal plasma setup.[85] The authors found that the conductivity of the catalyst influenced both plasma characteristics and catalytic activity by altering discharge types, enhancing the electric field, as well as enabling micro-discharge formation. For example, lower ionic conductivity enhances CO_2 conversion, as high ionic conductivity reduces plasma density and hinders CO_2 dissociation. Among the tested materials, the non-conductive Al_2O_3 support achieved the highest CO_2 conversion rate of 34.0%, while the zeolite-supported Cu-ZnO catalyst produced the maximum methanol yield of 0.5%. Similar systems were also studied by Neyts and Bogaerts using simulation and modeling.[86] Different from solely using plasma, Joshi and colleagues introduced a heating furnace around the plasma discharge area to study the synergistic effect among catalyst, temperature (by heating), and non-thermal plasma.[87] The authors found that plasma induced the formation of a mixture of spinel compounds (NiO-Fe_2O_3, $NiFe_2O_4$, and Fe_3O_4) that largely outperformed pure oxide catalysts; meanwhile, the synergy between plasma and thermal heating generated notable morphological changes for the catalyst, leading to a substantial improvement in the overall catalytic performance.

In connection with reactor and process studies, catalyst development has drawn the most attention. A range of transition metals such as Cu, Ni, Co, and Pt have been explored as catalysts for plasma-assisted CO_2 hydrogenation to methanol. For example, Wang and colleagues used CuO/Al_2O_3 catalysts in their non-thermal plasma reactor, achieving 21.2% CO_2 conversion with 53.7% methanol selectivity at room temperature and under ambient pressure.[84] Feliz and coworkers employed Cu with Zn as promoter, where the effect of support was noted.[85] Joshi and colleagues applied a binary NiO-Fe_2O_3 metal oxide catalyst, suggesting Ni to be a potential catalyst candidate other than Cu.[87] Ronda-Lloret and coworkers employed a CoO_x/MgO catalyst, achieving over 35% CO_2 conversion and 10% methanol yield.[88] In the works conducted by Ronda-Lloret and colleagues, the substrate effect was also investigated. The surface basicity of MgO was found to facilitate more effective CO_2 adsorption compared to the acidic γ-Al_2O_3, leading to improved catalytic performance. Cui and coworkers conducted a combined experimental and computational study on a size-selected Cu_{13}/γ-Al_2O_3 catalyst.[89] In this work, the catalyst's surface sensitivity to water was reported. The authors found that water induced desorption of methanol from the catalyst surface due to competitive absorption, boosting methanol selectivity to 65%.[89] Men and coworkers studied Pt/film/In_2O_3 catalysts in a chilled DBD reactor.[90] Different from the aforementioned approach by Wang and colleagues,[84] Men and coworkers used the water electrode as both the ground electrode and the cooling medium. This setup, together with Pt-based catalysts, outperformed the conventional Cu/ZnO/Al_2O_3 catalyst without additional cooling. In addition to the aforementioned systems, MnO_x/ZrO_2 as well as other Cu- and Ni-based catalysts have also been explored along the way for plasma-assisted conversion of CO_2 into methanol.[91–93]

3.3.2 Reverse Water-Gas Shift

The RWGS reaction provides a means to adjust the H_2 and CO ratio in syngas. It also converts CO_2 and H_2 into value-added CO. The generated CO can then be further processed into valuable chemicals, such as hydrocarbons for liquid fuels via

Fischer-Tropsch synthesis, as well as other oxygenated compounds using a range of well-established industrial methods.[94,95] The RWGS reaction is endothermic, requiring significant energy input. As a result, high temperatures, sometimes exceeding 1000 K, are necessary to achieve meaningful conversion rates (e.g., 50%). Additionally, selective catalysts are crucial to prevent the occurrence of the exothermic methanation reaction, which would otherwise produce CH_4 instead of CO (see Section 3.3.3).

$$\text{RWGS}: CO_2 + H_2 \leftrightarrow CO + H_2O \quad \Delta H^{\circ}_{298} = +41\,\text{kJ/mol}$$
$$CO + 3H_2 \leftrightarrow CH_4 + H_2O \quad \Delta H^{\circ}_{298} = -206\,\text{kJ/mol}$$

3.3.2.1 Electrothermal Reverse Water-Gas Shift Using Resistive Heating

In 2022, Wismann and colleagues introduced a novel electrified reverse water-gas shift (eRWGS) catalyst, advancing a zero-emissions method for producing syngas with any desired H_2/CO ratio. This innovation enables the sustainable production of carbon-based chemicals from CO_2 and renewable electricity, achieving high efficiency in H_2 and carbon utilization. The fully electrified process was implemented in a pressurized bench-scale reactor, utilizing a porous zirconia-based washcoat on aFeCrAl alloy that was impregnated with a nickel catalyst and heated through ohmic heating. Operating under industrially relevant conditions, the system processed a feed of H_2 and CO_2 in a 2.25 ratio at 10 bar and 1050 °C, producing syngas with a 2.0 H_2/CO ratio and no detectable CH_4, making it ideal for Fischer-Tropsch synthesis of sustainable fuels. It was observed that facilitating the RWGS through CH_4 as an intermediate was more effective than the selective RWGS route, as the route demonstrated higher activity and reduced carbon formation. The eRWGS™ reactor supports high-temperature operation, achieving a gas composition of 23% CO, 46% H_2, and 8% CO_2, with CH_4 levels below detection limits above 1050 °C.[96]

Zheng and coworkers introduced a promising approach for the direct eRWGS process using wash-coated structured reactors. These reactors employed catalytically activated open-cell foams, which provided optimal heat and mass transfer properties to serve as Joule heating substrates for the catalytic conversion of CO_2 with H_2. Specifically, the proposed reactor system utilized Joule-heated Rh/Al_2O_3-coated porous SiSiC foams, which acted both as catalyst supports and heating substrates. This setup demonstrated excellent catalytic and electrical stability for over 75 hours, operating at temperatures up to 800 °C and achieving near-equilibrium conversion at high space velocity for eRWGS. The reactor concept showed great potential for remarkably low specific energy demand in CO_2 valorization, with an optimized process configuration estimating approximately 0.7 kWh/Nm^3 CO_2 for eRWGS. Furthermore, during eRWGS processes above 675 °C, almost no CH_4 was detected in the product streams. Using an H_2-rich feed, the team achieved a high CO_2 conversion rate of 94.2%.[59]

Yu and colleagues made notable advancements by employing rapid pulse heating (RPH) for CO_2 hydrogenation over a Ni/Al_2O_3 catalyst at 1 bar.[97] This technique enhanced the reaction rate and shifted product selectivity toward CO over CH_4 at low reaction temperatures below 500 °C (Figure 3.7(A)), a range not achievable through conventional steady-state operation at similar average temperatures or conversions.

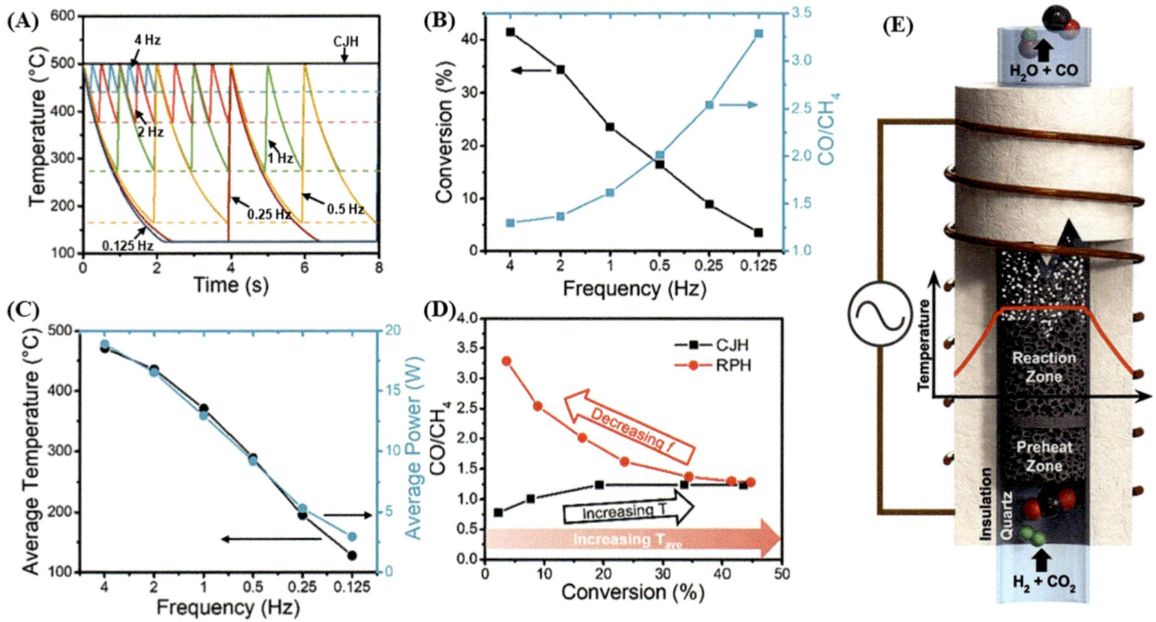

FIGURE 3.7 (A) CO_2 hydrogenation was investigated under two heating modes: continuous Joule heating (CJH) and rapid pulse heating (RPH). The temperature profiles demonstrate a steady 500 °C under CJH (solid black line) and oscillating temperatures under RPH at varying frequencies. (B) The impact of RPH frequency on CO_2 conversion and the CO/CH_4 product ratio. (C) The relationship plot between the average reactor temperature and frequency. (D) The CO/CH_4 product ratio is plotted against conversion to reveal how heating modes and frequencies influence reaction efficiency and product distribution. Adapted from Ref. 97, DOI: 10.1016/j.cej.2024.148528. (E) Schematic illustrating inductively heated reactor designed for the RWGS reaction. The silicon carbide foam baffle provides a high-surface-area support for catalyst pellets and functions as a volumetrically heated susceptor with three distinct heating zones. The temperature plot reveals the cross-sectional temperature distribution within the reactor's reaction zone. Adapted from Ref. 99, DOI: 10.1016/j.joule.2024.07.017.

The CO/CH_4 ratio was strongly influenced by pulse frequency (Figure 3.7(B)). As the frequency decreased and the average temperature dropped (Figure 3.7(C)), the CO/CH_4 ratio increased, indicating that dynamic operation favored CO production over CH_4. Additionally, RPH and continuous Joule heating exhibited different selectivity despite achieving similar conversion levels (Figure 3.7(D)). The authors identified *CO as a key surface intermediate, which either desorbs or undergoes further hydrogenation to form CH_4. The observed change in selectivity was attributed to the transient coverages of *CO and *H on the Ni surface during temperature pulsing. This dynamic process facilitated the desorption of *CO while suppressing its deeper hydrogenation to CH_4. Their findings revealed a dynamic catalyst surface with transient intermediate coverages, where *CO species desorbed more readily in the presence of *H. Pulse heating was proposed to enable the formation of these species by lowering the temperature and activating more *H, while high-temperature pulses (up to 500 °C) promoted the direct desorption of weakly bound *CO, resulting in enhanced CO selectivity.[97]

In another study, Basini and coworkers demonstrated an integrated process for converting CO_2 into synthetic methanol by coupling three components: a water electrolysis system, an eRWGS reactor, and a methanol synthesis reactor (see Section 3.3.1). This system utilized residual CO_2 from a biogas plant (300 Nm^3/h) and H_2 from a proton exchange membrane (PEM) electrolyzer. The authors found that the eRWGS reactor could operate at high temperatures (950 °C) with an energy efficiency exceeding 90%, significantly enhancing the CO yield. The process analysis revealed a specific energy consumption of 11 kWh/kg for methanol production, with a carbon efficiency of 90%. This innovative approach highlighted the potential for efficiently producing methanol from CO_2 and renewable H_2 through process intensification.[98]

3.3.2.2 Electrothermal Reverse Water-Gas Shift Using Induction Heating

An innovative study by Fan and colleagues introduced inductively heated reactors for eRWGS, achieving volumetric heating through three-dimensional metamaterial reactor baffles.[99] These baffles, designed as open-cell ceramic lattices, functioned as metamaterial susceptors with effective electrical and thermal conductivities, enabling efficient induction heating and high-temperature stability (Figure 3.7(E)). Experiments with lab-scale reactors demonstrated near-unity heating efficiencies, radially uniform heating profiles, and high CO_2 conversion rates that are close to equilibrium values at low gas flow rates. At higher flow rates, reduced residence times were found to reduce the conversion. The reactors achieved outstanding thermal efficiencies exceeding 85%, effectively converting electrical energy into useful heat, underscoring their potential for green chemical feedstock processing and clean energy applications. Notably, it was suggested that scaling up the reactor to industrially relevant sizes can be further obtained by customizing the metamaterial baffle. During scaling up, the system efficiency should increase with the reactor size toward the efficiency of power electronics (e.g., >90%), which is highly promising for industrially relevant electrified thermochemical processes.

3.3.3 Carbon Dioxide Methanation

Dou and coworkers investigated electric internal heating for CO_2 methanation using Ni-foam (NF) structured catalysts. This method, which heats the catalyst from inside the reactor, demonstrated largely enhanced catalytic activity and improved resistance to catalyst poisoning compared to conventional heating techniques. The RuNi/NF catalyst, characterized by densely packed ultrathin nanoplates on the NF skeleton, achieved good performance at 300 °C, with 73% CO_2 conversion and 99% CH_4 selectivity. Notably, its performance remained stable even after repeated heating and cooling cycles. In addition, the electric internal heating enabled rapid heating from 50 °C to over 350 °C within just 2 minutes, making it ideal for reactors requiring quick start-up and shut-down capabilities. The enhanced catalytic activity was attributed to the uniform temperature distribution, minimized temperature gradients, and optimal catalyst utilization provided by the electric internal heating approach.[100]

3.4 ELECTROTHERMAL BIOMASS CONVERSION

The global production of biomass, its derivatives, and primary waste is estimated to reach approximately 100 billion metric tons annually.[101,102] Such abundant and highly available resource holds immense potential for chemical synthesis and manufacturing to complement industrial processes that use fossil fuels as feedstock. Like many other applications discussed above, biomass conversion to high-value chemicals using electrothermal methods offers a promising route to decarbonize the chemical industry. Despite being at an early stage, several electrothermal methods have shown promising effects in achieving improved reaction performance and enhanced energy efficiency.

3.4.1 Electrothermal Biomass Conversion Using Microwave

Among various electrothermal methods used for biomass conversion, microwave heating has gained notable attention due to its ability to selectively heat materials using an external electromagnetic field, which has profound impact when working with a solid feedstock, especially with the large variety of biomass species. Specifically, this "inside-out" heating mechanism offers several advantages, including rapid heating rates, uniform temperature distribution, quick process regulation, minimal energy dissipation, selective heating, and high energy conversion efficiency. Owing to these benefits, microwave heating has been used to make biofuels, bio-oils, as well as various compounds such as methanol, ethanol, glycerol, glucose, furfural, and 5-hydroxymethylfurfural, and many more.

Among various pyrolysis products from biomass, biofuels and bio-oils are widely explored by the community. For microwave-assisted processes, large variations in biofuel and bio-oil compositions were reported based on various raw biomass used. Gautam and coworkers analyzed and compared bio-oil production from macroalgae using microwave-assisted pyrolysis and analytical fast pyrolysis.[103] Microwave-assisted pyrolysis exhibited reduced anhydrosugar contents and high selectivity to aromatics (20%) and furan derivatives (40%) compared to conventional methods owing to higher localized temperatures, which promoted their secondary breakdown into low-molecular-weight organic compounds. Microwave-assisted pyrolysis also led to the formation of monocyclic and polycyclic aromatic compounds (15–20%) through mechanisms such as intramolecular addition of cyclopentadienyl radicals and β-scission of C-C and C-H bonds.[103] Note that the dielectric properties of lignocellulosic biomass are generally poor, necessitating the addition of microwave absorbents or susceptors for efficiency improvement. Borges and coworkers reported enhanced bio-oil yields through fast microwave-assisted pyrolysis using feedstocks like wood sawdust and maize stover with SiC as microwave adsorbent. The authors also found that the optimum temperature for high bio-oil yield falls between 480 and 560 °C depending on the substrate.[104] Wang and colleagues later reported the pyrolysis of soapstock and found the optimal temperature for aromatic production was approximately 550 °C, agreeing with the prior results.[105] In addition to these results, Kostas and coworkers reported that bio-oil derived from olive pomace contained a higher concentration of acetic acid, whereas Abas and colleagues reported that bio-oil obtained from microwave-assisted pyrolysis of oil palm fiber had increased levels of water and various organic acids.[106,107]

To improve the product value and enhance the practicality of the technology, microwave-assisted co-pyrolysis or co-processing using biomass and other materials or multiple biomass species has been explored. For instance, An and coworkers reported increased yields of liquid oil and pyrolytic gas when biomass was co-processed with lignite.[108] This feedstock mixture favored the generation of aliphatic hydrocarbons, reducing the production of single-ring aromatic compounds, carboxylic acids, alcohols, aldehydes, and ketones. Suriapparao and colleagues found that combining biomass with polypropylene and polystyrene produced bio-oil with reduced water content and a higher heating value of 38–42 MJ/kg. Aromatic and aliphatic hydrocarbons were the major products with 80% and 62% maximum yields in bio-oil from polystyrene mixed with rice husk and polypropylene mixed with rice husk, respectively.[109] Li and colleagues further investigated the microwave-assisted co-pyrolysis of oil sands and sawdust, which increased the aromatic hydrocarbon content in bio-oil. The authors also observed that microwave-generated chars and residues exhibited microwave-absorbing ability, potentially facilitating the heating process.[110]

One of the issues with conventional microwave reactors is that the electromagnetic field tends to scatter randomly within the cavity, leading to a relatively low field intensity and an unpredictable microwave energy distribution. Previous research has used microwave susceptors, such as carbon materials and SiC, to improve the heating of biomass. Recently, conventional magnetron-based devices were replaced by semiconductor generators, which can provide a sharp microwave spectrum with accurate control over frequency. In addition, it is observed that spatial separation of electric (E) and magnetic (H) fields is more efficient as the E-field heats raw biomass through dielectric heating, while the H-field induces heating in conductive carbonized biomass formed during pyrolysis. Tsubaki and coworkers applied this method to bamboo biomass using a semiconductor microwave generator. The authors showed that the E-field facilitated gasification and carbonization in 30 seconds with substantial production of H_2, CO, and tar. After carbonization, the reactants were more efficiently heated by the H-field, producing amorphous biochar with more defects (Figure 3.8(A)).[111]

Catalytic microwave-assisted pyrolysis has also shown promise. Wang and coworkers reported that a modified HZSM-5 catalyst increased phenol selectivity

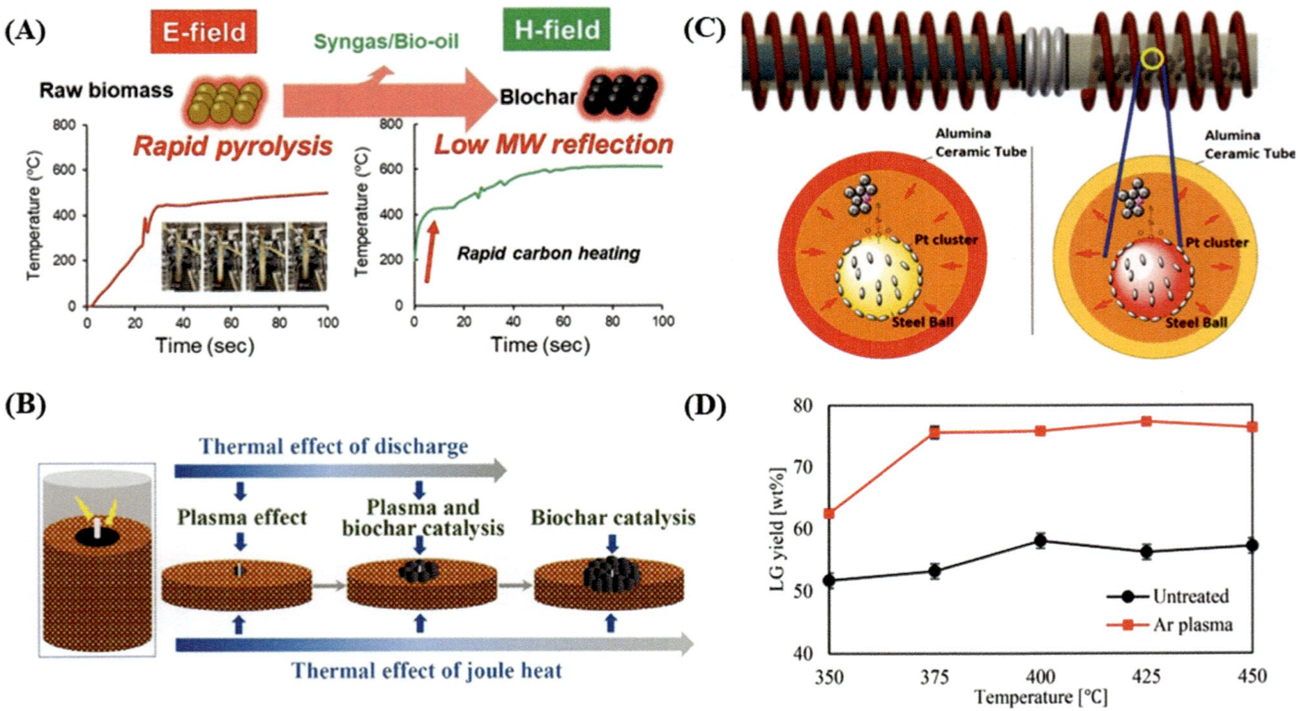

FIGURE 3.8 (A) Combination of the spatially separated E-field (for extensive carbonization) and H-field (for low reflection from carbonized bamboo and effective heating) for efficient pyrolysis of bamboo biomass. Adapted from Ref. 111, DOI: 10.1016/j.cej.2024.156260. (B) Triggering effect of microwave-assisted heating coupled with tip-metal. Adapted from Ref. 115, DOI: 10.1016/j.renene.2022.12.076. (C) Schematic of different heat transfer modes, namely outside-to-inside and inside-to-outside, via different reactor designs. Adapted from Ref. 125, DOI: 10.1016/j.enconman.2019.01.025. (D) LG yields were obtained from the untreated and Ar plasma-pre-treated cellulose using different pyrolysis temperatures. Adapted from Ref. 129, DOI: 10.1039/D0GC00255K.

and bio-oil yield during cellulose pyrolysis, which is unusual as phenols are typically derived from lignin instead of cellulose.[112] The authors found that the Fe/HZSM-5 catalyst increased bio-oil yields by 11.4% and reduced coke formation by about 24% compared to unmodified HZSM-5. On the other hand, the Fe-Ni catalyst almost fully converted saccharides in the bio-oil into phenols and low-molecular-weight compounds. The transformation of furans into aromatics within the pores of HZSM-5, followed by further conversion to phenols on the catalyst's external surface mainly accounted for the unconventional selectivity towards phenol. Zhu and colleagues achieved a high hydrocarbon concentration of 52.77% in bio-oil over a biochar catalyst. In comparison, only 0.72% hydrocarbon content was obtained for the bio-oil from non-catalytic runs.[113] The biochar helped in the deoxygenation of raw bio-oil, forming hydrocarbon and phenol products efficiently. Shi and coworkers explored the influence of biochar, ZSM-5, and metal-based catalysts (e.g., Zn, Zr) in microwave pyrolysis using bamboo as the starting material. The authors discovered distinct activities and product distributions using these catalysts.[114] For example, ZSM-5 showed the highest bio-oil yield (42.16 wt%), while biochar catalyst achieved the highest gas yield (43.10 wt%). The Zn-based catalyst was found to enhance aromatic hydrocarbon formation. In comparison, Zr-based catalyst led to improved bio-oil and syngas yield.

In previous approaches, the mixing of catalyst powders with biomass during pyrolysis presented challenges in separating the catalysts from the final solid products. To address this issue, researchers have explored the use of microwave irradiation in combination with immobilized catalytic components, such as a tip-metal. In addition to facilitating product and catalyst separation, the discharge induced by microwaves at the metal tip could enhance the pyrolysis process. For instance, Li and coworkers observed a synergistic interaction between microwave irradiation and an iron-based metal tip during the pyrolysis of a binary lignin and xylan mixture.[115] This combination resulted in a remarkable 180% increase in efficiency compared to microwave pyrolysis without the metal tip (Figure 3.8(B)). In another study, Zhang and colleagues studied the effects of microwaves combined with a needle-shaped iron wire during the pyrolysis of pine sawdust and a biomass mixture. This design increased the maximum pyrolysis rate by 92% and 56%, respectively, while also reducing the activation energy by 52% and 49%.[116] In comparison to microwave-assisted pyrolysis without metal, this method significantly altered the product distribution and properties. Key improvements included increased graphite content in biochar, enhanced aromatic structure, and the decomposition of phenols in bio-oil into benzene and olefins. Moreover, a notable rise in syngas production was observed. These enhancements were attributed to the combined plasma and thermal effects generated by microwave discharge and the continuous Joule heating caused by the presence of metal. Together, these findings highlight the potential of this process for advancing microwave-assisted pyrolysis technologies.

3.4.2 Electrothermal Biomass Conversion Using Induction Heating

Induction heating provides another pathway for biomass conversion, utilizing electromagnetic fields to generate eddy currents within materials, thereby heating them internally without direct contact. This method has been applied to various chemical synthesis processes, as discussed above, as well as for biomass pyrolysis. In one of the initial works, Tsai and coworkers employed induction heating in an externally heated fixed-bed reactor to quickly pyrolyze some of the most available agricultural wastes like rice straw, sugarcane bagasse, and coconut shell to produce value-added products. The technique allowed for fine control of heating rates between 100 and 500 C/min, yielding a maximum of 50% bio-oil.[117,118] After these early studies, rapid pyrolysis of different biomass substrates without catalysts has been the subject of numerous investigations employing an inductive heating reactor setup. Boldor and colleagues conducted several studies on using induction heating for the pyrolysis of various biomass materials such as pinewood sawdust, energy cane, Chinese tallow tree, cottonwood, and switchgrass. The authors reported an energy conversion efficiency of up to 80% using inductively heated steel reactor.[119–121] Wu and colleagues demonstrated that the addition of stainless-steel metal particles to an inductively heated reactor substantially enhanced biofuel production. Using reed straw as the feedstock, induction heating reduced complete pyrolysis times by up to 28.9%, increased combustible gas and bio-oil yields by 13.2% and 7.0%, respectively, and improved both the lower heating value of gas and the higher heating value of bio-oil. Furthermore, the energy consumption was reduced by up to 26.5%.[122]

Biomass conversion using induction heating together with catalysts represents another major line of research. Muley and coworkers studied the use of zeolite for biomass pyrolysis powered by induction heating, where the catalyst bed was also heated by induction at 290 °C, 330 °C, and 370 °C. Compared to conventional heated catalysts, induction heating showed high-quality bio-oil with decent yields of aromatic hydrocarbons (C_5-C_{10}) and reduced oxygen content. These results can be explained by the lower amount of carbon deposition with inductive heating, whereas pronounced thermal gradients in conventional reactors cause greater condensation of molecules on the catalyst surface at lower temperatures.[123] Laban and coworkers demonstrated the pyrolysis of sawdust into bio-oil using induction heating, achieving partial deoxygenation using Pt/Al_2O_3 pellets as a catalyst mixed with steel balls.[124] For comparison, the authors applied a traditional heating method using an electric heating tape as an external heat source. In this control experiment, the catalyst experienced rapid fouling from coke production and showed little to no deoxygenation when the bed was heated to the same target temperature (230–240 °C). With induction heating, however, the catalyst acted as a heat source rather than a heat sink, as is typical in conventional heating methods. By directly heating the catalyst and metallic susceptor mixture through radiofrequency-based induction, cooling and condensation of pyrolysis vapors on the catalyst surface were minimized, significantly reducing coke formation. The team later functionalized stainless-steel spheres with Pt-nanoparticles using a silane linker for the upgrading of biofuel in an induction reactor. The induction heating was found to be highly efficient, which requires a minimum power of only 150 W to reach a temperature of 525 °C and is capable of effectively reducing the oxygen content in bio-oil (Figure 3.8(C)).[125]

Notably, in addition to using various raw biomass, induction heating has been applied to power up reactions to convert biomass-derived molecules such as furfural and 5-hydroxymethylfurfural. For example, Chaudret and colleagues achieved 100%

selective hydrodeoxygenation of these molecules into alkylfurans—specifically 2-methylfuran and 2,5-dimethylfuran—using $Fe_{2.2}C$@Ru nanoparticles as catalysts. This approach leveraged the high heating efficiency of $Fe_{2.2}C$ NPs and the catalytic activity of Ru. Notably, inductive heating enabled the reactions to proceed with low catalyst loadings and under low H_2 pressures (e.g., 3 bar), highlighting the scalability potential of this technology.[126] It is expected that induction heating together with heterogeneous catalysts will have a range of applications in similar applications, transforming biomass-derived intermediate products into fine chemicals.

3.4.3 Electrothermal Biomass Conversion Using Plasma

Plasma technologies have been employed on biomass conversion reactions across a wide temperature range, with both "low-" and "high-temperature" plasma serving different purposes. "Low-temperature" plasmas are widely used for biomass pretreatment, such as delignification, prior to acid or enzyme-catalyzed hydrolysis. In contrast, "high-temperature" plasma (different from fusion plasma) facilitates the conversion of biomass in inert or oxidizing atmospheres, aiming at producing biofuels and syngas or other valuable chemicals.

For the pretreatment of biomass precursors. Benoit and coworkers studied non-thermal atmospheric plasma for cellulose pretreatment. The authors demonstrated superior reaction performance compared to other methods, such as ionic liquid treatment or ball-milling. Specifically, the non-thermal plasma process led to nearly 22 wt% glucose yield after hydrolysis.[127] Bandulasena and colleagues developed a DBD plasma microbubble reactor that generated reactive oxygen and nitrogen species like ozone for the effective pretreatment of lignocellulosic biomass.[128] Lusi and coworkers reported a high 78.6% yield of levoglucosan produced from cellulose just by pretreating it with DBD plasma for 10–60 s followed by pyrolysis at 350–450 °C. The authors found that homolytic cleavage of glycosidic bonds occurs during the pretreatment forming free radicals, which further helps with the depropagation processes (Figure 3.8(D)).[129]

Plasma has been applied to gasification, targeting both raw biomass and its derivatives as feedstocks. Mallinson and colleagues demonstrated that concentrated glycerol and pure glycerol could be effectively converted into syngas with over 80% selectivity using a DC pulse plasma system operating at ambient temperature and pressure.[130] Je-Lueng and coworkers used a torch plasma reactor for pyrolyzing biomass waste of rice straw. The authors found that the plasma induced fast reaction rate, where almost 90% of gaseous product was released within just 3 minutes.[131] An'Shakov and colleagues used a jet-transferred-arc plasma torch for sawdust and rice hull gasification. Interestingly, the authors found that H_2 production decreased with dried sawdust. The result indicates that H_2O also contributes to H_2 formation.[132]

Other than pretreatment and gasification, plasma-assisted liquefaction has also gained attention. Ye and colleagues treated bamboo shoot shells using in-liquid plasma discharges within a polyethylene glycol/ethylene glycol solvent mixture and with an acid catalyst. This method required no external heating and achieved faster liquefaction than conventional methods, yielding approximately 97% liquid product in just 3 minutes.[133] Zhou and coworkers demonstrated rapid and efficient conversion of both dry and wet algal biomass into value-added chemicals and solid carbon under mild plasma conditions.[134] The authors used H_2SO_4 and KOH as homogeneous catalysts under mild conditions, which helped reduce the liquefication time from hours to minutes. In just 7 minutes, the acid and basic catalyst produced a liquid yield of 73.95 wt% and 69.80 wt%, respectively.

3.5 ELECTROTHERMAL AMMONIA SYNTHESIS

3.5.1 Electrothermal Ammonia Synthesis Using Hydrogen

NH_3 is the second-most produced chemical globally. Its synthesis plays a pivotal role in the production of fertilizers, the development of clean fuels, and the storage of hydrocarbon-free energy. Three Nobel laureates have been recognized for their groundbreaking contributions to NH_3 synthesis: Fritz Haber (1918) and Carl Bosch (1931) for inventing and industrializing the Haber-Bosch process, and Gerhard Ertl (2007) for applying surface science to elucidate the reaction mechanism.[135,136] Notably, the Haber-Bosch process has revolutionized global food production over the past century, supporting a substantial increase in the global population. Today, approximately 170 million metric tons of NH_3 are produced every year, with nearly 80% used in fertilizers. Beyond agricultural applications, NH_3 is gaining attention as a carbon-free energy and H_2 storage material due to its high energy density (3 kWh/kg) and substantial H_2 content (17.6 wt%).[137,138] It can be readily liquefied either by compressing to 1 MPa at room temperature (298 K) or by cooling to 240 K at ambient pressure (0.1 MPa), enabling efficient and cost-effective storage, transportation, and distribution.[139] Established infrastructure for such purposes, including shipping routes and pipelines, broadly exists in regions such as the United States, Australia, and Eastern Europe.

While playing a critical role in the chemical industry for more than a century, the Haber-Bosch process remains extremely energy-intensive and carbon-heavy, consuming 1–2% of global energy, utilizing 3–5% of the world's natural gas supply, and contributing 1–3% of global CO_2 emissions.[140–142] These challenges underscore the importance of developing more sustainable methods for NH_3 production.[143] In addition to being energy intensive and carbon heavy, the Haber-Bosch process is constrained to centralized, large-scale operations due to the need for high temperatures (700–800 K) and pressures (up to 250 atm) for economic feasibility. With the growing interest in making NH_3 for H_2 storage and local fertilizer production with renewable energy, smaller scale and distributed synthesis of NH_3 draws significant attention. Compared to electrochemical NH_3 synthesis methods, electrothermal NH_3 synthesis potentially offers better scale and faster rate. It is expected that distributed electrothermal NH_3 synthesis will likely play an indispensable role in the near future in the NH_3 supply chain, complementing the centralized and large-scale Haber-Bosch process.

3.5.1.1 Electrothermal Ammonia Synthesis Using Resistive Heating

While distributed and electrified NH_3 synthesis has been explored via electrochemical and photochemical methods, production rates remain limited due to the slow kinetics of N_2 activation and low NH_3 yield. NH_3 synthesis from N_2 and H_2 is an exothermic reaction (see reaction below). Although lower temperatures favor NH_3 yield, higher temperatures (or other activation techniques) are still necessary to achieve a high reaction rate[144]

$$N_2 + 3H_2 \leftrightarrow NH_3 \quad \Delta H^{\circ}_{298} = -91.8\,kJ\,/\,mol$$

Recently, Dong and colleagues have invented a unique concept to conduct electrothermal chemical synthesis under non-equilibrium conditions.[35,145] By using a catalyst-loaded, porous carbonaceous Joule heating element, this method allows for periodic rapid heating and quenching (e.g., 110 ms on, 990 ms off; $T_{high} = 1200$ K, $T_{low} = 700$ K) according to a programmed pattern by modulating the heating time and pulse temperature, in stark contrast to continuous heating at constant temperatures used in conventional thermochemical NH_3 synthesis. Promising results toward energy-efficient synthesis of NH_3 under ambient conditions have been demonstrated. In comparison to conventional continuous heating under near-equilibrium conditions, programmable heating and quenching allows the authors to access a non-equilibrium regime by accelerating the rate-limiting N_2 dissociation reaction at high temperatures and then rapidly quenching the reactor to avoid catalyst agglomeration, which is particularly beneficial for nanostructured catalysts (Figure 3.9(A)). Furthermore, by using periodic heating, the approach can reduce the energy cost compared to continuous heating for a comparable product yield.

With this programmable heating and quenching process enabled by pulsed Joule heating, Dong and colleagues demonstrated the ability to achieve temperature swing between near-room temperature and up to 1400 K in milliseconds using a flow reactor. Using simple catalyst compositions, such as Ru and Fe nanoparticles, the authors achieved high NH_3 synthesis rates under ambient pressure. Specifically, the Fe nanoparticle catalyst led to a high NH_3 synthesis rate of > 6000 $\mu mol/g_{Fe}/h$ under ambient pressure, which is among the highest in literature reports. The performance using pulsed Joule heating is 100 times better than that of continuous heating. More importantly, the Fe nanoparticle catalyst exhibited stable performance for over 100 h owing to the fast temperature quenching, whereas the same catalyst showed severe agglomeration and performance decay under continuous heating conditions (Figure 3.9(B)).

3.5.1.2 Electrothermal Ammonia Synthesis Using Plasma

One of the earliest industrial approaches to N_2 fixation before the Haber-Bosch process was the Birkeland-Eyde process, which was developed based on a thermal plasma reactor where an electrical arc was used to react N_2 and O_2 to produce nitric acid and water. Since the first reported plasma-driven NH_3 synthesis by Morren and Perrot in 1859 using a DBD reactor,[146] the field received intermittent attention until the 1980s, when low-pressure glow discharges became a major focus, particularly for studying surface effects and the roles of ions such as N_2^+ and N_2H^+. After that, numerous reports have demonstrated novel approaches for utilizing plasma technology in NH_3 synthesis.[147]

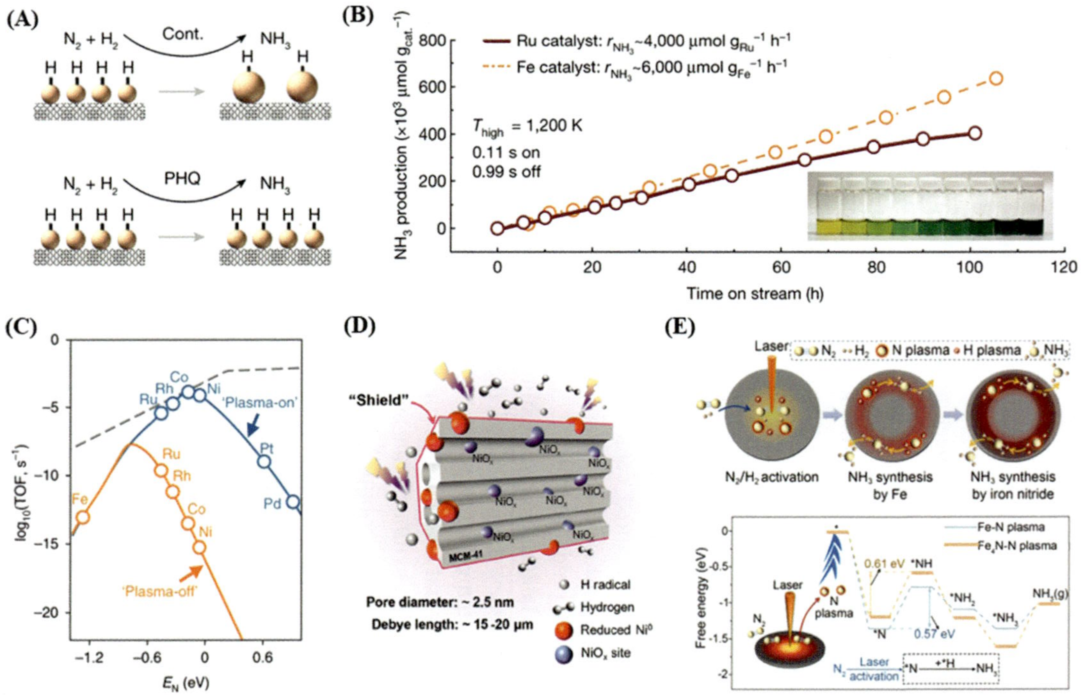

FIGURE 3.9 (A) Morphological evolution of nanocatalyst during NH_3 synthesis by continuous heating and pulsed Joule heating. (B) NH_3 production as a function of reaction time by pulsed Joule heating using non-optimized Ru and Fe catalysts. The inset shows the testing solutions for quantifying the NH_3 yield using the Berthelot method. Adapted from Ref. 35, DOI: 10.1038/s41586-022-04568-6. (C) N_2 vibrational-distribution-weighted (plasma-on) and thermal (plasma-off) NH_3 synthesis rates on step sites. Adapted from Ref. 148, DOI: 10.1038/s41929-018-0045-1. (D) The shielding effect of mesoporous MCM-41 on Ni species within its mesopores in an H_2 plasma discharge. Adapted from Ref. 158, DOI: 10.1021/jacs.2c01950. (E) Top: schematic of the catalytic mechanism of a Laser-induced ammonia synthesis (LIAS) process. Bottom: the energy profiles of the NH_3 synthesis process based on density functional theory (DFT) calculations (N plasma as the feed gas). Adapted from Ref. 163, DOI: 10.1016/j.nanoen.2023.108855.

Although plasma reactors enable NH_3 production under non-thermal conditions, the underlying mechanisms remain insufficiently understood. Much like traditional thermocatalysis, vibrational excitation is the primary step in plasma catalysis. Mehta and coworkers demonstrated that non-thermal plasma uniquely activates molecular degrees of freedom via vibrational excitation, which decreases the dissociation barrier of N_2.[148] Using density-functional-theory-based microkinetic models, the authors highlighted the potential for improved reaction rates through precise plasma property control and catalyst optimization. The authors computed the plasma-on reaction rates and compared this to the thermal catalytic rates (plasma-off), where they showed that the plasma-on reaction rates are several orders of magnitude higher than those observed in thermal catalysis. In addition, Co and Ni show optimal performance at lower nitrogen binding energies, comparable to Ru step reaction rate during Haber–Bosch process under high-temperature and high-pressure conditions (Figure 3.9(C)).

Rouwenhorst and colleagues explored the mechanistic NH_3 synthesis process and found that catalytic plasma reactors support four potential pathways for NH_3 formation: (1) Plasma-phase NH_3 synthesis: gaseous phase recombination of excited N^* and H^* radicals. (2) Surface-enhanced plasma-driven NH_3 synthesis: adsorbed radicals recombining and undergoing hydrogenation on the catalyst surface. (3) Plasma-enhanced semi-catalytic NH_3 synthesis: catalyst facilitating H_2 dissociation but not participating in N_2 dissociation. (4) Plasma-enhanced catalytic NH_3 synthesis: N_2 undergoing vibrational excitation in the plasma stream, while H_2 dissociating directly on the catalyst surface and subsequent hydrogenation proceeds independently of the plasma environment.[149]

A number of other studies also enriched the fundamental understanding and optimized NH_3 synthesis through tailored catalytic strategies.[150–155] Notably, an interesting study by Gómez-Ramírez and colleagues showed that applying ferroelectric materials between the electrodes of a DBD plasma reactor can enhance the overall energy efficiency of NH_3 synthesis. The authors varied different parameters as well as the type of ferroelectric material ($BaTiO_3$ versus lead zirconate titanate). The optimal condition led to an N_2 conversion rate of over 7%.[156] This approach represents a new version of the DBD plasma reactor with ferroelectric layers, which may find broader application for plasma catalysis and electrified manufacturing.

Plasma-assisted NH_3 synthesis needs to consider the influence of afterglow. Van't Veer and colleagues examined this mechanism using a microkinetic model for filamentary micro-discharges. The study revealed that the synthesized NH_3 can be decomposed during the afterglow. Despite this side effect, the authors found a net gain of NH_3 production by the electron-induced dissociation of N_2 and H_2 and the following catalytic processes.[157] To better control this potent decomposition pathway, Wang and coworkers applied a mesoporous MCM-41-supported Ni catalyst, where the mesopores facilitate the diffusion of desorbed NH_3, protecting it from decomposition by the extensive plasma discharge (Figure 3.9(D)).[158] Peng and colleagues employed metal halides such as $MgCl_2$ to absorb NH_3 from the gas phase to avoid decomposition.[159] Wu and colleagues reported an NaH-mediated strategy, where plasma activates N_2 and NaH both, forming intermediates that undergo thermal hydrogenation, completing a stepwise chemical cycle that efficiently produces NH_3 with minimal reverse reactions.[160]

Several plasma-assisted electrochemical processes have also been reported recently. Sharma and coworkers integrated electrocatalysis with plasma in a proton-conducting solid oxide electrolyzer. This system used plasma-activated N_2 and H_2 from water oxidation to achieve NH_3 production rates of up to 26.8 nmol NH_3 s^{-1} cm^{-2}, with Faradaic efficiencies of up to 88%.[161] Zhang and coworkers developed a simpler process where H_2O and N_2 activation can happen at the same time via non-thermal plasma. The authors demonstrated an NH_3 production rate of 2.67 mmol g_{cat}^{-1} h^{-1} with an optimized Ru/MgO catalyst.[162] Wu and colleagues reported using laser irradiation to ionize N_2 into plasma. Assisted by Fe, the plasma migrated to an annular region, where Fe and iron nitride catalyzed the NH_3 synthesis. The process initially achieved NH_3 yields of 70.8 μmol g^{-1} min^{-1} (28.32 μmol min^{-1}), later stabilizing at 23.63 μmol g^{-1} min^{-1} (9.45 μmol min^{-1}) (Figure 3.9(E)).[163]

3.5.1.3 Electrothermal Ammonia Synthesis Using Microwave Heating

Continuous-flow microwave reactors operating under atmospheric pressure offer a promising platform for NH_3 synthesis. Hu and coworkers studied microwave-mediated NH_3 synthesis with Ru catalysts.[164] At mild temperature (280 °C) and pressure, the authors achieved ~1 mmol $NH_3/g_{cat}/h$. The authors found that promoter ions such as K, Ce, and Ba can enhance the synthesis of Ru-based catalysts by increasing dielectric loss and improving microwave absorption. Wang and colleagues then examined the role of promoters such as CeO_2 and Cs, which enhance Ru's activity more effectively than the MgO supports alone. This can be ascribed to their higher electron-donating abilities and lower electronegativities.[165] With the $CsRu/CeO_2$ catalyst under microwave irradiation, the authors found a positive relationship between the applied pressure (in the range of 0.1–1.13 MPa) and the NH_3 production rate, largely due to the enhanced collision probability at higher molecule concentrations. For example, under a 5:1 H_2 to N_2 ratio, raising the pressure from ambient to 0.65 MPa increases the NH_3 production rate from 0.022 to 0.031 $g/h/g_{cat}$. For a 1:1 H_2 to N_2 ratio, the same pressure increase results in a rise of NH_3 production rate from 0.026 to 0.072 $g/h/g_{cat}$.[166] The authors further explored the effect of temperature (in the range of 280–400 °C), GHSV, and particle size of the CeO_2 support (25 nm) on NH_3 production. For the CeO_2 support, smaller particle size was found to exhibit higher NH_3 synthesis activity.[167]

To check the scalability of electrothermal NH_3 synthesis using microwave heating, Melkote and coworkers recently explored a 10× expanded microwave-assisted NH_3 synthesis system through computational modeling and experimental testing. The study led to the largest microwave-powered NH_3 synthesis reactor documented to date according to the authors. The influence of the catalyst's dielectric properties on field distribution and the energy inefficiencies despite adequate attenuation are examined by modeling. A production rate of 56.6 g NH_3/day and an energy efficiency of 45.6 g NH_3/kWh were achieved, representing the highest values reported for small-scale operations.[168]

3.5.2 Electrothermal Ammonia Synthesis Using Hydrocarbons

Over the years, alternative approaches have explored the use of hydrocarbons, particularly CH_4, as a hydrogen source for NH_3 synthesis under electrothermal conditions.

These methods often yield hydrocarbon coproducts, some of which are comparably valuable as NH_3. Traditionally, the hydrogen required for NH_3 synthesis is derived from natural gas reforming or coal gasification, processes that contribute to approximately 50% of production costs in NH_3 plants. As such, the direct utilization of natural gas (e.g., CH_4 and C_2H_6) as the hydrogen source presents an attractive, cost-effective alternative that potentially simplifies the NH_3 synthesis process.

$$N_2 + CH_4 \leftrightarrow NH_3 + \text{hydrocarbon} \quad \Delta H^{\circ}_{298} = -91.8\,\text{kJ/mol}$$

An early study investigated CH_4 conversion co-fed with N_2 by microwave plasma, where only a trace amount of NH_3 was detected.[169] Despite the low yield, the concept underscores the potential for further research into using alternative hydrogen sources for NH_3 synthesis. Recently, Bai and colleagues developed a γ-Al_2O_3-supported Co and Co-Fe catalyst.[170] The γ-Al_2O_3 support proved to be an excellent absorber of the microwave energy. Under microwave irradiation, the activation energy for CH_4 decomposition was reduced, generating dihydrogen species on the catalyst surface. However, the carbon deposition gradually deactivated the catalyst. The addition of Fe promoter facilitated carbon nanotube (CNT) formation, while activated nitrogen atoms combined with hydrogen to form NH_3 on the catalyst surface. The study found that CH_4 conversion and NH_3 yield increased significantly when microwave plasma was employed. Later, Hu and coworkers demonstrated the concurrent activation of CH_4 and N_2 in a microwave-assisted catalytic reaction, yielding NH_3 along with valuable C_2 hydrocarbon coproducts.[171] Detailed investigations using density functional theory and microkinetic modeling revealed that Ru-based catalysts with step sites are the most effective for NH_3 synthesis as well as CH_4 dissociation. K-promoted Ru terrace sites demonstrated prolonged activity, enabling the formation of value-added C_2 hydrocarbons despite some coking on the catalyst surface. Experimental validation showed that microwave irradiation lowered the activation energy for intermediates, with NH_3 formation attributed to non-thermal effects and C_2 hydrocarbon coproduct formation driven by thermal effects. The same team later addressed selectivity challenges using a post-plasma catalytic system.[172] Activated species generated in plasma were passed through a thermally heated Ag-Pd/CeO_2 catalyst. This process reduced the selectivity for hydrogen cyanide by 9% while ensuring consistent NH_3 production with a selectivity of 9%. C_2H_4 selectivity improved markedly, rising from 3% to 37%, surpassing conventional thermal heating for C_2 hydrocarbon production. Wang and colleagues employed a non-thermal plasma catalytic system to synthesize NH_3 directly from CH_4 and N_2, with value-added C_2 hydrocarbons as coproducts as well.[173] Upon down-selecting various 3d metal and main-group element oxides, the authors identified that Al and Fe are most effective in enhancing NH_3 production and CH_4 conversion. Under optimized conditions, the team achieved an NH_3 concentration of 3534.32 ppm and an energy efficiency of 836.61 mg/kWh. These findings highlight the feasibility and economic potential of direct NH_3 synthesis from hydrocarbon and N_2 under mild conditions. Future research may prioritize scaling up the process and enriching the understanding of the underlying mechanisms.

3.6 SUMMARY AND OUTLOOK

This chapter reviewed the progress and technological advancements in electrothermal techniques, including resistive heating, microwave heating, induction heating, and plasma-assisted reactions, with a focus on key industrial processes such as hydrocarbon conversion, CO_2 conversion, biomass conversion, and NH_3 synthesis. As renewable electricity becomes increasingly available and the demand for industrial decarbonization grows, the field of electrothermal chemical synthesis and manufacturing is poised for significant expansion. These techniques and transformations are expected to evolve rapidly also in response to sustainability challenges like carbon emissions and resource scarcity.

Despite their promise, most electrothermal technologies remain at an early stage, with technology readiness levels of 3–4. Addressing critical challenges—such as scaling up, cost reduction, and improving system durability—will require collaborative efforts among industry, academia, and policymakers to realize practical applications. Looking ahead, the successful integration of electrothermal processes into industrial systems will not only accelerate the transition toward sustainable chemical manufacturing but also inspire new pathways for a greener energy landscape. Continued investment in research, infrastructure, and cross-disciplinary partnerships will be crucial to unlocking the full potential of these transformative technologies.

REFERENCES

(1) Wismann, S. T.; Engbæk, J. S.; Vendelbo, S. B.; Bendixen, F. B.; Eriksen, W. L.; Aasberg-Petersen, K.; Frandsen, C.; Chorkendorff, I.; Mortensen, P. M. Electrified methane reforming: A compact approach to greener industrial hydrogen production. *Science (1979)* 2019, *364* (6442), 756–759. https://doi.org/10.1126/science.aaw8775

(2) Neumann, B.; Jacob, K. Die Gleichgewichtsverhältnisse Bei Der Methanbildung Aus Kohlenoxyd Und Wasserstoff, Bzw. Kohlendioxyd Und Wasserstoff. *Zeitschrift für Elektrochemie und angewandte physikalische Chemie* 1924, *30* (23–24), 557–576. https://doi.org/10.1002/bbpc.192400208

(3) Zhang, H.; Sun, Z.; Hu, Y. H. Steam reforming of methane: Current states of catalyst design and process upgrading. *Renewable and Sustainable Energy Reviews*. Elsevier Ltd October 1, 2021. https://doi.org/10.1016/j.rser.2021.111330

(4) Spath, P. L.; Mann, M. K. *Life Cycle Assessment of Hydrogen Production via Natural Gas Steam Reforming*; United States, 2000. https://doi.org/10.2172/764485

(5) Kim, Y. T.; Lee, J. J.; Lee, J. Electricity-driven reactors that promote thermochemical catalytic reactions via joule and induction heating. *Chemical Engineering Journal*. Elsevier B.V. August 15, 2023. https://doi.org/10.1016/j.cej.2023.144333

(6) Wang, W.; Zhao, S.; Tang, X.; Chen, C.; Yi, H. Electrothermal catalysis for heterogeneous reaction: Mechanisms and design strategies. *Chemical Engineering Journal*. Elsevier B.V. January 1, 2023. https://doi.org/10.1016/j.cej.2022.140272

(7) Zheng, L.; Ambrosetti, M.; Tronconi, E. Joule-heated catalytic reactors toward decarbonization and process intensification: A review. *ACS Engineering Au* 2024, *4* (1), 4–21. https://doi.org/10.1021/acsengineeringau.3c00045

(8) Spagnolo, D. A.; Cornett, L. J.; Chuang, K. T. Direct electro-steam reforming: A novel catalytic approach; *International Journal of Hydrogen Energy* 1992, *17*.

(9) Zhang, Q.; Nakaya, M.; Ootani, T.; Takahashi, H.; Sakurai, M.; Kameyama, H. Simulation and experimental analysis on the development of a co-axial cylindrical methane steam reformer using an electrically heated alumite catalyst. *International Journal of Hydrogen Energy* 2007, *32* (16), 3870–3879. https://doi.org/10.1016/j.ijhydene.2007.05.031
(10) Zhou, L.; Guo, Y.; Zhang, Q.; Yagi, M.; Hatakeyama, J.; Li, H.; Chen, J.; Sakurai, M.; Kameyama, H. A novel catalyst with plate-type anodic alumina supports, $Ni/NiAl_2O_4/\gamma$-Al_2O_3/alloy, for steam reforming of methane. *Applied Catalysis A: General* 2008, *347* (2), 200–207. https://doi.org/10.1016/j.apcata.2008.06.007
(11) Zhou, L.; Guo, Y.; Yagi, M.; Sakurai, M.; Kameyama, H. Investigation of a novel porous anodic alumina plate for methane steam reforming: Hydrothermal stability, electrical heating possibility and reforming reactivity. *International Journal of Hydrogen Energy* 2009, *34* (2), 844–858. https://doi.org/10.1016/j.ijhydene.2008.10.057
(12) Guo, Y.; Zhou, L.; Kameyama, H. Thermal and hydrothermal stability of a metal monolithic anodic alumina support for steam reforming of methane. *Chemical Engineering Journal* 2011, *168* (1), 341–350. https://doi.org/10.1016/j.cej.2011.01.036
(13) Wismann, S. T.; Engbæk, J. S.; Vendelbo, S. B.; Eriksen, W. L.; Frandsen, C.; Mortensen, P. M.; Chorkendorff, I. Electrified methane reforming: Elucidating transient phenomena. *Chemical Engineering Journal* 2021, *425*. https://doi.org/10.1016/j.cej.2021.131509
(14) Ledesma, E. B.; Idamakanti, M.; Bollini, P.; Harold, M. P.; Ratnakar, R. R. Decarbonizing steam-methane reforming: Enhancing activity through joule heating of a Ni/ZrO_2-coated FeCrAl coil. *ChemCatChem* 2024, *16* (8). https://doi.org/10.1002/cctc.202301110
(15) Renda, S.; Cortese, M.; Iervolino, G.; Martino, M.; Meloni, E.; Palma, V. Electrically driven SiC-based structured catalysts for intensified reforming processes. *Catalysis Today* 2022, *383*, 31–43. https://doi.org/10.1016/j.cattod.2020.11.020
(16) Zheng, L.; Ambrosetti, M.; Marangoni, D.; Beretta, A.; Groppi, G.; Tronconi, E. Electrified methane steam reforming on a washcoated SiSiC foam for low-carbon hydrogen production. *AIChE Journal* 2023, *69* (1). https://doi.org/10.1002/aic.17620
(17) Zheng, L.; Ambrosetti, M.; Zaio, F.; Beretta, A.; Groppi, G.; Tronconi, E. Direct electrification of Rh/Al_2O_3 washcoated SiSiC foams for methane steam reforming: An experimental and modelling study. *International Journal of Hydrogen Energy* 2023, *48* (39), 14681–14696. https://doi.org/10.1016/j.ijhydene.2022.12.346
(18) Ambrosetti, M.; Beretta, A.; Groppi, G.; Tronconi, E. A numerical investigation of electrically-heated methane steam reforming over structured catalysts. *Frontiers in Chemical Engineering* 2021, *3*. https://doi.org/10.3389/fceng.2021.747636
(19) Balakotaiah, V.; Ratnakar, R. R. Modular reactors with electrical resistance heating for hydrocarbon cracking and other endothermic reactions. *AIChE Journal* 2022, *68* (2). https://doi.org/10.1002/aic.17542
(20) Peloquin, J. F.; Francoeur, D.; Leclerc, W.; Mehanovic, D.; Dufault, J. F.; Camus, P.; Castellanos-Beltran, I. J.; Braidy, N.; Fréchette, L. G.; Picard, M. Electrified steam methane reforming microreactor. *International Journal of Hydrogen Energy* 2024, *49*, 907–915. https://doi.org/10.1016/j.ijhydene.2023.07.343
(21) Dong, Q.; Hu, S.; Hu, L. Electrothermal synthesis of commodity chemicals. *Nature Chemical Engineering* 2024, *1* (11), 680–690. https://doi.org/10.1038/s44286-024-00134-1
(22) Song, H.; Liu, Y.; Bian, H.; Shen, M.; Lin, X. Energy, environment, and economic analyses on a novel hydrogen production method by electrified steam methane reforming with renewable energy accommodation. *Energy Conversion and Management* 2022, *258*. https://doi.org/10.1016/j.enconman.2022.115513

(23) Vinum, M. G.; Almind, M. R.; Engbæk, J. S.; Vendelbo, S. B.; Hansen, M. F.; Frandsen, C.; Bendix, J.; Mortensen, P. M. Dual-function cobalt–nickel nanoparticles tailored for high-temperature induction-heated steam methane reforming. *Angewandte Chemie* 2018, *130* (33), 10729–10733. https://doi.org/10.1002/ange.201804832
(24) Scarfiello, C.; Bellusci, M.; Pilloni, L.; Pietrogiacomi, D.; La Barbera, A.; Varsano, F. Supported catalysts for induction-heated steam reforming of methane. *International Journal of Hydrogen Energy* 2021, *46* (1), 134–145. https://doi.org/10.1016/j.ijhydene.2020.09.262
(25) Poletto Dotsenko, V.; Bellusci, M.; Masi, A.; Pietrogiacomi, D.; Varsano, F. Improving the performances of supported NiCo catalyst for reforming of methane powered by magnetic induction. *Catalysis Today* 2023, *418*. https://doi.org/10.1016/j.cattod.2023.114049
(26) Almind, M. R.; Vendelbo, S. B.; Hansen, M. F.; Vinum, M. G.; Frandsen, C.; Mortensen, P. M.; Engbæk, J. S. Improving performance of induction-heated steam methane reforming. *Catalysis Today* 2020, *342*, 13–20. https://doi.org/10.1016/j.cattod.2019.05.005
(27) Ma, J.; Jiang, B.; Gao, Y.; Yu, K.; Lv, Z.; Si-ma, W.; Yuan, D.; Tang, D. A compact and high-efficiency electrified reactor for hydrogen production by methane steam reforming. *International Journal of Hydrogen Energy* 2022, *47* (98), 41421–41431. https://doi.org/10.1016/j.ijhydene.2022.04.281
(28) Ma, J.; Jiang, B.; Li, L.; Yu, K.; Zhang, Q.; Lv, Z.; Tang, D. A high temperature tubular reactor with hybrid concentrated solar and electric heat supply for steam methane reforming. *Chemical Engineering Journal* 2022, *428*. https://doi.org/10.1016/j.cej.2021.132073
(29) Meloni, E.; Martino, M.; Palma, V. Microwave assisted steam reforming in a high efficiency catalytic reactor. *Renewable Energy* 2022, *197*, 893–901. https://doi.org/10.1016/j.renene.2022.07.157
(30) Crynes, B. L. *Survey of Recent Methane Pyrolysis Literature a Survey of Methane Pyrolysis Data is Presented and Discussed*. https://pubs.acs.org/sharingguidelines
(31) Frenklach, M. Reaction mechanism of soot formation in flames. In *Physical Chemistry Chemical Physics* 2002, *4*, 2028–2037. https://doi.org/10.1039/b110045a
(32) Porsin, A. V.; Kulikov, A. V.; Amosov, Y. I.; Rogozhnikov, V. N.; Noskov, A. S. Acetylene synthesis by methane pyrolysis on a tungsten wire. *Theoretical Foundations of Chemical Engineering* 2014, *48* (4), 397–403. https://doi.org/10.1134/S0040579514040241
(33) Sigaeva, S. S.; Likholobov, V. A.; Tsyrul'Nikov, P. G. Pyrolysis of methane on a heat-treated FeCrAl coil heated with electric current. *Kinetics and Catalysis* 2013, *54* (2), 199–206. https://doi.org/10.1134/S0023158413010126
(34) Shekunova, V. M.; Aleksandrov, Y. A.; Tsyganova, E. I.; Filofeev, S. V. Cracking of light hydrocarbons in the presence of electrically heated metal wires. *Petroleum Chemistry* 2017, *57* (5), 446–451. https://doi.org/10.1134/S0965544117050097
(35) Dong, Q.; Yao, Y.; Cheng, S.; Alexopoulos, K.; Gao, J.; Srinivas, S.; Wang, Y.; Pei, Y.; Zheng, C.; Brozena, A. H.; Zhao, H.; Wang, X.; Toraman, H. E.; Yang, B.; Kevrekidis, I. G.; Ju, Y.; Vlachos, D. G.; Liu, D.; Hu, L. Programmable heating and quenching for efficient thermochemical synthesis. *Nature* 2022, *605* (7910), 470–476. https://doi.org/10.1038/s41586-022-04568-6
(36) Spivey, J. J.; Hutchings, G. Catalytic aromatization of methane. *Chemical Society Reviews*. Royal Society of Chemistry February 7, 2014, 792–803. https://doi.org/10.1039/c3cs60259a
(37) Deng, Y.; Bai, X.; Abdelsayed, V.; Shekhawat, D.; Muley, P. D.; Karpe, S.; Mevawala, C.; Bhattacharyya, D.; Robinson, B.; Caiola, A.; Powell, J. B.; van Bavel, A. P.; Hu, J.; Veser, G. Microwave-assisted conversion of methane over H-(Fe)-ZSM-5: Evidence for formation of hot metal sites. *Chemical Engineering Journal* 2021, *420*. https://doi.org/10.1016/j.cej.2021.129670

(38) Bai, X.; Muley, P. D.; Musho, T.; Abdelsayed, V.; Robinson, B.; Caiola, A.; Shekhawat, D.; Jiang, C.; Hu, J. A combined experimental and modeling study of microwave-assisted methane dehydroaromatization process. *Chemical Engineering Journal* 2022, *433*. https://doi.org/10.1016/j.cej.2021.134445
(39) Julian, I.; Pedersen, C. M.; Jensen, A. B.; Baden, A. K.; Hueso, J. L.; Friderichsen, A. V.; Birkedal, H.; Mallada, R.; Santamaria, J. From bench scale to pilot plant: A 150x scaled-up configuration of a microwave-driven structured reactor for methane dehydroaromatization. *Catalysis Today* 2022, *383*, 21–30. https://doi.org/10.1016/j.cattod.2021.04.013.
(40) Yang, H.; Nuran Zaini, I.; Pan, R.; Jin, Y.; Wang, Y.; Li, L.; Caballero, J. J. B.; Shi, Z.; Subasi, Y.; Nurdiawati, A.; Wang, S.; Shen, Y.; Wang, T.; Wang, Y.; Sandström, L.; Jönsson, P. G.; Yang, W.; Han, T. Distributed electrified heating for efficient hydrogen production. *Nature Communications* 2024, *15* (1). https://doi.org/10.1038/s41467-024-47534-8
(41) Gao, Y.; Zhang, S.; Sun, H.; Wang, R.; Tu, X.; Shao, T. Highly efficient conversion of methane using microsecond and nanosecond pulsed spark discharges. *Applied Energy* 2018, *226*, 534–545. https://doi.org/10.1016/j.apenergy.2018.06.006
(42) Fulcheri, L.; Rohani, V. J.; Wyse, E.; Hardman, N.; Dames, E. An energy-efficient plasma methane pyrolysis process for high yields of carbon black and hydrogen. *International Journal of Hydrogen Energy* 2023, *48* (8), 2920–2928. https://doi.org/10.1016/j.ijhydene.2022.10.144
(43) Scapinello, M.; Delikonstantis, E.; Stefanidis, G. D. Direct Methane-to-ethylene conversion in a nanosecond pulsed discharge. *Fuel* 2018, *222*, 705–710. https://doi.org/10.1016/j.fuel.2018.03.017
(44) Amghizar, I.; Vandewalle, L. A.; Van Geem, K. M.; Marin, G. B. New trends in olefin production. *Engineering* 2017, *3* (2), 171–178. https://doi.org/10.1016/J.ENG.2017.02.006
(45) Sattler, J. J. H. B.; Ruiz-Martinez, J.; Santillan-Jimenez, E.; Weckhuysen, B. M. Catalytic dehydrogenation of light alkanes on metals and metal oxides. *Chemical Reviews*. American Chemical Society October 22, 2014, 10613–10653. https://doi.org/10.1021/cr5002436
(46) Docherty, S. R.; Rochlitz, L.; Payard, P. A.; Copéret, C. Heterogeneous alkane dehydrogenation catalysts investigated via a surface organometallic chemistry approach. *Chemical Society Reviews*. Royal Society of Chemistry May 7, 2021, 5806–5822. https://doi.org/10.1039/d0cs01424a
(47) Liu, L.; Bhowmick, A.; Cheng, S.; Blazquez, B. H.; Pan, Y.; Zhang, J.; Zhang, Y.; Shu, Y.; Tran, D. T.; Luo, Y.; Ierapetritou, M.; Zhang, C.; Liu, D. Alkane dehydrogenation in scalable and electrifiable carbon membrane reactor. *Cell Reports Physical Science* 2023, *4* (12). https://doi.org/10.1016/j.xcrp.2023.101692
(48) Chen, W.; Malhotra, A.; Yu, K.; Zheng, W.; Plaza-Gonzalez, P. J.; Catala-Civera, J. M.; Santamaria, J.; Vlachos, D. G. Intensified microwave-assisted heterogeneous catalytic reactors for sustainable chemical manufacturing. *Chemical Engineering Journal* 2021, *420*. https://doi.org/10.1016/j.cej.2021.130476
(49) Kwak, Y.; Wang, C.; Kavale, C. A.; Yu, K.; Selvam, E.; Mallada, R.; Santamaria, J.; Julian, I.; Catala-Civera, J. M.; Goyal, H.; Zheng, W.; Vlachos, D. G. *Microwave-Assisted, Performance-Advantaged Electrification of Propane Dehydrogenation*, 2023. https://www.science.org
(50) Jung, H.; Yoon, W. L.; Lee, H.; Park, J. S.; Shin, J. S.; La, H.; Lee, J. D. Fast start-up reactor for partial oxidation of methane with electrically-heated metallic monolith catalyst. *Journal of Power Sources* 2003, *124* (1), 76–80. https://doi.org/10.1016/S0378-7753(03)00604-9

(51) Ramirez, A.; Hueso, J. L.; Abian, M.; Alzueta, M. U.; Mallada, R.; Santamaria, J. Escaping undesired gas-phase chemistry: Microwave-driven selectivity enhancement in heterogeneous catalytic reactors. *Science Advances* 2024, *5* (3), eaau9000. https://doi.org/10.1126/sciadv.aau9000
(52) Ramirez, A.; Hueso, J. L.; Mallada, R.; Santamaria, J. Microwave-activated structured reactors to maximize propylene selectivity in the oxidative dehydrogenation of propane. *Chemical Engineering Journal* 2020, *393*. https://doi.org/10.1016/j.cej.2020.124746
(53) Yang, L.; Ge, X.; Wan, C.; Yu, F.; Li, Y. Progress and perspectives in converting biogas to transportation fuels. *Renewable and Sustainable Energy Reviews*. Elsevier Ltd 2014, 1133–1152. https://doi.org/10.1016/j.rser.2014.08.008
(54) Schwab, E.; Milanov, A.; Schunk, S. A.; Behrens, A.; Schödel, N. Dry reforming and reverse water gas shift: Alternatives for syngas production? *Chemie-Ingenieur-Technik*. Wiley-VCH Verlag April 1, 2015, 347–353. https://doi.org/10.1002/cite.201400111
(55) Wittich, K.; Krämer, M.; Bottke, N.; Schunk, S. A. Catalytic dry reforming of methane: Insights from model systems. *ChemCatChem*. Wiley Blackwell April 20, 2020, 2130–2147. https://doi.org/10.1002/cctc.201902142
(56) Pakhare, D.; Spivey, J. A review of dry (CO_2) reforming of methane over noble metal catalysts. *Chemical Society Reviews*. Royal Society of Chemistry November 21, 2014, 7813–7837. https://doi.org/10.1039/c3cs60395d
(57) Labrecque, R.; Lavoie, J. M. Dry reforming of methane with CO_2 on an electron-activated iron catalytic bed. *Bioresource Technology* 2011, *102* (24), 11244–11248. https://doi.org/10.1016/j.biortech.2011.09.088
(58) Rieks, M.; Bellinghausen, R.; Kockmann, N.; Mleczko, L. Experimental study of methane dry reforming in an electrically heated reactor. *International Journal of Hydrogen Energy* 2015, *40* (46), 15940–15951. https://doi.org/10.1016/j.ijhydene.2015.09.113
(59) Zheng, L.; Ambrosetti, M.; Beretta, A.; Groppi, G.; Tronconi, E. Electrified CO_2 valorization driven by direct joule heating of catalytic cellular substrates. *Chemical Engineering Journal* 2023, *466*. https://doi.org/10.1016/j.cej.2023.143154
(60) Yu, K.; Wang, C.; Zheng, W.; Vlachos, D. G. Dynamic electrification of dry reforming of methane with in situ catalyst regeneration. *ACS Energy Letters* 2023, *8* (2), 1050–1057. https://doi.org/10.1021/acsenergylett.2c02666
(61) Kameshima, S.; Tamura, K.; Ishibashi, Y.; Nozaki, T. Pulsed dry methane reforming in plasma-enhanced catalytic reaction. *Catalysis Today* 2015, *256* (Part 1), 67–75. https://doi.org/10.1016/j.cattod.2015.05.011
(62) Lahijani, P.; Zainal, Z. A.; Mohammadi, M.; Mohamed, A. R. Conversion of the greenhouse gas CO_2 to the fuel gas CO via the Boudouard reaction: A review. *Renewable and Sustainable Energy Reviews*. Elsevier Ltd 2015, 615–632. https://doi.org/10.1016/j.rser.2014.08.034
(63) Chung, W. C.; Chang, M. B. Review of catalysis and plasma performance on dry reforming of CH_4 and possible synergistic effects. *Renewable and Sustainable Energy Reviews*. Elsevier Ltd September 1, 2016, 13–31. https://doi.org/10.1016/j.rser.2016.04.007
(64) Tu, X.; Whitehead, J. C. Plasma dry reforming of methane in an atmospheric pressure AC gliding arc discharge: Co-generation of syngas and carbon nanomaterials. *International Journal of Hydrogen Energy* 2014, *39* (18), 9658–9669. https://doi.org/10.1016/j.ijhydene.2014.04.073
(65) Wang, L.; Yi, Y.; Wu, C.; Guo, H.; Tu, X. One-step reforming of CO_2 and CH_4 into high-value liquid chemicals and fuels at room temperature by plasma-driven catalysis. *Angewandte Chemie* 2017, *129* (44), 13867–13871. https://doi.org/10.1002/ange.201707131.

(66) Ahasan, M. R.; Hossain, M. M.; Ding, X.; Wang, R. Non-equilibrium plasma-assisted dry reforming of methane over shape-controlled CeO_2 supported ruthenium catalysts. *Journal of Materials Chemistry A Material* 2023, *11* (20), 10993–11009. https://doi.org/10.1039/d3ta01196h
(67) Ahasan, M. R.; Hossain, M. M.; Barlow, Z.; Ding, X.; Wang, R. Low-temperature plasma-assisted catalytic dry reforming of methane over CeO_2 nanorod-supported NiO catalysts in a dielectric barrier discharge reactor. *ACS Applied Materials & Interfaces* 2023, *15* (38), 44984–44995. https://doi.org/10.1021/acsami.3c09349
(68) Li, Y.; Liu, X.; Wu, T.; Zhang, X.; Han, H.; Liu, X.; Chen, Y.; Tang, Z.; Liu, Z.; Zhang, Y.; Liu, H.; Zhao, L.; Ma, D.; Zhou, W. Pulsed laser induced plasma and thermal effects on molybdenum carbide for dry reforming of methane. *Nature Communications* 2024, *15* (1). https://doi.org/10.1038/s41467-024-49771-3
(69) Gangurde, L. S.; Sturm, G. S. J.; Valero-Romero, M. J.; Mallada, R.; Santamaria, J.; Stankiewicz, A. I.; Stefanidis, G. D. Synthesis, characterization, and application of ruthenium-doped $SrTiO_3$ perovskite catalysts for microwave-assisted methane dry reforming. *Chemical Engineering and Processing Process Intensification* 2018, *127*, 178–190. https://doi.org/10.1016/j.cep.2018.03.024.
(70) Sharifvaghefi, S.; Shirani, B.; Eic, M.; Zheng, Y. Application of microwave in hydrogen production from methane dry reforming: Comparison between the conventional and microwave-assisted catalytic reforming on improving the energy efficiency. *Catalysts* 2019, *9* (7). https://doi.org/10.3390/catal9070618
(71) Lei, Y.; Mehmood, F.; Lee, S.; Greeley, J.; Lee, B.; Seifert, S.; Winans, R. E.; Elam, J. W.; Meyer, R. J.; Redfern, P. C.; Teschner, D.; Schlögl, R.; Pellin, M. J.; Curtiss, L. A.; Vajda, S. Increased silver activity for direct propylene epoxidation via subnanometer size effects. *Science (1979)* 2010, *328* (5975), 224–228. https://doi.org/10.1126/science.1185200
(72) Burgoyne, J. H.; Burden, F. A. Ethylene oxide explosions. *Nature* 1948, *162* (4109), 181–182. https://doi.org/10.1038/162181a0
(73) Yuan, Y. H.; Zhou, X. G.; Wu, W.; Zhang, Y. R.; Yuan, W. K.; Luo, L. Propylene epoxidation in a microreactor with electric heating. *Catalysis Today* 2005, *105*, 544–550. https://doi.org/10.1016/j.cattod.2005.06.047
(74) Ramírez, A.; Hueso, J. L.; Mallada, R.; Santamaría, J. Ethylene epoxidation in microwave heated structured reactors. *Catalysis Today* 2016, *273*, 99–105. https://doi.org/10.1016/j.cattod.2016.01.007.
(75) Ramirez, A.; Hueso, J. L.; Mallada, R.; Santamaria, J. In situ temperature measurements in microwave-heated gas-solid catalytic systems. Detection of hot spots and solid-fluid temperature gradients in the ethylene epoxidation reaction. *Chemical Engineering Journal* 2017, *316*, 50–60. https://doi.org/10.1016/j.cej.2017.01.077
(76) Ono, K.; Erhard, A. *Ullmann's Encyclopedia of Industrial Chemistry*, 2011. https://doi.org/10.1002/14356007.o17_o02
(77) Wender, I. Reactions of synthesis gas. *Fuel Processing Technology* 1996, *48* (3), 189–297. https://doi.org/10.1016/S0378-3820(96)01048-X
(78) Supp, E. *How to Produce Methanol from Coal*. Springer: Imprint: Springer: Berlin, Heidelberg, 1990. https://doi.org/10.1007/978-3-662-00895-9
(79) Chorkendorff, I.; Niemantsverdriet, H. Concepts of modern catalysis and kinetics I. Chorkendorff, J. W. Niemantsverdriet. *SERBIULA (sistema Librum 2.0)*, 2005. https://doi.org/10.1002/3527602658.ch3
(80) Yang, Y.; Mims, C. A.; Mei, D. H.; Peden, C. H. F.; Campbell, C. T. Mechanistic studies of methanol synthesis over Cu from $CO/CO_2/H_2/H_2O$ mixtures: The source of C in methanol and the role of water. *Journal of Catalysis* 2013, *298*, 10–17. https://doi.org/10.1016/j.jcat.2012.10.028

(81) Arena, F.; Mezzatesta, G.; Spadaro, L.; Trunfio, G. *Latest Advances in the Catalytic Hydrogenation of Carbon Dioxide to Methanol/Dimethylether*, 2014, 103–130. https://doi.org/10.1007/978-3-642-44988-8_5
(82) Kanuri, S.; Roy, S.; Chakraborty, C.; Datta, S. P.; Singh, S. A.; Dinda, S. An insight of CO hydrogenation to methanol synthesis: Thermodynamics, catalysts, operating parameters, and reaction mechanism. *International Journal of Energy Research* 2022, *46* (5), 5503–5522. https://doi.org/10.1002/er.7562
(83) Azhari, N. J.; Erika, D.; Mardiana, S.; Ilmi, T.; Gunawan, M. L.; Makertihartha, I. G. B. N.; Kadja, G. T. M. Methanol synthesis from CO_2: A mechanistic overview. *Results in Engineering* 2022, *16*, 100711. https://doi.org/10.1016/j.rineng.2022.100711
(84) Wang, L.; Yi, Y.; Guo, H.; Tu, X. Atmospheric pressure and room temperature synthesis of methanol through plasma-catalytic hydrogenation of CO_2. *ACS Catalysis* 2018, *8* (1), 90–100. https://doi.org/10.1021/acscatal.7b02733
(85) Feliz, M. Q.; Polaert, I.; Ledoux, A.; Fernandez, C.; Azzolina-Jury, F. Influence of ionic conductivity and dielectric constant of the catalyst on DBD plasma-assisted CO_2 hydrogenation into methanol. *Journal of Physics D: Applied Physics* 2021, *54* (33), 334003. https://doi.org/10.1088/1361-6463/abfddd
(86) Neyts, E. C.; Bogaerts, A. Understanding plasma catalysis through modelling and simulation—A review. *Journal of Physics D: Applied Physics* 2014, *47* (22), 224010. https://doi.org/10.1088/0022-3727/47/22/224010
(87) Joshi, N.; Sivachandiran, L. Exploring the feasibility of liquid fuel synthesis from CO_2 under cold plasma discharge: Role of plasma discharge in binary metal oxide surface modification. *RSC Advances* 2021, *11* (44), 27757–27766. https://doi.org/10.1039/d1ra04852j
(88) Ronda-Lloret, M.; Wang, Y.; Oulego, P.; Rothenberg, G.; Tu, X.; Shiju, N. R. CO_2 hydrogenation at atmospheric pressure and low temperature using plasma-enhanced catalysis over supported cobalt oxide catalysts. *ACS Sustainable Chemistry & Engineering* 2020, *8* (47), 17397–17407. https://doi.org/10.1021/acssuschemeng.0c05565
(89) Cui, Z.; Meng, S.; Yi, Y.; Jafarzadeh, A.; Li, S.; Neyts, E. C.; Hao, Y.; Li, L.; Zhang, X.; Wang, X.; Bogaerts, A. Plasma-catalytic methanol synthesis from CO_2 hydrogenation over a supported Cu cluster catalyst: Insights into the reaction mechanism. *ACS Catalysis* 2022, *12* (2), 1326–1337. https://doi.org/10.1021/acscatal.1c04678.
(90) Men, Y.-L.; Liu, Y.; Wang, Q.; Luo, Z.-H.; Shao, S.; Li, Y.-B.; Pan, Y.-X. Highly dispersed Pt-based catalysts for selective CO_2 hydrogenation to methanol at atmospheric pressure. *Chemical Engineering Science* 2019, *200*, 167–175. https://doi.org/10.1016/j.ces.2019.02.004
(91) Du, J.; Zong, L.; Zhang, S.; Gao, Y.; Dou, L.; Pan, J.; Shao, T. Numerical investigation on the heterogeneous pulsed dielectric barrier discharge plasma catalysis for CO_2 hydrogenation at atmospheric pressure: Effects of Ni and Cu catalysts on the selectivity conversions to CH_4 and CH_3OH. *Plasma Processes and Polymers* 2022, *19* (2), 2100111. https://doi.org/10.1002/ppap.202100111
(92) Joshi, N.; Loganathan, S. Methanol synthesis from CO_2 using Ni and Cu supported Fe catalytic system: Understanding the role of nonthermal plasma surface discharge. *Plasma Processes and Polymers* 2021, *18* (5), 2000104. https://doi.org/10.1002/ppap.202000104
(93) Zhang, X.; Sun, Z.; Shan, Y.; Pan, H.; Jin, Y.; Zhu, Z.; Zhang, L.; Li, K. Boosting methanol production via plasma catalytic CO_2 hydrogenation over a $MnOx/ZrO_2$ catalyst. *Catalysis Science & Technology* 2023, *13* (8), 2529–2539. https://doi.org/10.1039/D2CY02015G

(94) Daza, Y. A.; Kuhn, J. N. CO_2 conversion by reverse water gas shift catalysis: Comparison of catalysts, mechanisms and their consequences for CO_2 conversion to liquid fuels. *RSC Advances* 2016, *6* (55), 49675–49691. https://doi.org/10.1039/c6ra05414e
(95) Mortensen, P. M.; Dybkjær, I. Industrial scale experience on steam reforming of CO_2-rich gas. *Applied Catalysis A: General*. Elsevier B.V. April 5, 2015, 141–151. https://doi.org/10.1016/j.apcata.2015.02.022
(96) Thor Wismann, S.; Larsen, K. E.; Mølgaard Mortensen, P. Electrical reverse shift: Sustainable CO_2 valorization for industrial scale. *Angewandte Chemie, International Edition* 2022, *61* (8). https://doi.org/10.1002/anie.202109696
(97) Yu, K.; Sourav, S.; Zheng, W.; Vlachos, D. G. Dynamic electrification steers the selectivity of CO_2 hydrogenation. *Chemical Engineering Journal* 2024, *481*. https://doi.org/10.1016/j.cej.2024.148528
(98) Basini, L. E.; Furesi, F.; Baumgärtl, M.; Mondelli, N.; Pauletto, G. CO_2 capture and utilization (CCU) by integrating water electrolysis, electrified reverse water gas shift (E-RWGS) and methanol synthesis. *Journal of Cleaner Production* 2022, *377*. https://doi.org/10.1016/j.jclepro.2022.134280
(99) Lin, C. H.; Wan, C.; Ru, Z.; Cremers, C.; Mohapatra, P.; Mantle, D. L.; Tamakuwala, K.; Höfelmann, A. B.; Kanan, M. W.; Rivas-Davila, J.; Fan, J. A. Electrified thermochemical reaction systems with high-frequency metamaterial reactors. *Joule* 2024. https://doi.org/10.1016/j.joule.2024.07.017
(100) Dou, L.; Yan, C.; Zhong, L.; Zhang, D.; Zhang, J.; Li, X.; Xiao, L. Enhancing CO_2 methanation over a metal foam structured catalyst by electric internal heating. *Chemical Communications* 2019, *56* (2), 205–208. https://doi.org/10.1039/c9cc07525a
(101) Kumar, R.; Singh, S.; Singh, O. V. Bioconversion of lignocellulosic biomass: Biochemical and molecular perspectives. *Journal of Industrial Microbiology & Biotechnology* 2008, *35* (5), 377–391. https://doi.org/10.1007/s10295-008-0327-8
(102) Singh, N.; Singhania, R. R.; Nigam, P. S.; Dong, C.-D.; Patel, A. K.; Puri, M. Global status of lignocellulosic biorefinery: Challenges and perspectives. *Bioresource Technology* 2022, *344*, 126415. https://doi.org/10.1016/j.biortech.2021.126415
(103) Gautam, R.; Shyam, S.; Reddy, B. R.; Govindaraju, K.; Vinu, R. Microwave-assisted pyrolysis and analytical fast pyrolysis of macroalgae: Product analysis and effect of heating mechanism. *Sustainable Energy & Fuels* 2019, *3* (11), 3009–3020. https://doi.org/10.1039/C9SE00162J
(104) Borges, F. C.; Du, Z.; Xie, Q.; Trierweiler, J. O.; Cheng, Y.; Wan, Y.; Liu, Y.; Zhu, R.; Lin, X.; Chen, P.; Ruan, R. Fast microwave assisted pyrolysis of biomass using microwave absorbent. *Bioresource Technology* 2014, *156*, 267–274. https://doi.org/10.1016/j.biortech.2014.01.038
(105) Wang, Y.; Dai, L.; Wang, R.; Fan, L.; Liu, Y.; Xie, Q.; Ruan, R. Hydrocarbon fuel production from soapstock through fast microwave-assisted pyrolysis using microwave absorbent. *Journal of Analytical and Applied Pyrolysis* 2016, *119*, 251–258. https://doi.org/10.1016/j.jaap.2016.01.008
(106) Kostas, E. T.; Durán-Jiménez, G.; Shepherd, B. J.; Meredith, W.; Stevens, L. A.; Williams, O. S. A.; Lye, G. J.; Robinson, J. P. Microwave pyrolysis of olive pomace for bio-oil and bio-char production. *Chemical Engineering Journal* 2020, *387*, 123404. https://doi.org/10.1016/j.cej.2019.123404
(107) Abas, F. Z.; Ani, F. N.; Zakaria, Z. A. Microwave-assisted production of optimized pyrolysis liquid oil from oil palm fiber. *Journal of Cleaner Production* 2018, *182*, 404–413. https://doi.org/10.1016/j.jclepro.2018.02.052

(108) An, Y.; Tahmasebi, A.; Yu, J. Mechanism of synergy effect during microwave co-pyrolysis of biomass and lignite. *Journal of Analytical and Applied Pyrolysis* 2017, *128*, 75–82. https://doi.org/10.1016/j.jaap.2017.10.023

(109) Suriapparao, D. V.; Boruah, B.; Raja, D.; Vinu, R. Microwave assisted co-pyrolysis of biomasses with polypropylene and polystyrene for high quality bio-oil production. *Fuel Processing Technology* 2018, *175*, 64–75. https://doi.org/10.1016/j.fuproc.2018.02.019

(110) Li, H.; Li, J.; Fan, X.; Li, X.; Gao, X. Insights into the synergetic effect for co-pyrolysis of oil sands and biomass using microwave irradiation. *Fuel* 2019, *239*, 219–229. https://doi.org/10.1016/j.fuel.2018.10.139

(111) Tsubaki, S.; Fukushima, J.; Yamaguchi, A.; Nishioka, M.; Wada, Y. Process intensification of the ultra-rapid pyrolysis of bamboo by spatially separated microwave electric and magnetic fields. *Chemical Engineering Journal* 2024, *499*, 156260. https://doi.org/10.1016/j.cej.2024.156260

(112) Wang, W.; Wang, M.; Huang, J.; Tang, N.; Dang, Z.; Shi, Y.; Zhaohe, M. Microwave-assisted catalytic pyrolysis of cellulose for phenol-rich bio-oil production. *Journal of the Energy Institute* 2019, *92* (6), 1997–2003. https://doi.org/10.1016/j.joei.2018.10.012

(113) Zhu, L.; Zhang, Y.; Lei, H.; Zhang, X.; Wang, L.; Bu, Q.; Wei, Y. Production of hydrocarbons from biomass-derived biochar assisted microwave catalytic pyrolysis. *Sustainable Energy & Fuels* 2018, *2* (8), 1781–1790. https://doi.org/10.1039/C8SE00096D

(114) Shi, X.; Li, P.; Wang, X.; Song, J.; Fang, S.; Chang, C.; Pang, S. Enhancement of the production of aromatics and bio-syngas from microwave ex-situ pyrolysis based on Zn/Zr modified biochar and multi-catalysts. *Energy* 2022, *261*, 125307. https://doi.org/10.1016/j.energy.2022.125307

(115) Li, L.; Cai, D.; Zhang, L.; Zhang, Y.; Zhao, Z.; Zhang, Z.; Sun, J.; Tan, Y.; Zou, G. Synergistic effects during pyrolysis of binary mixtures of biomass components using microwave-assisted heating coupled with iron base tip-metal. *Renewable Energy* 2023, *203*, 312–322. https://doi.org/10.1016/j.renene.2022.12.076

(116) Zhang, L.; Tan, Y.; Cai, D.; Sun, J.; Zhang, Y.; Li, L.; Zhang, Q.; Zou, G.; Song, Z.; Bai, Y. Enhanced pyrolysis of woody biomass under interaction of microwave and needle-shape metal and its production properties. *Energy* 2022, *249*, 123667. https://doi.org/10.1016/j.energy.2022.123667

(117) Lee, M.-K.; Tsai, W.-T.; Tsai, Y.-L.; Lin, S.-H. Pyrolysis of Napier grass in an induction-heating reactor. *Journal of Analytical and Applied Pyrolysis* 2010, *88* (2), 110–116. https://doi.org/10.1016/j.jaap.2010.03.003

(118) Tsai, W. T.; Lee, M. K.; Chang, Y. M. Fast pyrolysis of rice straw, sugarcane bagasse and coconut shell in an induction-heating reactor. *Journal of Analytical and Applied Pyrolysis* 2006, *76* (1), 230–237. https://doi.org/10.1016/j.jaap.2005.11.007

(119) Muley, P. D.; Henkel, C.; Abdollahi, K. K.; Marculescu, C.; Boldor, D. A critical comparison of pyrolysis of cellulose, lignin, and pine sawdust using an induction heating reactor. *Energy Conversion and Management* 2016, *117*, 273–280. https://doi.org/10.1016/j.enconman.2016.03.041

(120) Henkel, C.; Muley, P. D.; Abdollahi, K. K.; Marculescu, C.; Boldor, D. Pyrolysis of energy cane bagasse and invasive Chinese tallow tree (Triadica Sebifera L.) biomass in an inductively heated reactor. *Energy Conversion and Management* 2016, *109*, 175–183. https://doi.org/10.1016/j.enconman.2015.12.013

(121) Daniel, D. J.; Ellison, C. R.; Bursavich, J.; Benbow, M.; Favrot, C.; Blazier, M. A.; Marculescu, C.; Nokes, S. E.; Boldor, D. An evaluative comparison of lignocellulosic pyrolysis products derived from various parts of populus deltoides trees and panicum virgatum grass in an inductively heated reactor. *Energy Conversion and Management* 2018, *171*, 710–720. https://doi.org/10.1016/j.enconman.2018.06.026

(122) Wu, L.; Ma, H.; Mei, J.; Li, Y.; Xu, Q.; Li, Z. Low energy consumption and high quality bio-fuels production via in-situ fast pyrolysis of reed straw by adding metallic particles in an induction heating reactor. *International Journal of Hydrogen Energy* 2022, *47* (9), 5828–5841. https://doi.org/10.1016/j.ijhydene.2021.11.229
(123) Muley, P. D.; Henkel, C.; Abdollahi, K. K.; Boldor, D. Pyrolysis and catalytic upgrading of pinewood sawdust using an induction heating reactor. *Energy & Fuels* 2015, *29* (11), 7375–7385. https://doi.org/10.1021/acs.energyfuels.5b01878
(124) Abu-Laban, M.; Muley, P. D.; Hayes, D. J.; Boldor, D. Ex-situ up-conversion of biomass pyrolysis bio-oil vapors using Pt/Al_2O_3 nanostructured catalyst synergistically heated with steel balls via induction. *Catalysis Today* 2017, *291*, 3–12. https://doi.org/10.1016/j.cattod.2017.01.010
(125) Bursavich, J.; Abu-Laban, M.; Muley, P. D.; Boldor, D.; Hayes, D. J. Thermal performance and surface analysis of steel-supported platinum nanoparticles designed for bio-oil catalytic upconversion during radio frequency-based inductive heating. *Energy Conversion and Management* 2019, *183*, 689–697. https://doi.org/10.1016/j.enconman.2019.01.025
(126) Asensio, J. M.; Miguel, A. B.; Fazzini, P.-F.; van Leeuwen, P. W. N. M.; Chaudret, B. Hydrodeoxygenation using magnetic induction: High-temperature heterogeneous catalysis in solution. *Angewandte Chemie International Edition* 2019, *58* (33), 11306–11310. https://doi.org/10.1002/anie.201904366
(127) Benoit, M.; Rodrigues, A.; Zhang, Q.; Fourré, E.; De Oliveira Vigier, K.; Tatibouët, J.-M.; Jérôme, F. Depolymerization of cellulose assisted by a nonthermal atmospheric plasma. *Angewandte Chemie International Edition* 2011, *50* (38), 8964–8967. https://doi.org/10.1002/anie.201104123.
(128) Wright, A.; Bandulasena, H.; Ibenegbu, C.; Leak, D.; Holmes, T.; Zimmerman, W.; Shaw, A.; Iza, F. Dielectric barrier discharge plasma microbubble reactor for pretreatment of lignocellulosic biomass. *AIChE Journal* 2018, *64* (11), 3803–3816. https://doi.org/10.1002/aic.16212
(129) Lusi, A.; Hu, H.; Bai, X. Producing high yield of levoglucosan by pyrolyzing nonthermal plasma-pretreated cellulose. *Green Chemistry* 2020, *22* (6), 2036–2048. https://doi.org/10.1039/D0GC00255K
(130) Zhu, X.; Hoang, T.; Lobban, L. L.; Mallinson, R. G. Plasma reforming of glycerol for synthesis gas production. *Chemical Communications* 2009, *20*, 2908–2910. https://doi.org/10.1039/B823410H
(131) Shie, J.-L.; Tsou, F.-J.; Lin, K.-L.; Chang, C.-Y. Bioenergy and products from thermal pyrolysis of rice straw using plasma torch. *Bioresource Technology* 2010, *101* (2), 761–768. https://doi.org/10.1016/j.biortech.2009.08.072
(132) An'shakov, A.; Faleev, V.; Danilenko, A.; Urbakh, E.; Urbakh, A. Investigation of plasma gasification of carbonaceous technogeneous wastes. *Thermophysics and Aeromechanics* 2007, *14*, 607–616. https://doi.org/10.1134/S0869864307040105
(133) Zhou, R.; Zhou, R.; Wang, S.; Lan, Z.; Zhang, X.; Yin, Y.; Tu, S.; Yang, S.; Ye, L. Fast liquefaction of bamboo shoot shell with liquid-phase microplasma assisted technology. *Bioresource Technology* 2016, *218*, 1275–1278. https://doi.org/10.1016/j.biortech.2016.07.042
(134) Zhou, R.; Zhou, R.; Zhang, X.; Fang, Z.; Wang, X.; Speight, R.; Wang, H.; Doherty, W.; Cullen, P. J.; Ostrikov, K. (Ken); Bazaka, K. High-performance plasma-enabled biorefining of microalgae to value-added products. *ChemSusChem* 2019, *12* (22), 4976–4985. https://doi.org/10.1002/cssc.201901772
(135) Erisman, J. W.; Sutton, M. A.; Galloway, J.; Klimont, Z.; Winiwarter, W. How a century of ammonia synthesis changed the world. *Nature Geoscience* 2008, *1* (10), 636–639. https://doi.org/10.1038/ngeo325

(136) Bowker, M. The 2007 nobel prize in chemistry for surface chemistry: Understanding nanoscale phenomena at surfaces. *ACS Nano* 2007, *1* (4), 253–257. https://doi.org/10.1021/nn700356g

(137) Wang, F.; Ma, J.; He, G.; Chen, M.; Zhang, C.; He, H. Nanosize effect of Al_2O_3 in Ag/Al_2O_3 catalyst for the selective catalytic oxidation of ammonia. *ACS Catalysis* 2018, *8* (4), 2670–2682. https://doi.org/10.1021/acscatal.7b03799

(138) Klerke, A.; Christensen, C. H.; Nørskov, J. K.; Vegge, T. Ammonia for hydrogen storage: Challenges and opportunities. *Journal of Materials Chemistry* 2008, *18* (20), 2304–2310. https://doi.org/10.1039/B720020J

(139) Valera-Medina, A.; Xiao, H.; Owen-Jones, M.; David, W. I. F.; Bowen, P. J. Ammonia for power. *Progress in Energy and Combustion Science* 2018, *69*, 63–102. https://doi.org/10.1016/j.pecs.2018.07.001

(140) MacFarlane, D. R.; Cherepanov, P. V.; Choi, J.; Suryanto, B. H. R.; Hodgetts, R. Y.; Bakker, J. M.; Ferrero Vallana, F. M.; Simonov, A. N. A roadmap to the ammonia economy. *Joule* 2020, *4* (6), 1186–1205. https://doi.org/10.1016/j.joule.2020.04.004

(141) Smith, C.; Hill, A. K.; Torrente-Murciano, L. Current and future role of Haber–Bosch ammonia in a carbon-free energy landscape. *Energy & Environmental Science* 2020, *13* (2), 331–344. https://doi.org/10.1039/C9EE02873K

(142) Suryanto, B. H. R.; Du, H.-L.; Wang, D.; Chen, J.; Simonov, A. N.; MacFarlane, D. R. Challenges and prospects in the catalysis of electroreduction of nitrogen to ammonia. *Nature Catalysis* 2019, *2* (4), 290–296. https://doi.org/10.1038/s41929-019-0252-4.

(143) Hatzell, M. C. The colors of ammonia. *ACS Energy Letters* 2024, *9* (6), 2920–2921. https://doi.org/10.1021/acsenergylett.4c01391

(144) Ponikvar, Ž.; Likozar, B.; Gyergyek, S. Electrification of catalytic ammonia production and decomposition reactions: From resistance, induction, and dielectric reactor heating to electrolysis. *ACS Applied Energy Materials* 2022, *5* (5), 5457–5472. https://doi.org/10.1021/acsaem.1c03045

(145) Liu, D.; Yao, Y.; Hu, L.; Dong, Q. High-temperature shock heating for thermochemical reactions (U.S. Patent 20230144856A1). U.S. Patent and Trademark Office.

(146) Alsfeld, M.; Wilhelmy, E. Über die Bildung von Ammoniak Aus Seinen Elementen Durch Elektrische Gasentladungen. *Annalen der Physik* 1931, *400* (1), 89–123. https://doi.org/10.1002/andp.19314000105

(147) Hong, J.; Prawer, S.; Murphy, A. B. Plasma catalysis as an alternative route for ammonia production: Status, mechanisms, and prospects for progress. *ACS Sustainable Chemistry & Engineering* 2018, *6* (1), 15–31. https://doi.org/10.1021/acssuschemeng.7b02381

(148) Mehta, P.; Barboun, P.; Herrera, F. A.; Kim, J.; Rumbach, P.; Go, D. B.; Hicks, J. C.; Schneider, W. F. Overcoming ammonia synthesis scaling relations with plasma-enabled catalysis. *Nature Catalysis* 2018, *1* (4), 269–275. https://doi.org/10.1038/s41929-018-0045-1.

(149) Rouwenhorst, K. H. R.; Kim, H.-H.; Lefferts, L. Vibrationally excited activation of N_2 in plasma-enhanced catalytic ammonia synthesis: A kinetic analysis. *ACS Sustainable Chemistry & Engineering* 2019, *7* (20), 17515–17522. https://doi.org/10.1021/acssuschemeng.9b04997.

(150) Engelmann, Y.; van't Veer, K.; Gorbanev, Y.; Neyts, E. C.; Schneider, W. F.; Bogaerts, A. Plasma catalysis for ammonia synthesis: A microkinetic modeling study on the contributions of Eley–Rideal reactions. *ACS Sustainable Chemistry & Engineering* 2021, *9* (39), 13151–13163. https://doi.org/10.1021/acssuschemeng.1c02713

(151) Mehta, P.; Barboun, P. M.; Engelmann, Y.; Go, D. B.; Bogaerts, A.; Schneider, W. F.; Hicks, J. C. Plasma-catalytic ammonia synthesis beyond the equilibrium limit. *ACS Catalysis* 2020, *10* (12), 6726–6734. https://doi.org/10.1021/acscatal.0c00684.

(152) Rouwenhorst, K. H. R.; Burbach, H. G. B.; Vogel, D. W.; Núñez Paulí, J.; Geerdink, B.; Lefferts, L. Plasma-catalytic ammonia synthesis beyond thermal equilibrium on Ru-based catalysts in non-thermal plasma. *Catalysis Science & Technology* 2021, *11* (8), 2834–2843. https://doi.org/10.1039/D0CY02189J

(153) Yamijala, S. S. R. K. C.; Nava, G.; Ali, Z. A.; Beretta, D.; Wong, B. M.; Mangolini, L. Harnessing plasma environments for ammonia catalysis: Mechanistic insights from experiments and large-scale ab initio molecular dynamics. *Journal of Physical Chemistry Letters* 2020, *11* (24), 10469–10475. https://doi.org/10.1021/acs.jpclett.0c03021

(154) Barboun, P. M.; Daemen, L. L.; Waitt, C.; Wu, Z.; Schneider, W. F.; Hicks, J. C. Inelastic neutron scattering observation of plasma-promoted nitrogen reduction intermediates on Ni/γ-Al_2O_3. *ACS Energy Letters* 2021, *6* (6), 2048–2053. https://doi.org/10.1021/acsenergylett.1c00643

(155) Barboun, P.; Mehta, P.; Herrera, F. A.; Go, D. B.; Schneider, W. F.; Hicks, J. C. Distinguishing plasma contributions to catalyst performance in plasma-assisted ammonia synthesis. *ACS Sustainable Chemistry & Engineering* 2019, *7* (9), 8621–8630. https://doi.org/10.1021/acssuschemeng.9b00406

(156) Gómez-Ramírez, A.; Cotrino, J.; Lambert, R. M.; González-Elipe, A. R. Efficient synthesis of ammonia from N_2 and H_2 alone in a ferroelectric packed-bed DBD reactor. *Plasma Sources Science and Technology* 2015, *24* (6), 065011. https://doi.org/10.1088/0963-0252/24/6/065011

(157) van't Veer, K.; Engelmann, Y.; Reniers, F.; Bogaerts, A. Plasma-catalytic ammonia synthesis in a DBD plasma: Role of microdischarges and their afterglows. *The Journal of Physical Chemistry C* 2020, *124* (42), 22871–22883. https://doi.org/10.1021/acs.jpcc.0c05110

(158) Wang, Y.; Yang, W.; Xu, S.; Zhao, S.; Chen, G.; Weidenkaff, A.; Hardacre, C.; Fan, X.; Huang, J.; Tu, X. Shielding protection by mesoporous catalysts for improving plasma-catalytic ambient ammonia synthesis. *Journal of the American Chemical Society* 2022, *144* (27), 12020–12031. https://doi.org/10.1021/jacs.2c01950

(159) Peng, P.; Chen, P.; Addy, M.; Cheng, Y.; Anderson, E.; Zhou, N.; Schiappacasse, C.; Zhang, Y.; Chen, D.; Hatzenbeller, R.; Liu, Y.; Ruan, R. Atmospheric plasma-assisted ammonia synthesis enhanced via synergistic catalytic absorption. *ACS Sustainable Chemistry & Engineering* 2019, *7* (1), 100–104. https://doi.org/10.1021/acssuschemeng.8b03887

(160) Wu, H.; Yang, L.; Wen, J.; Xu, Y.; Cai, Y.; Gao, W.; Wang, Q.; Guan, Y.; Feng, S.; Cao, H.; He, T.; Liu, L.; Zhang, S.; Guo, J.; Chen, P. Plasma-driven nitrogen fixation on sodium hydride. *Advanced Energy Materials* 2023, *13* (27), 2300722. https://doi.org/10.1002/aenm.202300722

(161) Sharma, R. K.; Patel, H.; Mushtaq, U.; Kyriakou, V.; Zafeiropoulos, G.; Peeters, F.; Welzel, S.; van de Sanden, M. C. M.; Tsampas, M. N. Plasma activated electrochemical ammonia synthesis from nitrogen and water. *ACS Energy Letters* 2021, *6* (2), 313–319. https://doi.org/10.1021/acsenergylett.0c02349

(162) Zhang, T.; Zhou, R.; Zhang, S.; Zhou, R.; Ding, J.; Li, F.; Hong, J.; Dou, L.; Shao, T.; Murphy, A. B.; Ostrikov, K. (Ken); Cullen, P. J. Sustainable ammonia synthesis from nitrogen and water by one-step plasma catalysis. *Energy & Environmental Materials* 2023, *6* (2), e12344. https://doi.org/10.1002/eem2.12344

(163) Wu, T.; Chang, B.; Li, Y.; Zhang, X.; Zhao, X.; Liu, Z.; Zhang, G.; Liu, X.; Zhao, L.; Zhang, Y.; Zhang, H.; Liu, H.; Zhou, W. Laser-induced plasma and local temperature field for high-efficiency ammonia synthesis. *Nano Energy* 2023, *116*, 108855. https://doi.org/10.1016/j.nanoen.2023.108855

(164) Hu, J.; Wildfire, C.; Stiegman, A. E.; Dagle, R. A.; Shekhawat, D.; Abdelsayed, V.; Bai, X.; Tian, H.; Bogle, M. B.; Hsu, C.; Luo, Y.; Davidson, S. D.; Wang, Y. Microwave-driven heterogeneous catalysis for activation of dinitrogen to ammonia under atmospheric pressure. *Chemical Engineering Journal* 2020, *397*, 125388. https://doi.org/10.1016/j.cej.2020.125388
(165) Wang, Y.; Wildfire, C.; Khan, T. S.; Shekhawat, D.; Hu, J.; Tavadze, P.; Quiñones-Fernández, R.; Moreno, S. Effects of support and promoter on Ru catalyst activity in microwave-assisted ammonia synthesis. *Chemical Engineering Journal* 2021, *425*, 130546. https://doi.org/10.1016/j.cej.2021.130546
(166) Wang, Y.; Khan, T. S.; Wildfire, C.; Shekhawat, D.; Hu, J. Microwave-enhanced catalytic ammonia synthesis under moderate pressure and temperature. *Catalysis Communications* 2021, *159*, 106344. https://doi.org/10.1016/j.catcom.2021.106344
(167) Araia, A.; Wang, Y.; Robinson, B.; Jiang, C.; Brown, S.; Wildfire, C.; Shekhawat, D.; Hu, J. Microwave-assisted ammonia synthesis over Cs-Ru/CeO_2 catalyst at ambient pressure: Effects of metal loading and support particle size. *Catalysis Communications* 2022, *170*, 106491. https://doi.org/10.1016/j.catcom.2022.106491
(168) Melkote, S. G.; Muley, P.; Dutta, B.; Wildfire, C.; Weiss, R.; Hu, J. Developing a microwave-driven reactor for ammonia synthesis: Insights into the unique challenges of microwave catalysis. *Catalysis Science & Technology* 2023, *13* (8), 2393–2406. https://doi.org/10.1039/D2CY02181A
(169) Oumghar, A.; Legrand, J. C.; Diamy, A. M.; Turillon, N.; Ben-Aïm, R. I. A kinetic study of methane conversion by a dinitrogen microwave plasma. *Plasma Chemistry and Plasma Processing* 1994, *14* (3), 229–249. https://doi.org/10.1007/BF01447080.
(170) Bai, X.; Tiwari, S.; Robinson, B.; Killmer, C.; Li, L.; Hu, J. Microwave catalytic synthesis of ammonia from methane and nitrogen. *Catalysis Science & Technology* 2018, *8* (24), 6302–6305. https://doi.org/10.1039/C8CY01355A
(171) Tiwari, S.; Khan, T.; Tavadze, P.; Hu, J. Activation of two highly stable molecules – Nitrogen and methane to co-produce ammonia and ethylene. *Chemical Engineering Journal* 2020, *413*, 127501. https://doi.org/10.1016/j.cej.2020.127501
(172) Tiwari, S.; Ibrahim, S. A.; Robinson, B.; Brown, S.; Wang, Q.; Che, F.; Hu, J. Post-plasma catalysis: Charge effect on product selectivity in conversion of methane and nitrogen plasma to ethylene and ammonia. *Catalysis Science & Technology* 2023, *13* (10), 2966–2981. https://doi.org/10.1039/D2CY02077G.
(173) Wang, X.; Qian, C.; Tang, S.-Y.; Zhou, S. Ammonia synthesis via nitrogen-coupled methane conversion at ambient temperature and plasma conditions. *Industrial and Engineering Chemistry Research* 2023, *62* (44), 18416–18426. https://doi.org/10.1021/acs.iecr.3c02943

4 Electrochemical Processes for Sustainable Chemical Manufacturing

Rong Xia, Yuanjun Chen, Cong Tian, Jianan (Erick) Huang, and Aditya Menon

4.1 INTRODUCTION

Atmospheric CO_2 concentrations have escalated to a record high of 426 ppm, underscoring the urgency of transitioning from fossil fuels to renewable energy sources to mitigate emissions and achieve a carbon-neutral economy by 2050.[1] Over the past decade, the economic landscape of energy production has shifted dramatically; renewable energy, primarily solar and wind, has become more cost-effective than fossil fuel. From 2010 to 2022, the levelized cost of electricity (LCOE) from offshore wind, onshore wind, and solar PV has dramatically decreased by 59%, 69%, and 87%, while the cost of coal has approximately remained constant (Figure 4.1). The solar LCOE has plummeted from \$392 per MWh to \$50 per MWh in 2024—making it 56% cheaper than the weighted average cost of fossil-fueled alternatives.[2] This trend is projected to continue, with solar electricity prices expected to potentially reach as low as \$0.03 per kWh by 2030.[3]

This reduction in the cost of renewable energy has spurred notable advancements in electrification across various sectors, exemplified by the rapid development of electric vehicles, a clear sign of progress in replacing fossil fuels with cleaner alternatives. Similarly, the industrial sector—which accounts for 21% of global greenhouse gas emissions, with chemical, mineral, and metal manufacturing responsible for over 93% of these—presents significant opportunities for transformation through electrification.[4] Traditional manufacturing processes in the industry sector rely heavily on fossil fuels and operate at high temperatures and pressures, presenting a prime opportunity for decarbonization through electrochemical manufacturing.

Electrochemical manufacturing, powered by renewable electricity, offers a viable pathway to transform the chemical industry by enabling the conversion of non-fossil feedstocks—such as carbon dioxide, nitrogen, and biomass derivatives—into value-added fuels, commodity chemicals, and specialty chemicals that are traditionally associated with a high carbon footprint. Electrochemical processes are particularly suited for modular systems and distributed productions, which can operate at smaller scales with high throughput, thus enabling more localized and flexible

DOI: 10.1201/9781003655442-6

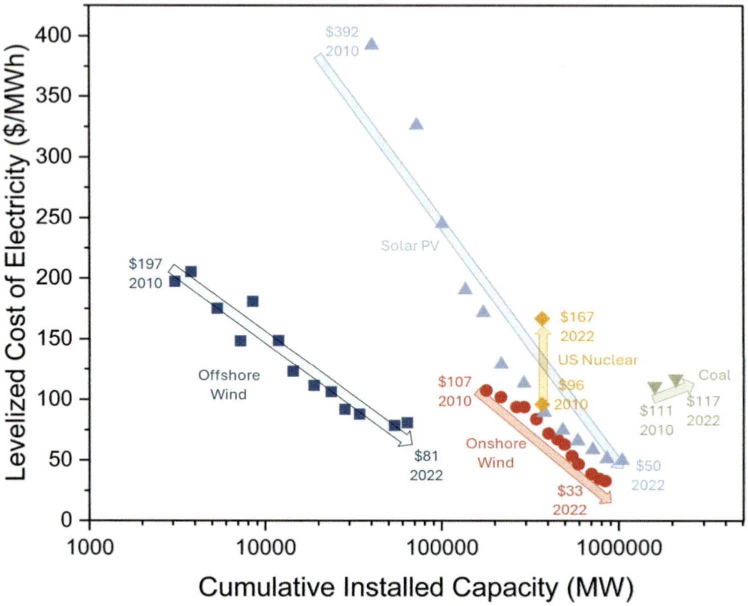

FIGURE 4.1 Change in levelized cost of electricity from 2010 to 2022 of renewable and fossil-powered energy sources.

(Renewable energy data is from IRENA 2022 Report, nuclear and coal LCOE is from Lazard 2023 Report, and nuclear and coal capacities are from IAEA and globalenergymonitor.org, respectively.)

manufacturing setups. Despite its promise, scaling electrochemical processes to meet global sustainability and energy needs remains a formidable challenge. To date, only a few processes, such as the Chlor-Alkali and Hall-Heroult, have achieved commercial scalability. Consequently, a critical focus needs to be placed on transitioning these technologies from laboratory-scale experiments to industrial-scale production.

This chapter aims to discuss the latest advancements in electrochemical manufacturing, with a particular focus on the synthesis of chemicals and fuels from sustainable feedstocks such as carbon dioxide, nitrogen, and biomass derivatives. We will delve into key challenges in electrochemical manufacturing such as energy efficiency, mass transfer limitations, reactant/product solubility, electrode stability, and ion conductivity. The discussion will highlight the pivotal role of catalyst development and the implementation of innovative reactor designs for advancing electrochemical manufacturing techniques. Furthermore, we will evaluate the potential to electrify chemical manufacturing, particularly through the production of high-demand commodity chemicals with substantial CO_2 footprints. The chapter concludes by discussing the broad impacts of widespread electrochemical manufacturing, stressing the pivotal research and development—from lab to industrial scale—needed to shift toward renewable energy and effectively decarbonize society.

4.2 ELECTROCHEMICAL CO_2 REDUCTION TO CHEMICALS AND FUELS

As atmospheric CO_2 levels continue to rise due to fossil fuel combustion and industrial emissions, the development of innovative technologies for CO_2 capture and conversion is critical to combating climate change.[5] Electrochemical CO_2 reduction (CO_2R) has emerged as a promising solution, transforming captured CO_2 into a diverse array of chemicals and fuels such as syngas, methane, methanol, formic acid, ethanol, ethylene, acetic acid, and propanol.[6,7] When powered by renewable electricity, CO_2R offers a viable approach to close the anthropogenic carbon cycle. This approach not only recycles CO_2 into products that re-enter industrial and consumer supply chains but also reduces reliance on fossil fuel-derived chemicals and fuels, thereby fostering a sustainable carbon economy.[8]

Over the past decade, CO_2R technology has seen notable advancements in catalyst design and system optimization, resulting in significantly enhanced Faradaic efficiency (the fraction of total charge passed that leads to target product), and energy efficiency (the ratio of theoretical energy input to the total energy consumed). Significant progress has been made in engineering the surface structure and electronic properties of catalysts to enhance catalytic selectivity toward desired carbon-based products while suppressing the competing hydrogen evolution reaction (HER). Despite these advancements, several key challenges remain: improving catalyst selectivity to reduce the formation of undesired byproducts, decreasing the high energy requirements associated with substantial electrocatalytic overpotentials, and enhancing the long-term stability of catalysts during continuous operation.

The development of advanced electrolyzer configurations, such as flow cells and membrane electrode assemblies (MEAs), is crucial for achieving higher current densities with reduced energy losses. Additionally, enhancing CO_2 single-pass conversion efficiency and addressing issues related to the formation of carbonate/bicarbonate and CO_2 crossover are vital for reducing the costs associated with CO_2 regeneration and product separation. To advance CO_2R toward economic viability, future efforts should concentrate on developing cost-effective catalysts, designing energy-efficient systems, and achieving seamless integration with renewable energy sources.

4.2.1 Fundamental Insight of Electrochemical CO_2R

The thermodynamic potential of CO_2R and HER are comparable; for instance, the equilibrium potentials of converting CO_2 to carbon monoxide and ethylene are -0.11 V and +0.07 V, respectively, relative to the reversible hydrogen electrode (RHE).[9] However, the actual energy input required for CO_2R often exceeds the theoretical minimum due to kinetic barriers associated with multi-step proton and electron transfer processes.

The initial step of CO_2 reduction involves the adsorption and activation of CO_2 on the catalyst surface. There are two adsorptive configurations of CO_2 that lead to distinct reactive intermediates: *OCHO, which results in the formation of formate, and *COOH, leading to the production of carbon monoxide. On metals such as copper, *CO can remain adsorbed sufficiently long to enable further chemical transformations. If *CO undergoes hydrogenation, methane is produced, while dimerization leads to C–C bond formation, resulting in the production of multicarbon products like ethylene, ethanol, and propanol.[10]

The requirement for proton donors in CO_2R adds significant complexity to the process, since protons can easily reduce to hydrogen. This complexity is further compounded by the numerous available reaction pathways, each yielding different products and thereby reducing the overall selectivity of the process. Additionally, linear scaling relations that connect the binding energies of key intermediates like *COOH and *CO further constrain the optimization of selective pathways. Breaking these scaling relations is essential, especially for the synthesis of multicarbon products, as it allows for more precise control over reaction selectivity and enhances the efficiency of CO_2 conversion into designated products.[11]

To enhance CO_2R activity and selectivity, significant progress has been made in catalyst development, reactor design, and system optimization. These advancements include modifying catalyst surface and structure, controlling local CO_2 and proton concentrations, and engineering the electrode-electrolyte interface. Techniques such as gas diffusion electrodes (GDEs) in flow-cell reactor systems increase CO_2 availability at the catalyst surface, improving CO_2R selectivity over HER. The solid-liquid-gas interface is also crucial; solvated cations like Na^+, K^+, and Cs^+ in the electrolyte stabilize CO_2R intermediates at the reaction interface through coordination and solvation effects.[12] Understanding these interfacial interactions and tailoring catalyst surfaces accordingly are essential for advancing CO_2R selectivity and efficiency.

4.2.2 Catalyst Design for Electrochemical CO_2R

The development of efficient catalysts with high activity, selectivity, and durability is crucial for the commercial feasibility of electrochemical CO_2 reduction. The performance of CO_2R catalysts heavily depends on their composition and structure, as these factors determine their interactions with intermediates. Based on the binding strength with intermediates, catalysts can be divided into three groups: Group I metals (e.g., Ag, Au, Pd, Zn) prefer the *COOH intermediate and mainly produce CO; Group II metals (e.g., Ga, In, Sn, Bi) favor *OCOH, leading to formate. Copper, belonging to Group III, is unique in its ability to stabilize both *COOH and *CO, facilitating the production of multicarbon (C_{2+}) products like ethylene and ethanol. At the nanoscale, the geometric and electronic structures of catalysts, such as crystal facets, grain boundaries, and defects, significantly influence the performance of CO_2R. On a mesoscopic scale, features like particle size and porosity affect reactant mass transfer and the electrochemically active surface area. Additionally, controlling the macroscopic morphology of materials provides an effective strategy to maximize the number of active sites.[13]

The electrochemical CO_2 to CO involves two-electron and two-proton transfer. Initially, CO_2 adsorbs onto the catalyst surface and is reduced to a *COOH intermediate via a concerted proton-electron transfer. Another proton and electron interaction yields H_2O and *CO, which then desorbs to produce CO. Forming *COOH and desorbing *CO are rate-limiting steps, as weak *COOH binding hinders formation, and strong *CO binding impedes desorption.[14] Metal nanoparticles such as Ag, Au, Pd, and Zn have demonstrated over 90% Faradaic efficiency in CO production. Additionally, earth-abundant single-atom catalysts like Ni, Fe, Co, Sn, and Bi, anchored on nitrogen-doped carbon, are of interest for enhancing the utilization of active sites. Molecular catalysts based on porphyrins and phthalocyanines, such as iron porphyrin and cobalt phthalocyanine, have also been explored, featuring precise control over steric and electronic properties near the catalytic sites.

For CO_2-to-formic acid conversion, CO_2 binds to the electrode surface through oxygen atoms, forming a bidentate *OCHO intermediate.[15] Through proton-coupled electron transfer, this intermediate converts to HCOOH or $HCOO^-$. An optimal *OCHO binding energy enhances the CO_2-to-HCOOH efficiency. Oxophilic metals such as Sn, Bi, In, and Pb favor the adsorption of *OCHO intermediates, making them selective catalysts for formic acid production. Additionally, their weak binding to *H effectively suppresses the competing hydrogen evolution reaction.

Copper-based catalysts are highly effective for converting CO_2 into multicarbon products like ethylene, ethanol, acetate, and propanol.[16,17] Strategies to enhance performance focus on engineering the size, crystal facets, and morphology. Incorporating secondary metals into Cu to form Cu-based alloys is another effective strategy by leveraging geometric ensemble effects and electronic effects. Geometric ensemble effects involve changes in the atomic arrangement of active sites, which alters the interaction between intermediates and the active sites. Electronic effects result from modifications in the electronic structure of Cu when doped with heteroatoms such as Au, Ag, Pd, Al, Zn, or Sn, which adjust the binding strength of intermediates on the catalyst surface, thereby influencing the kinetics of CO_2R.

4.2.3 Electrolyzer Configuration for Electrochemical CO_2R

The three common configurations of electrochemical electrolyzers used for CO_2 reduction (CO_2R) are (i) H-cells, (ii) flow cells with a liquid catholyte and GDE, and (iii) MEAs.

The H-cell is the simplest and most effective setup for lab-scale studies. In this configuration, the cathode is fully immersed in an aqueous electrolyte, and CO_2 gas is bubbled into the cathodic chamber to maintain a CO_2-saturated catholyte. However, the low solubility of CO_2 in aqueous electrolytes—34 mM under ambient conditions—limits the operating current density to several mA cm^{-2}. Although some high-performing CO_2R catalysts have been reported in H-cells, their practical application is limited due to scalability and operational constraints. Moreover, the performance of CO_2R is significantly influenced by the local reaction environment, which is highly sensitive to variations in current density and applied voltage. Consequently, results obtained from H-cell configurations often do not translate directly to industrial conditions. To bridge this gap, advanced testing systems are required to evaluate catalysts under industrially relevant conditions.

Employing GDEs in a flow electrolyzer represents an advancement over the H-cell configuration. In this configuration, the gaseous chamber and the catholyte chamber are separated by a hydrophobic carbon-based gas diffusion layer, which enables the reaction to occur at the gas-liquid-catalyst triple-phase boundary, effectively overcoming the mass transport limitations and allowing CO_2R to operate at industrial-level current densities. However, this approach also poses challenges, such as the flooding issue, where the catholyte can fully saturate the gas diffusion electrode, limiting access to the gaseous reactant.

MEA-based electrolyzers are more industrially relevant for improving the energy efficiency and durability of CO_2 reduction systems. In traditional flow cells, the liquid electrolyte layer between the cathode and the anode increases internal resistance, resulting in high cell voltage and poor energy efficiency. In MEA-based zero-gap electrolyzers, cathodes and anodes are in direct contact with the membrane. Electrolyte is introduced through flow channels embedded in the endplate, minimizing the distance between the cathode and anode and significantly reducing ohmic losses. Additionally, the membrane separates the liquid electrolyte from the gas diffusion electrode, enhancing durability and largely preventing flooding issues. However, achieving thousands of hours of operation remains a significant challenge for industrial applications. Ongoing advancements in membrane technology and optimization of the cathode environment will be crucial for realizing the full industrial potential of MEA-based CO_2R electrolyzers.

4.2.4 System Engineering for Electrochemical CO_2R

To date, most CO_2 reduction research has utilized alkaline and neutral electrolytes, which create a highly alkaline local environment due to the in-situ production of OH^-, advantageous for suppressing the competing HER and promoting CO_2R performance. However, the use of alkaline or neutral electrolytes often leads to carbonate formation. This occurs because every mole of electron transfer is accompanied by the generation of one mole of hydroxide, and the CO_2 reacts with the electrochemically generated hydroxides to form carbonate. Such carbonate formation significantly limits the single-pass conversion (SPC) of CO_2, with only a small fraction of the CO_2 being converted to target products. In alkaline conditions, over 70% of the consumed CO_2 is converted to carbonate, and the process of regenerating CO_2 from carbonate, that is, calcium caustic loop, is very energy-intensive.[18] Meanwhile, carbonate formation also leads to salt precipitation on the gas diffusion electrode, which negatively impacts its durability. Additionally, the carbonate crossover to the anode, driven by the electric field, leads to the release of CO_2 in a mixture with O_2. Separating CO_2 from O_2 further increases energy consumption. To prevent CO_2 loss due to carbonate formation and increase the single-pass conversion of CO_2, emerging strategies include acidic CO_2 reduction (CO_2R), cascade CO_2R, and electrochemical reactive capture (Figure 4.2).

One promising approach is to replace the commonly used alkaline and neutral electrolytes with acidic ones, which could prevent the carbonate formation.[19] However, in acidic media, the hydrogen evolution reaction tends to dominate over CO_2R, leading to poor selectivity, especially for C_{2+} products. Although some progress has been made in enhancing CO_2R performance in acidic environments through modification of catalyst structure and tuning of the local reaction environment, further advancements in catalyst design are necessary to suppress HER and improve both product selectivity and energy efficiency.

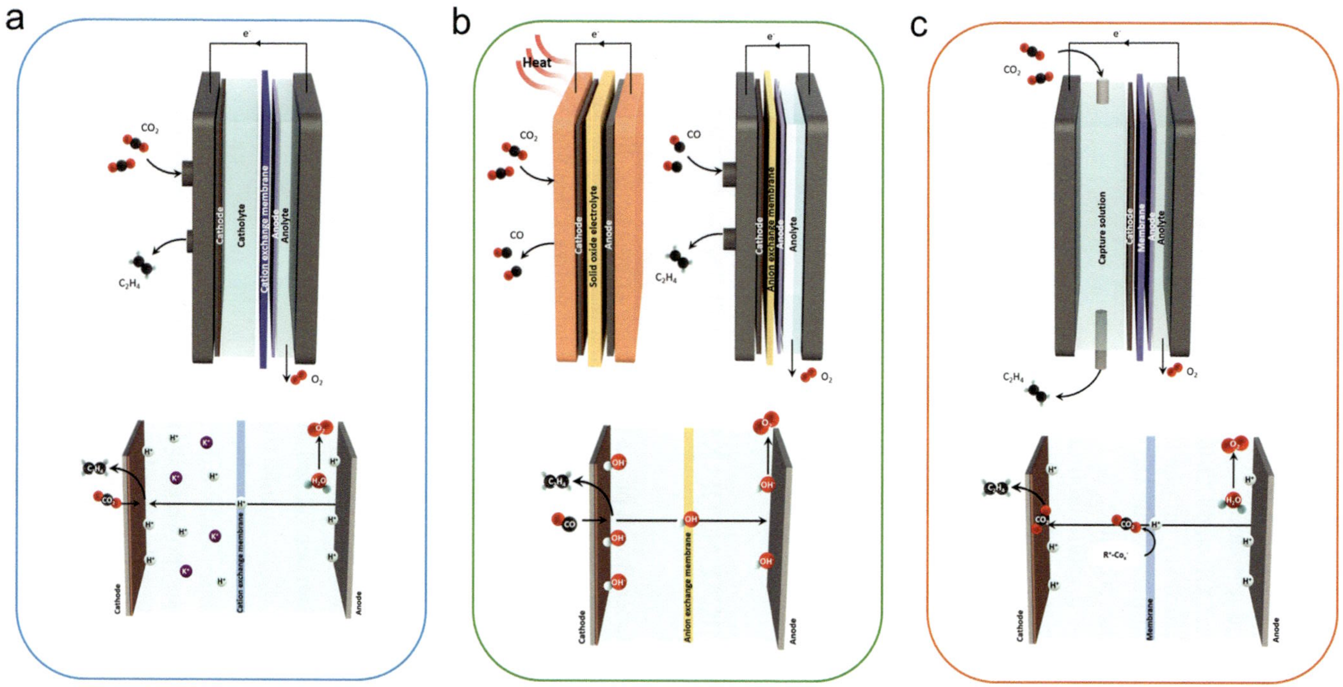

FIGURE 4.2 Strategies to prevent CO_2 loss due to carbonate formation and increase the single-pass conversion of CO_2. (a) Schematic illustration of acidic CO_2R. (b) Schematic illustration of a cascade CO_2R-COR system. CO_2 is reduced to CO in an SOEC, followed by CO reduction to value-added products in an MEA electrolyzer. (c) Schematic illustration of reactive capture: in (bi)carbonate solutions, protons react with (bi)carbonate to generate CO_2 *in situ* at the catalyst surface, initiating CO_2R; in amine-based solutions, the amine-CO_2 adduct ($R^+–CO_x^-$) is directly reduced at the catalyst surface.

(The figure is reprinted from "Catalyst design for electrochemical CO_2 reduction to ethylene," by Y. Chen, et al., 2023, *Matter*, Volume 7, Issue 1, p. 25. Copyright 2024 with permission from Elsevier.)

Another promising strategy is the cascade system, in which CO_2 is first converted to CO, followed by CO electroreduction (COR) to C_{2+} products.[20] This approach leverages recent advances in solid-oxide electrolysis cell (SOEC) technology,[21] which efficiently converts CO_2 to CO at low energy costs and with high SPC. An additional advantage is that CO, unlike CO_2, does not react with hydroxide ions, thereby avoiding carbonate formation. Given the similarities of reaction intermediates and pathways between COR and CO_2R, existing CO_2R catalyst design strategies can be adapted for COR. However, unique factors like local CO coverage and distinct local reaction environment necessitate the development of new catalysts tailored to COR to further enhance performance.

Electrochemical reactive capture facilitates the direct conversion of capture solutions (e.g., carbonate from direct air capture or carbamide from point-source carbon capture) into target product, that is, carbon monoxide, methane, and ethylene, eliminating the need for traditional CO_2 regeneration processes.[22,23] This method bypasses the need for gaseous CO_2, thus avoiding the CO_2 loss due to carbonate formation. The carbonate or carbamide is electrochemically regenerated to CO_2, which is then reduced to the target product. The challenges lie in enhancing the local CO_2 concentration at the catalyst surface to ensure effective CO_2 adsorption and activation, even at low concentrations, while simultaneously suppressing the hydrogen evolution reaction.

4.3 ELECTROCHEMICAL N_2 TRANSFORMATION FOR FERTILIZER PRODUCTION

Since its inception, the Haber-Bosch process—reacting atmospheric dinitrogen with hydrogen in the presence of iron catalyst under high pressures and temperatures—has been instrumental in producing synthetic nitrogen fertilizers that feed nearly half of the world's population. Despite a century of optimization, this method remains deeply dependent on fossil fuels and is energy-intensive, primarily due to the harsh reaction conditions needed to break the stable N-N triple bond. It accounts for 2 percent of global energy consumption and produces approximately 420 million tons of CO_2 emissions annually.[24] Additionally, the conventional fertilizer production process is infrastructure-intensive and vulnerable to supply chain disruptions, compounded by the high costs of long-distance transportation. These factors underscore the need for transitioning from centralized industrial production to a more decentralized configuration. In light of these challenges, there is an urgent need for greener alternatives to decarbonize fertilizer production. Promising methods include enzyme catalysis, photocatalysis, plasma-assisted catalysis, and electrochemical catalysis. Among these, electrocatalysis, powered by renewable energy sources, emerges as a particularly promising solution. It not only significantly reduces the carbon footprint associated with ammonia production but also supports a modular operational mode that can decentralize fertilizer production, thereby enhancing food security and fostering economic resilience in developing regions.

4.3.1 Lithium-Mediated Electrochemical N_2 Reduction

Electrifying ammonia production through N_2 electroreduction presents a compelling alternative, which uses water as the proton source instead of hydrogen, operating at ambient temperature and pressure. Despite significant research efforts dedicated to direct N_2 electroreduction in aqueous electrolytes, challenges persist. The selectivity for ammonia remains low, and the production rate typically does not exceed 20 nmol cm^{-2} s^{-1}, largely due to the difficulty in activating the robust N-N triple bond of dinitrogen.[25]

Li-mediated N_2 electroreduction is emerging as a promising method for ammonia synthesis, utilizing metallic lithium to dissociate N_2 and form lithium nitride, which is subsequently protonated to produce ammonia and regenerate Li^+. The initial lithium-mediated nitrogen reduction was documented in 1930 by Fichter, who utilized ethanol as the proton source and achieved a 10% Faradaic efficiency under 1,000 bar of N_2.[26] However, the thermodynamic instability of electrodeposited metallic lithium in contact with alcohol leads to its dissolution and hydrogen evolution. Further developments were reported by Tsuneto and co-authors in the 1990s, who conducted lithium-mediated nitrogen reduction in a tetrahydrofuran electrolyte containing 0.2 N $LiClO_4$ with 1 vol% ethanol as the proton source.[27] This aprotic electrolyte significantly mitigates metallic lithium dissolution, achieving a 48.7% ammonia Faradaic efficiency under 50 atm of N_2 pressure.

McEnaney and colleagues report a molten-salt lithium-mediated N_2 reduction process.[28] Lithium hydroxide is reduced to metallic lithium at 450°C in a molten-salt electrolyzer. The resulting metallic lithium reacts with nitrogen at ambient pressure to form lithium nitride (Li_3N), which upon hydrolysis in water releases ammonia and recovers LiOH. This process achieves a Faradaic efficiency of approximately 88.5% at a current density of 500 mA cm^{-2}, substantially higher than any previously reported figures. Nonetheless, the high cell voltage of over 3.1 V leads to low energy efficiency, and the transferring of solid reactants introduces significant challenges in downstream separation.

Recent developments in low-temperature lithium-mediated nitrogen reduction using aprotic electrolytes are attracting considerable attention. Research indicates that forming a surface passivation layer, known as a solid-electrolyte interphase (SEI), can mitigate the interaction between metallic lithium and the proton source that typically leads to lithium dissolution and hydrogen evolution—a phenomenon also observed in lithium-ion batteries.[29] Optimizing this SEI is crucial for enhancing the efficiency of low-temperature lithium-mediated nitrogen reduction. Li et al. demonstrated that utilizing a lithium tetrafluoroborate electrolyte with a highly porous copper electrode forms a uniform, LiF-enriched SEI layer.[30] This configuration promotes even lithium deposition and significantly reduces uncontrolled electrolyte degradation, demonstrating a 95% Faradaic efficiency under 20 bar of N_2. Subsequent studies revealed that adding a small amount of oxygen (0.6–0.8% molar fraction) to the system enhances the uniformity of the SEI during lithium deposition, which in turn boosts the N_2 reduction to a 78% Faradaic efficiency at a current density of 4 mA cm^{-2}.[31]

One significant challenge in the process is the use of a sacrificial proton source. Initially, ethanol was used as the proton source, which could be oxidized at the anode to produce CO_2. To overcome this, Lazouski introduced a non-aqueous gas diffusion electrode that facilitates hydrogen oxidation while paired with lithium-mediated nitrogen reduction, effectively preventing solvent oxidation and CO_2 generation at the anode.[32] Suryanto et al. reported using a phosphonium cation as a proton shuttle, which has demonstrated a 69% ammonia Faradaic efficiency at a current density of 22.5 mA cm^{-2}, equivalent to a production rate of 53 nmol cm^{-2} s^{-1}.[33] Further advancements by Fu et al. involved the development of anodic hydrogen oxidation catalysts that reduce the overall cell voltage, thereby improving the energy efficiency to 13% at a current density of 6 mA cm^{-2}.[34]

Another significant challenge is the high overpotential required for lithium plating, which leads to low energy efficiency in the process. The thermodynamic equilibrium potential for nitrogen reduction is 0.057 V versus RHE, whereas lithium-mediated nitrogen reduction depends on reducing lithium to metallic lithium, necessitating a potential of −2.63 V versus RHE at pH 7. Experimentally, lithium plating involves an overpotential exceeding 3 V, which economically challenges the viability of lithium-mediated nitrogen reduction. Although integrating hydrogen oxidation can lower the full cell voltage, the necessity of lithium plating caps the theoretical maximum energy efficiency at just 27.8%.[34]

4.3.2 Alternative Approaches for Ammonia Production

Alternative approaches for electrifying ammonia production include the use of high-temperature protonic ceramic fuel cells and plasmatic-electrocatalytic nitrogen reduction. Vasileios Kyriakou has developed a protonic ceramic membrane reactor, featuring VN-Fe, $BaZr_{0.8}Ce_{0.1}Y_{0.1}O_{3-x,}$ and Ni-$BaZr_{0.7}Ce_{0.2}Y_{0.1}O_{3-x}$ as the cathode, membrane, and anode respectively.[35] This reactor integrates hydrogen production with ammonia synthesis at 550°C, achieving an ammonia Faradaic Efficiency of 14% at a full cell voltage of 0.3 V. Recent studies have shown that plasma-induced vibrational activation of N_2 can reduce the activation energy by one-third in thermocatalytic nitrogen reduction.[36] Kumari et al. demonstrated a 47% increase in ammonia production in plasma-aided electroreduction compared with conventional electrochemical reduction and plasmatic nitrogen reduction.[37] Qiao and collaborators successfully coupled low-temperature plasma oxidation with electrochemical reduction, achieving N_2 conversion to ammonia via nitrate as an intermediate. Nickel boride was applied as the electrocatalyst to reduce plasma-derived NO_x^- to NH_3 with approximately 100% selectivity.[38] Liang et al. reported the cascade plasmatic-electrocatalytic N_2 reduction to ammonia and demonstrated over 90% Faradaic efficiency and 45.45 mg h^{-1} ammonia production rate.[39]

The energy efficiency of the conventional Haber–Bosch process has improved from 36% to 62% after a century of development.[29] The US Department of Energy's REFUEL program has set an ambitious target for electrochemical ammonia synthesis: a 90% Faradaic efficiency at a current density exceeding 300 mA cm^{-2}, with an overall energy efficiency above 60%.[40] Despite significant research efforts, achieving these

Faradaic efficiency and current density targets remains challenging. Implementing flow electrolyzer integrated with gas diffusion electrode could enhance gas transport to the electrode surface and allow for higher operating current densities. Managing pressure and temperature effectively is crucial for optimizing nitrogen transport and achieving high Faradaic efficiencies. However, these advancements introduce new challenges, such as electrode flooding and changes in SEI chemistry, necessitating further advances in understanding and optimizing the catalytic SEI and refining kinetic and transport models. Additionally, the lifetime of these systems is a critical factor for their industrial relevance.

4.4 ELECTROCHEMICAL PATHWAY FOR BIOMASS VALORIZATION

Biomass is globally available and one of the largest sources of renewable carbon. It can be derived from diverse sources, from edible crops to algae, crop residues, and bio-wastes. Nowadays, the estimated global supply of dry biomass from agriculture and forestry reaches approximately 11.9 gigatons per year (Gt/y), and there is an additional availability of over 100 Gt/y from industrial, municipal, and food wastes.[41] This total supply of biomass feedstocks provides abundant renewable carbon, surpassing the growing demand for fossil-based fuels and chemicals. Therefore, biomass valorization holds immense potential as a cornerstone for the sustainable production of biofuels, bio-based chemicals, and bioproducts.

The electrochemical biomass conversion holds great potential for carbon- and energy-efficient valorization of biomass feedstocks, which have traditionally been dominated by expensive and toxic redox reagents[42] or carbon-emitting fermentation processes.[43] Additionally, electrochemical valorization can lead to new carbon-reducing strategies and products that are challenging to obtain through conventional thermos-chemical valorization processes.[44]

Recent advances in electrolyzer design, electrocatalyst development, and reaction mechanism investigation reveal the potential of using an electrochemical approach to address the limitations in biomass valorization.[6,45] Electrochemically driven biomass valorization benefits from several advantages: it uses renewable electricity as the driving force, reducing carbon intensity and enabling intermittent energy storage; it operates under ambient or near-ambient reaction conditions, which enhance process efficiency and avoid undesired polymerization reactions; and it allows for tunable oxidation/reduction driving force by tuning the potential without the need for toxic or expensive oxidizing/reducing agents, leading to improved overall process sustainability.

To address the emerging need for sustainable chemical transformations with biomass feedstocks, we highlight three key priorities in electrochemical biomass valorization to meet the requirement for carbon- and energy-efficient biomass transformation: (1) biomass-derived electrochemically valorized products from platform chemicals, (2) enhancing the energy efficiency of electrochemical transformation of biomass into energy-dense fuels, and (3) advancing direct electrochemical biogas valorization and industrial reactor design (Figure 4.3).

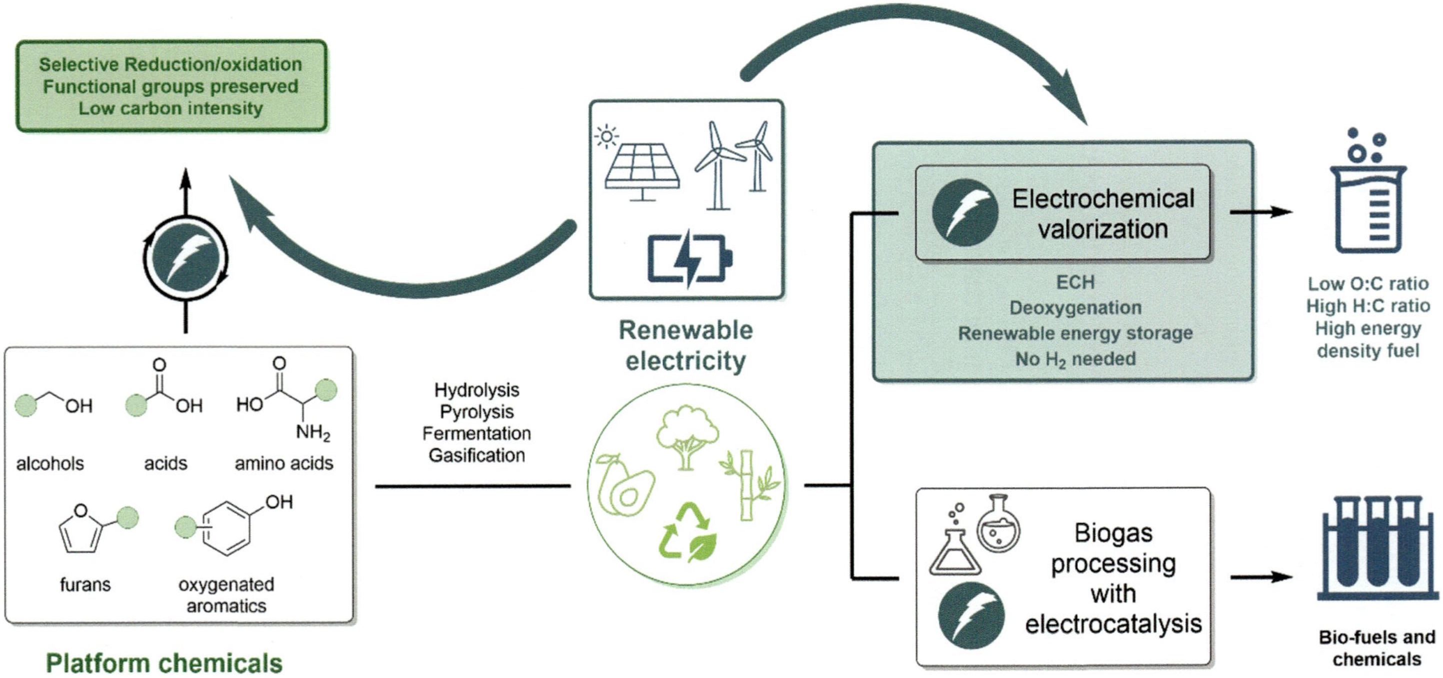

FIGURE 4.3 Electrified biomass valorization pathways.

4.4.1 Biomass-Derived Electrochemically Valorized Products from Platform Chemicals

Biomass platform chemicals[46] are building block molecules derived from various biomass pretreatments including dehydration, fermentation, pyrolysis, and hydrolysis. Unlike fossil-derived chemical feedstocks, these components retain the rich oxygen-containing functional groups in biomass, thereby eliminating the need for capital-intensive oxidation processes downstream.[47] Another advantage of platform chemicals is that preserving the aromaticity of lignocellulosic-derived molecules allows for the direct production of aromatic products, eliminating the need for costly and energy-intensive thermocatalytic processes used in the petroleum industry.[48] Similarly, the preservation of amino groups found in biomass can circumvent energy-intensive downstream amination processes.

In the exploration of potential industrial electrochemical biomass transformations, three categories of feedstocks—bio-alcohols, bio-acids, and furans—are of particular interest as platform chemicals. These are chosen due to their industrial applicability, market size, and versatility in producing various derivatives. Valorization of these platform chemicals offers an alternative approach to either compete with or complement fossil-derived chemical production, while maintaining comparable physical and chemical properties.

Although electrochemical biomass valorization holds significant potential, controlling the electrochemical selectivity of multi-functional groups in biomass feedstocks presents substantial challenges, especially when targeting partial oxidation to yield valuable products rather than overoxidation to CO_2. To address these challenges, research efforts should focus on implementing sequential oxidation-reduction reactions within a single electrochemical reactor, advancing catalyst development and gaining mechanistic insights through theoretical calculations and operando spectroscopy, developing surface engineering techniques for novel and efficient/selective heterogeneous catalysts, and optimizing electrolyzer design to minimize side reactions and direct chemical transformations in the solution phase. This integrated approach can enhance both the efficiency and selectivity of the biomass valorization process.

4.4.2 Improved Energy Efficiency in the Electrochemical Transformation of Biomass to Energy-Dense Fuels

Biomass-derived fuels, or biofuels, serve as low-carbon alternatives to traditional energy sources, with bioethanol and biodiesel already utilized as transportation fuels and additives.[49,50] For biofuels to be economically viable, they must have a high energy density to offset production costs. Additionally, since biomass feedstocks are distributed in nature, biofuels with higher energy density can reduce transportation costs per unit of energy. Compounds such as cellulose and hemicellulose-derived carboxylic acids, like lactic acid and levulinic acid, show promise for conversion into fuel-grade long-chain alkanes and cycloalkanes, which can be used as additives to current fossil fuels.

Energy-dense biofuels require a high hydrogen-to-carbon (H:C) ratio and a low oxygen-to-carbon (O:C) ratio.[51] Consequently, oxygenated biomass feedstocks need an increased H:C ratio and a decreased O:C ratio. To achieve this, thermocatalytic hydrogenation is commonly used to remove the abundant oxygen functional groups or convert highly oxidized hydrocarbons in biomass feedstocks into energy-dense hydrocarbons. This process demands substantial amounts of hydrogen as a reductant and operates under harsh conditions, including high temperatures and high pressures. It is crucial to select biofuel production processes based on the target energy density, specifically the H/C ratio, and to consider the processing energy input (PEI) as a critical factor in the development of industrial electrochemical biomass valorization strategies.

Unlike thermocatalytic hydrogenation processes that rely on H-H cleavage of H_2, electrocatalytic hydrogenation (ECH) can utilize H_2O directly to produce H* or a proton-coupled electron transfer (PCET) from the solution.[52, 53] ECH reactions have been demonstrated for transformation of biomass platform chemicals to fuels[51] and can benefit from better PEI compared with the thermocatalytic approach. Bio-acids, lipids, and oxygenated platform chemicals can also be treated directly with electrochemistry to enable an energy-efficient and modular transformation process. Bio-oil, a liquid mixture of hydrocarbons derived from pyrolysis, can be readily converted into suitable fuel with an appropriate octane number.

This can be achieved via optimally developed ECH strategy that bypasses the separation cost and produces a stream of energy-dense fuel.[54]

Electrochemically derived biofuel production offers several key advantages, including modular operation, the use of water as a proton source rather than pressurized hydrogen, and inherently milder reaction conditions. However, significant technological challenges remain that hinder its commercialization. For instance, using aqueous-based electrolytes as a proton source introduces the HER as a competing side reaction. This necessitates a dedicated catalyst design to suppress HER and enable selective electrochemical hydrogenation. Additionally, it is crucial to design electrochemical systems capable of processing complex biomass mixtures. Industrial applications require higher reaction rates (higher current densities), which in turn demand advancements in both catalyst development and electrolyzer engineering.

4.4.3 Direct Electrochemical Biogas Upgrading and Industrial Electrocatalytic System Design

Biogas, a major product of biomass processing, is primarily produced through the methanation of biomass and organic wastes.[55] It mainly consists of methane, carbon dioxide, water vapor, and other impurities. Biomass gasification followed by Fischer-Tropsch synthesis can produce energy-dense product streams.[56] Apart from direct ECH for fuel production, electrocatalysis can also play a crucial role in the upstream and downstream processes of biogas production. Given that CO_2 is a major component of biogas, it can be directly utilized for the electrochemical reduction of CO_2 to syngas, which can then be integrated into Fischer-Tropsch upgrading.

Strategies such as alloying, catalyst surface modification, and electrolyte tuning have been employed to enhance the electrochemical transformation of biomass. The main challenge in this field is preventing overoxidation and improving product selectivity. Methods like direct hydrogenation/oxidation, cascade reactions, and redox-mediated reactions have been explored to selectively convert biomass into value-added products. Among these, redox-mediated electrooxidation stands out as a particularly promising but underexplored method capable of upgrading larger molecules, which are typically difficult to adsorb onto catalyst surfaces due to their size. Additionally, the development of specialized electrochemical reactors, such as membrane reactors, provides opportunities for challenging reactions, including the cleavage of polymer linkages.

4.5 OXIDATIVE ELECTROCHEMICAL UPGRADING OF ALKENES AND AROMATICS

The hydrogen evolution reaction and electrochemical CO_2 reduction, powered by renewable energy, are promising cathodic reactions extensively studied to decarbonize the production of important fuels and chemicals like H_2, CO, methane, ethylene, and ethanol. These reactions are typically paired with the oxygen evolution reaction (OER) at the anode, which produces oxygen and imposes a high thermodynamic potential (1.23 V vs. RHE) to the full cell voltage and offers little economic benefit. An alternative anodic reaction that could reduce energy consumption and simultaneously produce value-added products is highly desirable. One promising avenue is the selective partial oxidation of hydrocarbons into valuable commodity chemicals typically associated with large carbon footprints, such as ethylene oxide, ethylene glycol, propylene oxide, adipic acid, and terephthalic acid. Currently, these oxygenates are primarily produced by partial oxidation of hydrocarbons through energy-intensive thermal processes at high temperatures and pressures, driven by fossil fuel combustion. The overoxidation of CO_2, due to limited selectivity, further contributes to carbon emissions. Electrifying hydrocarbon oxidation under ambient conditions offers the potential for better control over product selectivity and avoiding overoxidation.

4.5.1 Electro-Oxidation of Ethylene and Propylene

Ethylene oxide and glycol (EO and EG), along with propylene oxide and glycol (PO and PG), are the most demanded chemicals derived from ethylene and propylene globally. These chemicals are essential in the production of plastics, textiles, solvents, and detergents. The current manufacturing processes for these substances contribute significantly to greenhouse gas (GHG) emissions, releasing approximately 2 and 4 tons of CO_2 per ton of EG and PG produced, respectively. These emissions stem both from the combustion of fossil fuels and from overoxidation to CO_2. Electrochemical oxidation of ethylene and propylene presents a significant opportunity to decarbonize the production of epoxide and glycol, offering a cleaner alternative to traditional methods.[57, 58]

In current industrial practices, ethylene and propylene are oxidized through distinct thermal catalytic processes due to their different chemical structures. Ethylene is directly oxidized with oxygen at high temperatures and pressures to produce ethylene oxide, while propylene generally undergoes either the chlorohydrin process or the hydrogen peroxide to propylene oxide (HPPO) process. Direct oxidation of propylene with O_2 typically results in overoxidation and poor selectivity towards propylene oxide. Both ethylene glycol and propylene glycol are subsequently produced through the hydrolysis of EO and PO under acidic conditions or at elevated temperatures. Therefore, this discussion primarily focuses on the production of EO and PO.

4.5.1.1 Redox Mediator Pathway

The redox-mediated pathway is among the earliest and most extensively studied methods for electrocatalytic hydrocarbon oxidation due to their high similarity with the existing industrial thermal methods. By using halide ions (Cl^- or Br^-) and H_2O_2 as mediators, the reaction benefits from high reactivity and controlled oxidation under ambient conditions. The mediated oxidation works through the generation of reactive halogen intermediates or H_2O_2, which then selectively oxidize ethylene or propylene to form EO or PO.

In the halogen-mediated reaction pathway, halide ions are oxidized to halogens, which react with water to form oxidizing agents like hypochlorous acid and hypobromous acid. For example, in chloride-mediated ethylene oxidation, chloride ions electrochemically oxidize to Cl_2, which reacts with water to produce HOCl. The hypochlorous acid then reacts with ethylene to form 2-chloroethanol, subsequently reacting with hydroxide ions generated at the cathode to produce ethylene oxide.[59] While this pathway offers high Faradaic efficiency, it poses challenges due to the generation of chlorinated byproducts and complex downstream separation.

Bromide-mediated partial oxidation follows a similar reaction mechanism to chloride-mediated oxidation but presents some distinct advantages. The oxidation of bromide has a lower thermodynamic potential (1.09 V vs. RHE) compared to chlorine evolution (1.36 V vs. RHE). Additionally, bromide-mediated oxidation exhibits superior selectivity and Faradaic efficiency for PO production. In a study by Zeng et al., bromide-mediated electrolysis also demonstrated a reduction in side reactions compared to chloride, achieving Faradaic efficiency values up to 91% and selectivity near 99.9% in propylene oxidation.[60]

The hydrogen peroxide-mediated pathway utilizes electrochemically produced H_2O_2 from oxygen reduction, making it environmentally friendly since water is the only byproduct. However, electrochemically producing H_2O_2 at high concentrations (>5wt%) is challenging due to the tendency of H_2O_2 to further reduce to water.

4.5.1.2 Direct Alkane Oxidation

Despite the high efficiency and selectivity of redox-mediated methods, they require complex waste handling and product separation processes, which can be both carbon and energy-intensive. This has driven the development of direct oxidation pathways to achieve similar efficiencies with less environmental impact. Direct oxidation methods seek to eliminate the need for redox mediators by oxidizing ethylene or propylene directly on the catalyst surface using water as the oxygen source, thus simplifying the reaction process and reducing waste. In this pathway, the electrocatalyst itself provides active sites for hydrocarbon adsorption and oxygenation, leading to selective formation of EO or EG, and PO or PG.

Pd was discovered as the first effective catalyst material for activating olefin at relatively low potentials. It shows high selectivity for ethylene oxidation to EG. For propylene, it has been discovered that the metallic Pd sites tend to oxidize the methyl group and lead to the formation of the allyl alcohol and acrolein[61] At higher oxidation potential where the surface is fully or partially oxidized, Pd again shows a much higher selectivity toward the π-bond activation and produces PG as the main products, with similar pattern observed in ethylene to EG process.

Metal dopants have proven to be an effective method for enhancing catalytic performance. Au and Rh doping in Pd improve the faradaic efficiency for EG and PG from 50% to over 70%, respectively. These dopants have been demonstrated to tune the electronic structure of Pd and inherently change the adsorption energy of surface hydroxyls and oxygenate intermediates.[62] The modified structure generates a more favorable pathway toward EG and PG compared to that on pristine Pd. These catalysts only have less than 5% doping and Pd still serves as the active site for the reaction. Pt-Pd alloys with a 1:1 ratio currently are the most promising catalysts for the selective oxidation of propylene-to-propylene oxide under mild conditions. The $PdPtO_x$ catalyst uniquely addresses the challenge of balancing selectivity and stability at high current densities, which are essential for industrially relevant applications. This catalyst leverages the combined properties of Pd and Pt, where Pd enhances hydrocarbon adsorption and activation, and Pt contributes to stabilizing oxygen intermediates, thereby suppressing competing OER. The oxide layer ($PdPtO_x$) formed on the catalyst surface plays a crucial role by providing reactive oxygen species specifically for the epoxidation of propylene at the last step of OER[63] Additionally, a mixture of acetonitrile and water as electrolyte can increase the solubility of olefins and suppress the OER. This synergy in the $PdPtO_x$ catalyst and system achieves 66% FE for PO at 50 mA/cm^2.

4.5.2 Electrochemical Oxidation of Organic Substrates

While the field of oxidative electrochemistry has been rapidly expanding to transform high-volume commodity chemicals such as ethylene and propylene, there remains tremendous opportunity in yet under-explored transformations of organic substrates. Table 4.1 lists three examples of highly desirable oxidative transformations where electrochemical approaches may improve productivity and selectivity.

An oxidative transformation that has recently gained attention is the oxidation of KA oil (a mixture of cyclohexanol and cyclohexanone) to adipic acid (AA), a precursor for the production of Nylon 6,6. The current route of adipic acid synthesis requires large amounts of concentrated nitric acid, which decomposes into toxic NOx species.[64] Altogether, this accounts for 5–8% of global N_2O emissions.[65] Instead, the use of nitric acid can be circumvented if the partially oxidized cyclohexane product, KA oil, is electrochemically oxidized to AA. To mitigate the limited solubility of KA oil in water, Liu et al. synthesized a nickel hydroxide catalyst with a hydrophobic sodium dodecyl sulfonate layer.[66] This enabled efficient molecular-level mixing of water and organic substrate at the catalyst surface, which was supported by molecular dynamics simulations. However, achieving high current densities along with high Faradaic efficiencies remained challenging, as in water the oxygen evolution reaction exists as a competing reaction. As nickel hydroxide is also known to be an OER catalyst in alkaline conditions, altering the electrocatalyst surface can tune the surface chemistry to promote selectivity to AA. By doping vanadium into the nickel hydroxide electrocatalyst, Liu et al. demonstrated that the catalyst surface was reconstructed to increase the kinetics of the rate-determining oxidation of nickel hydroxide and

TABLE 4.1
The List of Hydrocarbon Oxidation that Produces Highly Demanded Commodity Chemicals

	Hydrocarbon	Electron Transfer	E^0 (V vs. RHE)	Products
Reported	Methane	$2e^-$	0.63	Methanol
	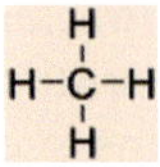			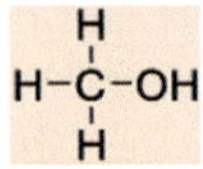
	Ethylene	$2e^-$	0.51	Ethylene glycol/oxide
				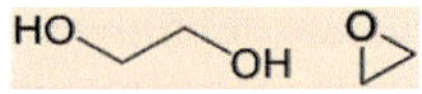
	Propylene	$2e^-$	0.57	Propylene glycol/oxide
				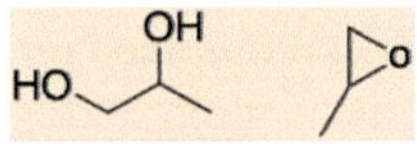
		$4e^-$	0.356	Acrolein
				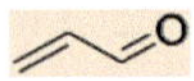
Not yet reported	"KA" oil (cyclohexanone & cyclohexanol)	$6e^-$	1.15	Adipic acid
	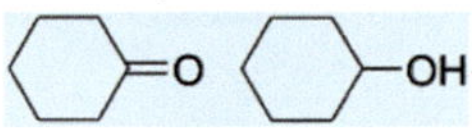			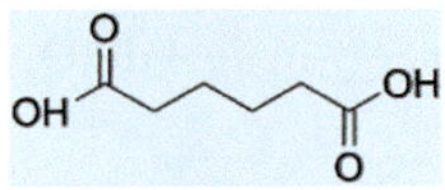
	Benzene	$2e^-$	0.34	Phenol
				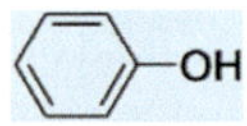
	p-Xylene	$6e^-$	0.36	Terephthalic acid
	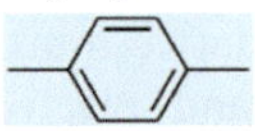			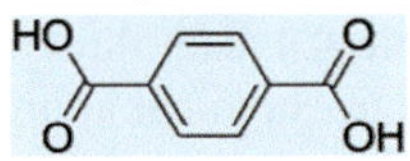

mitigate oxygen evolution.[67] This resulted in an enhanced performance of 82% FE at 300 mA cm^{-2}, parameters at which techno-economic analysis indicated such a technology may be profitable at scale.

Finally, a longstanding goal in the field of oxidation chemistry is the selective activation of C-H bonds, which remains challenging due to high C-H bond stability and competing oxidation pathways. As an important example, the oxidation of p-xylene to terephthalic acid (TPA) is the largest liquid phase oxidative reaction in the chemicals industry. Ding et al. demonstrated a two-step electrosynthesis, whereby p-xylene is first halogenated by anodically evolving bromine.[68] The bromine is hydrolyzed to form a di-alcohol species, which can then be electrochemically oxidized in a second step to give the carboxylic acid TPA. This system highlights the ability of halogenating organic molecules to provide a reactive intermediate to further products, a strategy similarly employed in organic chemistry. Further, the in-situ generation of halogen species significantly decreases safety risks involved with handling large quantities of liquid bromine at elevated temperatures and pressures. However, selectively brominating the side-chain instead of the aromatic ring has required the use of a two-phase system in which the reactant p-xylene was dissolved in chloroform. The produced bromine forms reactive Br_2O intermediates, which partition into the organic phase and form radical species.[69] However, the rate of this reaction is limited. Further, the use of a co-solvent increases system complexity. Alternatively, a direct approach to the oxidation of p-xylene to TPA may be beneficial in reducing the complexity of the reaction scheme. This remains a significant challenge from a catalysis standpoint due to the stability of the pX methyl group. Lv et al. developed a MoO_2/MoP catalyst and attributed high performance to a surface reconstruction that favored the adsorption of partially oxidized TPA intermediates.[70] In an H-cell configuration, 76.9% FE and 94.8% selectivity to TPA was achieved. However, the reaction was limited to low current densities due to the limited solubility of pX in a mixed aqueous-acetonitrile electrolyte. The further development of highly active and selective catalysts, paired with novel reactor architectures employing an aqueous-organic interface, may improve the performance of TPA electrosynthesis toward industrial viability.

4.6 CHALLENGES AND OPPORTUNITIES IN THE INDUSTRIAL APPLICATION OF ELECTROCHEMICAL TECHNOLOGIES

The declining cost of renewable electricity and heightened awareness of climate change are making electrochemical processes increasingly attractive for industrial production, especially for decarbonizing commodity chemicals known for their substantial carbon footprints. Traditionally, chemical manufacturing has relied heavily on petrochemical refineries. Considering global capacity and greenhouse gas emissions, significant decarbonization efforts are needed for commodities such as ethylene, propylene, ammonia, benzene/toluene/xylene (BTX), methanol, terephthalic acid, styrene, and ethylene oxide (Figure 4.4). The electrochemical production of these chemicals opens up promising new avenues for addressing the significant carbon footprints associated with essential industrial chemical transformations. While laboratory-scale research has primarily concentrated on reactions like hydrogen evolution, nitrogen reduction, and CO_2 reduction, exciting opportunities exist to electrify the production of other carbon-intensive chemicals. Yet, there are few examples that demonstrate high selectivity at industrially relevant current densities, particularly for commodities such as benzene/toluene/xylene and terephthalic acid. Substantial efforts are required to advance the electrochemical production of these value-added chemicals through innovative catalyst development and reactor engineering.

There are significant opportunities in organic electrolysis, particularly for processing molecules like biomass or aromatics that exhibit low water solubility. Organic electrosynthesis is typically conducted in non-aqueous solvents such as acetonitrile or dimethylformamide, utilizing costly and hazardous supporting electrolytes like tetrabutylammonium hexafluorophosphate. These systems often face challenges such as high internal resistance due to low conductivity, which leads to high full cell voltage and low energy efficiency. The absence of water in these setups also precludes the use of water as a proton or oxygen source, which is a major benefit in traditional organic electrosynthesis. To address these issues, there is significant potential in developing reactors that leverage the aqueous-organic interface at the surface of a heterogeneous catalyst, such as through the introduction of a liquid

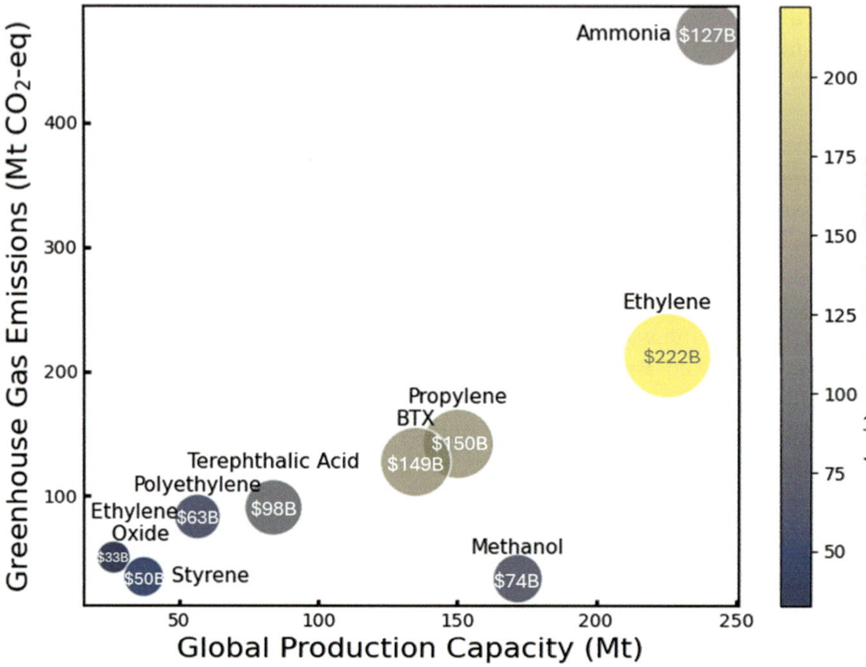

FIGURE 4.4 Greenhouse gas emissions from the production of commodity chemicals and associated global production capacities and market values.

(Global production capacity and market value data from Statista. Greenhouse gas emissions were estimated from the product of production capacity and emissions of each chemical from the US DOE GREET lifecycle model.)

diffusion electrode. At this interface, reactive species from both phases can adsorb onto the catalyst surface to participate in the electrochemical reaction—overcoming the solubility limitations of organic substrates in aqueous electrolytes—but also enable operations at higher current densities, enhancing overall efficiency.

Moreover, the integration of thermochemical and electrochemical methodologies, a synergistic approach, holds substantial promise in overcoming challenges that electrochemistry alone may not be able to address. By leveraging the unique advantages of each method—thermocatalysis' ability to activate difficult reactions at high temperatures and electrocatalysis' precision in driving reactions under mild conditions—a more efficient, selective, and sustainable pathway for challenging reaction such as C-H bonds activation and N_2 reduction could be realized. Additionally, coupling electrochemical and thermochemical technologies can broaden the spectrum of producible chemicals. For example, electrochemical reactions can generate building blocks such as syngas, which can then be upgraded via thermocatalytic processes like the Fischer-Tropsch synthesis to produce diesel and jet fuels (C_{12}-C_{16} hydrocarbons) directly from CO_2 captured from the air, offering a sustainable pathway for the production of carbon-neutral fuels.

It's crucial to transition electrochemical technologies from lab-scale experiments to industrial applications to meet the pressing challenges of energy and sustainability.

Although scaling up any new technology presents challenges, the successful implementation of large-volume electrochemical processes—such as those used for the production of chlorine, alkali, aluminum, and adiponitrile—demonstrates the inherent scalability of electrochemistry. Currently, only a few emerging electrochemical technologies have reached high Technology Readiness Levels (TRLs), signifying their maturity for full commercialization. For example, water electrolysis, including both alkaline and proton exchange membrane (PEM)-based water electrolyzer, has achieved TRL 9. In contrast, low-temperature CO_2 electroreduction to carbon monoxide is at TRL 5–6, and its reduction to more complex products stands at TRL 2–4. Likewise, electrochemical nitrogen reduction and electrochemical biomass conversion are in the early stages, with TRLs ranging from 2 to 4. This disparity underscores the need for focused development efforts to enhance the readiness of lower-TRL technologies for sustainable energy solutions.

Despite considerable advances, electrolysis technology still faces significant challenges that hinder its industrial application. Four key metrics determine the economic feasibility of electrochemical processes: Faradaic efficiency, cell voltage, single-pass conversion, and operating current density. Faradaic efficiency is crucial as it indicates the proportion of electrical charge that results in the desired product. Cell voltage directly affects energy consumption and operational costs, making it a critical parameter for optimization. Single-pass conversion influences the costs associated with the compression and recycling of unreacted reactants, requiring optimizations to minimize material handling. Additionally, achieving industrial-level current densities above 100 mA cm^{-2} remains a significant hurdle, as many electrochemical technologies have yet to meet this standard. Enhancing these performance metrics is essential to transition electrolysis from laboratory research to economically viable, large-scale industrial applications.

Stability and lifetime present another significant challenge. Taking CO_2 electroreduction as an example, the current designs typically employ a hydrophobic gas diffusion electrode, adapted from water electrolyzer technologies, which facilitates reactions at the gas-catalyst-liquid triple-phase boundary. However, the issue of flooding significantly undermines the lifetime of these electrodes as the porous diffusion layer becomes saturated with water, often causing the competing hydrogen evolution reaction to dominate. Transitioning to a membrane-electrode assembly design has effectively mitigated flooding by having gas diffusion electrodes directly contact the ion-exchange membrane rather than the liquid electrolyte. Despite this improvement, prolonged electrolysis still results in water crossover from the anode to the cathode, and the electrochemical reduction process creates high-alkaline conditions at the surface of the gas diffusion electrode. Since these electrodes are predominantly made of carbon materials, they are vulnerable to degradation by OH ions that attack the carbon surface, reducing its hydrophobicity during electrolysis. As a result, the longest continuous electrolysis reported to date lasted for 5,200 hours, but for industrial applications, the target lifetime is typically 60,000–80,000 hours, or 7–10 years.

Product separation in electrochemical processes also requires enhanced attention, particularly due to the necessity of using supporting electrolytes to facilitate ion transfer. Commonly used salt solutions like KOH, K_2SO_4, or H_2SO_4 complicate the separation process when the target product is highly soluble in water, resulting in a complex mixture of salt and water. The removal of salt often requires expensive dialysis processes, and the typically low concentration of the target product further necessitates costly separation to remove water. In addition to focusing on electrolyzer performance metrics such as Faradaic efficiency and cell voltage, it is crucial to consider the entire process, including downstream separation. A promising future direction involves leveraging ion-exchange membranes to eliminate the need for salts, operating in pure-water conditions, and achieving higher product concentrations. This approach could streamline the separation process and reduce operational costs.

The advancement of electrochemical technology requires integrating Techno-Economic Analysis (TEA) and Life Cycle Assessment (LCA) studies. For electrochemical manufacturing to gain traction in existing markets, it requires policy-driven economic incentives alongside technological advancements. A comprehensive TEA is crucial as it assesses economic viability and offers strategic guidance for prioritizing system development. Additionally, detailed LCAs are critical for evaluating the global warming impacts of alternative electrochemical manufacturing routes from cradle to gate. Performing quantitative analyses of emerging electrochemical technologies is essential to assess potential CO_2 emission reductions from a lifecycle perspective, identify the most effective electrification strategies, and evaluate their environmental impacts relative to traditional industry practices.

Finally, electrifying chemical production also depends on the development of renewable energy. Globally, there are 615 crude oil refineries, each with an average capacity of 150,000 barrels per day (bpd). To fully electrify these refineries using CO_2 captured from the air and subsequently reduced to syngas via a CO_2 electrolyzer, or converted thermocatalytically to syngas using hydrogen produced from a water electrolyzer, massive scale-ups are required: about 1,600 times for direct air capture technology, 700 times for CO_2 electrolysis, and 300 times for water electrolysis.[71] Additionally, powering an average-sized refinery would require about 5.5 GW of renewable electricity, totaling approximately 3,400 GW for all global refineries. For comparison, by 2023, the global renewable energy capacity stood at 2,080 GW, which is projected to increase to 5,520 GW by 2030.[72] This suggests that the energy required to electrify chemical manufacturing alone exceeds the current global annual renewable electricity generation. However, given the rapid advancement of renewable energy technologies, it is plausible that renewable electricity will eventually meet the energy demands of electrochemical manufacturing. Furthermore, the growth of electrification in the chemical industry could in turn stimulate the expansion of renewable energy capacity.

4.7 SUMMARY

In this chapter, we discuss the state-of-the-art advancements in electrochemical technologies, focusing on their potential to reduce industrial carbon emissions and advance sustainable manufacturing. We cover four key areas: the electrochemical reduction of CO_2 to chemicals and fuels, the transformation of N_2 into fertilizers, biomass valorization, and the oxidative upgrading of alkenes and aromatics. Each section underscores the potential of electrochemical technologies to convert greenhouse gases and non-fossil feedstock into valuable chemicals and fuels, thereby contributing to a greener industrial landscape. The chapter also discusses the key challenges and opportunities within the industrial application of these technologies, particularly the hurdles in scaling processes from lab to industrial scales. It highlights the need for innovative catalysts, reactor designs, and integrated systems to overcome these obstacles. Furthermore, we explore the opportunities electrochemical technologies offer in reducing CO_2 emissions in the production of commodity chemicals with substantial carbon footprints. Through detailed analysis, this chapter aims to illuminate how electrochemical technologies can revolutionize the industrial sector, propelling it toward a sustainable and economically viable future.

REFERENCES

1. Inc. SS Average monthly carbon dioxide (CO_2) levels in the atmosphere worldwide from 1990 to 2024. 2024. [cited] Available from: https://www.statista.com/statistics/1091999/atmospheric-concentration-of-co2-historic/
2. IRENA Renewable power generation costs in 2023. 2024 [cited] Available from: https://www.irena.org/-/media/Files/IRENA/Agency/Publication/2024/Sep/IRENA_Renewable_power_generation_costs_in_2023.pdf
3. Peplow M. A new kind of solar cell is coming: Is it the future of green energy? *Nature* 2023, **623**(7989): 902–905.
4. Xia R, Overa S, Jiao F. Emerging electrochemical processes to decarbonize the chemical industry. *JACS Au* 2022, **2**(5): 1054–1070.
5. Hepburn C, Adlen E, Beddington J, Carter EA, Fuss S, Mac Dowell N, et al. The technological and economic prospects for CO_2 utilization and removal. *Nature* 2019, **575**(7781): 87–97.
6. De Luna P, Hahn C, Higgins D, Jaffer SA, Jaramillo TF, Sargent EH. What would it take for renewably powered electrosynthesis to displace petrochemical processes? *Science* 2019, **364**(6438): eaav3506.
7. Shin H, Hansen KU, Jiao F. Techno-economic assessment of low-temperature carbon dioxide electrolysis. *Nature Sustainability* 2021, **4**(10): 911–919.
8. Jordaan SM, Wang C. Electrocatalytic conversion of carbon dioxide for the Paris goals. *Nature Catalysis* 2021, **4**(11): 915–920.
9. Nitopi S, Bertheussen E, Scott SB, Liu X, Engstfeld AK, Horch S, et al. Progress and perspectives of electrochemical CO_2 reduction on copper in aqueous electrolyte. *Chemical Reviews* 2019, **119**(12): 7610–7672.
10. Wang Y, Liu J, Zheng G. Designing copper-based catalysts for efficient carbon dioxide electroreduction. *Advanced Materials* 2021, **33**(46): e2005798.
11. Ross MB, De Luna P, Li Y, Dinh C-T, Kim D, Yang P, et al. Designing materials for electrochemical carbon dioxide recycling. *Nature Catalysis* 2019, **2**(8): 648–658.
12. Birdja YY, Pérez-Gallent E, Figueiredo MC, Göttle AJ, Calle-Vallejo F, Koper MTM. Advances and challenges in understanding the electrocatalytic conversion of carbon dioxide to fuels. *Nature Energy* 2019, **4**(9): 732–745.
13. Li L, Li X, Sun Y, Xie Y. Rational design of electrocatalytic carbon dioxide reduction for a zero-carbon network. *Chemical Society Reviews* 2022, **51**(4): 1234–1252.
14. Jin S, Hao Z, Zhang K, Yan Z, Chen J. Advances and challenges for the electrochemical reduction of CO_2 to CO: From fundamentals to industrialization. *Angewandte Chemie, International Edition in English* 2021, **60**(38): 20627–20648.
15. Nankya R, Elgazzar A, Zhu P, Chen F-Y, Wang H. Catalyst design and reactor engineering for electrochemical CO_2 reduction to formate and formic acid. *Materials Today* 2024, **76**: 94–109.
16. Ma W, He X, Wang W, Xie S, Zhang Q, Wang Y. Electrocatalytic reduction of CO_2 and CO to multi-carbon compounds over Cu-based catalysts. *Chemical Society Reviews* 2021, **50**(23): 12897–12914.

17. Chen Y, Miao RK, Yu C, Sinton D, Xie K, Sargent EH. Catalyst design for electrochemical CO_2 reduction to ethylene. *Matter* 2024, **7**(1): 25–37.
18. Xia R, Lv J-J, Ma X, Jiao F. Enhanced multi-carbon selectivity via CO electroreduction approach. *Journal of Catalysis* 2021, **398**: 185–191.
19. Huang JE, Li F, Ozden A, Sedighian Rasouli A, García de Arquer FP, Liu S, et al. CO_2 electrolysis to multicarbon products in strong acid. *Science* 2021, **372**(6546): 1074–1078.
20. Ozden A, Wang Y, Li F, Luo M, Sisler J, Thevenon A, et al. Cascade CO_2 electroreduction enables efficient carbonate-free production of ethylene. *Joule* 2021, 5, 706–719.
21. Hauch A, Kungas R, Blennow P, Hansen AB, Hansen JB, Mathiesen BV, et al. Recent advances in solid oxide cell technology for electrolysis. *Science* 2020, **370**(6513).
22. Lee G, Li YC, Kim J-Y, Peng T, Nam D-H, Sedighian Rasouli A, et al. Electrochemical upgrade of CO_2 from amine capture solution. *Nature Energy* 2020, **6**(1): 46–53.
23. Sullivan I, Goryachev A, Digdaya IA, Li X, Atwater HA, Vermaas DA, et al. Coupling electrochemical CO_2 conversion with CO_2 capture. *Nature Catalysis* 2021, **4**(11): 952–958.
24. Chen JG, Crooks RM, Seefeldt LC, Bren KL, Bullock RM, Darensbourg MY, et al. Beyond fossil fuel–driven nitrogen transformations. *Science* 2018, **360**(6391).
25. Chen Y, Liu H, Ha N, Licht S, Gu S, Li W. Revealing nitrogen-containing species in commercial catalysts used for ammonia electrosynthesis. *Nature Catalysis* 2020, **3**(12): 1055–1061.
26. Fichter F, Girard P, Erlenmeyer H. Elektrolytische Bindung von komprimiertem Stickstoff bei gewöhnlicher Temperatur. *Helvetica Chimica Acta* 1930, **13**(6): 1228–1236.
27. Tsuneto A, Kudo A, Sakata T. Efficient electrochemical reduction of N_2 to NH_3 catalyzed by lithium. *Chemistry Letters* 1993(5): 851–854.
28. McEnaney JM, Singh AR, Schwalbe JA, Kibsgaard J, Lin JC, Cargnello M, et al. Ammonia synthesis from N_2 and H_2O using a lithium cycling electrification strategy at atmospheric pressure. *Energy & Environmental Science* 2017, **10**(7): 1621–1630.
29. Chang W, Jain A, Rezaie F, Manthiram K. Lithium-mediated nitrogen reduction to ammonia via the catalytic solid–electrolyte interphase. *Nature Catalysis* 2024, **7**(3): 231–241.
30. Li S, Zhou Y, Li K, Saccoccio M, Sažinas R, Andersen SZ, et al. Electrosynthesis of ammonia with high selectivity and high rates via engineering of the solid-electrolyte interphase. *Joule* 2022, **6**(9): 2083–2101.
31. Li K, Andersen SZ, Statt MJ, Saccoccio M, Bukas VJ, Krempl K, et al. Enhancement of lithium-mediated ammonia synthesis by addition of oxygen. *Science* 2021, **374**(6575): 1593–1597.
32. Lazouski N, Chung M, Williams K, Gala ML, Manthiram K. Non-aqueous gas diffusion electrodes for rapid ammonia synthesis from nitrogen and water-splitting-derived hydrogen. *Nature Catalysis* 2020, **3**(5): 463–469.
33. Suryanto BH, Matuszek K, Choi J, Hodgetts RY, Du H-L, Bakker JM, et al. Nitrogen reduction to ammonia at high efficiency and rates based on a phosphonium proton shuttle. *Science* 2021, **372**(6547): 1187–1191.
34. Fu X, Pedersen JB, Zhou Y, Saccoccio M, Li S, Sažinas R, et al. Continuous-flow electrosynthesis of ammonia by nitrogen reduction and hydrogen oxidation. *Science* 2023, **379**(6633): 707–712.
35. Kyriakou V, Garagounis I, Vourros A, Vasileiou E, Stoukides M. An electrochemical Haber-Bosch process. *Joule* 2020, **4**(1): 142–158.
36. Rouwenhorst KHR, Kim H-H, Lefferts L. Vibrationally excited activation of N_2 in plasma-enhanced catalytic ammonia synthesis: A kinetic analysis. *ACS Sustainable Chemistry & Engineering* 2019, **7**(20): 17515–17522.

37. Kumari S, Pishgar S, Schwarting ME, Paxton WF, Spurgeon JM. Synergistic plasma-assisted electrochemical reduction of nitrogen to ammonia. *Chemical Communications* 2018, **54**(95): 13347–13350.
38. Li L, Tang C, Cui X, Zheng Y, Wang X, Xu H, et al. Efficient nitrogen fixation to ammonia through integration of plasma oxidation with electrocatalytic reduction. *Angewandte Chemie International Edition* 2021, **60**(25): 14131–14137.
39. Liang WP, Zhang X-M, Bai P-W, Zhang Z, Chen J-H, Liu W, et al. Cascade N_2 reduction process with DBD plasma oxidation and electrocatalytic reduction for continuous ammonia synthesis. *Environmental Science & Technology* 2023, **57**(39): 14558–14568.
40. Soloveichik G Future of ammonia production: Improvement of Haber-Bosch process or electrochemical synthesis. *AIChE annu. Meet. Top. Conf. Minneapolis: NH_3 Energy*. 2017.
41. Energy USDo 2016 Billion-ton Report: Advancing Domestic Resources for a Thriving Bioeconomy. 2016 [cited 2023 May] Available from: https://www.energy.gov/sites/prod/files/2.16/12/f34/2.16_billion_ton_report_12.2.16_0.pdf
42. Meyer TH, Choi I, Tian C, Ackermann L. Powering the future: How can electrochemistry make a difference in organic synthesis? *Chem* 2020, **6**(10): 2484–2496.
43. Tuck CO, Pérez E, Horváth IT, Sheldon RA, Poliakoff M. Valorization of biomass: Deriving more value from waste. *Science* 2012, **337**(6095): 695–699.
44. Lucas FWS, Grim RG, Tacey SA, Downes CA, Hasse J, Roman AM, et al. Electrochemical routes for the valorization of biomass-derived feedstocks: From chemistry to application. *ACS Energy Letters* 2021, **6**(4): 1205–1270.
45. Seh ZW, Kibsgaard J, Dickens CF, Chorkendorff I, Nørskov JK, Jaramillo TF. Combining theory and experiment in electrocatalysis: Insights into materials design. *Science* 2017, **355**(6321): eaad4998.
46. Bozell JJ, Petersen GR. Technology development for the production of biobased products from biorefinery carbohydrates—the US Department of Energy's "Top 10" revisited. *Green Chemistry* 2010, **12**(4): 539–554.
47. Zhou H, Li Z, Ma L, Duan H. Electrocatalytic oxidative upgrading of biomass platform chemicals: From the aspect of reaction mechanism. *Chemical Communications* 2022, **58**(7): 897–907.
48. Rahimpour MR, Jafari M, Iranshahi D. Progress in catalytic naphtha reforming process: A review. *Applied Energy* 2013, **109**: 79–93.
49. Shahid EM, Jamal Y. Production of biodiesel: A technical review. *Renewable and Sustainable Energy Reviews* 2011, **15**(9): 4732–4745.
50. Huber GW, Iborra S, Corma A. Synthesis of transportation fuels from biomass: Chemistry, catalysts, and engineering. *Chemical Reviews* 2006, **106**(9): 4044–4098.
51. Akhade SA, Singh N, Gutiérrez OY, Lopez-Ruiz J, Wang H, Holladay JD, et al. Electrocatalytic hydrogenation of biomass-derived organics: A review. *Chemical Reviews* 2020, **120**(20): 11370–11419.
52. Ma J, Wang Z, Majima T, Zhao G. Role of Ni in PtNi alloy for modulating the proton–electron transfer of electrocatalytic hydrogenation revealed by the in situ Raman–rotating disk electrode method. *ACS Catalysis* 2022, **12**(22): 14062–14071.
53. Tyburski R, Liu T, Glover SD, Hammarström L. Proton-coupled electron transfer guidelines, fair and square. *Journal of the American Chemical Society* 2021, **143**(2): 560–576.
54. Lam CH, Das S, Erickson NC, Hyzer CD, Garedew M, Anderson JE, et al. Towards sustainable hydrocarbon fuels with biomass fast pyrolysis oil and electrocatalytic upgrading. *Sustainable Energy & Fuels* 2017, **1**(2): 258–266.
55. Awe OW, Zhao Y, Nzihou A, Minh DP, Lyczko N. A review of biogas utilisation, purification and upgrading technologies. *Waste and Biomass Valorization* 2017, **8**(2): 267–283.

56. Zhang L, Rao TU, Wang J, Ren D, Sirisommboonchai S, Choi C, et al. A review of thermal catalytic and electrochemical hydrogenation approaches for converting biomass-derived compounds to high-value chemicals and fuels. *Fuel Processing Technology* 2022, **226**: 107097.
57. Bartolini N, Casasso A, Bianco C, Sethi R. Environmental and economic impact of the antifreeze agents in geothermal heat exchangers. *Energies* 2020, **13**(21): 5653.
58. Nijhuis TA, Makkee M, Moulijn JA, Weckhuysen BM. The production of propene oxide: Catalytic processes and recent developments. *Industrial & Engineering Chemistry Research* 2006, **45**(10): 3447–3459.
59. Leow WR, Lum Y, Ozden A, Wang Y, Nam D-H, Chen B, et al. Chloride-mediated selective electrosynthesis of ethylene and propylene oxides at high current density. *Science* 2020, **368**(6496): 1228–1233.
60. Chi M, Ke J, Liu Y, Wei M, Li H, Zhao J, et al. Spatial decoupling of bromide-mediated process boosts propylene oxide electrosynthesis. *Nature Communications* 2024, **15**(1): 3646.
61. Winiwarter A, Silvioli L, Scott SB, Enemark-Rasmussen K, Sariç M, Trimarco DB, et al. Towards an atomistic understanding of electrocatalytic partial hydrocarbon oxidation: Propene on palladium. *Energy & Environmental Science* 2019, **12**(3): 1055–1067.
62. Lum Y, Huang JE, Wang Z, Luo M, Nam D-H, Leow WR, et al. Tuning OH binding energy enables selective electrochemical oxidation of ethylene to ethylene glycol. *Nature Catalysis* 2020, **3**(1): 14–22.
63. Chung M, Maalouf JH, Adams JS, Jiang C, Román-Leshkov Y, Manthiram K. Direct propylene epoxidation via water activation over Pd-Pt electrocatalysts. *Science* 2024, **383**(6678): 49–55.
64. Yang J, Liu J, Ge Y, Huang W, Neumann H, Jackstell R, et al. Direct and selective synthesis of adipic and other dicarboxylic acids by palladium-catalyzed carbonylation of allylic alcohols. *Angewandte Chemie International Edition* 2020, **59**(46): 20394–20398.
65. Sato K, Aoki M, Noyori R. A "green" route to adipic acid: Direct oxidation of cyclohexenes with 30 percent hydrogen peroxide. *Science* 1998, **281**(5383): 1646–1647.
66. Li Z, Li X, Zhou H, Xu Y, Xu S-M, Ren Y, et al. Electrocatalytic synthesis of adipic acid coupled with H_2 production enhanced by a ligand modification strategy. *Nature Communications* 2022, **13**(1): 5009.
67. Liu X, Zhu Y-Q, Li J, Wang Y, Shi Q, Li A-Z, et al. Electrosynthesis of adipic acid with high faradaic efficiency within a wide potential window. *Nature Communications* 2024, **15**(1): 7685.
68. Ding H, Orazov M, Oliveira N, Yan Y. A benign synthesis route to terephthalic acid via two-step electrochemical oxidation of P-xylene. *Journal of the Electrochemical Society* 2024, **171**(5): 053510.
69. Thasan R, Kumarasamy K. Electrochemical bromination and oxidation of alkyl aromatic compounds by two-phase electrolysis. *Korean Journal of Chemical Engineering* 2014, **31**(3): 365–373.
70. Lv Y, Peng M, Yang W, Liu M, Kong A, Fu Y, et al. Surface reconstruction enabling MoO_2/MoP hybrid for efficient electrocatalytic oxidation of p-xylene to terephthalic acid. *Applied Catalysis B: Environmental* 2024, **340**: 123229.
71. Vogt ETC, Weckhuysen BM The refinery of the future. *Nature* 2024, **629**(8011): 295–306.
72. iEA I. Renewables 2024. 2024 [cited] Available from: https://www.iea.org/reports/renewables-2024

5 Microwave Heating for Energy-Efficient Chemical Production

Xinkang Peng, Xiaohua Lu, and Tuo Ji

5.1 INTRODUCTION

Microwave energy is a natural phenomenon and has existed ever since the beginning of the universe. Unlike other natural things, microwaves cannot be detected by man's eyes if not using an instrument. Its discovery traces back to Hans Christian Oersted's wire and compass experiment in 1819. The development of microwave technology advanced significantly during World War 2 with radar. This led to the invention of microwave ovens. Throughout history, many scientists and engineers contributed to understanding and applying microwave energy for heating. Nowadays, microwave energy has been considered a powerful and non-conventional energy source for performing materials synthesis, organic synthesis, fuel engineering, biomass conversion, etc. [1–5]. Since the first publication using microwave heating, correlational studies in this new technique have grown exponentially and have been extensively reviewed. The increasing studies reveal microwave heating brings rapid reaction rate, cleaner products, and mild reaction conditions for chemistry [6–8].

Microwave heating is also demonstrated to be an energy-efficient approach in chemical engineering for the production of chemicals. Figure 5.1 illustrates the characteristics, advantages, and effects of microwave heating in chemical engineering. Briefly, it offers several advantages such as rapid heating, selective heating, and the ability to achieve efficient chemical reactions [9, 10]. This technique has shown great potential in various chemical processes, allowing for more sustainable and efficient manufacturing of a wide range of chemicals. Since a large number of chemical engineering processes involve heterogeneous reactions, microwave can selectively heat the target material. This selective heating can significantly expedite the interfacial reaction. It can make the reaction at the interface occur more rapidly and efficiently. At the same time, it can also improve the mass transfer process. These characteristics of microwave would have a profound impact on and effectively enhance the interfacial reaction process. It would bring about local hot-spot occurrence and the development of chemical reactions in various chemical engineering processes.

This chapter mainly focuses on presenting state-of-the-art research related to microwave heating for energy-efficient chemical production. It delves into detailing how microwave technology operated and its importance within these reactions.

DOI: 10.1201/9781003655442-7

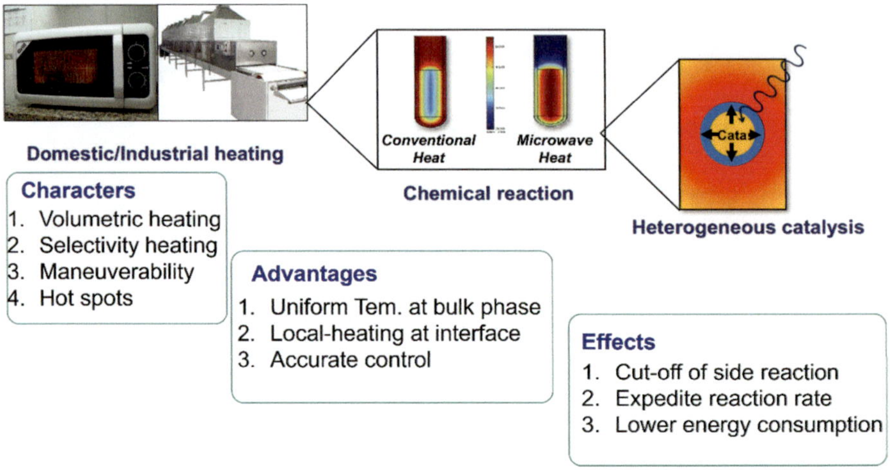

FIGURE 5.1 The characters, advantages, and effects of microwave heating for energy-efficient chemical production.

The fundamental mechanism of microwave heating is thoroughly illustrated and explained to provide a clear understanding of how it functions and exerts its influence in the reaction process. The principle and evaluation method of microwave-heating reaction were discussed, exploring various aspects and considerations. At the same time, the strategies for efficiency enhancement in different reactions were also discussed in depth. The chapter aims to provide a comprehensive overview and understanding of the subject matter, offering valuable insights and knowledge in the field of microwave-assisted reaction and their applications in heterogeneous catalyst reactions.

5.2 THE PRINCIPLE OF MICROWAVE HEATING

The range of microwaves lies within the frequencies from 300 to 0.3 GHz (wavelengths from 1 mm to 1 m), situated between infrared and radio frequencies. Generally, radar transmissions employ frequencies between 1 and 30 GHz, while telecommunications utilize the remaining frequencies. In the event of interference with telecommunication devices, household and industrial microwave ovens operate at either 915 or 2450 MHz, which are set aside by the Federal Communications Commission (FCC) [11]. Notably, the majority of common domestic microwave ovens and lab-scale microwave reactors commonly use a frequency of 2.45 GHz. The selection of such a frequency can likely be attributed to factors like suitable transmission depth and an appropriate heating rate.

Conventional heating is a conduction and convection procedure, through which the heat is transferred from an external heat source like a heating plate, steam, or resistive heater toward the surface of the material. Consequently, the heating rate is dependent on the thermal conductivity of the material and convection currents [9, 12]. In contrast, microwave heating is not merely heating but rather a form of energy conversion process, where electromagnetic energy is transformed into thermal energy. Microwaves can penetrate materials and deposit energy directly within them. This volumetric heating is a fundamental difference from a conduction process. As a result, microwave energy directly acts on the polar molecules of the entire reaction mixture, causing a rapid increase in temperature [13].

Microwave heating depends on the capacity of materials to absorb and convert microwave energy into heat. According to the diverse interactions of microwave with materials, they can be classified into three categories: conductors (such as metals and graphite), insulators (like quartz glass, porcelain, and ceramics), and dielectrics (e.g., water, carbon, Fe_3O_4), as shown in Figure 5.2. When microwaves irradiate a conductor, the majority of microwaves are reflected from its surface. In contrast, most microwaves can penetrate through the insulator without significant losses. Both these types of materials cannot effectively generate heat. When microwaves are absorbed or lost during penetration and generate heat (a process known as dielectric heating), these materials are referred to as dielectric materials (dielectrics). Dielectrics have two crucial properties: (1) the materials have very few free-charge carriers and (2) they exhibit a dipole movement [14]. A dipole moment is defined as two equal and opposite charges separated by a finite distance, which can either naturally or temporarily exist in dielectrics [15]. In the presence of an external electric field, non-polar molecules can be induced to form a temporary dipole moment. Whether they are natural or temporary dipoles, they are sensitive to the external electric field and attempt to align themselves through rotation. Due to the high frequency of microwaves (i.e., 2.45 GHz), this rotation occurs 2.45×10^9 times per second, which then leads to friction within dielectrics and the subsequent dissipation of energy as heat. Both the rotated molecules and the oscillatory migration of ions in the system can generate heat under microwave radiation.

According to Maxwell's equations, microwave is a kind of electromagnetic wave that contains two mutually perpendicular components of electric field and magnetic field, and the electric field and the magnetic field continuously change and are interrelated in space and time. The permittivity and permeability of dielectric materials affect their interaction with the electric field and the magnetic field and thus may absorb microwaves. The permittivity determines the degree of polarization of the material in the electric field, while the permeability reflects the response-ability to the magnetic field. When microwaves interact with dielectric materials, these properties will have an impact on the propagation, absorption, etc. The complex permittivity can be described by equation (5.1):

$$\varepsilon = \varepsilon' - j \cdot \varepsilon'' \tag{5.1}$$

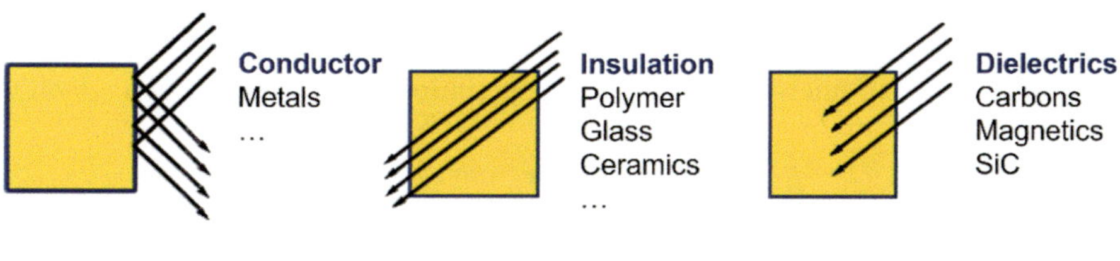

FIGURE 5.2 Interaction of microwave with three kinds of materials.

where the real part of the permittivity (ε') indicates the ability to store electric energy, while the imaginary part, known as the dielectric loss (ε''), represents the ability to convert electric energy. The ability of dielectrics to convert microwaves into heat energy is calculated using

$$\tan\delta = \varepsilon'' / \varepsilon' \tag{5.2}$$

where δ is the dissipation factor of dielectrics. A large value of the dissipation factor δ implies a high efficiency in generating heat from microwaves [16]. It is notable that the dielectric properties of dielectrics also depend on the system temperature, microwave frequency, and the properties of the dielectrics (purity, chemical state, structure) [17]. Generally, dielectrics can be classified into two categories: dielectric capacitor materials and dielectric loss materials. The former with relatively high permittivity (ε') is often used as capacitor dielectric materials, while dielectrics with a high dissipation factor (δ) are utilized as microwave absorption materials, similar to carbon-based materials, conducting polymers, semiconductors, and their composites. The amount of power that can be absorbed by a substance or the conversion of microwave energy to heat is expressed by the following equations [18]:

$$P = \sigma E^2 \tag{5.3}$$

$$\sigma = 2\pi\varepsilon_0\varepsilon'' f \tag{5.4}$$

where P is the power absorbed (W/m^3), E is the electric field intensity, σ is the dielectric conductivity, f is the frequency of the energy source (Hz), and ε_0 is the permittivity of vacuum or free space (8.85×10^{-12} F/m). In case the materials are influenced by oscillated magnetic fields, the magnetic loss can be described similarly

$$\mu = \mu' - j \cdot \mu'' \tag{5.5}$$

where the real part of permeability (μ') represents the ability to store magnetic energy, and the imaginary part (μ'') represents the ability to convert magnetic energy. Therefore, the dissipated power, which can be converted into thermal energy, can be expressed as [19]

$$Q = 2\pi\varepsilon_0\varepsilon'' fE^2 + 2\pi\mu_0\mu'' fH^2 \tag{5.6}$$

where H is the magnetic field intensity, and μ_0 is the permeability of vacuum or free space ($\mu_0 = 4\pi \times 10^{-7}$ H/m).

Another effect of microwave heating is penetration depth, which is given by [20]

$$D_p = \frac{1}{2\pi f}\left(\frac{2}{\mu'\mu_0\varepsilon_0\varepsilon'\kappa'}\right)^{1/2}\left[\left(1+\tan^2\delta\right)^{\frac{1}{2}} - 1\right]^{-1/2} \tag{5.7}$$

where the penetration depth at microwave frequencies is on the order of meters, for samples with a penetration depth greater than the system dimension, the heating effect may be disregarded. But when the penetration depth is smaller than the system dimension, the penetration of microwave energy is restricted, resulting in non-uniform heating. Understanding how microwaves interact with materials and the unique way they can deposit energy directly within substances is crucial for optimizing processes that utilize microwave heating. Consequently, an ideal material for microwave-heating reactions should minimize reflection and maximize absorption toward microwave irradiation, as seen in Figure 5.3.

Microwave heating shows great appeal in various catalytic reactions such as oxidation, hydrogenation, pyrolysis, dehydration, esterification, etc. [21–26]. It offers higher yields, mild reaction conditions, and shorter reaction times for these reactions, and can even initiate reactions that scarcely occur with conventional heating. This prompts scientists to explore mechanisms that go beyond just rapid heating [27]. According to past decades of literature, so-called non-thermal effects (microwave effect) are hypothesized to explain microwave-assisted reactions [9]. When in a highly polarizing field like microwaves, it may affect the mobility and diffusion of reactants and products and increase the likelihood of effective contacts. However, the existence of these effects remains controversial, and most current studies tend to explain the improvements through thermal effect approaches. For example, rapid microwave heating can effectively reduce temperature fluctuations during the reaction, especially when part of the reactants are added dropwise [28].

Heterogeneous catalytic reactions under microwaves have been paid more attention recently [29, 30]. In this scenario, microwaves can selectively heat the reaction interface, further accelerating the reaction rate and enhancing mass transfer. For instance, the temperature of the catalyst reached 140 °C within 2 minutes and was higher than that of the toluene phase [31]. In the case of 1-octanal oxidation, such selective heating promoted the yield to reach 67% in 5 minutes, which is higher than the yield in conventional heating.

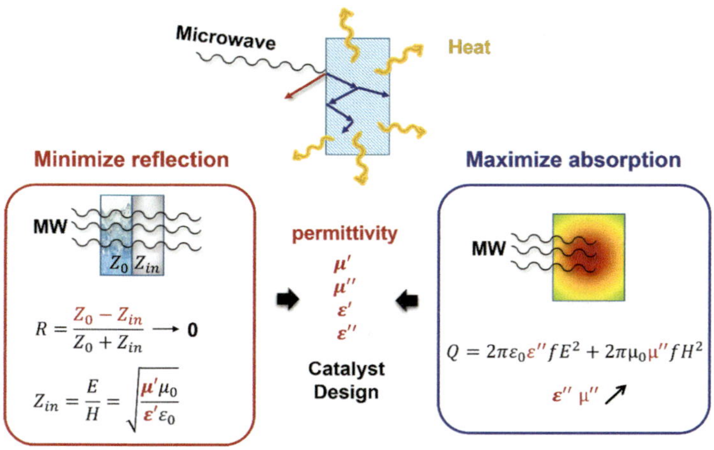

FIGURE 5.3 The principle for designing catalysts for microwave-heating reactions.

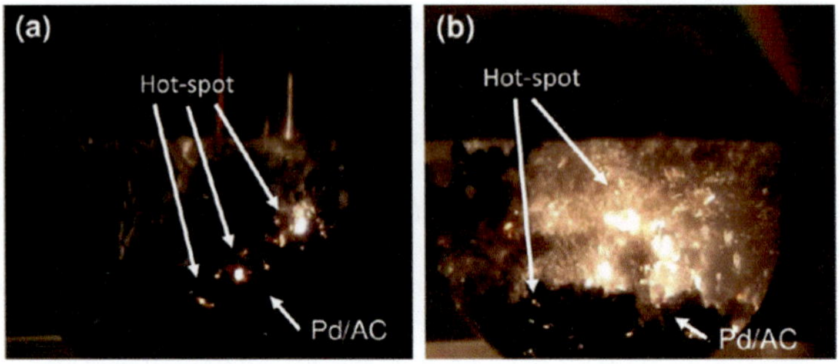

FIGURE 5.4 High-speed camera photos of hot spot generation on the Pd/activated carbon (AC) catalyst surface in the Suzuki-Miyaura coupling reaction: (a) 40 s and (b) 120 s irradiation [32].

When the catalysts are selectively heated, several studies have detected or postulated the existence of "hot spots" on the surface of heterogeneous catalysts. This is a thermal effect where the temperature in certain areas is significantly higher than the bulk temperature. Zhang et al. estimated that the temperature of hot spots is about 100–200 °C higher than the bulk temperature, and the size is about 100 μm in the H_2S decomposition reaction [33]. Satoshi et al. studied the Suzuki-Miyaura coupling reaction using Pd/AC and recorded the formation of hot spots on the catalyst surface in real time using a high-speed camera [32]. As shown in Figure 5.4(a), microscopic electric discharges (whitish-orange arcing) are generated on the catalyst surface when microwave irradiation lasts for 40 s. As the time increases to 120 seconds, the incidence of electric discharges increases. Many studies believe that hot spots make a major contribution to accelerating the reaction [34, 35]. The overheated reactants near hot spots form free radicals, which trigger the reaction to occur. For example, in the aqueous phase oxidation reaction, hot spots may pyrolyze H_2O molecules into hydroxyl radicals and hydrogen radicals around the hot spots. To effectively achieve such selective heating and the generation of hot spots on the surface of heterogeneous catalysts, scientists have developed microwave-responsive catalysts to enhance their response to microwaves. Thus, the alignment of hot spots and the reaction center would fully exert microwave energy in the heterogeneous catalyst reaction [36].

5.3 STATE-OF-THE-ART STUDIES

5.3.1 The Design Concept of Catalysts for the Microwave-Heating Reaction

To further improve the energy efficiency in the reaction, the dielectric properties of the catalysts are adjusted to enhance the ability of microwave absorption. Since the current research mainly focuses on microwave energy in the 2.45 GHz frequency band, the mainstream research is aimed at the response of the catalysts to microwaves in this frequency band. At the same time, the catalyst still needs to maintain the original sufficient reactive sites for efficient catalytic reactions. One approach is to construct a composite by coupling a microwave absorber with conventional catalyst supports. For example, CNT/TiO_2 core/shell composites were developed in saccharide dehydration reaction by microwave heating [36]. The carbon nanotubes (CNTs) are used as microwave receptor cores and sulfonic TiO_2 were used as catalytic active shells. High-resolution transmission electron microscopy (HRTEM) in the inset of Figure 5.5(a) showed the scheme of the CNT/TiO_2 core-shell structure. The catalytic performance showed such a hybrid catalyst promotes reaction temperature to 162 °C, which is 30% higher than the pure TiO_2 system (125 °C) at 15 W. Benefited from the better microwave absorption efficiency, the energy efficiency of 4.2 mol·KJ^{-1} L^{-1} is obtained by TiO_2/CNT, which is 6 times higher than pure TiO_2 and 4 times higher than CNTs. It benefits from its core-shell structure that offers sufficient active sites on the shell. When microwave radiation is applied, the internal CNT core gets heated and provides localized thermal energy to speed up the reaction. Additionally, by directly converting dielectrics into supports suitable for loading active sites, it can be used in microwave-assisted reactions. For example, solid-acid catalysts using carbon support were developed in microwave-assisted acid hydrolysis reaction or pyrolysis [38–40]. In this regard, microwave energy can directly heat carbon materials to drive the reaction on the interfacial active sites and enhance efficiency. Moreover, metal particles are doped on carbon materials for microwave-assisted ammonia decomposition to produce hydrogen [41]. Results showed that mesoporous carbons act as microwave receptors for heating, and the enhanced thermal effect enables ammonia to be completely converted at 450 °C over the 7.7 wt% Fe/mesoporous carbon, which only occurs in a conventional heating reactor at 600 °C.

Another approach is to regulate the dielectric properties at the interface, as presented in Figure 5.5(b). Research has focused on surface modification and crystal defect engineering approaches to boost the microwave response of traditional metal

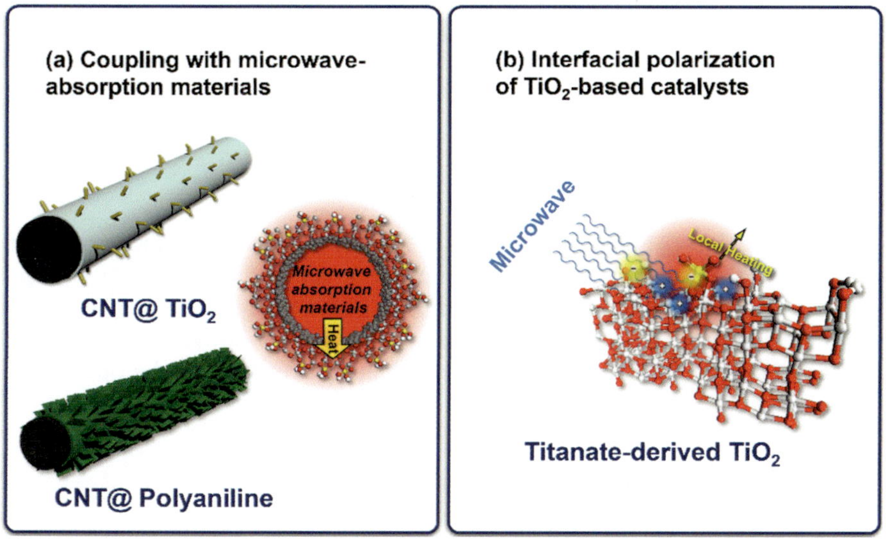

FIGURE 5.5 The methodologies of design materials for microwave-heating catalytic reactions [36, 37].

oxides. Xia et al. investigated the hydrogenated TiO_2 within the range of 2 to 18 GHz microwave [42]. They discovered that the amorphous TiO_2 layer on the surface after hydrogenation served as the dielectric loss source from the microwave. The crystal defects at the interface initiated microwave absorption at the grain boundary. A similar phenomenon was also noted on sulfated TiO_2 nanomaterials [37]. When ion-exchanged titanates ($(H_2Ti_nO_{2n+1}$) were directly heated at 400 °C, a large amount of Ti-O-SO_4^{2-} acid sites formed between the crystal layers of titanates. Both sulfonic groups accelerated charge transfer and formed electrical dipole moments on the surface. Consequently, such a polarized surface becomes highly active in the electromagnetic field, which assists in generating a considerable amount of heat. The results indicated that sulfonic groups increased 9 times in energy efficiency (5.6 mmol·(kJ·L)$^{-1}$) than other TiO_2-based solid acid (0.6 mmol·(kJ·L)$^{-1}$) in 5-hydroxymethylfurfural (HMF) synthesis.

Generally speaking, the microwave absorption capacity of the catalyst can be manifested through the maximum system temperature and the heating rate when the reaction is carried out under the fixed-power microwave. To quantify the microwave absorption characteristics, the open-ended coaxial probe method can be utilized to measure the catalysts. The catalyst powders are first dispersed in an ether solution along with paraffin wax, after which the solvent is evaporated. The dried samples are then gathered and compressed into a toroidal form. The relative complex permittivity and permeability of the wax-based composites will be determined on a vector network analyzer. This technique is capable of assisting in analyzing the microwave frequency at which the maximum absorption occurs.

Regarding the overall catalytic performance, activity, selectivity, and stability are three crucial indices in the evaluation of catalytic performance. In addition to these, energy efficiency is also a factor that needs to be taken into account in the microwave-heating catalytic reaction. Since microwave energy is generated by electricity, the consumption of such high-value energy directly determines its practicability in industrial applications. The energy efficiency in the microwave-heating catalytic reaction can be assessed by utilizing the following equation:

$$\eta = \frac{n_p}{P \cdot t \cdot n_c} \tag{5.8}$$

where η is the energy efficiency factor (J^{-1}), P is the input power of the microwave, t is elapsed time, n_p and n_c is the molar mass of the target product and catalyst, respectively. It directly reflects the overall performance of the catalyst including microwave absorption ability as well as activity. Based on this, a lower power input is favorable in terms of energy efficiency. Therefore, the input power needs to be optimized to maximize the yield of the target product.

5.3.2 Microwave Heating for Liquid-Solid Phase Reactions

A large number of studies have also targeted other liquid-solid phase reactions by microwave heating, aiming to enhance the process efficiency, especially in biomass conversions. For instance, Kong et al. employed a hydrothermal carbonization

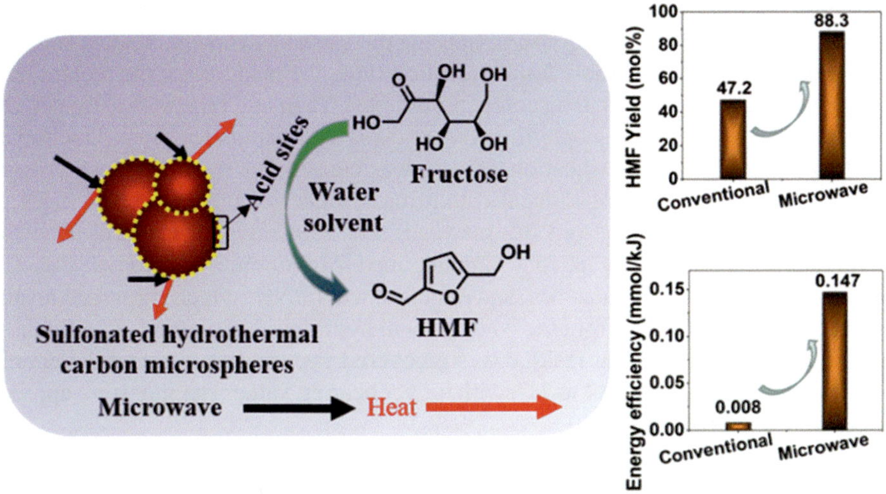

FIGURE 5.6 The process of fructose dehydration to produce HMF on sulfonated carbon microspheres [43].

technique to synthesize sulfonated carbon catalysts from glucose under reflux conditions (Figure 5.6) [43]. They investigated various reaction parameters to assess their impact on the conversion of fructose to HMF. Based on this evaluation equation (5.8), HMF yield reached 88.3 mol% with an energy efficiency of 0.147 mmol kJ^{-1}. It indicated that these catalyst attributes effectively enhanced the energy efficiency of fructose dehydration reactions. This was attributed to the good microwave absorption properties of carbon materials, with spherical morphology being advantageous for enhancing microwave energy efficiency. Additionally, the conversion of fructose to HMF required efficient transfer of acidic functional groups during the process. Similarly, Sandro L. Barbosa et al. employed hydrophilic sulfonated silica catalysts for the microwave-assisted dehydration of D-fructose to 5-hydroxymethylfurfural in dimethylsulfoxide (DMSO). Under microwave heating at 150 °C, D-fructose achieved 100% conversion to HMF within 10 minutes, with a selectivity of 100% for HMF. Compared to conventional heating, reactions under microwave irradiation required shorter reaction times and achieved higher yields. Microwave heating enhanced the energy efficiency of the reactions, and the catalyst may increase the specificity of the DMSO toward the dehydration of the hexose [44].

In addition to being applied in the dehydration reaction of saccharides, microwave heating is also applied in other processes of biomass conversion, aiming to improve the production efficiency of chemicals. Daeho Kim and others presented the energy efficiency aspect of microwave-accelerated esterification of free fatty acids with 5 wt% sulfated zirconia serving as the catalysts. In the reaction carried out at 60 °C and under atmospheric pressure, more than 90% of the esterification conversion was attained within 20 minutes through microwave heating, whereas it would take approximately 130 minutes with conventional heating. This remarkable reduction in reaction time led to microwave heating consuming merely 67% of the projected minimum thermal energy necessity for this accelerated esterification procedure. When comparing the energy consumption for the same conversion, as conventional heating requires a considerably longer reaction time, the heat loss of the reactor during the extended period has to be compensated [45]. Yuan and others investigated the usage of brown seaweed Ascophyllum nodosum as a feedstock for the production of bioethanol, and the saccharification process was carried out by microwave-assisted acid hydrolysis. The findings revealed that microwave heating led to 127 mg/g of monosaccharides released from the seaweed, and 20.8 mg/g of ethanol when the temperature was maintained at 150 °C maintained for 1 minute. The conversion efficiency based on glucose, galactose, and mannose was 60.7%, which surpassed results obtained using hydrolysis from the conventional hydrothermal method. Additionally, more than 50% of the algal residue was recovered post-hydrolysis, and the energy densification ranged from 1.4 to 1.7, with higher heating values ranging from approximately 19 to 24 MJ/kg [46].

For instance, Joseph V.L. Ruatpuia et al. explored the production of green biodiesel from Jatropha curcas oil (JCO) using a carbon-based solid-acid catalyst. They employed a response surface methodology combined with a central composite design (RSM-CCD) to optimize reaction parameters. Their analysis predicted a maximum JCO conversion of 98.24% after 50 minutes at 120 °C, with a catalyst loading of 9 wt% and a MeOH molar ratio of 20:1. In contrast to conventional heating methods, which achieved less than 40% conversion of JCO to fatty acid methyl esters, experimental results under microwave heating showed a remarkable JCO conversion rate of 98.7 ± 0.6%. This significant enhancement in performance under microwave heating highlights its efficacy in biodiesel production from JCO [47]. Besides, Qi. et al. reported the degradation of sulfamethoxazole (SMX) in a microwave-activated persulfate (MW/PS) system. The finding reveals that the degradation of SMX in the MW/PS system follows pseudo-first-order kinetics. Microwave heating shows a particular impact on the degradation of SMX compared to conventional heating, featuring a higher reaction rate and a shorter process duration. Complete degradation of SMX can be achieved within 4 minutes and 16 minutes when the reaction is conducted at 130 °C and 90 °C respectively. This is a typical phenomenon in many thermal-activated persulfate oxidation reactions, as more SO_4^- are generated at higher temperatures, leading to a quicker degradation of organic contaminants. These results imply that persulfate can be transformed into SO_4^- through microwave-induced heating, significantly facilitating the degradation of SMX [48].

5.3.3 Microwave Heating for Gas-Solid Phase Reactions

For gas-solid phase reactions, microwave heating is more widely applied to dry reforming of methane (DRM). For example, Shi et al. studied DRM under microwave heating and employed Ni/CeO_2-SiC as catalysts. The Ni-sintering performance was evaluated under conventional microwave heating. The results showed that carbon deposits reached 5.0 wt% by conventional heating, where the ratio decreased to 1.4 wt% under microwave heating. The authors suggested that Ni and SiC absorbed microwaves, significantly increasing the surface temperature of the catalyst, and enhancing DRM performance meanwhile reducing carbon deposition.

The microwave treatment rates for CH_4 and CO_2 reached 16.5 mmol/(g_{Ni}·s) and 18.6 mmol/(g_{Ni}·s), respectively. Experimental findings indicated that microwave heating exhibited lower apparent E_a. The possible mechanism involved that the absorption of microwave interaction changed the internal energy levels of CH_4 and CO_2 by interaction between microwave SiC with Ni, thereby reducing their activation barriers and accelerating their rates. This contrasted with the temperature effects promoting rates in traditional furnace heating, indicating that microwave heating reduced activation energies compared to conventional methods [49]. Sharifvaghefi et al. also investigated the application of microwave-assisted DRM on Ni-doped activated carbon (AC)-supported (Ni/AC) and Ni/MgO catalysts. The energy consumption between conventional and microwave heating was compared at the same catalyst bed. Among the Ni/AC catalysts, the 15% Ni loading showed the highest conversion rate, reaching 76% for CH_4 and 79% for CO_2. Meanwhile, microwave heating increased the energy efficiency by 10% using 10% Ni/AC catalyst, and the energy efficiency of the Ni/MgO/AC was 8%. This improvement in energy efficiency ascribed to microwave energy is directly transferred to the catalyst without any heat flux. This causes the temperature inside the catalyst to be higher than the temperature of the atmosphere around the surface, which can be translated into a more uniform heat distribution compared with traditional heating. Microwave heating also creates hot spots within the catalyst bed at temperatures higher than the measured average temperature of the bulk catalyst bed [50].

Zhang et al. explored the microwave discharge characteristics under different carbon materials, particle sizes, particle structures, and microwave electric/magnetic fields, and the following effects on DRM [51]. They achieved reactant conversions surpassing 80%, along with a syngas yield of 92% and an approximately 1:1 H_2/CO ratio. Figure 5.7(a–c) shows that the microplasma discharge of carbon materials in microwave fields effectively activated reaction molecules. The proposed mechanism, as seen in Figure 5.7(d), indicated that microwave discharge brought a synergistic effect combining surface thermal catalysis and microwave plasma catalysis, where the adsorbed CH_4 and CO_2 molecules were activated or even dissociated states. This activation enhances their adsorption and dissociation on the catalyst surface. Subsequently, ionic fragments recombine to form new molecules, significantly lowering the reaction energy barrier and facilitating the overall reaction process, thereby accelerating the reaction kinetics.

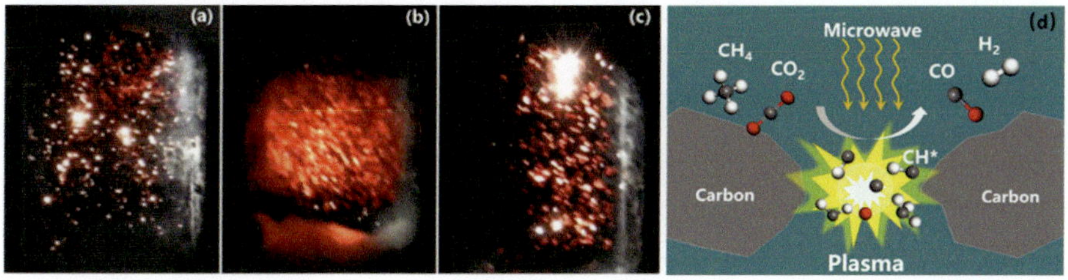

FIGURE 5.7 (a–c) Discharge phenomenon on three different carbon materials, (d) scheme of microwave-assisted DRM driven by plasma [51].

Hu et al. studied ammonia synthesis for the Haber-Bosch process by microwave heating [52]. The result showed that $CsRu/CeO_2$ catalysts achieved an ammonia yield of 0.025 mmol/g-cat/h using under 0.1 MPa and 320 °C, which is comparable to the performance of conventional heating at 370 °C. Furthermore, it has been demonstrated that the addition of promoter ions like potassium, cerium, and barium substantially enhances the ammonia production rate compared to undoped ruthenium-based catalysts. This enhancement stems from the increase of dielectric loss, resulting in more efficient microwave absorption by the catalyst, Figure 5.8. The authors suggested that the oscillating electric fields of the radiation can interact with adsorbed N_2 on the catalyst surface, enhancing its following dissociation. Given that the dissociation of dinitrogen on the surface is typically the rate-limiting step, this mechanism effectively accelerates the overall reaction.

Zhou and his coworkers studied microwave-assisted pyrolysis of waste plastics for fuel production processes [53]. Talc, used as a plastic filler in polypropylene, exhibited significant cracking activity. When ZSM-5 catalysts were introduced at a weight hourly space velocity of 10 h and a pyrolysis temperature of 620 °C, the process yielded a liquid product amounting to 48.9%. This liquid consisted predominantly of gasoline-range hydrocarbons (73.5), with high aromatic content (45.0%) and isomerized aliphatic compounds (24.6%). Energy balance analysis revealed that processing 1 kg of high-density polyethylene (HDPE) with the microwave-assisted pyrolysis system required 5 MJ of electrical energy, achieving an impressive energy efficiency of 89.6%. The gas products alone could generate 6.1 MJ of electrical energy, rendering the process self-sufficient in energy. Overall, the microwave-assisted pyrolysis system enables continuous production, avoiding the sweep-out of volatiles by carrier gases from the reactor. It combines microwave heating with a mixed SiC ball bed, offering several advantages: higher energy efficiency and lower temperature requirements compared to traditional fluidized bed reactors.

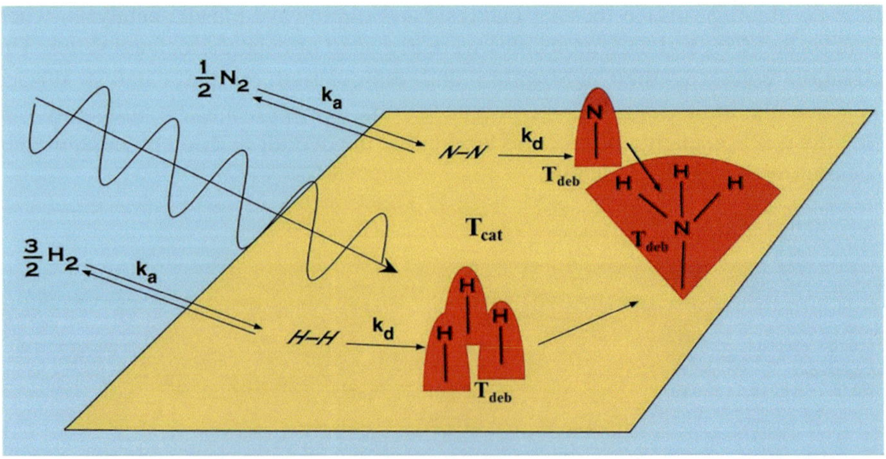

FIGURE 5.8 The dissociation of nitrogen and ammonia on the catalyst surface gives rise to polar hydride and nitride species. The interaction with microwave radiation selectively heats these species through Debye-type processes [52].

5.3.4 Microwave Heating for Absorbent Regeneration in CO_2 Capture

Microwave heating has also been carried out in the CO_2 capture process to improve regeneration energy efficiency. McGurk et al. compared the regeneration of 30% Monoethanolamine (MEA) solution for both microwave and conventional heating, and they found microwave regeneration exhibited rapid heating, low regeneration temperatures (70–90 °C), and fast recovery, resulting from the reduction of activation energy [54]. Tao et al. developed a triethylenetetramine (TETA) – polyethyleneglycol 200 (PEG200) solvent with a phase-changing feature for CO_2 capture [55]. CO_2-absorbed TETA became insoluble in PEG200, which caused ineffective regeneration by using conventional heating. On the contrary, CO_2-absorbed TETA can be quickly regenerated under microwave heating, and it showed good stability in the cycling tests.

At present, microwave regeneration is still less studied in the solvent-based CO_2 capture system. The feature of selective heating has not been fully excavated to maximize energy efficiency. Microwave energy can selectively heat the CO_2 absorption site rather than the whole solvent. Thus, the required regeneration energy may significantly be reduced. Recently, the National Energy Technology Laboratory (NETL, USA) developed a slurry of amine sorbent and silicone oil for a CO_2 adsorption/desorption process aided by the Microwave-Accelerated Regeneration of a Slurry (MARS), as presented in Figure 5.9 [56]. Using non-aqueous silicone oil can minimize water consumption and reduce the energy duty for slurry regeneration because silicone oil has a much lower heat capacity but a higher boiling point than water. Selective MW heating of CO_2-adsorbed amine-based sorbents can accelerate the CO_2 desorption flux by 10 times in comparison with thermal heating and consume much less energy for CO_2 desorption, even at process temperatures of 65–85 °C. These results indicated that the MW-accelerated swing method is highly energy efficient.

Microwave regeneration is also employed in solid sorbent regeneration for CO_2 capture. To date, only a few studies of microwave regeneration were exclusively aimed at CO_2 capture. The advantage of microwave regeneration is similar to electrical regeneration, where energy is directly transferred to the sorbent for expedited CO_2 desorption. Chronopoulos et al. compared the regeneration of microporous activated carbon (AC) using microwave heating and conventional heating [57].

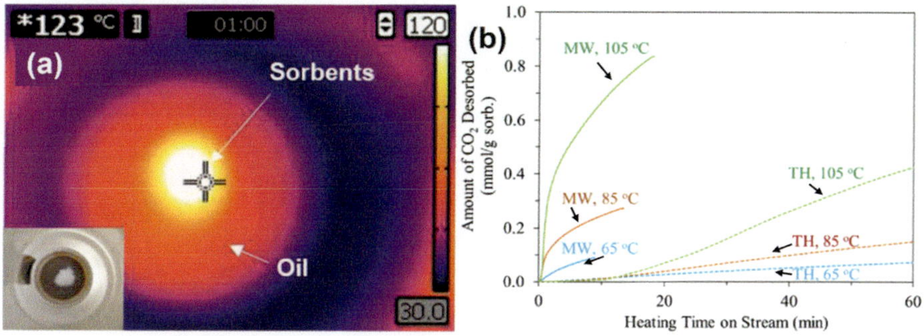

FIGURE 5.9 (a) Thermal images of CO_2-adsorbed slurry, (b) CO_2 desorption flow rates of slurry under microwave heating and conventional heating [56].

Experimental results showed that microwave heating increased CO_2 desorption rates, without the drawback of slow desorption of the residual CO_2. It also proved that microwave irradiation did not change the sorbent's morphology significantly.

Yang et al. studied microwave regeneration of perfluorinated silica-stabilized dry alkanolamines for CO_2 capture [58]. They found that conventional heating (hot plate) spent 3 h and consumed about 0.38 kW·h per cycle for sorbent regeneration, while microwave heating spent 1 h and consumed about 0.034 kW·h per cycle on average. This value for microwave is less than 10% of the energy consumption by conventional hotplate heating. Nigar et al. investigated microwave swing regeneration of three amine-functionalized mesoporous silicas (MCM-48) by grafting mono-amine, di-amine, and tri-amine molecules [59]. A tri-amino silane derivative gave the highest CO_2 adsorption and the fastest microwave heating, as CO_2 adsorption capacity decreased only 0.23% after 20 cycles.

NETL developed microwave-accelerated regeneration of sorbent for dilute carbon capture from air, Figure 5.10 [60]. The experimental studies indicated that the CO_2

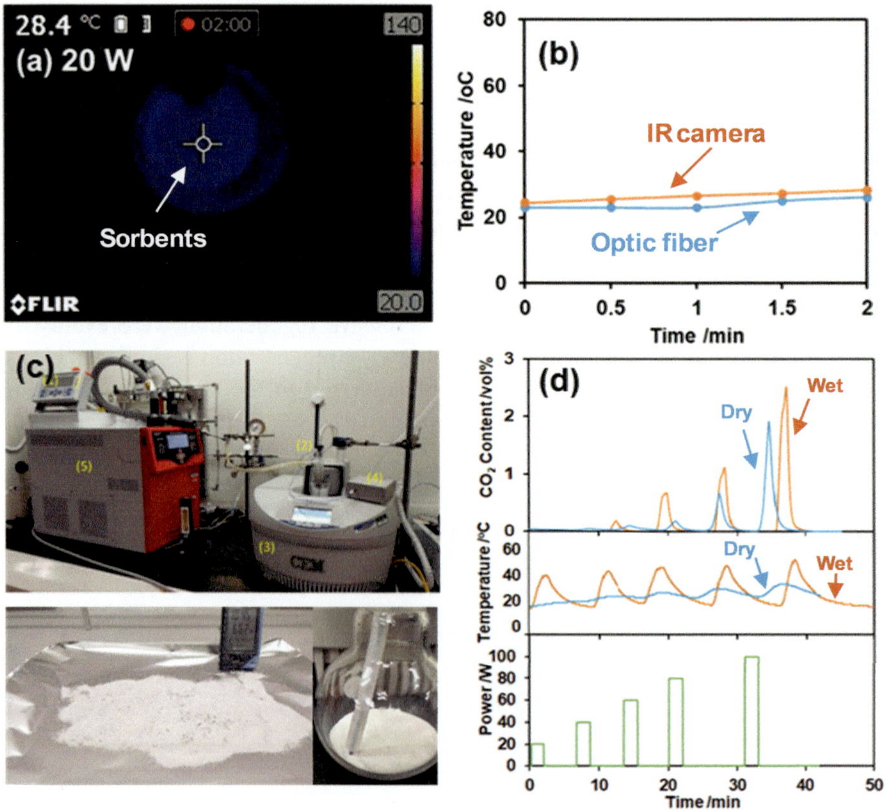

FIGURE 5.10 (a) IR images and (b) temperature profiles of sorbent under 20 W MW irradiations; (c) MW experimental setup and sorbent exposed in ambient air; (d) CO_2 desorption kinetics and temperature profiles [60].

desorption rate from the sorbents increased with microwave power, as seen in Figure 5.10(d). The reduced activation energy indicated enhanced CO_2 desorption kinetics by microwave-induced rotational vibrational (rovibrational) coupling transitions, primarily due to the highly polarized feature of the carbamate. The instant and selective delivery of microwave energy to the targeted polarized CeN bonds at room temperature is particularly advantageous for energy-efficient direct air capture. The experimental results also demonstrated a good working capacity of 0.6–1.4 mmol/g and a promising rapid MARS process with microwave swing. As it does not require steam regeneration and heat exchanger, a simple MARS process is attractive for CO_2 capture in water-stressed regions.

5.3.5 Mechanism Study of Microwave Heating in Reactions

As for the influence of the non-thermal effect on the reaction process, a lot of work attributes the non-thermal effect to the reduction of the activation energy in the reaction process. For example, Ma et al. studied oxidative dehydrogenation of propane by CO_2 in vanadium-lanthanum-doped mesoporous silica nanosphere (V-La-MSNS) [61]. Contrast experiments showed that the propane conversion increased from 2% in conventional reaction mode (CRM) to 13% in microwave catalytic reaction mode (MCRM) at 500 °C, with the propylene selectivity remaining at a high level. According to a kinetics study, the apparent activation energy of the V-La-MSNS with SiC microwave catalyst drops from 132.7 kJ/mol in the CRM to 71.4 kJ/mol in the MCRM, indicating a significant microwave direct catalytic effect. The authors proposed that the reduction in activation energy could be ascribed to the assistance of microwave irradiation in activating polar molecules (such as propane) to expedite its activation procedure. Qu et al. carried out a study on model lignin β-O-4 dimer compounds that were treated with copper oxide and H_2O_2 under microwave irradiation [62]. The products obtained through microwave heating and oil-bath heating in the same reaction vessel and with the same temperature profile were quantitatively compared. Microwave irradiation enhanced the cleavage of the side chain and promoted the oxidation of vanillin to vanillic acid during the reaction of the nonphenolic dimer, leading to the generation of veratric acid, veratraldehyde, and guaiacol. Nevertheless, no acceleration was noticed in the oxidation of veratraldehyde to veratric acid or the cleavage of aromatic rings for the production of dicarboxylic acids. The mechanism of promotion can be accounted for by selective heating, a thermal effect, and/or a MW-specific effect (non-thermal effect). Microwave heating produced larger Arrhenius values than CH during the degradation of both the phenolic and nonphenolic dimers. This outcome is in line with the non-thermal effects on elementary reactions. The authors also assert that the conclusion requires further validation as the slight fluctuation in temperature could impact the Arrhenius factors.

Shen et al. synthesized Cu-MnO_2 loaded onto nickel foam (Cu-MnO_2/NF) as catalysts for the removal of p-nitrophenol (PNP) through a microwave-assisted Fenton-like process [63]. By integrating with a microwave-enhanced Fenton-like system, it achieves a remarkable 92% removal efficiency of p-nitrophenol within 10 minutes. The involvement of Cu-MnO_2/NF substantially increased the energy efficiency of the system from 3.58 mg·kWh^{-1} to 44.35 mg·kWh^{-1}. This enhanced efficiency can be attributed to the strong microwave absorption capability of $Cu\text{-}MnO_2/NF$, which facilitates catalytic reactions more effectively. Additionally, the non-thermal effects of microwaves excite the reactant molecules, thereby weakening the chemical bonds of polar molecules.

Kubota et al. delved into the influence of microwave irradiation on the Nazarov cyclization process using an aryl vinyl ketone derivative [64]. The Nazarov cyclization was accelerated, and the reaction kinetics under microwave irradiation was 5.8 times quicker as compared to the conventional heating methods. They performed catalytic asymmetric Nazarov cyclization to further study the microwave-specific effects. Since enantioselectivity correlates with reaction temperature, maintaining consistent enantioselectivity implied similar reaction temperatures under both microwave and conventional heating. Interestingly, when employing a chiral copper catalyst for asymmetric Nazarov cyclization, microwave irradiation not only enhanced the reaction rate but also preserved the enantioselectivity. This enhancement, without compromising enantioselectivity, suggests that it was driven by a microwave-specific effect rather than thermal influences on the enantioselective reaction.

Xie et al. investigated microwave-assisted conversion of acrolein to 3-hydroxypropionaldehyde (3-HPA) [65]. They employed a simultaneous cooling method to ensure continuous microwave irradiation while maintaining a stable temperature during the reaction. Kinetic studies revealed a reduction in activation energy under microwave irradiation. Furthermore, theoretical calculations (DFT) were performed and found the increase in electric field intensity led to increasing bond length. The increase in bond length usually means a reduction in bond strength, indicating that the chemical bonds can be broken at lower energy consumption. Therefore, microwaves enhanced the adsorption behavior of acrolein on the catalyst by increasing the electrostatic potential charge and adsorption energy. Additionally, it weakened key chemical bonds in the reactants and intermediates under the microwave field, thereby reducing the overall energy barrier for acrolein hydration reaction and enhancing the reaction energy efficiency.

Tang et al. investigated the non-thermal effects of microwave irradiation on the removal of organic sulfur from coal using peracetic acid [66]. X-ray absorption near edge structure (XANES) analysis was employed to track changes in sulfur forms during the desulfurization process of high organic sulfur coal under microwave and conventional heating conditions. Quantum chemical calculations were used to study how microwave fields influence molecular properties, bond strengths, and reaction kinetics of organic sulfur-containing groups. Their findings revealed that microwave heating improved the desulfurization rate by 26% compared to conventional heating at the same temperature. XANES analysis suggested that microwave radiation facilitated the conversion of low-valence sulfur to high-valence sulfur, resulting in heterocyclic sulfur and oxidized sulfur predominating in the treated coal. Theoretical calculations indicated that the non-thermal microwave effect enhanced the nucleophilic reactivity of sulfur-containing groups and weakened the strength of C-S bonds under additional electric fields. Moreover, this effect reduced the activation energy and increased the reaction rate, thereby promoting the formation of oxidized sulfur in coal.

5.4 FUTURE PROSPECTS

Considerable attention has been directed toward the utilization of microwave heating. Nevertheless, as one of the new heating methods for future industrial electrification, there is still a significant amount of work that needs to be explored for microwave heating. Currently, studies mainly focus on promoting fast reactions through the introduction of microwave energy rather than on its energy efficiency. Note that microwave energy is high-value energy, which directly determines the feasibility of industrial applications. Although microwave heating accelerates the reaction rate, more energy might be input compared to the conventional heating process. Therefore, future research should take into account the energy efficiency of the reaction to establish a more economically viable microwave-assisted reaction process. New approaches and technologies need to be developed that can reduce greenhouse gas emissions in the industrial sector while enabling growth and enhancing productivity.

Besides, numerous works have demonstrated the crucial role of hot spots in the catalyst reaction. They can generate free radicals, enhance mass transfer, alter the reaction route, etc. However, the mechanism is still being explored due to their short lifespan and microscopic scale. Scientists have attempted to measure them using infrared pyrometers, but such macroscale measurements make it difficult to track them. Current research has found that the formation of hot spots may be related to the medium and surface properties of the catalysts. This mechanism can be studied in future research to gain a better understanding of microwave heating.

Finally, while most process models have incorporated conventional heating processes, few models have been developed for microwave heating, and those that exist are not easily integrated into complete process flows for total plant optimization. System-level optimization is particularly crucial for high-efficiency thermal processes, which can be thoroughly integrated into manufacturing plants to maximize the utilization of waste heat. As such, there is an urgency for the Institute to develop the necessary tools to support process design, optimization, system integration, and safe operation to facilitate the adoption of this technology.

5.5 SUMMARY

In conclusion, microwave heating can remarkably enhance energy efficiency in catalytic reactions. A great deal of work has already been done to apply microwave heating to multiple chemical reaction processes, including liquid-solid phase reactions and gas-solid phase reactions. In order to achieve higher energy efficiency, most works promote dielectric properties toward microwave energy, and many dielectric loss materials such as carbon materials and SiC have been engaged. Compared to conventional heating, many mechanism studies have shown that microwave enables the reduction of the activation energy of the reaction. Nevertheless, there is a lack of solid evidence of the non-thermal effect of microwaves, and the complex mechanism of thermal effects (e.g., hot spots, superheating) is awaiting exploration. More experimentation, computational simulation, and the development of fundamental theories are still required.

REFERENCES

[1] S. Faraji, F.N. Ani, Microwave-assisted synthesis of metal oxide/hydroxide composite electrodes for high power supercapacitors–A review, *J. Power Sources*, 263, (2014), 338–360.

[2] D. Dallinger, C.O. Kappe, Microwave-assisted synthesis in water as solvent, *Chem. Rev.*, 107, (2007), 2563–2591.

[3] C. Si, J. Wu, Y. Wang, Y. Zhang, X. Shang, Drying of low-rank coals: A review of fluidized bed technologies, *Drying Technol.*, 33, (2015), 277–287.

[4] F.J. Barba, Z. Zhu, M. Koubaa, A.S. Sant'Ana, V. Orlien, Green alternative methods for the extraction of antioxidant bioactive compounds from winery wastes and by-products: A review, *Trends Food Sci. Technol.*, 49, (2016), 96–109.

[5] F. Mushtaq, R. Mat, F.N. Ani, A review on microwave assisted pyrolysis of coal and biomass for fuel production, *Renew. Sustain. Energy Rev.*, 39, (2014), 555–574.

[6] S. Nomanbhay, M. Ong, A review of microwave-assisted reactions for biodiesel production, *Bioengineering*, 4, (2017), 57.

[7] F. Motasemi, M.T. Afzal, A review on the microwave-assisted pyrolysis technique, *Renew. Sustain. Energy Rev.*, 28, (2013), 317–330.

[8] R.N. Baig, R.S. Varma, Alternative energy input: Mechanochemical, microwave and ultrasound-assisted organic synthesis, *Chem. Soc. Rev.*, 41, (2012), 1559–1584.

[9] N. Remya, J.-G. Lin, Current status of microwave application in wastewater treatment—A review, *Chem. Eng. J.*, 166, (2011), 797–813.

[10] A. de la Hoz, A. Diaz-Ortiz, A. Moreno, Microwaves in organic synthesis. Thermal and non-thermal microwave effects, *Chem. Soc. Rev.*, 34, (2005), 164–178.

[11] A.K. Datta, P.M. Davidson, Microwave and radio frequency processing, *J. Food Sci.*, 65, (2000), 32–41.

[12] M. Oghbaei, O. Mirzaee, Microwave versus conventional sintering: A review of fundamentals, advantages and applications, *J. Alloys Compd.*, 494, (2010), 175–189.

[13] E. Thostenson, T.-W. Chou, Microwave processing: Fundamentals and applications, *Compos. Part A*, 30, (1999), 1055–1071.

[14] J.M. Osepchuk, A history of microwave heating applications, *IEEE Trans. Microw. Theory Tech.*, 32, (1984), 1200–1224.

[15] R. Kelly, N. Rowson, Microwave reduction of oxidised ilmenite concentrates, *Miner. Eng.*, 8, (1995), 1427–1438.

[16] H. Kingston, L. Jassie, Microwave energy for acid decomposition at elevated temperatures and pressures using biological and botanical samples, *Anal. Chem.*, 58, (1986), 2534–2541.

[17] F. Içier, T. Baysal, Dielectrical properties of food materials-1: Factors affecting and industrial uses, *Crit. Rev. Food Sci. Nutr.*, 44, (2004), 465–471.

[18] R.V. Decareau, *Microwaves in the food processing industry*, Academic Press, 1985.

[19] D.D. Dinčov, K.A. Parrott, K.A. Pericleous, Heat and mass transfer in two-phase porous materials under intensive microwave heating, *J. Food Eng.*, 65, (2004), 403–412.

[20] A.A. Metaxas, R.J. Meredith, *Industrial microwave heating*, IET, 1983.

[21] C. González-Arellano, J.M. Campelo, D.J. Macquarrie, J.M. Marinas, A.A. Romero, R. Luque, Efficient microwave oxidation of alcohols using low-loaded supported metallic iron nanoparticles, *ChemSusChem*, 1, (2008), 746–750.

[22] S.S. Lam, H.A. Chase, A review on waste to energy processes using microwave pyrolysis, *Energies*, 5, (2012), 4209–4232.

[23] T. Ji, R. Tu, L. Mu, X. Lu, J. Zhu, Structurally tuning microwave absorption of core/shell structured CNT/polyaniline catalysts for energy efficient saccharide-HMF conversion, *Appl. Catal., B*, 220, (2018), 581–588.

[24] T. Lieu, S. Yusup, M. Moniruzzaman, Kinetic study on microwave-assisted esterification of free fatty acids derived from Ceiba pentandra seed oil, *Bioresour. Technol.*, 211, (2016), 248–256.

[25] R. Babu, S.-H. Kim, A.C. Kathalikkattil, R.R. Kuruppathparambil, D.W. Kim, S.J. Cho, D.-W. Park, Aqueous microwave-assisted synthesis of non-interpenetrated metal-organic framework for room temperature cycloaddition of CO_2 and epoxides, *Appl. Catal., A*, 544, (2017), 126–136.

[26] R. Bernini, E. Mincione, M. Barontini, G. Provenzano, L. Setti, Obtaining 4-vinylphenols by decarboxylation of natural 4-hydroxycinnamic acids under microwave irradiation, *Tetrahedron*, 63, (2007), 9663–9667.

[27] J. Jacob, L. Chia, F. Boey, Thermal and non-thermal interaction of microwave radiation with materials, *J. Mater. Sci.*, 30, (1995), 5321–5327.

[28] T. Ji, C. Liu, X. Lu, J. Zhu, Coupled chemical and thermal drivers in microwaves toward ultrafast HMF oxidation to FDCA, *ACS Sustain. Chem. Eng.*, 6, (2018), 11493–11501.

[29] H. Will, P. Scholz, B. Ondruschka, Microwave-assisted heterogeneous gas-phase catalysis, *Chem. Eng. Technol.*, 27, (2004), 113–122.

[30] N. Sharma, U.K. Sharma, E.V. Van der Eycken, Microwave-assisted organic synthesis: Overview of recent applications, *Green techniques for organic synthesis and medicinal chemistry*, (2018), 441–468.

[31] D. Bogdal, M. Lukasiewicz, J. Pielichowski, A. Miciak, S. Bednarz, Microwave-assisted oxidation of alcohols using Magtrieve™, *Tetrahedron*, 59, (2003), 649–653.

[32] S. Horikoshi, A. Osawa, M. Abe, N. Serpone, On the generation of hot-spots by microwave electric and magnetic fields and their impact on a microwave-assisted heterogeneous reaction in the presence of metallic Pd nanoparticles on an activated carbon support, *J. Phys. Chem. C*, 115, (2011), 23030–23035.

[33] X. Zhang, D.O. Hayward, D.M.P. Mingos, Apparent equilibrium shifts and hot-spot formation for catalytic reactions induced by microwave dielectric heating, *Chem. Commun.*, (1999), 975–976. https://doi.org/10.1039/a901245a

[34] M. Guler, T. Dogu, D. Varisli, Hydrogen production over molybdenum loaded mesoporous carbon catalysts in microwave heated reactor system, *Appl. Catal., B*, 219, (2017), 173–182.

[35] A.L. Garcia-Costa, J.A. Zazo, J.J. Rodriguez, J.A. Casas, Microwave-assisted catalytic wet peroxide oxidation. Comparison of Fe catalysts supported on activated carbon and gamma-alumina, *Appl. Catal., B*, 218, (2017), 637–642.

[36] T. Ji, R. Tu, L. Mu, X. Lu, J. Zhu, Enhancing energy efficiency in saccharide–HMF conversion with core/shell structured microwave responsive catalysts, *ACS Sustain. Chem. Eng.*, 5, (2017), 4352–4358.

[37] T. Ji, R. Tu, L. Li, L. Mu, C. Liu, X. Lu, J. Zhu, Localizing microwave heat by surface polarization of titanate nanostructures for enhanced catalytic reaction efficiency, *Appl. Catal., B*, 227, (2018), 266–275.

[38] T. Chen, C. Fan, One-pot generation of mesoporous carbon supported nanocrystalline $H_3PW_{12}O_{40}$ heteropoly acid with high performance in microwave esterification of acetic acid and isoamyl alcohol, *J. Porous Mater.*, 20, (2013), 1225–1230.

[39] S. Horikoshi, T. Minagawa, S. Tsubaki, A. Onda, N. Serpone, Is selective heating of the sulfonic acid catalyst $ACSO_3H$ by microwave radiation crucial in the acid hydrolysis of cellulose to glucose in aqueous media?, *Catalysts*, 7, (2017).

[40] Y. Wu, Z. Fu, D. Yin, Q. Xu, F. Liu, C. Lu, L. Mao, Microwave-assisted hydrolysis of crystalline cellulose catalyzed by biomass char sulfonic acids, *Green Chem.*, 12, (2010), 696–700.

[41] D. Varisli, C. Korkusuz, T. Dogu, Microwave-assisted ammonia decomposition reaction over iron incorporated mesoporous carbon catalysts, *Appl. Catal., B*, 201, (2017), 370–380.

[42] T. Xia, C. Zhang, N.A. Oyler, X. Chen, Hydrogenated TiO_2 nanocrystals: A novel microwave absorbing material, *Adv. Mater.*, 25, (2013), 6905–6910.

[43] X. Kong, S. Vinju Vasudevan, M. Cao, J. Cai, H. Mao, Q. Bu, Microwave-assisted efficient fructose–HMF conversion in water over sulfonated carbon microsphere catalyst, *ACS Sustain. Chem. Eng.*, 9, (2021), 15344–15356.

[44] S.L. Barbosa, M. de S. Freitas, W.T.P. dos Santos, D.L. Nelson, S.I. Klein, G.C. Clososki, F.J. Caires, A.C.M. Baroni, A.P. Wentz, Dehydration of d-fructose to 5-hydroxymethyl-2-furfural in DMSO using a hydrophilic sulfonated silica catalyst in a process promoted by microwave irradiation, *Sci. Rep.*, 11, (2021), 1919.

[45] D. Kim, J. Choi, G.-J. Kim, S.K. Seol, Y.-C. Ha, M. Vijayan, S. Jung, B.H. Kim, G.D. Lee, S.S. Park, Microwave-accelerated energy-efficient esterification of free fatty acid with a heterogeneous catalyst, *Bioresour. Technol.*, 102, (2011), 3639–3641.

[46] Y. Yuan, D.J. Macquarrie, Microwave assisted acid hydrolysis of brown seaweed ascophyllum nodosum for bioethanol production and characterization of alga residue, *ACS Sustain. Chem. Eng.*, 3, (2015), 1359–1365.

[47] J.V.L. Ruatpuia, B. Changmai, A. Pathak, L.A. Alghamdi, T. Kress, G. Halder, A.E.H. Wheatley, S.L. Rokhum, Green biodiesel production from Jatropha curcas oil using a carbon-based solid acid catalyst: A process optimization study, *Renew. Energy*, 206, (2023), 597–608.

[48] C. Qi, X. Liu, C. Lin, X. Zhang, J. Ma, H. Tan, W. Ye, Degradation of sulfamethoxazole by microwave-activated persulfate: Kinetics, mechanism and acute toxicity, *Chem. Eng. J.*, 249, (2014), 6–14.

[49] Y. Shi, L. Wang, M. Wu, F. Wang, Order-of-magnitude increase in rate of methane dry reforming over Ni/CeO_2-SiC catalysts by microwave catalysis, *Appl. Catal., B*, 337, (2023), 122927.

[50] S. Sharifvaghefi, B. Shirani, M. Eic, Y. Zheng, Application of microwave in hydrogen production from methane dry reforming: Comparison between the conventional and microwave-assisted catalytic reforming on improving the energy efficiency, *Catalysts*, 9, (2019), 618.

[51] F. Zhang, X. Zhang, Z. Song, X. Li, X. Zhao, J. Sun, Y. Mao, X. Wang, W. Wang, Promotion of microwave discharge over carbon catalysts for CO_2 reforming of CH_4 to syngas, *Fuel*, 331, (2023), 125914.

[52] J. Hu, C. Wildfire, A.E. Stiegman, R.A. Dagle, D. Shekhawat, V. Abdelsayed, X. Bai, H. Tian, M.B. Bogle, C. Hsu, Y. Luo, S.D. Davidson, Y. Wang, Microwave-driven heterogeneous catalysis for activation of dinitrogen to ammonia under atmospheric pressure, *Chem. Eng. J.*, 397, (2020), 125388.

[53] N. Zhou, L. Dai, Y. Lv, H. Li, W. Deng, F. Guo, P. Chen, H. Lei, R. Ruan, Catalytic pyrolysis of plastic wastes in a continuous microwave assisted pyrolysis system for fuel production, *Chem. Eng. J.*, 418, (2021), 129412.

[54] S.J. McGurk, C.F. Martín, S. Brandani, M.B. Sweatman, X. Fan, Microwave swing regeneration of aqueous monoethanolamine for post-combustion CO_2 capture, *Appl. Energy*, 192, (2017), 126–133.

[55] M. Tao, J. Gao, W. Zhang, Y. Li, Y. He, Y. Shi, A novel phase-changing nonaqueous solution for CO_2 capture with high capacity, thermostability, and regeneration efficiency, *Ind. Eng. Chem. Res.*, 57, (2018), 9305–9312.

[56] T. Ji, H. Zhai, C. Wang, J. Culp, C.M. Marin, H.P. Paudel, W.C. Wilfong, Y. Duan, R. Xia, F. Jiao, B. Kail, Q. Wang, Y. Soong, F. Shi, M. Gray, Microwave-accelerated regeneration of a non-aqueous slurry for energy-efficient carbon sequestration, Mater. Today Sustain., 19, (2022), 100168.

[57] T. Chronopoulos, Y. Fernandez-Diez, M.M. Maroto-Valer, R. Ocone, D.A. Reay, CO_2 desorption via microwave heating for post-combustion carbon capture, *Microporous Mesoporous Mater.*, 197, (2014), 288–290.

[58] J. Yang, H. Tan, Q. Low, B. Binks, J. Chin, CO_2 capture by dry alkanolamines and an efficient microwave regeneration process, *J. Mater. Chem. A*, 3, (2015), 6440–6446.

[59] H. Nigar, B. Garcia-Baños, F.L. Peñaranda-Foix, J.M. Catalá-Civera, R. Mallada, J. Santamaría, Amine-functionalized mesoporous silica: A material capable of CO_2 adsorption and fast regeneration by microwave heating, *AIChE J.*, 62, (2016), 547–555.

[60] T. Ji, H. Zhai, C. Wang, C.M. Marin, W.C. Wilfong, Q. Wang, Y. Duan, R. Xia, F. Jiao, Y. Soong, F. Shi, M. Gray, Energy-efficient and water-saving sorbent regeneration at near room temperature for direct air capture, *Mater. Today Sustain.*, 21, (2023), 100321.

[61] Z. Ma, W. Xu, Q. Wang, Q. Zhou, J. Zhou, Highly effective microwave catalytic oxidative dehydrogenation of propane by CO_2 over V-La-doped dendritic mesoporous silica-based microwave catalysts, *Chem. Eng. J.*, 435, (2022), 135081.

[62] C. Qu, K. Ito, I. Katsuyama, T. Mitani, K. Kashimura, T. Watanabe, Directly microwave-accelerated cleavage of C–C and C–O bonds of lignin by copper oxide and H_2O_2, *ChemSusChem*, 13, (2020), 4510–4518.

[63] T. Shen, K. Tian, M. Cao, L. Li, F. Shi, J. Qu, Q. Zheng, G. Zhang, Application of nickel foam supported Cu–MnO_2 in microwave enhanced Fenton-like process for p-nitrophenol removal: Degradation, synergy and mechanism insight, *J. Clean. Prod.*, 397, (2023), 136442.

[64] Y. Kubota, J. Fukuzumi, K. Saito, T. Yamada, Experimental investigation of microwave-specific effect on nazarov cyclization, *Bull. Chem. Soc. Jpn.*, 95, (2022), 1730–1740.

[65] Q. Xie, Y. Zhou, Y. Wang, T. Pan, Y. Duan, S. Yu, W. Ji, Y. Nie, Microwave enhanced catalytic hydration of acrolein to 3-hydroxypropionaldehyde using simultaneous cooling: Experimental and theoretical studies, *Chem. Eng. Sci.*, 269, (2023), 118493.

[66] L. Tang, X. Chen, X. Tao, Non-thermal effect of microwave on organic sulfur removal from coal by microwave with peroxyacetic acid, *Fuel*, 338, (2023), 127262.

6 Industrial-Scale Sustainable Electrochemical Energy Conversion Systems for Chemical Transformations

Saket S. Bhargava

6.1 INTRODUCTION

Electrochemical energy conversion systems for chemical transformations that are powered by electricity derived from green energy sources are emerging as vectors for decarbonization by enabling sustainable industrial production of chemicals, fuels, and intermediates. More than a decade of predominantly academic research and more recently some industrial research has matured electrochemical systems for chemical transformations (hereafter referred to as "electrochemical systems") into reaching technology readiness levels (TRLs) of 4–5 for low-temperature electrochemical systems and TRLs of 6–8 for high-temperature electrochemical systems.[1] With so much improvement in the field of electrochemical systems, attention now needs to be diverted towards electrochemical process systems engineering approaches to de-risk the formulation of sustainable and scalable electrochemical systems so that TRLs of 6–9 can be achieved rapidly. Key performance metrics that will influence the system considerations are current density (total – j_{total} and partial – $j_{partial}$), Faradaic efficiency ($FE_{product}$), full cell energy efficiency ($EE_{product}$), reactant conversion per pass ($Conv_{product}$), and the durability/performance stability over time of the system. System considerations include the choice of reactor design, reactor operating conditions, chemistries, and catalysts at the cathode and the anode, and also include process and operational decisions to enhance the system performance stability, deal with process impurities, and inherently make the process more sustainable and easier to scale.

In this chapter, we explore reactor designs and operating conditions that are scalable and can make the process more sustainable. We also explore the choice of chemistries at both the electrodes, catalysis approaches, and catalysts for operation at industrially relevant scales. We also discuss factors that can affect performance stability over time (e.g., electrode design, impurities). Lastly, we close out our discussion by exploring process development and integration considerations (e.g., process

 DOI: 10.1201/9781003655442-8

network synthesis, heat management, scaling rules, process economics, lifecycle impact) to guide the development of electrochemical systems that can be operated sustainably at industrial scales. At several places in the chapter, we will use examples such as CO_2 electrolysis as case studies to elaborate on the discussion further.

6.2 REACTOR DESIGN

In this section, we briefly discuss the design, advantages, and disadvantages of the most commonly employed reactors and highlight the most appropriate reactor choice(s) for scaling electrochemical systems. We will limit our discussion to H-Cells or W-Cells (HWCs), Liquid Electrolyte Cells (LECs), Membrane-based Cells (MCs), and Solid Oxide Electrolyte Cells (SOECs). While there are also other types of reactors studied in the literature (e.g., semi-flow electrolyzers, rotating disk electrode-based electrolyzers, molten carbonate electrolysis cells, proton ceramic electrolyte cells, packed bed reactors), we will not be focusing on such reactors here as they are either too challenging to scale or are far from being considered as viable candidates at this time (due to low TRL). A comparison-at-a-glance of the four relevant reactor types is shown in Table 6.1.

TABLE 6.1
A Table Comparing the Commonly Employed Reactor Types for Electrochemical Systems

	Reactor Type			
Characteristic	**HWC**	**LEC**	**MC**	**SOEC**
General Usability Considerations				
Barrier-to-use/sophistication of equipment	Very low	High	High	Very high
Usable for *quick* testing/ characterization of catalysts?	Yes	No	No	No
Performance evaluation under catalytically relevant conditions?	No	Yes	Yes	Yes
Design Considerations				
Flexible in operating pH?	Somewhat	Yes	Yes	Not applicable
Encounter mass transfer limitations	Low currents/ potentials	High currents/ potentials	Moderate currents/ potentials	High currents/ potentials
Diffusion limitations	Severe	Low	Low	Low
Operational Considerations				
Direct gaseous feed?	Typically not	Yes	Yes	Yes
Electrolyte	Aqueous solutions/ Organic solvents	Aqueous solutions/ Organic solvents	Membranes	Solid oxides

(Continued)

TABLE 6.1 (Continued)

	Reactor Type			
Characteristic	**HWC**	**LEC**	**MC**	**SOEC**
iR-loss	High	Low to moderate	Low	Moderate
Process Considerations				
Can high rates be achieved?	Very difficult	Very easy	Moderately easy	Very easy
Faradaic side reactions possible?	Yes	Yes	Yes	Yes
Non-Faradaic side reactions possible?	Yes	Yes	Yes	No
Ease of product separation	Very difficult	Difficult	Moderately easy	Easy
Miscellaneous Considerations				
Durability and/or performance stability issues?	Low to moderate	Severe	Severe	Moderate
Product crossover	Very high	Moderate	High	Low
Pressure balance (across electrode) issues?	No	Yes	Yes	Yes
Ease of operation at high-temperature conditions	Moderately easy	Easy	Moderately easy	Very easy
Ease of operation at high-pressure conditions	Very easy	Difficult	Very easy	Easy
Cell energy efficiency at relevant reaction rates	Very low	High	Moderate to high	High to very high

6.2.1 HWCs

HWCs are the most commonly used reactors in research studies, especially when a quick/efficient catalytic performance characterization is needed. However, as listed in Table 6.1, HWCs quickly run into mass transfer limitations, often resulting in very low rates, and thus, the catalysts are not evaluated under industrially relevant conditions. So, while HWCs might be good for quick studies under a limited number/range of conditions, they are not scalable.

6.2.2 LECs

LECs are the second-most commonly used reactors in research studies because their design and use of a gas diffusion electrode (GDE) allow for direct gaseous and liquid feeds, which minimizes the mass transfer limitations encountered in HWCs significantly, resulting in industrially relevant rates and selectivities. Moreover, LECs are pH flexible and have small iR-losses because of small electrode-to-electrode distance. However, use of aqueous and organic electrolytes leads to unwanted chemistries, both Faradaic and non-Faradaic, which are discussed in detail later. Furthermore, the durability and performance stability over time of GDEs is very poor, warranting frequent replacement for prolonged operation. So, while LECs might be a good step up from HWCs, they are also generally not scalable.

6.2.3 MCs

MCs are the most promising candidates for scaling low-temperature electrolysis systems. MCs, like LECs, allow for direct gaseous humidified feeds and/or direct liquid feeds, thereby minimizing mass transfer limitations, but in addition, they also allow for a potentially liquid electrolyte-free operation by using membranes for charge transport. This liquid electrolyte-free operation also minimizes the Faradaic and the non-Faradaic side reactions associated with liquid electrolytes. The zero-gap arrangement in MCs results in a very small iR-loss, and it is also easier to pressurize these systems and assemble them in stacks. MCs are also pH flexible and the pH of the reaction(s) at the individual electrodes influences the membrane choice. MCs have also demonstrated industrially relevant performance for certain chemistries (e.g., CO_2 electrolysis); however, there is still scope for improvement in terms of performance metrics, especially, the durability and performance stability over time of membrane electrode assemblies (MEAs).

Membranes are of three types: (i) proton exchange membrane (PEM) or cation exchange membrane (CEM), (ii) anion exchange membrane (AEM) – of which the hydroxide exchange membrane (HEM) is of special interest, and (iii) bipolar membranes (BPM). PEM-based or CEM-based MCs are only useful if reactions at the cathode and at the anode can deliver high performance under acidic conditions. The choice of BPM-based MCs would typically make sense if reactions at the cathode and the reactions at the anode need to operate under different pH. AEM-based MCs should be chosen if the reactions at the cathode and at the anode are favored in alkaline pH conditions. HEMs are an emerging class of membranes that help to maintain alkaline pH near the electrodes and these need to be studied in more detail. For chemistries where water molecules are consumed in the reaction, the feeds need to be humidified else the membrane dries out.

6.2.4 SOECs

SOECs are the most promising candidates for scaling high-temperature electrolysis systems. SOECs, as the name suggests, use solid oxides as electrolytes instead of liquid electrolytes. This eliminates the non-Faradaic side reactions associated with liquid electrolytes. SOECs also use a direct gaseous feed and typically operate at elevated temperatures resulting in very high rates, but sometimes, the selectivities suffer due to the enhancement of Faradaic side reactions (e.g., HER). Furthermore, catalyst delamination is a major concern in terms of electrode durability and performance stability over time. Despite this, some pilot/commercial scale demos for SOECs are already underway (e.g., Haldor-Topsoe sells commercial SOECs for CO production *via* CO_2 electrolysis).

Based on this discussion, it is abundantly clear that for low-temperature electrolysis, HEM-based MCs and for high-temperature electrolysis, SOECs are the best options for scaling. Of course, for HEM-based MCs, their operation needs to be demonstrated at scale and the requisite stable performance needs to be delivered for thousands of hours with minimal to no reactant and/or product crossover. Similarly, for SOECs, the selectivities need to be enhanced and the durability of the electrodes, in particular, the catalyst delamination issue needs to be resolved.

6.3 REACTOR OPERATING CONDITIONS

Following the choice of reactor design, it is important that the reactor operating conditions be optimized to maximize the electrochemical performance. In this section, we briefly discuss the implications of manipulating several process parameters on the process performance for various reactor types (LECs, MCs, and SOECs) and how such process considerations would affect process integration, scaling, and sustainability.

6.3.1 Catalyst Loading

Representative of the number of electrochemically active catalyst sites, catalyst loading is generally expressed as the amount of catalyst in a given electrode area. However, the thickness of the catalyst layer and the method of deposition also influence the access to the catalyst sites that are electrochemically active and, hence, the electrochemical performance. When catalyst loadings are effectively optimized, the economic impacts are not significant, but for precious metal-based catalysts, supply chain issues might pose a significant risk as we show later in Calculation 6.2. To maximize catalyst mass activities, lower loadings are recommended if the rates ($j_{partial}$) do not decrease.

6.3.2 Electrolyte Properties

Depending on the nature of the electrolyte (solid oxide, aqueous, or organic), there are several considerations – such as conductivity or pH – that would influence the reactor operation. We will discuss solid oxide electrolytes, aqueous, and organic electrolytes separately.

1. **Solid Oxide Electrolytes**: For solid oxide electrolytes, ionic conductivity is the most important consideration other than composition. The solid oxides have relatively lower conductivities than aqueous solutions and thus the resistance due to solid oxides will be the governing resistance in SOECs at least at moderately high current densities. Higher conductivity solid oxides need to be developed to reduce resistance and facilitate ion transport – however, leakage currents become significant at high conductivities. Despite extensive research, there are no systematic/rational design rules for solid oxide electrolytes. This is still an ongoing area of research with an intuition-guided trial-and-error approach.
2. **Aqueous Solutions**: For aqueous solutions, in addition to the conductivity and composition – the pH, concentration, and flow rate also influence the

electrochemical performance. While studies do not show a direct correlation between conductivity of the electrolyte and electrochemical performance, in general, higher conductivity is preferred for easier facilitation of movement of ions and lower resistance. Prior studies have also shown that pH strongly influences the process performance of electrochemical reactions. However, the choice of electrolyte can quickly get complicated depending on the Faradaic and non-Faradaic side reactions and the ideal pH required for the electrode reactions. Use of high electrolyte concentrations is preferred because in general, higher electrolyte concentrations reduce concentration polarizations and correspond to higher pH and higher conductivities, resulting in lower solution resistance (R_{sol}), and thus, lower iR-drop due to solution resistance and significantly higher EEs. Higher electrolyte concentrations can also have a promotional effect on electrokinetics (as can be tracked by the charge transfer resistance – R_{ct}), but high electrolyte concentrations are generally corrosive and thus, there needs to be a fine balance. Very few studies have also explored the use of pure water (or water containing only trace amounts of inorganic salts) as an electrolyte in MCs while maintaining high electrochemical performance – such systems would be very beneficial from a scaling point of view and should be investigated further.[2] Finally, higher electrolyte flow rates are needed for high rates as the flow of electrolyte flow is responsible for the convective flux of ions near the electrode surface and in the bulk of the electrolyte. Electrolyte flow also influences the migration flux (as can be inferred from the Nernst-Planck equation) and the reaction rates (due to Le Chatelier's principle) at both electrodes by removing or replenishing reactant species and/or product species. Electrolyte flow rate can also be a strong influence on rates and selectivities because it influences the mass transport flux and indirectly; it also influences the pH of the electrolyte near the electrode. However, use of electrolytes at high flow rates is not feasible from a practical point of view due to the high volumes required and process integration/design (e.g., pipe size, material of construction, corrosion, recycle loop) considerations. Recirculation of used electrolytes is an option; however, the reduced pH and conductivity of the spent electrolyte being recirculated should not hamper the electrochemical performance. Regeneration of electrolyte from the spent electrolyte is very energy-intensive and the energy economics do not support this as a viable option.[3]

3. **Organic Solvents**: For organic solvents, most of the considerations discussed for aqueous solutions apply, but in addition, factors such as pk_a, miscibility with reactants/products, amount of water in the electrolyte mixture, junction potential, and the stability window will also influence the electrolyte choice. Presence of water might be necessary for a reaction to proceed, but it can also be used to tune the $FE_{product}$ by changing the concentration of water in the electrolyte. Higher conductivity is preferred to reduce junction potential. The choice of electrolyte will also dictate the allowed operating window to prevent degradation of the electrolyte.

6.3.3 Feed Gas Flow Rate

Higher conversion per pass of reactants to products and higher atom efficiency are desirable from energetics and downstream separation points of view. The conversion per pass is inversely related to the reactant feed flow rate, and, thus, running at low feed flow rates while achieving a high $j_{partial}$ is desirable. The stoichiometric values of the feed flow rates required can be determined using Faraday's law, and the net feed flow rates can be estimated by accounting for any losses to the electrolyte.[4]

Let us look at an example – consider the CO_2 electrochemical reduction reaction. For CO_2 gas feed, the carbon efficiency is hampered by side reactions like carbonate and bicarbonate formation for aqueous systems like LECs or MCs, but since the carbonate and bicarbonate formation side reactions are not a concern for SOECs, the carbon efficiency is not hampered. In Calculation 6.1, we show sample calculations for maximum carbon efficiencies in aqueous electrolyte-based systems when CO_2 is used as the gas feed. Table 6.2 summarizes the maximum carbon efficiency values, net number of CO_2 molecules required, and the net number of water molecules required for several CO_2 electrolysis products.

TABLE 6.2
A Table Highlighting the Number of Electrons Transferred (z), the Limiting Carbon Efficiency (LCE) Values, the Number of CO_2 Molecules Required (nCO_2), and the Number of Water Molecules Consumed (nH_2O) to Make one Molecule of Product *via* CO_2 Electrolysis Accounting for the Carbonate and/or Bicarbonate Formation Reactions

		Carbonate Formation			Bicarbonate Formation		
Product	**z**	**LCE (%)**	**nCO_2**	**nH_2O**	**LCE (%)**	**nCO_2**	**nH_2O**
Gaseous Products							
Carbon Monoxide	2	50	2	0	33	3	1
Methane	8	20	5	2	11	9	6
Ethylene	12	25	8	2	14	14	8
Ethane	14	22	9	3	13	16	10
Liquid Products							
Formic Acid	2	50	2	1	33	3	2
Formate		67	1.5	0.5	50	2	1
Oxalic Acid	2	67	3	1	50	4	2
Oxalate		Not Applicable			Not Applicable		
Methanol	6	25	4	2	14	7	5
Acetic Acid	8	33	6	2	20	10	6
Acetate		36	5.5	1.5	22	9	5
Ethanol	12	25	8	3	14	14	9
n-Propanol	18	25	12	4	14	21	13

Calculation 6.1.

A calculation showing the limiting carbon efficiency values for different products *via* CO_2 electrochemical reduction in aqueous systems.

To calculate the maximum values for the carbon efficiency and hence the limiting values for the conversion per pass of CO_2 to the desired product(s), we account for the non-Faradaic side reactions of CO_2 with OH^- to form carbonate and/or bicarbonate. We also report the net number of water molecules consumed per molecule of CO_2 in each case.

$$\text{Carbonate Formation Reaction: } CO_2 + 2OH^- \rightarrow CO_3^{2-} + H_2O$$

$$\text{Bicarbonate Formation Reaction: } CO_2 + OH^- \rightarrow HCO_3^-$$

Let us consider the formation of n-propanol *via* the electrolysis of CO_2. The reactions involved are as follows:

1. **n-propanol and carbonate formation**:

$$3CO_2 + 13H_2O + 18e^- \rightarrow C_3H_7OH + 18OH^-$$

$$9CO_2 + 18OH^- \rightarrow 9CO_3^{2-} + 9H_2O$$

Net Reaction:

$$12CO_2 + 4H_2O + 18e^- \rightarrow C_3H_7OH + 9CO_3^{2-}$$

2. **n-propanol and bicarbonate formation**:

$$3CO_2 + 13H_2O + 18e^- \rightarrow C_3H_7OH + 18OH^-$$

$$18CO_2 + 18OH^- \rightarrow 18HCO_3^-$$

Net Reaction:

$$21CO_2 + 13H_2O + 18e^- \rightarrow C_3H_7OH + 18HCO_3^-$$

Thus, for every 3 molecules of CO_2 consumed to produce 1 molecule of n-propanol, 9 molecules of CO_2 are also consumed to form 9 carbonate ions, or 18 molecules of CO_2 are also consumed to form 18 bicarbonate ions. This limits the carbon efficiencies to 25% and 14% for carbonate formation and bicarbonate formation, respectively. Moreover, in the case of carbonate formation, 4 water molecules are net consumed per 12 molecules of CO_2 consumed, while in the case of bicarbonate formation, 13 water molecules are net consumed per 21 molecules of CO_2 consumed.

6.3.4 Reaction Temperature and Pressure

Higher temperatures and pressures generally result in higher rates and also significantly influence the selectivities and energy efficiencies of electrocatalytic processes – however, the effects in electrocatalytic processes get much more complicated especially when liquid electrolytes are involved. The effects of temperature and pressure on kinetics, gas solubility and diffusion, electrolyte conductivity, selectivity, and other process considerations are discussed in Table 6.3. In addition to the enhanced electrochemical performance, the choice of employing elevated temperatures and pressures would also facilitate easier and efficient integration with not only the balance-of-plant but also with the upstream (e.g., flue gas stacks) and the downstream (e.g., Fisher-Tropsch reactors) unit operations. It is also important to note that the use of elevated temperatures and pressures would make the handling of transient operations such as startup and shutdown more challenging.

TABLE 6.3
A Table Summarizing the Effects of Temperature and Pressure on Several Process Considerations for the Various Reactor Types

Process Consideration	Influence of Temperature	Influence of Pressure
Effect on *Kinetics*	Decreased activation energy barriers with increase in temperature: higher j_{total}, and lower overpotentials Difficult to predict $j_{partial}$ trends because $FE_{product}$ trends are unpredictable Rates of side reactions – both Faradaic and non-Faradaic – would also increase with increase in temperature	Higher j_{total} with increase in pressure if positive dependence in the rate law expression Faradaic side reactions are generally suppressed at higher pressures Difficult to predict $j_{partial}$ trends because $FE_{product}$ trends are still unpredictable
Effect on *Gas Solubility and Diffusivity*	The diffusion coefficient and the rate of diffusion increase with increase in temperature: reduced mass transfer limitations The solubility in aqueous solutions decreases as temperature increases: lower local concentration near the electrode, and thus, reduced rates and selectivities	The rate of diffusion increases with increase in pressure: reduced mass transfer limitations The solubility in aqueous solutions increases as pressure increases: higher local concentration near the electrode, and thus, enhanced rates and selectivities The diffusion coefficient is largely independent of pressure for low and medium pressures
Effect on *Electrolyte Conductivity*	The conductivity of electrolytes generally increases with increase in temperature: decreased ohmic resistances and polarization would result in lower cell potentials	Low to medium pressures have negligible impact on the conductivity of the electrolyte

(*Continued*)

TABLE 6.3 (Continued)

Process Consideration	Influence of Temperature	Influence of Pressure
Effect on *Selectivity*	Higher temperatures can provide enough energy to break and make difficult chemical bonds $FE_{product}$ trends can be phenomenologically rationalized but not predicted	Higher pressures would increase the surface coverage of key reaction intermediates: favoring the production of desired products over undesired ones and favoring complex/difficult reactions Decreasing partial pressures leads to a loss of selectivity and increasing partial pressures increases the selectivity $FE_{product}$ trends can be phenomenologically rationalized but not predicted
Other Considerations	Higher j_{total} or $j_{partial}$ with increase in temperature: higher conversion per pass values Decreased cell potential requirements: higher EEs Product separation could be easier if vaporized products at high temperatures condense into liquid products under ambient conditions: this would also minimize liquid product crossover Systems integration with upstream and downstream unit operations would be more energetically efficient at high temperatures Degradation of electrodes is enhanced under high-temperature conditions Very few commercial membranes – **no AEMs** – rated for operation at temperatures > 100 °C Balance-of-plant might get more complicated due to operation at high temperatures Handling of transient steps such as startup and shutdown, and the development of a control strategy for load following can get challenging when operating at high temperatures	Higher j_{total} or $j_{partial}$ with increase in pressure: higher conversion per pass values Decreased cell potential requirements in addition to an increase in FEs: higher EEs Mechanical pliancy of commercially available cloth-based GDEs is better, and they can sustain higher ΔPs The ΔP tolerance of commercially available membranes – especially AEMs – is quite good, but more robust membranes are still needed Product separation can benefit from high-pressure operation as long as the supercritical limit is not reached Use of higher pressures might reduce concentration polarizations in SOECs System integration with upstream and downstream unit operations would be more energetically and entropically efficient at high pressures Degradation of electrodes is enhanced under high-pressure conditions: possibility of failure due to high ΔP across the membrane Mechanical pliancy of commercially available non-cloth-based GDEs is very low: quickly break under a very small ΔP and flood the cell during operation For SOECs, catalyst delamination is enhanced under high pressures Balance-of-plant might get more complicated due to operation at high pressures Handling of transient steps such as startup and shutdown, and the development of a control strategy for load following can get challenging when operating at high pressures

The discussion in this section has significantly narrowed down the choice of optimal/recommended operating conditions. For LECs, MCs, and SOECs – high conductivity, temperature, and pressure are preferred – and low gas flow rate (possibly with recirculation), and catalyst loading are preferred. For LECs and MCs specifically – high electrolyte concentration and electrolyte flow rate are required but not preferred – low to moderate electrolyte concentration (and ideally, the use of pure water), and low electrolyte flow rate (with recirculation) should be targeted. The development of AEMs (especially HEMs) that can be used at temperatures greater than 100 °C is urgently needed.

6.4 REACTION CHEMISTRIES

The next important consideration is the choice of chemistries to be pursued at the cathode and the anode. For example, to synthesize a specific product at a particular electrode, there could be multiple direct and indirect reaction routes possible – the choice of the ideal route, however, would depend on both electrodes – to ensure electroneutrality, minimize Faradaic and non-Faradaic side reactions, decrease energy requirements, improve process economics, and intensify the process performance by choosing optimal operating conditions. In this section, we will briefly discuss the factors affecting the choice of reaction chemistries to be pursued at the cathode and the anode.

The products generated depend on the catalyst, the local reaction environment, and the number of electrons transferred. Chemistries favoring the selective production of a single chemical in a single phase should be preferred if possible, and if not, possibly two chemicals in two distinct phases – production of multiple chemicals in a single phase will require downstream separation unit operations, leading to increased costs and energy requirements.

6.4.1 Faradaic Side Reactions

Faradaic side reactions are also possible, which reduce the FEs toward the desired product(s). The hydrogen evolution reaction is the most common Faradaic side reaction in LECs, MCs, and SOECs. For organic electrolytes, if potentials beyond the stable operating window are used, the electrolyte could degrade. Thus, it is imperative that we understand the selectivity branching points in the reaction networks and direct our efforts to improve selectivity and suppress side reactions *via* manipulating the catalyst composition, reaction microenvironment, and/or reactor operating conditions.

6.4.2 Non-Faradaic Side Reactions

The presence of non-Faradaic side reactions is not so much an issue with SOECs but is a major challenge for LECs and MCs. For example, in the case of CO_2 electrolysis, in aqueous electrolytes, the carbonate and bicarbonate formation reactions are the most important non-Faradaic side reactions, which result in a loss of carbon efficiency, hampering the conversion per pass to the desired product, as explained in the previous section. Thus, the choice of chemistry should be such that it avoids or minimizes any non-Faradaic side reactions.

6.4.3 Tandem Reactions

If a single-step synthesis proves to be challenging, a prudent strategy would be to use the approach of tandem reactions. Using tandem reactions, the desired products can be produced indirectly from the starting material, and the reactor designs and operating conditions can be easily optimized for the individual reactors involved. For example, consider the production of methanol *via* CO_2 electrolysis. One approach could be the direct electrochemical reduction of CO_2 to methanol, as shown in Scheme 6.1. Another approach could be the electrochemical reduction of CO_2 to formic acid first, followed by the electrochemical reduction of formic acid to methanol. In either case, the total number of electrons transferred/required to make one molecule of methanol from CO_2 is the same. Using the approach of tandem reactions allows for individual reactor optimization resulting in higher process efficiency. The approach of using tandem reactions also opens avenues for hybrid processes – for example, if direct electrochemical reduction of N_2 to ammonia proves to be challenging, an alternative approach could be to source the hydrogen *via* water electrolysis and proceed with the conventional Haber-Bosch process to synthesize ammonia.[5] Using tandem reactions could also help in mitigating side reactions. For example, to eliminate the carbonate and bicarbonate formation reactions in CO_2 electrolysis, the electrochemical reduction of CO_2 to CO can be done in an SOEC, and the CO generated can be further reduced electrochemically to ethylene in an HEM-based MC. The approach of using tandem reactions can open a world of possibilities for CO_2 conversion to value-added fuels, chemicals, and intermediates – any of the stable reaction intermediates in the mechanistic reaction network for CO_2 electrolysis (shown in Figure 6.1) can be used as the saddle point for producing chemicals that are difficult to produce in a single step *via* CO_2 electrochemical conversion. In Figure 6.1, we only limit the reaction network to certain functional groups in molecules containing three carbon atoms, but the reaction network can be extended further to produce molecules containing more carbon atoms, heteroatoms, and/or multiple functional groups.

Scheme 6.1.

Production of Methanol *via* CO_2 Electrochemical Transformation.

Approach 1: Direct CO_2 Electrochemical Reduction to Methanol

Reaction:

$$CO_2 + 5H_2O + 6e^- \rightarrow CH_3OH + 6OH^-$$

Approach 2: Tandem Approach – CO_2 Electrochemical Reduction to Formic Acid followed by Formic Acid to Methanol

$$HCOOH + 3H_2O + 4e^- \rightarrow CH_3OH + 4OH^-$$

Reactions:

$$CO_2 + 2H_2O + 2e^- \rightarrow HCOOH + 2OH^-$$

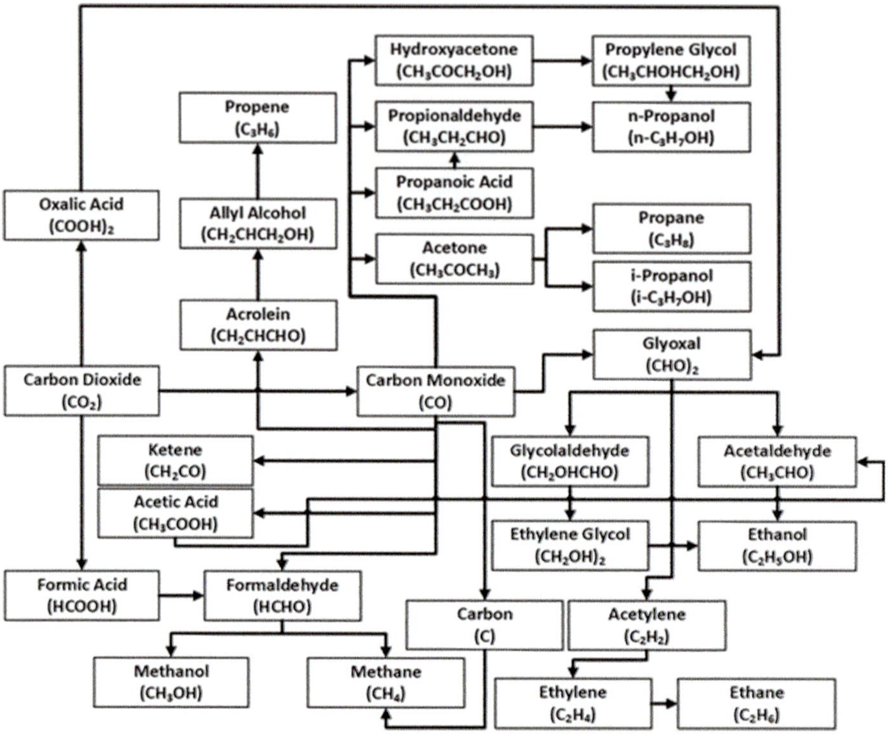

FIGURE 6.1 Comprehensive reaction network showing the several stable intermediates that can be produced *via* CO_2 electrochemical reduction. The interdependencies among several intermediates and the central role of CO are also highlighted. The direction of the arrow represents reduction. The mechanistic pathways might be different or more nuanced than those highlighted here.

6.4.4 Pairing of Reactions at the Cathode and Anode

An electrochemical system must satisfy electroneutrality, and thus, for reactions at the working electrode, there will always be reactions at the counter electrode as well. Thus, from an electrochemical systems perspective, unit operations and process optimization can get tricky because reactions at both electrodes need to be considered. However, there can be some advantages if the right reactions are paired at the cathode and the anode – for example, replacing oxygen evolution reaction at the anode with less energy-intensive reactions such as organic electrooxidation reactions can not only reduce the energy requirements for the system but also produce value-added products at both electrodes, potentially improving process economics and reducing carbon footprint of the system. The choice of chemistry will also be influenced by reactor design and reactor operating conditions – for example, it would be difficult to pursue the electrooxidation of organic wastes in SOECs but could be a possibility for LECs and MCs. By pairing the right chemistries at both electrodes and/or using tandem reaction schemes, ambitious chemistries like epoxidations, insertions, bond formations (e.g., C-C, C-H, C-N, or C-D) should be explored.[6]

6.5 CHOICE OF CATALYST

Catalysts play an important role in several chemical processes to accelerate rates and improve selectivities. Transition metals such as Ag, Au, Zn, Sn, and Cu are often the most common choice as electrocatalysts. In this section, we will briefly discuss the considerations that will influence catalyst selection for electrochemical systems. It is important to realize that a catalyst/catalytic insight that is apt for operation at 1 mA/cm^2 *will not translate* to operation at even 10 mA/cm^2 and definitely not for current densities at industrially relevant levels. Thus, it is *imperative* that electrocatalysts are evaluated under the relevant operating conditions.

6.5.1 Catalyst Morphology and Composition

In addition to the catalysts mentioned above, several other types/classes of catalysts have also been explored as electrocatalysts – these include metal oxide catalysts, abundant earth metal-based catalysts, bi- or multi-metallics (trimetallics or high entropy alloys), surface-modified catalysts (e.g., modifications include ionic liquid moieties, inorganic or organic graftings/coatings), defect-/vacancy-engineered catalysts (e.g., sulfur vacancies), non-metal catalysts (based on carbon and nitrogen), and hybrid catalysts (e.g., metal-nitrogen-carbon). Effects of catalyst morphology (i.e., shape, size, facet, step, edge) have also been investigated in detail. However, the rates may be independent of the above-mentioned characteristics but rather be heavily influenced by the reaction surface area provided by the catalyst.

6.5.2 Tolerance to Impurities

The choice of catalysts is also influenced by the tolerance to several kinds of impurities present in feeds or elsewhere. Impurities are discussed in detail in a separate section later. However, a key takeaway is that catalyst design needs to focus on developing impurity-tolerant electrodes.

6.5.3 Abundance and Manufacturing of Catalysts

It should be noted that the scalable synthesis of some catalysts, especially those that are not yet commercially available, might be tricky. Furthermore, a lot of the transition metals are not earth-abundant, and some are even critical in terms of their reserves.[7] Thus, it is imperative that while in the short-term we focus on improving the catalyst utilization or the catalyst mass activity (mA/mg$_{cat}$), in the long-term we move away from critical metal-based catalysts to earth-abundant catalysts to minimize the supply chain risk. To contextualize the supply chain threat of using critical metal-based catalysts such as IrO_2 in industrial electrolyzers, in Calculation 6.2, we calculate the annual consumption of IrO_2 in the first operational year for just one CO_2 electrolyzer that produces enough CO (assuming 100% FE) for further conversion to methanol (in a thermocatalytic reactor). We then compare the amount obtained to the world's annual production of IrO_2, which is estimated to be approximately 7000 kgs.[8] The calculation shows that about 12% of the world's annual production of IrO_2 will be required for just one electrolyzer in the first operational year assuming a constant annual production rate of Ir and highly optimal operating conditions (see assumptions in Calculation 6.2). Assuming a 90% recycle efficiency, we would still require roughly 4% of the world's annual production of IrO_2 for just one electrolyzer in the first operational year. Once non-idealities are factored in or if the multielectron products are targeted, the absolute amount of IrO_2 required will only go higher; however, after the first operational year, the absolute amount of IrO_2 required will be lower compared to the amounts needed in the first operational year. Improvements in catalyst loading and/or electrode durability will further reduce the amounts of IrO_2 required but when the global chemical production is envisioned using electrochemical systems, the IrO_2 supply falls noticeably short. Clearly, this shows that the use of critical metal-based catalysts could potentially be a significant supply chain issue threatening the scalability and sustainability of electrochemical systems. Furthermore, the iridium deposits are limited and thus, the production rate can decrease over time, if the current mines are depleted earlier than new mines are brought into operation. Moreover, recycling precious metal-based catalysts will require energy, and the CO_2 emissions associated with mining/producing precious metal-based catalysts or critical metal-based catalysts are very significant, threatening the sustainability of electrochemical systems even further.[9]

Calculation 6.2.

A calculation showing the amount of iridium oxide required annually in the first operational year for operating one CO_2 electrolyzer.

Assumptions:

1. Conversion of CO into methanol = 100%
2. Average electrical power consumption (EPC) for CO production = 10 kWh/Nm3
3. jCO = −1 A/cm^2 at a Vcell = −2.2 V
4. Operating year = 8400 hours
5. Catalyst loading = 0.5 mg/cm^2
6. Duration of stable performance = 2100 hours

Basis: 1 MMTPA (million metric tons per annum) production of methanol – it should be noted that the world's annual production of methanol in 2024 is approximately 110 MMTPA.[10]

Based on stoichiometry, the CO to methanol mol ratio = 1:1 (assuming no side reactions)

$$\text{Molar flow rate of methanol} = \frac{1000{,}000{,}000\,\dfrac{\text{kg}}{\text{year}}}{32\,\dfrac{\text{kg}}{\text{kgmol}}} = 31{,}250{,}000\,\frac{\text{kgmol}}{\text{year}}$$

$$\therefore \text{Volumetric flow of CO required}$$
$$= 31{,}250{,}000 \frac{\text{kgmol}}{\text{year}} * \frac{1\,\text{year}}{8400\,\text{hours}} * 0.0821 \frac{\text{atm} * \text{m}^3}{\text{kgmol} * \text{K}} * \frac{298.15}{1} \frac{\text{K}}{\text{atm}}$$
$$\approx 91{,}000 \frac{\text{Nm}^3}{\text{hour}}$$

$$P = V_{\text{cell}} * j_{CO} * \text{area required} = \text{Production Rate} * \text{EPC}$$
$$= 91{,}000 \frac{\text{Nm}^3}{\text{hour}} * 10 \frac{\text{kWh}}{\text{Nm}^3} = 910{,}000\,\text{kW} = 910\,\text{MW}$$

$$\therefore \text{Area required} = \frac{910{,}000{,}000\,(\text{W})}{-1 * -2.2 \left(\text{V} * \frac{\text{A}}{\text{cm}^2} \right)} \approx 413{,}636{,}364\,\text{cm}^2$$

$$\therefore \text{Amount of IrO}_2 \text{ required} = \frac{0.5 \frac{\text{mg}}{\text{cm}^2}}{1\,\text{cycle}} * 413{,}636{,}364\,\text{cm}^2 \approx 206.8 \frac{\text{kg}}{\text{cycle}}$$

$$\text{Now, number of cycles per year} = \frac{\text{Operating time}}{\text{Electrode durability}} = \frac{8400 \frac{\text{hours}}{\text{year}}}{2{,}100\,\text{hours}} = 4 \frac{\text{cycles}}{\text{year}}$$

$$\therefore \text{Total amount of IrO}_2 \text{ required} = 206.8 \frac{\text{kg}}{\text{cycle}} * 4 \frac{\text{cycles}}{\text{year}} = 827.2 \frac{\text{kg}}{\text{year}}$$

$$\therefore \text{Annual consumption rate of IrO}_2 = \frac{827.2 \frac{\text{kg}}{\text{year}}}{7000 \frac{\text{kg}}{\text{year}}} * 100 \approx 11.8\%$$

Assuming a 90% recycle efficiency for IrO_2,

$$\text{Total amount of IrO}_2 \text{ required} = 206.8 + 20.68 + 20.68 + 20.68 = 268.8 \frac{\text{kg}}{\text{year}}$$

$$\therefore \text{Annual consumption rate of IrO}_2 = \frac{268.8 \frac{\text{kg}}{\text{year}}}{7000 \frac{\text{kg}}{\text{year}}} * 100 \approx 3.8\%$$

6.6 ELECTRODE DURABILITY AND PERFORMANCE STABILITY OVER TIME

One of the biggest barriers to scaling electrochemical systems is the poor electrode durability and performance stability over time. This is an ongoing area of research that is making excellent strides in not only improving the fundamental understanding of electrode failure mechanisms but also engineering electrodes or the local environment for enhanced durability and performance stability over time. In this section, we discuss some key electrode structures and possible techniques to manipulate the local reaction environment to enhance electrode durability and demonstrate a stable performance during prolonged operation for LECs, MCs, and SOECs.

6.6.1 Electrode Structure

GDEs used in LECs and MCs can fail *via* a variety of failure modes, for example, flooding, salt deposits on catalyst surface, catalyst leaching, and, thus, the development of an optimal electrode structure is not straightforward. The binders and solvents present in catalyst ink solutions and the method of catalyst layer deposition (e.g., airbrushing, dip coating, electrodeposition, sputtering) on the electrode substrate have shown significant effects on electrochemical performance, but the effects on durability have still not been investigated in detail. Moreover, for electrode optimization, the materials for the electrode components are primarily responsible for improving the mechanical and chemical stability of the electrodes and still need to be optimized (e.g., using polytetrafluoroethylene (PTFE) as the carbon fiber substrate layer). For MEAs, in addition to the considerations for GDEs listed above, the membranes used should be mechanically stable and chemically resistant, and the ionic conductivity of the membranes should not decrease over time. The method of preparation of MEAs (i.e., catalyst-coated backings or catalyst-coated membranes) might also influence electrode durability and performance stability over time. Based on already commercialized systems (e.g., chlor-alkali process), the most ideal choice of GDE at the cathode would be the oxygen depolarization cathode (ODC)-type GDEs reinforced with metal mesh supports and/or foams of minimal thicknesses to improve the mechanical stability, and at the anode, the use of dimensionally stable anode (DSA)-type electrodes is recommended.

For SOECs, the electrodes are porous materials and the biggest issue with these electrodes is catalyst delamination. The other main issue is the formation of a poorly conducting oxide layer between the electrodes and the interconnect plates. Unfortunately, there is no easy fix, and these problems are an active area of research. One common issue across catalysts in all different types of electrodes is catalyst restructuring (*in situ* or *ex situ*), which causes catalyst properties to change that might result in a loss of stable performance and/or also lead to catalyst leaching.

6.6.2 Local Reaction Environment

While the structural and chemical stability of the electrodes is important, the reaction (micro)environment also plays a huge role in influencing the electrode durability and performance stability over time and tailoring the local environment *via* electrolyte engineering can be a very prudent strategy, but detailed durability and performance stability studies are still lacking. For SOECs, the solid oxide electrolytes are stable for >100s or 1000s of hours; however, their stability can be further improved by doping with certain materials or by manipulating the microstructure. This is still an active area of research, and the solution is not straightforward. For LECs and MCs, which use aqueous electrolytes, the situation gets tricky due to the ideal pH requirements and the Faradaic and non-Faradaic side reactions. When organic electrolytes are employed, junction potential and the stable operating window complicate things further. This conundrum might be solved using engineering approaches (e.g., periodic washing of electrode surfaces, self-cleaning systems, operando activation by intermittent doping), but these need to be investigated in more detail before they can be employed at industrial scales. Furthermore, some studies are also needed on employing multivalent cation-based salts and/or surfactants in electrolytes to minimize salt deposit formation and manipulate the hydrophobic-hydrophilic balance to minimize flooding.

Finally, one big challenge for all types of reactors that still remains is the absence of rigorous universal protocols and techniques to assess electrode durability and performance stability over time. The protocols and techniques existing for well-studied electrochemical systems like water electrolysis or fuel cells cannot be used for more complex systems (e.g., CO_2 or CO electrolysis processes) because the individual selectivities ($FE_{product}$) and rates ($j_{partial}$) also need to be tracked over time. Moreover, whenever cycling studies are recommended, an equivalence between the number of cycles and the operating time (in hours) needs to be established. It is also important that the electrode durability and performance stability tests be performed under industrially relevant operating conditions, and the degradation rates are calculated and reported.

6.7 CATALYSIS

The catalysis approach used in electrochemical systems has a significant impact on rates, selectivities, and overpotentials. Thus, it is important to understand the advantages and disadvantages of the different approaches and choose or design an optimal approach. In this section, we will briefly discuss heterogeneous, homogeneous/molecular, and emerging catalysis approaches without going into the mechanistic details.

6.7.1 Heterogeneous Electrocatalysis

Heterogeneous electrocatalysts typically provide a very high surface area for reaction, resulting in higher rates (and thus, higher throughput), but also uncontrolled selectivities. Several approaches have been investigated to tailor selectivities, but further research is still needed to achieve optimized selectivities.

6.7.2 Homogeneous/Molecular Electrocatalysis

Molecular electrocatalysts do not necessarily provide a high surface area for reaction, resulting in low rates in general, but they do provide a very good control over selectivities. Several studies have focused on optimizing the structure of the catalytic complexes; however, further optimization of the catalyst structure is needed to achieve higher rates while maintaining high selectivities.

6.7.3 Emerging Frontiers in Electrocatalysis

From a catalysis point of view for MCs, the way forward should involve combining the best of both worlds: the rate advantage of heterogeneous electrocatalysts and the selectivity advantage of molecular electrocatalysts. We highlight the two most promising approaches – single-atom catalysts and molecular catalysts immobilized on heterogeneous surfaces.

Single-atom catalysts, as the name suggests, are basically catalysts that are single atoms of one or more metals – and in the limit of single atoms, these catalysts mimic molecular catalysts even though they are heterogeneous in nature, giving the advantages of both types of catalysts. The other approach involves immobilizing molecular catalysts on heterogeneous surfaces. By doing so, the selectivity advantage of the molecular catalyst and the rate advantage due to the heterogeneous surface can intensify the electrochemical performance to the desired metrics. Further optimization studies are needed before this approach can be used at industrial scales.

An often-overlooked aspect of electrocatalysis is the proper characterization of mass transfer limitations – the formulation of theoretical principles and the development of experimental techniques to accurately quantify mass transfer in electrochemical systems are urgently needed.[11] Furthermore, an understanding of catalyst surface reconstruction and selectivity splits is especially important to designing catalysts for selective electrochemical transformations. The approach of using molecular catalysts immobilized on heterogeneous surfaces is likely the best approach for scaling MCs.

6.8 ROLE OF IMPURITIES

Several types of impurities (trace or bulk) can adversely affect the electrochemical performance and electrode durability and/or performance stability over time. Effects of impurities on electrochemical reactions have not been studied in great detail as most studies use pure feeds. Moreover, there might be some inconspicuous safety hazards and/or even the possibility of runaway reactions caused by or catalyzed by impurities! In this section, we briefly discuss how impurities play a role in influencing the scalability and sustainability of electrolysis processes.

6.8.1 Gaseous Impurities

Impurities in the gaseous phase, relevant for MCs and SOECs, can come from a variety of sources (e.g., inefficient purification, complete or incomplete combustion/oxidation of various gases together, cross-contamination with other gases/pollutants). Some of the most common gaseous impurities are CO, SO_x, NO_x, H_2S, and O_2. Several studies indicate that these gaseous impurities poison the catalyst surface by blocking the active sites and that this poisoning is not fully reversible. Furthermore, the catalyst can catalyze the electrochemical reduction/oxidation of these impurities, leading to a decline in the electrochemical process performance. One strategy to counteract the negative effects of impurities would be to engineer the catalysts/electrodes so that they are impurity-tolerant and prevent the adsorption of these impurities on active sites. Another strategy would be to manipulate the reaction environment/conditions to suppress the reduction/oxidation of impurities and/or their adsorption on the catalyst surface.

6.8.2 Liquid Impurities

Impurities in the liquid or solution phase, relevant for MCs, are somewhat tricky to deal with. Some of the liquid or solution phase impurities involve oils or organic acids, which will stay in liquid form and then there might be impurities like metals as salts or surfactants which will be in solution form but might precipitate/tailor the electrode surface. These impurities might lower the ionic conductivity of the electrolyte or alter the catalyst microenvironment in a way that might hamper electrochemical performance. The metals-as-salts impurities might even precipitate and form a passivating layer on the electrode surface thereby blocking active catalytic sites. Some chelating compounds might be able to remove these impurities as evidenced in some studies. Engineering the catalysts/electrodes for tolerance to impurities could also be another strategy.

6.8.3 Solid Impurities

Relevant for MCs and SOECs, the most common solid impurity could be solid waste, which needs to be removed *via* some form of pre-treatment. Elemental sulfur or sulfur compounds could also poison the catalyst by blocking the active catalytic sites. Impurities present in materials of construction could react with reactants or products and thus, pose safety risks – for example, Ni impurities present in stainless steel current collectors might react with CO to form metal carbonyls, which can cause severe material degradation among other safety concerns. Periodic washing of the electrode surfaces might be able to remove some macroscopic solid impurities but not the microscopic ones.

These discussions stress the importance of detailed studies on the effects of impurities on electrochemical process performance, especially under catalytically relevant conditions in MCs and SOECs. Such studies would help in assessing if the process can tolerate certain impurities or if their removal is necessary, which in some cases could be incredibly challenging, expensive, and energy intensive. Flue gas-tolerant and wastewater-tolerant electrodes could allow for electrolyzer operation directly with flue gases and wastewater, thereby significantly improving the economics and process integration aspects of electrochemical systems.

6.9 MISCELLANEOUS CONSIDERATIONS

Up until this point, we have discussed several considerations related to either the chemistry or the engineering aspects of electrochemical processes. In this section, we will focus on considerations that will influence system-level decisions for effective process integration when scaling electrochemical systems sustainably.

6.9.1 Scalability

System considerations for scalability are primarily guided by economics and ease of integration if the technology is mature enough to be scaled up. The scale of studies is particularly important because the electrochemical performance observed with a 1 cm^2 active area might not translate to a 100 cm^2 or a 1 m^2 active area.

1. **Intermittency**: In the long-term scenario, electrolyzers should be powered using electricity derived from clean or renewable energy sources. However, renewable energy sources suffer from intermittency due to their inherent nature (e.g., dependence on sun and wind). This means that either energy storage solutions need to be used to ensure continuous operation of electrolyzers, which adds to the costs significantly, or the electrolyzers be also operated intermittently which results in economic penalties due to throughput reduction and operational complications/inefficiencies (e.g., electrode degradation due to shunt currents and polarity reversal, or system efficiency loss due to residual ripples), and, thus, there is a tradeoff. Switchable fuel cell-electrolyzer configurations can be a potential solution for intermittency issues as such configurations would be load-flexible. An efficient strategy for load flexibility would become extremely important when temporal matching is needed.
2. **Modularity**: As discussed previously, it would be great to run several different chemistries at the cathode and/or at the anode, and we should be able to easily switch between different chemistries. This suggests that our electrolyzers be as modular as possible – which is an interesting and a challenging process design problem where operating conditions, choice of membrane, operating pH/pK_a, materials of construction, and balance-of-plant integration would need to be factored in for all the desired chemistries.
3. **(De-)Centralization**: The decision to go for centralized systems or distributed systems, interestingly at this point, is not limited by the state of technology readiness but by the economics, infrastructure, and logistics. For example, does the economics make sense to have distributed systems if we have easier access to feeds, the throughput demand is not high, and it is easy to sell the product off? Or would it make sense to have centralized systems if the throughput demand is high and plant integrations are easy and recommended? Since electrochemical systems have no significant economies-of-scale advantage, the choice of electrochemical systems is justified for cases where bulk production is not needed.
4. **Technoeconomics**: Several analyses have tried to ascertain the economic feasibility of electrochemical processes. However, several key questions still remain open – for example, are the proposed operating conditions feasible, what does the optimal product mix look like, what are the capital and operating expenses required, and what would be the recommended process train/sequence of unit operations, to name a few? Answering such questions will help in determining the requisite scale (or throughput) of operation, and therefore, help in the development of scaling rules.

5. **Scaling Rules**: Scaling rules need to be formulated to guide the scaling of an electrolyzer while maintaining industrially relevant electrochemical performance. For formulating scaling rules, we need to invoke the Buckingham-π theorem in concert with multiscale modeling to identify the key dimensionless parameters/numbers (e.g., Wagner number, transference number), guide the design of feed flow fields as well as select the scaling approach (e.g., scale-out (stacking or number scale-up), scale-up (area scale-up), or a combined scale-out/scale-up approach). Scaling rules should also account for heat management needs for the system due to the heat generated from energy efficiency losses.

6.9.2 Sustainability

System considerations for sustainability are primarily guided by lifecycle analysis (LCA) and associated carbon footprint, and adherence to the 12 basic principles of green chemistry and green engineering. Including sustainability considerations early in the process design influences decisions since the design stage, and thus, makes system integration and late-stage process development much easier and inherently sustainable.

1. **LCA and Carbon Footprint**: Rigorous LCA studies adhering to ISO 14000 standards need to be performed to identify the sources of highest carbon intensities (e.g., unit operations, transporting process feeds, embodied emissions) and other lifecycle impacts (e.g., water consumption, ecological impacts). The need to move away from critical or precious metal-based catalysts has already been stressed from a supply chain point of view earlier; however, the carbon footprint and water consumption associated with the production of these precious metal-based catalysts is also very high – in some cases, even orders of magnitude higher than for earth-abundant metal-based catalysts.[9] LCA studies should perform a detailed lifecycle inventory, and report scope 1, 2, and 3 emissions to enable the identification of avenues to achieve circularity, significantly reduce carbon footprint and lifecycle impacts, and potentially, even achieve carbon-neutrality. Finally, the carbon footprint associated with electricity usage from the grid must also be accounted for – the grid emission factor is heavily dependent on the fraction of renewable energy in the grid.
2. **Principles of Green Chemistry and Green Engineering**: Formulated by Anastas and co-workers, the 12 basic principles of green chemistry and green engineering serve as guiding principles for making a process inherently greener and more sustainable.[12,13] We summarize the principles, their interdependencies, and their influence on electrochemical processes in Tables 6.4 and 6.5. The key message from this discussion is that while tradeoffs will exist, as many principles as possible should concurrently guide sustainable reactor/process design in the early stages of process development. The most suited reactor choices based on green chemistry and engineering principles are MCs and SOECs.

TABLE 6.4

A Table Showing the Green Chemistry Principles and How They Influence System Considerations for Developing a Sustainable Electrolysis Process

Green Chemistry Principle	Remarks/System Considerations
Chemistry Principle 1 (CP1): Prevention of waste/byproduct(s) Also relates to engineering principles EP2 and EP4	Maximize selectivity toward the desired product(s) by changing the applied potential/current
Chemistry Principle 2 (CP2): Maximum incorporation of reactants into the final product(s) Also relates to engineering principle EP2	Thorough atom balance needs to be performed to account for atom incorporation in product(s) and highlight potential areas for improving the atom economy
Chemistry Principle 3 (CP3): Prevention or minimization of hazardous product(s)	Minimize selectivity toward the undesired product(s) by changing the applied potential/current
Chemistry Principle 4 (CP4): Design of safer chemical(s)	Not generalizable – depends on the target molecule(s)
Chemistry Principle 5 (CP5): Selection of appropriate solvent(s)	Electrolyte considerations discussed in previous section(s) apply here but moving toward a liquid electrolyte-free operation (e.g., MCs and SOECs) would be preferred
Chemistry Principle 6 (CP6): Energy requirement for any synthesis should be minimum Also relates to engineering principles EP3, EP4, and EP10	Perhaps, inarguably the most overlooked research aspect in the field of electrochemical systems Significant research to reduce the overpotential and/or increase $j_{partial}$ – but efforts need to simultaneously do both because the energy consumption depends on both Focus should also be on the production of easy-to-separate chemicals to minimize energy consumption associated with product separation
Chemistry Principle 7 (CP7): Use of renewable feedstock(s) Also relates to engineering principle EP12	Feeds like CO_2, seawater, or N_2 are non-depleting feedstocks, but freshwater availability is limited
Chemistry Principle 8 (CP8): Unnecessary derivatization should be avoided whenever possible	Not generalizable – depends on the synthesis pathway
Chemistry Principle 9 (CP9): Use of catalysts rather than stoichiometric reagents should be preferred wherever possible Also relates to engineering principle EP4	Discussions on catalysts and catalysis in prior sections summarize the considerations relevant to this principle
Chemistry Principle 10 (CP10): Products obtained should be biodegradable	Not applicable – depends on the biodegradability of the synthesized molecule
Chemistry Principle 11 (CP11): Use of accurate and precise analytical techniques, especially to control hazardous compounds	Methods for accurate determination of reactant and product crossover still need improvement The robustness of analytical methods should be evaluated before deployment at industrial scales Moreover, online analytical techniques should be preferred

(*Continued*)

TABLE 6.4 (Continued)

Green Chemistry Principle	Remarks/System Considerations
Chemistry Principle 12 (CP12): Substance(s) and the phase(s) of substance(s) should be so chosen to minimize process safety hazards	Operating temperature should be below the flash point/autoignition temperature of reactant(s) and product(s) Considerations like reaction runaways, incompatibilities with materials of construction, short circuiting, shunt currents, static discharge, fires (e.g., metal fires, electrical fires) and explosions (e.g., vapor cloud explosions, boiling liquid expanding vapor explosions) need to be explored Risk-based process safety management system should be employed

Note: Relation with green engineering principles is also highlighted.

TABLE 6.5

A Table Showing the Green Engineering Principles and How they Influence System Considerations for Developing a Sustainable Electrolysis Process

Green Engineering Principle	Remarks/System Considerations
Engineering Principle 1 (EP1): Use of inherently non-hazardous material and energy input(s) and output(s)	Renewably powered electricity is generally more benign than fossil fuel-based energy inputs
Engineering Principle 2 (EP2): Prevent waste(s) formation Also relates to chemistry principles CP1 and CP2	Atom economy, mass activity, and product selectivity should be maximized to avoid/minimize wastage and/or inefficient utilization
Engineering Principle 3 (EP3): Minimize energy consumption of separation and purification operations Also relates to chemistry principle CP6	Thermodynamically favorable and intensified separations should be preferred Process design should also help with this – for example, formation of products in different phases, maximum conversion in the reactor
Engineering Principle 4 (EP4): Mass, energy, space, and time efficiency should be maximized Also relates to chemistry principles CP1, CP6, and CP9	Mass efficiency: selectivity and conversion toward the desired products should be maximized by changing the applied potential or current; Faradaic and non-Faradaic side reactions should be minimized Energy efficiency: mass transfer limitations and the number of unit operations should be minimized through intentional design of the synthesis pathway and/or using process intensification approaches Space efficiency: using modular and distributed systems (in stack configurations) will reduce the physical area required for production Time efficiency: reactions should be completed as quickly as possible, and catalysts should be used to speed up the reactions

(Continued)

TABLE 6.5 (Continued)

Green Engineering Principle	Remarks/System Considerations
Engineering Principle 5 (EP5): Design should be output-driven rather than input-driven	The production output should be based on the demand rather than the availability of feedstock and energy, thus, minimizing excess capacity – modular and distributed configurations should be preferred
Engineering Principle 6 (EP6): Recycle, reuse, or disposition choices should be based on embedded entropy and complexity	Not generalizable – depends on the reaction product(s)
Engineering Principle 7 (EP7): Expendable component(s) should not have a lifespan significantly longer than their utility	Use of non-degradable or non-reusable materials for equipment and/or cell components should be avoided
Engineering Principle 8 (EP8): Design should be customized	For greenfield projects, overdesign should be minimized/avoided without compromising process safety For brownfield projects, process retrofits should be made
Engineering Principle 9 (EP9): Material diversity in multicomponent product(s) should be minimized	Not generalizable – depends on the reaction product(s)
Engineering Principle 10 (EP10): Design must include integration and interconnectivity with available material and energy flow(s) Also relates to chemistry principle CP6	Pinch analysis must be performed to synthesize heat exchanger networks and mass exchanger networks for efficient heat and mass integration
Engineering Principle 11 (EP11): Modularity, versatility, and reconfigurability should be preferred	Design cell stacks that are compatible with multiple chemistries for a range of throughputs
Engineering Principle 12 (EP12): Material and energy input(s) should be renewable/non-depleting Also relates to chemistry principle CP7	Feeds such as CO_2 or N_2 are non-depleting and electricity should be powered from renewable or clean energy sources

Note: Relation with green chemistry principles is also highlighted.

6.9.3 Process Intensification

The use of process intensification approaches (such as the use of plasma, or the application of magnetic and/or acoustic fields) can significantly increase the electrochemical system performance by improving mass transfer, which also improves the catalyst utilization and decreases the overall energy requirements.[14–16]

6.9.4 Process and Systems Modeling

Research studies often report optimization of a single process parameter at the expense of other process parameters – this approach is not suitable when considering scaling. For scaling, all process parameters must be optimized simultaneously. However, this is difficult because several functional relationships between process parameters and process performance are unknown. Here, use of hybrid modeling techniques (e.g., multiphysics-based machine learning models, or combined computational-kinetics-multiphysics models) can be a good approach for multi-objective optimization. Robust process models can allow for *in silico* or virtual experimentation, minimizing the number of experiments needed in the lab and providing additional confidence in the process design before scaling. Robust economic models and lifecycle models can provide accurate costing and lifecycle footprint of the proposed systems, and also help to understand the tradeoffs.

Furthermore, heat exchanger networks (HENs) and mass exchanger networks (MENs) need to be synthesized using pinch analysis to optimize the utility requirements, maximize heat recovery, and enhance selective separation of components from carrier process streams.[17] While we are now at the stage where detailed scaling and sustainability studies can guide the formulation of scaling rules, several key challenges still need to be overcome – for example, theoretical (based on the Buckingham-π theorem), modeling (both multiscale and multiphysics), and experimental (on bench-scale and pilot-scale) scaling studies are needed to de-risk scaling at the intended scale of operation (dictated by economics and supply chain constraints). Systems-at-scale need to be load-flexible. Sustainability studies need to be more rigorous and rely on lesser assumptions, check if green chemistry principles are followed, synthesize HENs and MENs using pinch analysis, and report a detailed lifecycle inventory. There is a great potential for significant energy savings using process intensification techniques, improving both economics and sustainability; however, detailed exploration is needed before these approaches can be used at scale. Process safety studies are needed to de-risk hazards associated with runaways and side reactions/unknown chemistries.

Studies have also shown that post-electrolysis separations are one of the most energy-intensive and carbon-positive steps. Electrifying separations and using process intensification approaches can significantly reduce the energy requirements and the associated carbon footprint if the electricity is from a clean source.

6.10 SUMMARY

Electrochemical systems have some distinct advantages over other kinds of systems – for example, reactions in electrochemical systems are not limited by equilibrium conversion, electrochemical systems can be modular, and electrochemical systems can be operated under milder reaction conditions by changing the potential (or current) to overcome the Gibbs free energy barrier. However, electrochemical systems also have some distinct disadvantages – for example, high capital and operating costs, high energy requirements, and the possibility of disproportionation is high when multielectron reactions are involved.

The key metrics that should be evaluated when electrochemical systems are being considered to replace incumbent technologies or as an alternative to emerging technologies are:

1. **Energy Intensity**: energy intensity is a measure of the amount of energy required for the chemical transformation to produce one ton of product (kWh/ton).

2. **Cost**: capital expenses and operating costs represent the total investment required to set up and operate the plant (US$), but the minimum selling price of a product is also an important measure (US$/ton).
3. **Lifecycle Footprint**: in the earlier stages, carbon footprint of the product ($kgCO_2$/ton) should be calculated, but, if possible, a detailed lifecycle assessment should be conducted.

From a technology deployment perspective, electrochemical systems should only be selected if they are better than the incumbent or emerging systems in *at least one of the three metrics* mentioned above.[18] Electrochemistry-at-Scale is a "*Materials Intensity Challenge*."

REFERENCES

1. Küngas, R. Review—Electrochemical CO_2 reduction for CO production: Comparison of low- and high-temperature electrolysis technologies. *J. Electrochem. Soc.* **167**, 044508 (2020).
2. Endrodi, B. et al. Operando cathode activation with alkali metal cations for high current density operation of water-fed zero-gap carbon dioxide electrolyzers. *Nat. Energy* **6**, 439–448 (2021).
3. Rabinowitz, J. A. & Kanan, M. W. The future of low-temperature carbon dioxide electrolysis depends on solving one basic problem. *Nat. Commun.* **11**, 5231 (2020).
4. Bhargava, S. S. et al. System design rules for intensifying the electrochemical reduction of CO_2 to CO on Ag nanoparticles. *Chemelectrochem* **7**, 2001–2011 (2020).
5. Overa, S., Feric, T. G., Park, A. H. A. & Jiao, F. Tandem and hybrid processes for carbon dioxide utilization. *Joule* **5**, 8–13 (2021).
6. Leow, W. R. et al. Chloride-mediated selective electrosynthesis of ethylene and propylene oxides at high current density. *Science* **368**, 1228–1233 (2020).
7. Bullock, R. M. et al. Using nature's blueprint to expand catalysis with Earth-abundant metals. *Science* **369**, eabc3183 (2020).
8. Clapp, M., Zalitis, C. M. & Ryan, M. Perspectives on current and future iridium demand and iridium oxide catalysts for PEM water electrolysis. *Catal. Today* **420**, 114140 (2023).
9. Glaister, B. J. & Mudd, G. M. The environmental costs of platinum-PGM mining and sustainability: Is the glass half-full or half-empty? *Miner. Eng.* **23**, 438–450 (2010).
10. The Methanol Industry. https://www.methanol.org/the-methanol-industry/
11. Jang, J., Ruscher, M., Winzely, M. & Morales-Guio, C. G. Gastight rotating cylinder electrode: Toward decoupling mass transport and intrinsic kinetics in electrocatalysis. *AIChE J.*, e17605 (2022).
12. Anastas, P. T. & Zimmerman, J. B. Design through the 12 principles of green engineering. *Environ. Sci. Technol.* **37**, 94A–101A (2003).
13. Anastas, P. T. & Kirchhoff, M. M. Origins, current status, and future challenges of green chemistry. *Acc. Chem. Res.* **35**, 686–694 (2002).
14. Islam, M. H. et al. The effects of power ultrasound (24 kHz) on the electrochemical reduction of CO_2 on polycrystalline copper electrodes. *Ultrason. Sonochem.* **72**, 105401 (2021).
15. Mota-Lima, A. The electrified plasma/liquid interface as a platform for highly efficient CO_2 electroreduction to oxalate. *J. Phys. Chem. C* **124**, 10907–10915 (2020).
16. Bhargava, S. S. et al. Decreasing the energy consumption of the CO_2 electrolysis process using a magnetic field. *ACS Energy Lett.* **6**, 2427–2433 (2021).

17. Al-Kalbani, H., Xuan, J., Garcia, S. & Wang, H. Z. Comparative energetic assessment of methanol production from CO_2: Chemical versus electrochemical process. *Appl. Energy* **165**, 1–13 (2016).
18. Grim, R. G. et al. Transforming the carbon economy: Challenges and opportunities in the convergence of low-cost electricity and reductive CO_2 utilization. *Energy Environ. Sci.* **13**, 472–494 (2019).

Part III

Decarbonization of the Iron and Steel Industry

7 Hydrogen for Iron and Steel Decarbonization*

Hanna Breunig and Fabian Rosner

7.1 GLOBAL STEELMAKING OVERVIEW

7.1.1 Definition, Products, and Services

Steel is arguably one of the most important materials to human development, with use in buildings, infrastructure, tools, vehicles, power generation, and machinery. It is a recyclable metal, an alloy of iron and carbon derived from iron ore, that can be modified with manufacturing processes and the addition of other elements to obtain a wide range of material characteristics. Iron by itself is weak and soft, while iron with too much carbon is brittle. Key products include pig iron in molten or solid form (with carbon contents higher than that needed for steel, which is typically 2.14%-wt-C), raw steel, wrought iron, ferroalloys, and steel products[1] in forms such as pipes, tubes, billets, meshes, bars, plates, rods, sheets, and wires. Steelmaking is a large global industry operating at a significant scale in over 60 countries as of 2020. Presently, China dominates global steel production, despite its dependency on imports for the iron feedstock. India, Japan, the United States, and Russia follow, with each producing well above 70,000 million metric tons (MMT) annually (World Steel Association).[2]

Steelmaking falls under the North American Industry Classification System (NAICS) classification *331110 as Iron and Steel Mills and Ferroalloy Manufacturing*. Industries under this classification are primarily engaged with the direct reduction of iron ore, manufacturing pig iron or converting it into steel by reducing carbon content, manufacturing electrometallurgical ferroalloys, or making steel and steel products, excluding bars for concrete reinforcement. Processes that involve adding elements to iron during steelmaking are included.

Early iron ore mining and forging processes in Europe were often done together and at small scales using furnaces for smelting, referred to as charcoal-powered bloomeries.[1] Larger ironworks for smelting at rates of 50 kg iron per day emerged in the early 13th century in Europe with the advent of water mills for energy (Figure 7.1). Prior to the industry's emergence in the Middle Ages in Europe, it is now believed that the appearance of iron in Ancient Egypt was possible due to its extraction from meteorites,[2] long before humans began smelting; however, some evidence

* This chapter is dedicated to Dr. Damien Jullié, whose ingenuity and stubbornness shone through as he taught himself to forge copper on his rooftop by watching YouTube during the pandemic. May we all embody that spirit in our research endeavors.

DOI: 10.1201/9781003655442-10

FIGURE 7.1 Remains of a water-powered iron smelting foundry "la ferrería" in Beasain, Spain. Iron was brought from nearby regions and stored, as can be seen on the left. Charcoal was used to heat the furnaces. The design is attributed to Francisco de Ibero in 1782.

(Photo: June 2024 by Hanna Breunig.)

suggests that primitive smelting practices were discovered and developed in many civilizations, including those in Anatolia or modern-day Turkey, South Asia, Africa, Western Tanzania, and elsewhere.

Following bloomeries were Spanish Catalan forges, and then taller furnaces (stuckofen) in Austria and Germany.[3] These developments ultimately led to the modern blast furnace, which uses air to intensify heat, with the first few examples identified in Sweden, Belgium, and France. The blast furnaces and bloomeries were limited by the availability of biomass for charcoal, rather than iron ore, and depletion of forests became an environmental challenge to the industry, along with contamination from waste. Areas with limited hardwood such as England were some of the first to switch from charcoal to bituminous coal. Similarly, iron ore deposits identified in other areas of the world, such as Australia were left untapped due to a lack of coal or adequate fuel for smelting.

Charcoal- and later anthracite coal-powered ironworks technologies were brought over to the United States, consisting of smelting ore into pig iron followed by refining in forges as done in Europe. Early ironworks were located near anthracite and bituminous coal seams in states formerly known as British Colonies, such as in Pennsylvania. The advent of the Bessemer furnace in the 1800s led to the replacement of old blast furnaces. However, the Bessemer process and variations of it were ultimately replaced by the Linz-Donawitz basic oxygen furnace (BOF) steelmaking process, which was significantly faster and more efficient at removing impurities. The breakthrough of the BOF involved being able to provide adequate quantities of concentrated oxygen at scale.

Since then, blast furnaces coupled with the BOF process (BF-BOF) have operated successfully, with modern processes described in detail in the literature.[4,5] Generally, BF-BOF begins with sintering and coke production, where sintering involves the combination of iron ore, coke breeze, and iron-bearing minerals. Coke is derived from coal through pyrolysis (~700-800 kg coke per metric ton of coal),[6] with some other sources of carbon also used, including recycled rubber tires. The pig iron resulting from the blast furnace is sent to the BOF along with ferroalloys, or combined with ferroalloys and scrap steel in an electric arc furnace (EAF). The blast furnace is not without its limitations, despite being a highly relied upon process. Aside from the large consumption of coal and associated emissions, it requires sintering and does not allow for powder ores to be used directly.

While BF-BOF continues to be used today in ironmaking, other furnaces have emerged for the reduction of iron ore in the United States, China, and elsewhere.[7,8] There are 97 shaft furnace modules designed by Midrex operating worldwide, comprising approximately 80% of today's natural gas-based DRI.[3] Other companies have developed shaft furnaces including Energiron. Direct reduction in a shaft furnace produces a metallic iron product referred to as "DRI" from iron ore without having the iron transition into its liquid phase, with ~92% iron content. It has low density and high porosity, with a melting temperature of 1300 °C implying that substantial energy input is required to melt the DRI further downstream in order to produce cast iron products. The DRI product has a different composition than pig iron (~96% iron content), which has high density and low porosity, and a slightly lower melting temperature of 1250 °C.

Shaft furnaces can be run on a diversity of gases, including natural gas, hydrogen, syngas from natural gas or coal, and blends of the three. In theory, iron ore can be reduced using a wide range of elements (Ellingham-Richardson),[9] but only CO (orange line) and H_2 (red line) are economical (Figure 7.2).[9] The use of H_2 for reducing iron ore is not a new concept, with investigations of H_2 and CO for reduction published in English more than five decades ago.[4] Evaluations of reduction conclude that operating at higher temperatures is critical when reducing with H_2 as the stability of iron increases with temperature, leading to improved gas utilization. Conversely, the stability of iron increases at lower temperatures when considering Fe-O-C systems, but the slow kinetics of CO-based reduction require that higher temperatures are used, sacrificing some stability. As such, DRI furnaces are operated at temperatures far above 570 °C.[9] Estimates of hydrogen consumption range from 0.08 to 0.12 metric tons H_2 per metric ton of iron ore.[10]

To produce a finished steel product, DRI product is delivered either hot or cold to an EAF and subsequent casting and finishing processes, to supplement or replace pig iron. Without further processing into briquettes for transport, DRI product is a hazardous and challenging material to handle, as it can self-combust. The product must be cooled, processed, and stored in an airproof manner until picked up for transport. The DRI product can be used in place of pig iron in EAF for processing into steel products. Figure 7.3 shows the process for a DRI-EAF system run on natural gas, where syngas entering the DRI is produced in a steam methane reformer, and DRI product is fed directly into an EAF. Facilities with a hotlink between the DRI furnace and the EAF feed DRI product directly from the shaft furnace without any necessary

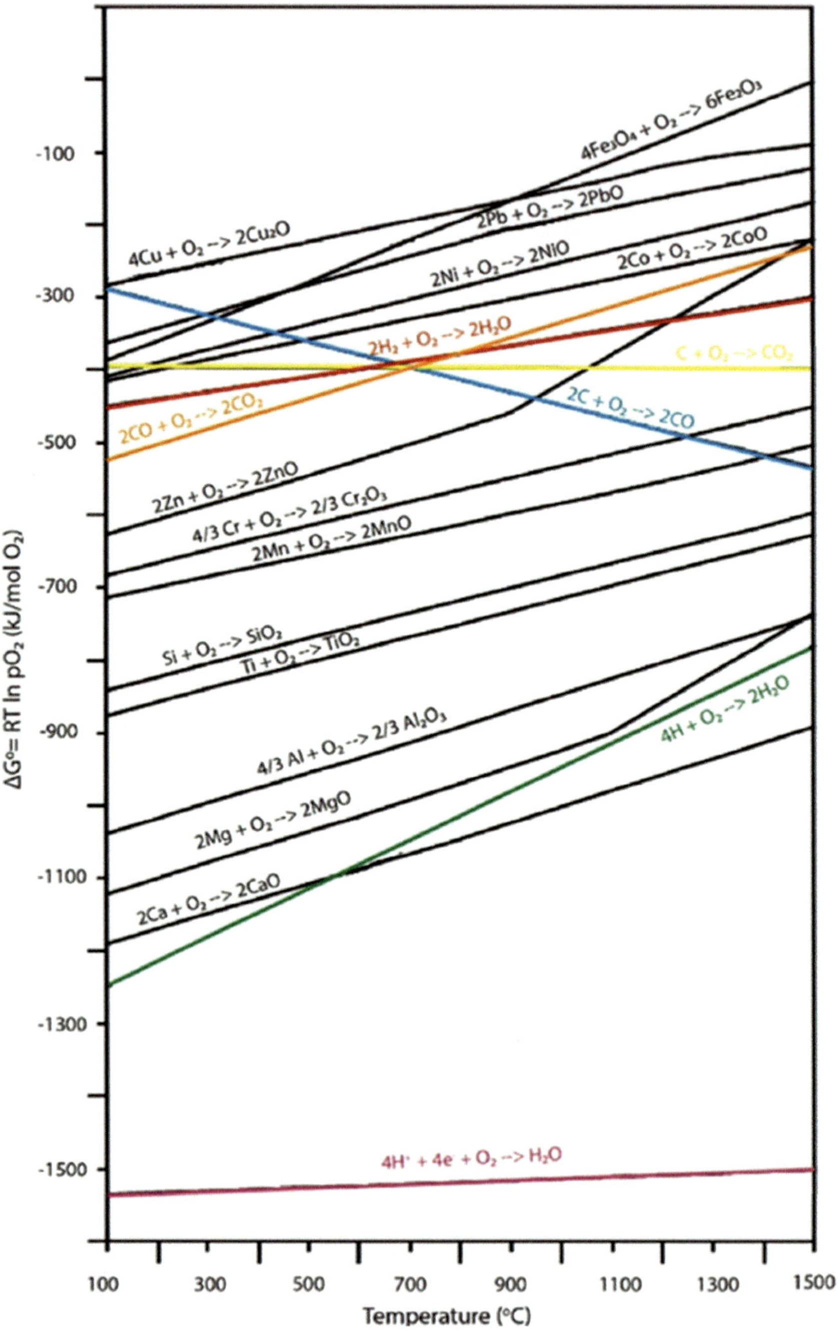

FIGURE 7.2 Ellingham-Richardson diagram illustrating standard free energy of oxide formation with temperature.

(Reproduced with permission 7/22/2024.[9])

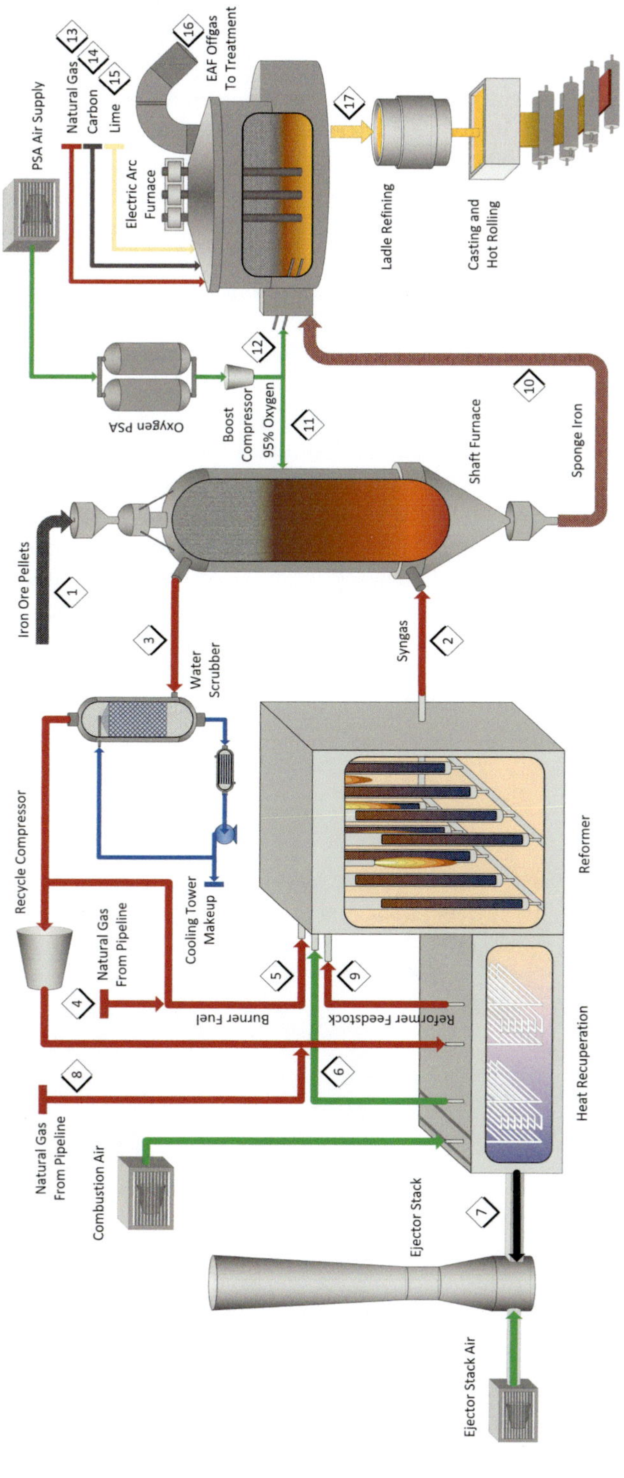

FIGURE 7.3 Steelmaking facility using natural gas DRI hotlinked to EAF and postprocessing.

(Reproduced with permission 7/29/2024.[11])

cooling or processing step. Carbon is added to the EAF, along with lime. Beyond leveraging the embodied heat in the DRI product, there are other synergies of tightly coupled DRI and EAF systems. As an example, cooling water used for the EAF jacket and for the EAF top gas cooling can be recirculated and processed with cooling water consumed at the primary steel DRI facility, including water recovered from the DRI top gas.[11] As of 2007, 97% of steel products from EAF and BOF went through continuous casting processes, followed by hot and cold rolling and finishing, representing another significant demand for thermal energy in the steelmaking sector.

Alternatively, EAF mini-mills typically receive pig iron and scrap steel as feedstocks, along with oxygen, carbon, and lime. Once at a facility with an EAF, the DRI product is reheated. Newly proposed oxyfuel combustion designs from companies such as Air Products are being developed for this reheating process (Figure 7.4). Each EAF operates as a batch process, but a facility may operate multiple EAFs in parallel to allow for more steady operation. Estimates for continuous-feed EAF processing of scrap steel in China suggest a carbon intensity of 0.911 kg CO_2e per kg steel product, with 73% EAF and 27% continuous casting and rolling. Air Products estimates the DRI and conversion process to be roughly 0.85 kg CO_2e per kg steel, at a capacity of 50 tons per hour.[12] The energy intensity of hot rolling alone is on the same order of magnitude as energy consumption for sintering and the steelmaking process (EAF) itself (between 50 and 75 kg coal equivalent (kg ce)/t-product).[13] One kg coal equivalent is 29.307 MJ and is a unit of energy intensity commonly used in China.

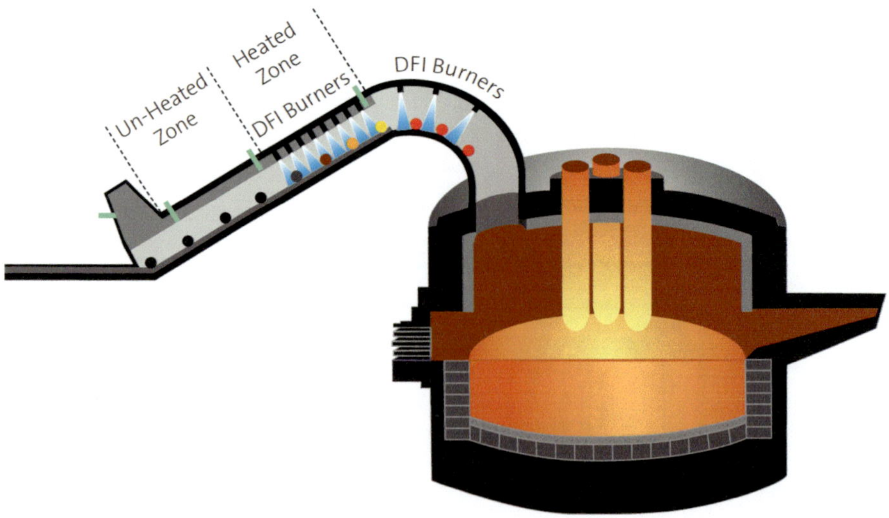

FIGURE 7.4 Preheating process concept from Air Products for DRI product entering an EAF.[12]

7.1.2 Trends

Decarbonization of steelmaking has gained momentum globally, with prominent companies announcing aggressive targets for carbon emissions reduction and sustainability. Energy efficiency improvements have advanced to lower sector risk related to energy reliability and price volatility, and recycling of scrap steel has accelerated in many countries, particularly those with limited domestic iron ore supply. The transition from blast furnaces to DRI furnaces, as discussed previously, is a major trend in the industry that affects both the iron and steelmaking industries. This has allowed for a transition away from coal, and the placement of integrated primary steel production in new geographic locations rich with natural gas and iron ore, rather than coal seams and iron ore. Research for the direct electrification of iron reduction remains an active area of research but is seen to be costly and not deployable in the near future. An emerging alternative is the use of electrolysis to produce iron, which is then converted to steel. One example is Boston Metal's molten oxide electrolysis technology, which has been demonstrated at gram scales, with the potential to scale up to tons per day of iron.[4] Saltwater and iron oxide can be converted into iron electrochemically, with byproducts being sodium hydroxide and chlorine.[14] Noble et al identify a process that leverages components from the mature chlor-alkali process, in an effort to generate iron using electricity without steel prices. Other research has focused on the potential of using slag as a source of carbon capture material,[15] or as a source of critical materials such as zinc.[16]

Carbon capture from top gas of shaft furnaces has so far been challenging due to the significant temperature and volume of gas to be handled but may emerge as a valuable point source of CO_2. Concentrations of CO_2 in waste gases from the steelmaking process can reach as high as 40%, and a wide number of technologies have emerged to capture CO_2 from different sources at a steel facility.[17] Notable reviews suggest separation is feasible using solid sorbents.[18,19] Rather than separate CO_2 from waste gas streams, waste heat from steelmaking processes has also been discussed as a source of heat for the regeneration step in direct air capture systems. A high-temperature heat pump can be leveraged to deliver heat via hot water boilers with high efficiency, avoiding consumption of natural gas for sorbent-based direct air capture systems.[20]

Beyond these strategies for decarbonization, the use of hydrogen and hydrogen-rich syngas for DRI furnace operation has gained increasing attention, as decarbonization roadmaps point to substantial opportunities for emission reduction in primary steel production, in relation to BF-BOF systems. In a subsequent section, we present opportunities for using hydrogen in steelmaking. Thanks to political pressure to decarbonize the steel industry, research funding across multiple countries has led to a number of experimental, pilot, analysis, and demonstration projects aimed at advancing hydrogen-based iron and steel production systems. We provide a few examples. First, pilots are being designed and operated that coupled DRI furnaces with hydrogen production and conditioning systems. Production systems under study include natural gas steam methane reforming (SMR), natural gas methane pyrolysis such as the technology under development by Molten Industries,[21] and electrolysis including proton exchange membrane (PEM)[22,23] and solid oxide electrolyzer cell

(SOEC)[5] hydrogen production systems. Hydrogen from natural gas/methane pyrolysis has a lower energy efficiency than SMR, consuming twice the natural gas as SMR, but has a similar energy efficiency when considering the penalty of carbon capture and sequestration on SMR efficiencies. Pyrolysis also produces a solid carbon product that may be useful in the steelmaking process. Second, new research has emerged suggesting that hydrogen can be used with or without supplemental syngas or natural gas in DRI furnaces, with tunable effects on iron ore quality.[24] Key parameters driving pore structure and reduction efficiencies include reaction rates, temperatures, and hydrogen-to-natural gas ratios. Notable challenges with a "carbon-free" DRI process are impacts on the downstream EAF steelmaking process, ranging from energy demand to stirring conditions.[25] Third, systems analysis of the technical, economic, and environmental implications of alternative process designs and supply chains has emerged and highlights key decisions that impact overall performance.[26–28] For example, electrolysis systems run on intermittent renewables require large hydrogen storage systems.[26] Steam methane reforming-based hydrogen can operate at high capacity factor, avoiding the need for storage.

7.2 GLOBAL IRON ORE MINING

7.2.1 Definition, Products, and Services

Meteorites and iron ore are the only sources that can be used to make primary iron. Iron ore mines typically require blasting using explosives and transportation of rock with electric shovels (e.g., capacity of 85 tons of rock) and large trucks (e.g., capacities of 240 tons).[6] Material is transported to a processing facility or loaded onto train cars. The iron ore mining industry delivers the ore raw or processes the ore into a concentrate, briquette, or pellet. By the NAICS, the iron ore mining industry falls under 212210. Under this classification, the industry is primarily engaged with developing mine sites, mining, and preparing iron ores and manganiferous ores, and, second, producing sinter iron ore and iron ore agglomerates (excluding products made in iron and steel mills).[7] Such agglomerates include taconite, which is a hard low-grade iron ore (over 15% iron) and the predominant iron ore remaining in the United States. The iron may be magnetite or hematite. In fact, the hematite iron ore Ngwenya mine in Mbabane, Eswatini is thought to be the oldest known mine, at over 40,000 years old, although not mined for smelting initially, and only closed in 2014. Prior to modern developments, with increasing demand, Europe transitioned from bog iron ore mining such as that conducted in Poland to underground mining. It was hundreds of years later that iron ore mining would begin in Australia in the late 1800s, despite discoveries of deposits due to limited energy for smelting. The discovery of effective processes for upgrading taconite by Dr. E.W. Davis in the 1940s catalyzed the mining of taconite deposits that were discovered in the 1870s in the United States but were too costly to leverage.[8]

Today, Australia produces the most ore per year, at 960 MMT annually in 2023, with exports to China representing possibly the largest-ever single product export between two countries in human history. From the 1960s and on, the iron ore mining industry in Australia has experienced tremendous growth, due to large deposits (the world's largest reserve at ~29%) and significant exporting to Japan and then China and South Korea. Brazil (440 MMT), China (280 MMT), and India (270 MMT) are the next largest producers, followed by countries with production that are substantial but under 100 MMT, most notably Russia, Iran, Canada, South Africa, Kazakhstan, the United States, Sweden, and Ukraine. The companies Vale, Rio Tinto, BHP, and Fortescue Metals mined nearly half of global iron ore in 2019.[9] In comparison, the U.S. iron ore mining industry is a 7 billion dollar revenue industry producing 44 MMT in 2023, employing over 4,500, and comprised of approximately 18 large businesses.[29] Cleveland-Cliffs (43.7%) and U.S. Steel Corp (27.55%) are the major players in the industry.[30] Iron ore production often operates in a vertically integrated manner regionally, where companies operating the steel mills have ownership or contracting with suppliers of ore and pellets. For example, Hibbing Taconite Company in Hibbing, Minnesota, United States, produces 29 million tons of ore annually, equivalent to 7.8 million tons of standard iron-bearing pellets and 5 million tons of iron ore pellets (2022). Pellets are transported by rail to a Cleveland-Cliffs facility in Indiana, United States. As of 2023, the mine is 30,670 acres and employs 730.[10] This region in Northeastern Minnesota, referred to as the Iron Range has a long history of iron ore mining, due to the high iron content of surface rocks. As mining of high quality ores ended in the 1980s, taconite mining is now the primary method, with pellets for blast furnaces transported by rail and the Port of Duluth to Ohio and other states for steelmaking.

Globally, emerging sources of iron ore are being considered for development, including the Simandou iron ore project in Guinea, which holds an estimated 2 billion tons of relatively high-quality iron ore reserves, where Chinese and Australian companies along with the Guinea government, have sought to begin production at a scale of 60 MMT per year by 2025. The Republic of Guinea boasts a large iron ore deposit that is currently under development by several domestic and international stakeholders, including Chinese- and Singapore-based companies. Other countries with significant fractions of global reserves of iron ore include Brazil 19%, Russia 15%, and China 12%.

Processing of mined materials includes comminution, or the reduction of particle size via crushing, grinding, milling, iron concentration, and pelletization. Nonferrous materials removed during the process of extracting iron ore are referred to as "gangue" and are disposed of through a number of methods at or near mines. Recovery of iron out of gangue has gained interest. Mills may use water to get the rock material down to a fine powder. The magnetic nature of ferrous materials is leveraged in separation to recover the concentrate, or taconite powder, while the gangue is deposited in mine tailing storage facilities or open piles. Taconite processing facilities in the United States produce pellets with 60% iron at 0.5-inch diameter.[11] The concentrate is still wet when rolled into briquette and heated into hardened pellets. Trains, trucks, and ore ships are used in the United States to transport pellets to steel mills. As of 2024, revenue in the U.S. iron industry is largely from the production of iron ore pellets (89%), with other products making up the remaining.[30] Pellets for direct reduction, or DR-grade pellets, have started to be produced by companies such as Cleveland-Cliffs and USSteel to prepare for DRI methods replacing blast furnaces. DR-grade pellets have higher iron and lower silica by %-wt. Silica and alumina are impurities in pellets that are removed in blast furnaces and DRI.

Materials handling is the largest consumer of energy at mines in the United States, with 87% from diesel (Figure 7.6). Comminution or the reduction of rock to finer diameter particle sizes, accounts for 75% of beneficiation and processing energy consumption. A review of energy consumption at mines by Engeco in 2021 (Figure 7.5) suggests energy intensities for magnetite processing of 0.3 GJ/t ore, attributed to mining diesel (41%), comminution (43%) and other processes, and 0.15 GJ/t ore for hematite processing, attributed to mining diesel (90%) and processing.[12] Ores of lower grade require more energy for processing; however, it is suggested across the full supply chain that magnetite offers lower emissions in blast furnaces (Figure 7.6).

7.2.2 Trends

Rising prices of oil increase the cost of ore transport, and fluctuations in steel production have affected revenues of the iron ore mining sector. Barriers to entry are considered very high, and the market can be significantly impacted by fluctuations in demand for steel products. The trend in the United States and globally has been consolidation and refocus on energy efficiency.

Despite these challenges, low interest rates and the U.S. Federal Reserve working to lower inflation can enable investment in new equipment and lowering energy consumption, as well as policy incentives that drive up consumption of steel, such as tax credits for electric vehicles, renewable energy equipment, and the CHIPS and Science Act driving up domestic production of electrical components.

Several regions have placed new emphasis on bolstering domestic resources and production of iron and steel, such as the United States. The Bipartisan Infrastructure Law stipulates all iron and steel used in projects must be U.S. made.[30] In comparison, iron ore produced in Australia and Brazil is of low-cost compared with U.S. ore products, due to massive economies of scale and due to the appreciation of the US dollar. However, exports to China have increased and are expected to remain high due to urbanization but may be affected by country relations.[12]

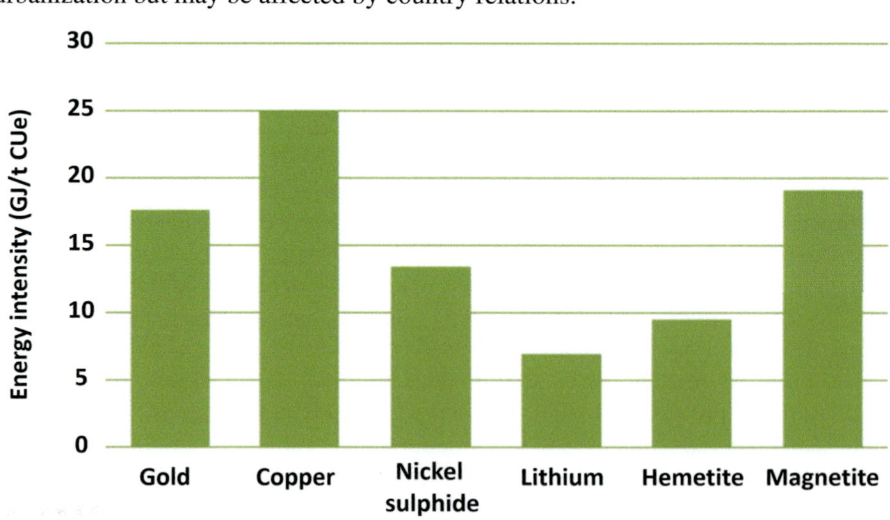

FIGURE 7.5 Energy intensities of different mined commodities (2021). Reproduced with permission.[12]

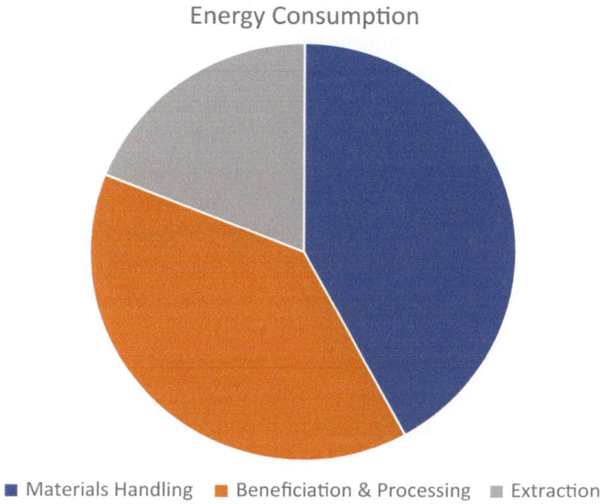

FIGURE 7.6 In 2003, the U.S. Department of Energy released a third Roadmap "Exploration and Mining Technologies," highlighting how materials handling consumes 42% of energy at mines, while beneficiation and processing consumes 39%, and extraction consumes 19%.[31]

A number of prominent iron ore companies have announced interest or partnerships with hydrogen developments in the past decade.[13,14] For example, in 2021, a conference organized by Energy & Mines, a Canadian company providing information and media related to decarbonizing mines, focused on outlining a vision for hydrogen and mines as part of a broader discussion on decarbonization in the mining sector. Mining companies such as Fortescue announced net zero emissions across their value chain by 2040. With mines often having large land available, it is possible that renewable generation and hydrogen production could be collocated on the mine to reduce costly electricity charges and hydrogen transportation, furthermore making the mining sector a potential early adopter for renewable hydrogen companies to pursue. At the same time, trends in increased demand for recycled steel pushed by green initiatives have further reduced demand for mining, especially with evidence of the high quality of recycled steel. This is correlated with the increased deployment of EAFs, which have lower emissions and less dependence on iron ore availability as blast furnaces.

Key in riding out the market's high-risk/high-reward nature is ensuring high output and efficiency, and ensuring products can meet changes in market demand. As an example, notable innovations on the iron mining side include the production of feedstocks for EAFs, such as hot briquette iron (HBI) and DRI.

7.3 HYDROGEN END USES IN IRON AND STEEL

7.3.1 Applications of Hydrogen and Demonstrations

Mining industries including iron ore companies and steel manufacturers have shown major interest in sustainability branding, with hydrogen featured prominently. Hydrogen has applications in the iron and steel sector as a reducing agent but also as a means of delivering renewable fuel, heat, and power to various end uses. Hauling and delivery vehicles, shipping vehicles, and equipment used for a variety of processes ranging from comminution (grinding down rocks) at mines to hot rolling can all be technically operated on the electricity generated from hydrogen-powered fuel cells during times of low renewable electricity generation. Particularly for high-quality, high-temperature heat, hydrogen can be utilized through combustion in place of natural gas or coal, which can have advantages over electrified heating methods such as resistive heating. Vehicles and ships can also operate on the combustion of fuels generated from renewable hydrogen such as compressed hydrogen or secondary hydrogen fuels like methanol. In this section, we detail the key applications of hydrogen in this sector, specifically the processes involved and the state of development.

7.3.1.1 Hydrogen at Iron Mines

Surface mining requires rugged, large pieces of equipment that can operate long shifts with little downtime. Equipment for collecting, loading, and transporting materials generated from blasts includes loaders and trucks, while equipment for processing rock material includes crushers, grinding, screens, mills, concentrators, and pelletizers.[32] All processes are aimed at separating gangue from iron ore concentrates. Much of the equipment is presently run on diesel, while grinders and mills are run on electricity.

Concepts for off-road vehicles include fuel cell systems with 700 bar onboard storage or metal hydride storage tanks. The company Australian Mines Limited submitted a patent on solid-state hydrogen storage leveraging a metal hydride, which is noted to be for "transport" applications.[33] Developing advanced onboard storage systems for vehicles and loading equipment, as well as mobile refuelers would further improve potential advantages of H_2 fuel in the mining sector. Whether a stationary refueling system is sensible will vary by mine, and in some cases, mobile refuelers may be more effective at reducing "downtime" associated with refueling the equipment. Hydrogen-powered drill rigs, excavators, large haul trucks, and bulk transportation systems ranging from ports to trucks to rail may all operate with shorter refueling times than battery-based models that require longer charge times. In South Africa, Anglo-American partnered with First Mode,[15] which built the largest hydrogen battery hybrid hauling truck "nuGen" in 2023. With a capacity of 290 tons and a total laden weight of 510 tons, the 2 MW fuel cell is a truly impressive example of hydrogen replacing a diesel system.[16] Forty haul trucks are currently being retrofitted at the Mogalakwena South Africa site. In 2024, First Mode also deployed a retrofitted Komatsu 830E ultra-class haul truck at TransAlta Centralia Mine, a former coal mine being reclaimed.

Another potentially large demand for hydrogen is the replacement of ammonium nitrate-based explosives with hydrogen-based explosives. This may simply mean replacing Haber Bosch ammonia with "green" ammonia synthesized from renewable hydrogen, or it may mean using hydrogen directly. Queensland Nitrates supplies mining in Queensland and partnered in 2023 on a feasibility study to produce 3,500 tons per year of electrolyzer hydrogen to produce 20,000 tons per year of green ammonia for explosives.[34] In 2023, Incitex Pivot Limited, a subsidy of Fortescue Metals Group in Australia, shut down natural gas ammonia production in Australia to produce 400,000 tons per year of ammonia from a large hydrogen electrolysis facility, indicating it would increase the cost of ammonium explosives until costs decrease, roughly by 2040.[17] In 2021, Enaex in Chili, a major producer of ammonium nitrate for explosives, moved forward with a green ammonia facility expected to produce 18,000 tons per year from electrolysis hydrogen.[18]

7.3.1.2 Hydrogen in Blast Furnaces Primary Steel Facility

Rather than completely replacing blast furnaces, it is technically possible to inject H_2-rich fuels such as syngas, coke oven gas, the blast furnace top gas, or town gas into a blast furnace for ironmaking as an auxiliary reducing agent (ARA), replacing CO, which is produced inside the furnace from pulverized coal or coke.[35] When natural gas is injected into the high-temperature section of the furnace, it readily converts to CO and H_2, avoiding the need for a natural gas SMR system upstream. Typical consumption rates of coke per ton of hot metal produced in conventional blast furnaces vary widely in practice (334–800 kg/tHM), as do theoretical limits (<200–300 kg/tHM).[36] To maintain temperatures, gas fuels may require preheating, or the blast furnace may need enriching with oxygen. Challenges exist, including incomplete conversion of the pulverized coal, demonstrating overall efficiency and heat transfer, temperature control, minimizing water condensation, and other factors that will impact the rate of production and quality of the product pig iron.[37] Initial work suggests higher hydrogen content fuels are most effective in reducing coke consumption. Older reports suggest 3 to 3.3 kg coke are replaced per kg H_2 at an operating temperature between 900 and 1000 °C, up to 27–44 kg per ton HM (Barnes 1975; Astier et al 1982),[38,39] while Yilmaz et al 2017 presents a process design achieving 3.93 kg coke per kg ARA.[36] Process models exploring the injection of pure H_2 along with pulverized coal suggest a 21.4% reduction of direct carbon dioxide emissions (Figure 7.7).

A number of prominent companies have performed successful demonstrations of the concept as of the year 2024. For example, Nippon Steel Corporation in Japan, piloted using heated hydrogen in 2023 with the hopes of reaching a 30% reduction in direct carbon dioxide emissions. Cleveland-Cliffs in Ohio, United States, piloted a hydrogen injection trial in one of its largest blast furnaces in 2023. That same year, Tata Steel in India announced a 40% injection of hydrogen into a blast furnace via standard injection systems, lowering coke consumption by 10%. Thyssenkrupp began transitioning a blast furnace in Duisburg from coal dust to hydrogen for up to 24 hours in 2019 before delays occurred during the coronovirus pandemic. Finally, German steel manufacturer Salzgitter also announced ambitious plans for replacing coke consumption in blast furnaces, relying on wind power to achieve renewable hydrogen generation.

7.3.1.3 Hydrogen in DRI Primary Steel Facility

As discussed in previous sections, an alternative to the blast furnace is the direct reduction shaft furnace or DRI process, which presently runs on reformed natural gas (a.k.a syngas), and can run with varying H_2:CO ratios through hydrogen injection without retrofit. Modified DRI designs have emerged that can operate on 100% hydrogen; however, in the near term it is expected that natural gas DRI will be retrofitted to replace closer to 30% of natural gas feedstock for reduction with renewable hydrogen.[10] The ability to lower direct CO_2 emissions from primary steel manufacturing is

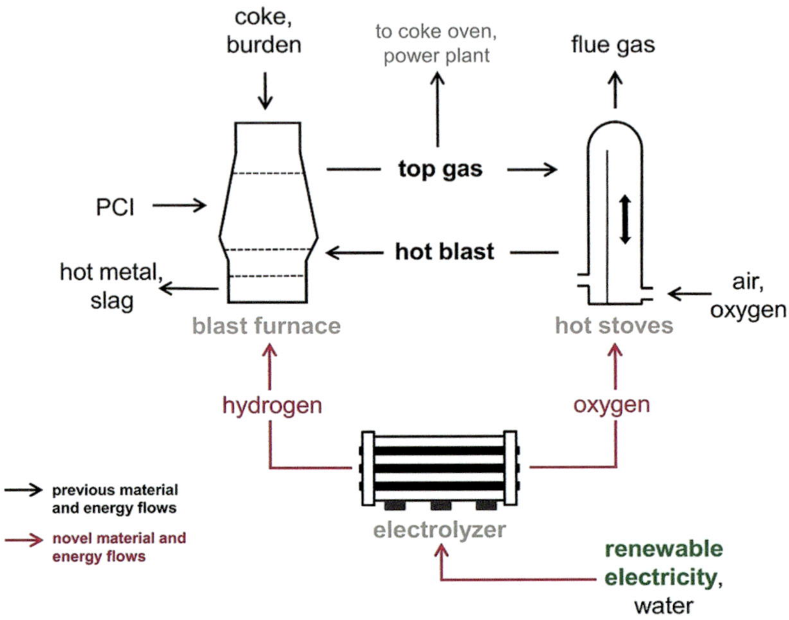

FIGURE 7.7 Hydrogen from water splitting delivered to a blast furnace for reduction. PCI: pulverized coal injection.

(Reproduced with permission 7/29/2024.[36])

even greater than the use of H2 in blast furnaces, and as such, hydrogen-based DRI has gained significant attention from government agencies, industry, and the research community. As noted by Spreitzer and Schenk,[40] there are two major challenges associated with the use of hydrogen for reducing iron oxides. The first challenge is simply producing enough hydrogen to run the process. If all crude steel production was converted to DRI run on 100% hydrogen, demand would be roughly 12 MMT per year in the United States.[10] Presently only 10 MMT of hydrogen are produced (mostly from fossil fuel sources), with dedicated offtakers including ammonia and petroleum refining. The second challenge is the proof of concept, demonstration, and optimization of high or 100% hydrogen furnace operation while ensuring high equipment utilization, feedstock conversion, and product quality.

Reduction of iron can be accomplished with pure hydrogen, and a number of reviews are available exploring key aspects related to the reduction process chemistry and its effect on iron ore pellet conversion.[40,41]

Rosner et al present a process and cost analysis of a DRI shaft furnace system coupled or "hot linked" with an EAF, where carbon is added to the DRI product exiting the DRI in the EAF.[11] Results suggest a DRI-EAF facility producing 1,162,100 metric tons of steel product annually has a product cost of United States Dollar (USD) 1018/kg steel when the price of hydrogen is USD 7/kg H_2. Using the same methodology, we estimate a facility of the same size, without a collocated EAF, would produce 1,385,200 metric tons of DRI product annually at USD 890/kg product, based on modifications described below. Companies linking DRI with EAF include HyBrix, Midrex, ArcelorMittal, SSAB, Liberty, and H2 Green Steel. Few companies are using ultra-high concentrations of 100% hydrogen feeds; however, Midrex and Energiron now offer high hydrogen content furnace options.

Configurations and operation of DRI run on 100% hydrogen vary; however, it is expected that some natural gas will still be used and injected in the lower part of the DRI furnace for cooling and carbon addition. Process designs have been published recently describing and benchmarking configurations and operation scenarios for the DRI (Figure 7.8 from Rosner et al). Hydrogen entering the DRI must be preheated and compressed to at least two to five bars to push through piping. Top gas from the DRI will contain unreacted hydrogen. This top gas must be sent through a treatment process before recirculating to drop out moisture and particulates before combining it with a fresh hydrogen feed stream, which is then heated and sent to the DRI. Research investigating the integration of all systems along with feed gas and recycled gas preheating has been funded by the US Department of Energy, including pilot projects and the University of California, Irvine, and Missouri State University. Critical data will emerge from the coupling of the DRI with intermittent electrolysis-based hydrogen. More than likely this hydrogen will need to be produced using more stable electricity such as grid, nuclear, or hydro. ArcelorMittal in Germany, Salzgitter AG in Germany, SSAB in Sweden, LKAB in Sweden, Liberty in France, and H2 Green Steel in Sweden claim to use some fraction of electrolysis-based hydrogen. Many electricity grids continue to use carbon-intensive fuels and hydrogen produced from grid-powered electrolysis would do little to lower overall carbon dioxide emissions.

The design of the facility changes when the EAF is not collocated. Aside from the obvious removal of the EAF itself, and downstream steel product processing systems including ladling and casting, there will be several other adjustments. Capital costs decrease by 57%.

First, the total cooling load and water withdrawal of the facility will be reduced by 20% without an EAF because the EAF requires wall and offgas cooling. This will also affect total surface water discharge, which decreases by 50%, and cooling tower capacity. Second, there will be no slag disposal from the EAF. Third, feedstocks to the EAF including carbon, natural gas, electricity, lime, nitrogen, and oxygen will not be needed. This will affect the purchase and flows at the facility level. Also, electricity will be reduced by 80% and natural gas by 13%. Fourth, related labor, maintenance, and construction costs associated with the EAF will be removed, which account for approximately half of these total costs. However, a system to process and store the DRI product will be required. DRI can auto-combust and has a high particulate emission risk, requiring specialized storage equipment. Companies such as Midrex are taking steps to compress the DRI into briquettes to simplify transportation. The facility's costs, energy efficiency, and CO_2 emissions will be adjusted as a result of the above adjustments. The steel facility generates an estimated 0.675 kg of greenhouse gas emission directly from the DRI-EAF process for every kg of steel product produced when natural gas is used in the DRI. Switching the DRI system to hydrogen results in a 76.3% reduction in onsite greenhouse gas emissions.[11] If the facility is modified to only make DRI products, and not final steel products, with the

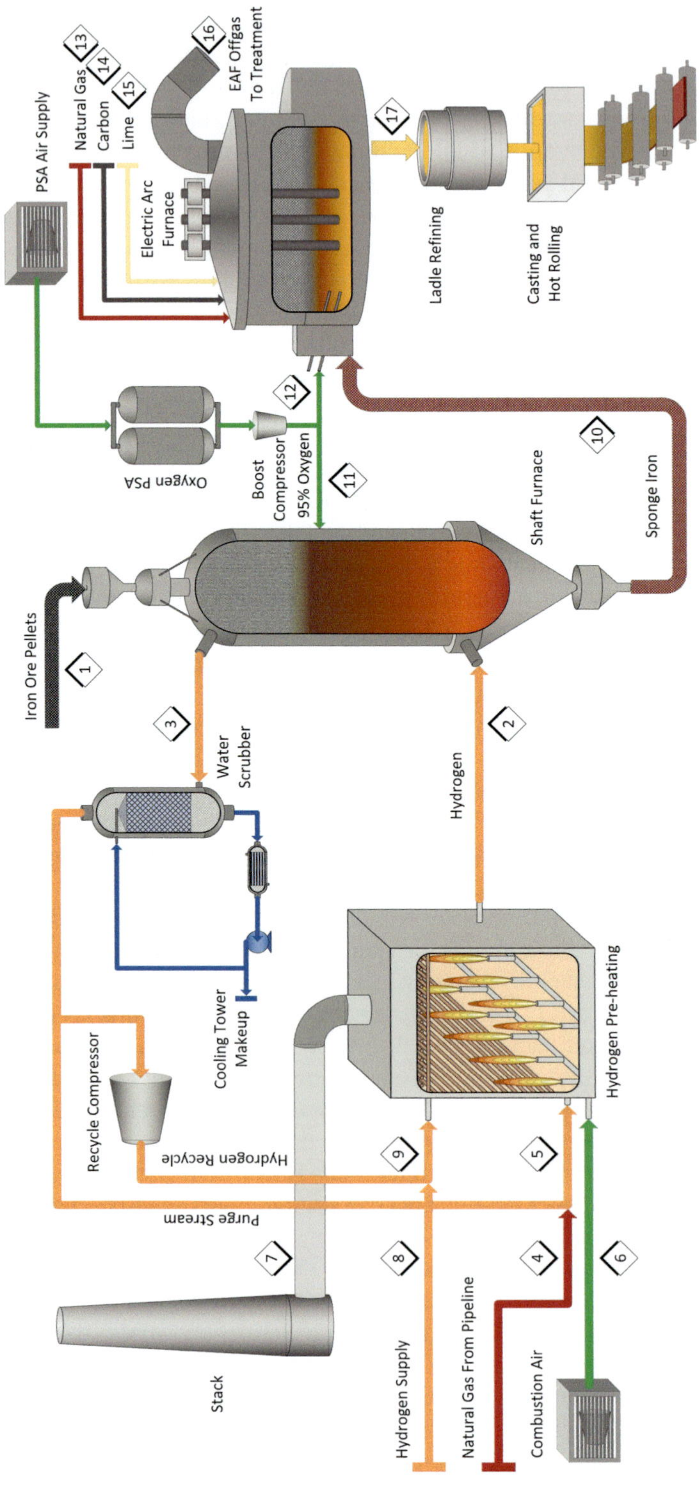

FIGURE 7.8 Steelmaking facility using pipeline hydrogen DRI hotlinked to an EAF. (Reproduced with permissions 7/29/2024.[11])

removal of the EAF system, direct emissions drop to 0.03 kg of greenhouse gas emissions per kg DRI product, reflecting the avoidance of emissions from carbon addition. While surface water discharge reduces by half and the facility no longer generates slag, such impacts will transfer to a degree to the EAF location where the DRI product is delivered.

Alternative solutions to a DRI-EAF primary steel process that still leverage hydrogen-based DRI may be possible, including:

- feeding DRI product into a blast furnace to reduce coke (e.g., Voestalpine in Linz and Leoben, Austria with DRI produced in Texas, USA; Liberty in Dunkirk, France; Salzgitter AG in Salzgitter and Wilhelmshaven, Germany by 2022)[19];
- coupling a DRI shaft furnace with a BOF using a melting unit (e.g., ThyssenKrupp in Duisburg, Germany by 2025);
- using a fluidized bed reactor in place of a shaft furnace. In 2021, Kove Steel, Ltd. in Japan announced a successful demonstration of feeding Midrex hot briquette iron (compressed DRI) into a blast furnace with a 20% reduction in direct carbon dioxide emissions.[20]

7.3.2 Economic Driving Factors in Steel Production

Steel is a globally traded commodity, with its prices subject to significant fluctuations driven by a variety of economic forces. For instance, in the United States, the price of hot-rolled coil steel traded as low as USD 351 per ton in November 2015 and as high as USD 1,930 per ton in August 2021.[21] Similarly, Chinese steel rebar prices varied from CNY 1,589 (USD 249) per ton in November 2015 to CNY 5,413 (USD 840) per ton in September 2021.[22] These price movements are influenced by several key factors, including supply and demand dynamics, cost of raw materials, cost of energy, capital expenditure, and labor cost.

When economic activity is robust and demand for steel surges, prices tend to rise if supply cannot meet the increased demand, often resulting in inflated profit margins. However, this price volatility does not always reflect the actual production cost of steel, leading to a disconnect between market pricing and the fundamental economics of steelmaking. The true cost of steel production is more accurately reflected in factors such as the cost of raw materials (iron ore, scrap metal), energy requirements (typically coal or natural gas for traditional processes, though increasingly hydrogen and electricity for EAF), labor expenses, and capital investment in equipment and facilities. These are the fixed and variable costs that steel producers must bear, regardless of market conditions. Consequently, in economic and engineering analysis, the cost of production is arguably a more critical metric, as it establishes the minimum sales price required to achieve profitability. Market prices, while crucial, are often influenced by speculative trading, geopolitical factors, and short-term disruptions (e.g., tariffs, supply chain bottlenecks), which can distort the perception of long-term economic sustainability.

In addition, the global steel industry faces increasing pressure to decarbonize, which will likely have profound economic implications. The shift to greener production methods – such as hydrogen-based steelmaking or EAF powered by renewable energy – could drive up capital and operational costs in the short term. However, as these technologies scale, they may reduce dependency on volatile fossil fuel prices and bring down long-term production costs, potentially reshaping the cost structure of the industry.

To more accurately predict the steel production cost, a holistic approach that spans the lifetime of a steel facility is essential. Such a method must account for infrequent, often costly, one-time investments as well as ongoing maintenance activities. This approach is often referred to as the levelized cost of production (LCOP), which calculates the average cost per unit of product by factoring in all expenses and investments over the facility's operational lifetime. A simple approach to calculating the LCOP typically includes the capital charge factor (CCF) derived from the financial structure of the project and reflects the cost of capital, the total overnight capital (TOC) expenditure, including the plant cost and other capital expenditures incurred at the beginning of the project, the fix operating costs (OC_{fix}) covering employee salaries, insurance, and other consistent expenditures, and the variable operating costs (OC_{var}) comprised of feedstock, maintenance, energy and disposal costs. These expenditures are then balanced against the plant's capacity factor (CF) representing the percentage of annual operational uptime, and the plant's production capacity (PC), or designed output of the facility.[11]

$$\mathrm{LCOP} = \frac{\mathrm{CCF} \cdot \mathrm{TOC} + \mathrm{OC}_{\mathrm{fix}} + \mathrm{CF} \cdot \mathrm{OC}_{\mathrm{var}}}{\mathrm{CF} \cdot \mathrm{PC}}$$

Cost of raw materials: Iron ore, a primary feedstock in steel production, experiences similar price fluctuations as steel due to the direct correlation between their supply and demand. Since iron ore has few other industrial applications outside of steelmaking, its prices are highly sensitive to shifts in the steel market. As of August 2024, global iron ore prices average around USD 101 per metric ton, with historical highs of USD 216 in June 2021 and lows of USD 41 in December 2015.[23]

Iron ore used in steelmaking typically contains at least 62% iron, and under ideal conversion and recovery rates, the cost contribution from iron ore alone amounts to approximately USD 162 per ton of steel produced. It is important to note that many steel production pathways rely on iron ore pellets, and procurement costs must account for pellet premiums. For example, DRI pellets command a premium of around USD 53 per ton, while BF pellets typically carry a slightly lower premium of about USD 45 per ton.[24] In the case of steel recycling via mini-mills, scrap metal serves as the primary feedstock. Scrap prices in the United States range from USD160 to 280 per ton.[25] Unsurprisingly, this aligns closely with the cost of producing steel from iron ore due to market equilibrium.

Cost of energy: Steel production is highly energy-intensive, with significant energy requirements for the reduction of iron ore (or iron oxide) and the subsequent melting of iron, regardless of the production route – whether through the BF-BOF process or the DRI-EAF route. In the BF-BOF process, coke is used to produce carbon monoxide (CO), which reduces iron ore to elemental iron. Theoretically, producing one ton of iron requires approximately 322.5 kg of coke, translating to around

2.6 MWh of energy. As of 2024, coking coal in the U.S. traded at USD 243 per ton as of 2024,[26] implying iron ore reduction cost of USD 78 per ton of steel. However, actual coke consumption is typically more than double this theoretical value due to the additional energy required for heating and melting, bringing energy costs closer to USD 178 per ton of steel, or about 5.8 MWh of energy per ton of steel produced.[42]

The energy balance in the DRI process is more complex because a portion of the top gas is recycled and burned in a reformer to drive the endothermic natural gas reforming reactions. A practical figure is about 3.0 MWh of natural gas required per ton of steel,[11] which at a natural gas price of USD 44.4 per MWh[27] corresponds to USD 133 per ton of steel. In a DRI scenario, however, additional energy in the form of electricity, ~475 kWh per ton,[28] is needed in the EAF to melt the sponge iron adding a cost of USD 38 per ton of steel (assuming an electricity price of 8.14 ¢/kWh).[29] Mini-mills, which use recycled scrap as feedstock, exhibit similar energy consumption to EAFs, as the scrap does not require reduction but only melting.

Capital expenditure: Building a steel facility is a capital-intensive, long-term investment with operational lifetimes extending beyond 30 years. At the end of this period, while the facility may not necessarily shut down, major reinvestments – such as reactor relining – are required to continue operations, marking the end of its economic cycle. These capital expenditures lock in technologies and processes for decades. Given that a significant portion of U.S. steel facilities will reach the end of their operational life by 2030, it is critical to invest in sustainable steel technologies during the 2020s.[43]

BF-BOF steelmaking requires a capital investment of around USD 1100 per ton of installed capacity,[30] DRI-EAF around USD500–700 per ton[11,44] of installed capacity, and EAF/mini-mill investments are typically in the range of USD190–400 per ton of capacity.[31,32] These figures are subject to economies of scale and vary depending on the size of the facility. To incorporate these capital expenditures into LCOP, the plant cost is multiplied by the CCF, which typically ranges from 5 to 15% for most projects.

Labor Cost: Steel production remains labor-intensive, despite increased automation and technological advancements. In 2023 around 101,300 workers were employed in the US steel manufacturing industry (NAICS 331110),[33] a significant decline from the 230,000 workers in the 1980s. The reduction in the workforce can be attributed in part to bankruptcies, foreign competition, and high labor costs, but more recent trends point to productivity improvements, particularly the shift toward mini-mills and more efficient production processes.[45] Comparing this to the annual steel production in the United States of 80 million tons,[34] this means that every worker is responsible for producing about 790 tons of steel per year. Considering an average salary of USD 65,000 annually, this translates to a labor cost of approximately USD 82 per ton of steel.[35]

These ballpark figures highlight the major cost drivers in steel production, illustrating the industry's competitiveness and the tight profit margins typical of a globally traded commodity. While various steel companies – accounting for 17% of global production[46] – have pledged to reduce CO_2 emissions, the steel market is highly sensitive to economic fluctuations, and introducing the additional complexities of CO_2 emission reductions presents a significant challenge. To succeed, these efforts must be economically viable. Two mainstream approaches dominate the discussion on decarbonizing the steel industry: 1) carbon capture and storage (CCS) – capturing CO_2 emissions from steel production processes and storing them to prevent their release into the atmosphere, and 2) hydrogen as fuel and reductant – using hydrogen, instead of carbon-based fuels like coke, both as a fuel and as a reductant in the steelmaking process. Both approaches require significant investment and careful economic justification.

The current understanding is that BF-BOF scenarios with CCS as well as alternative low-carbon drop in solutions such as biochar, renewable natural gas, and green electricity lead to a substantial increase in the levelized cost of steel production.[47] In the BF-BOF route, the BF can also be operated with H_2 injection to reduce coke usage. The H_2 breakeven price for this approach is around USD 3; however, only marginal CO_2 reduction of 10–20% is achievable.[48] A more cost-effective strategy to decarbonize BF-BOF steel production is to transition to natural gas-based DRI-EAF. Moreover, DRI shaft furnaces with H_2 offer a cheaper pathway to full decarbonization compared to BF-BOF with CCS,[47] assuming H_2 prices around USD 1.30 per kg. A recent comparison between state-of-the-art DRI-EAF with CCS and hydrogen-based DRI-EAF indicates while DRI-EAF with CCS results in lower premiums for decarbonized steel, the H_{2}-DRI route achieves a higher degree of CO_2 reduction. It is important to note that the analysis shows that the current cost of green H_2 remains prohibitively high but highlights that if H_2 production costs decline, H_{2}-DRI could eventually offer a lower levelized cost of steel (LCOS) compared to CCS-based solutions.[49] The average breakeven H_2 price for H_{2}-DRI to become economically competitive with natural gas-based DRI in the U.S. averages around USD 1.70 per kg of H_2, though significant regional differences exist across states.[11] This suggests that once the US Department of Energy's H_2 production target of USD 1.00 per kg of H_2 is achieved, H_{2}-DRI could provide an economical route to steel production without imposing any premiums for low CO_2 emissions. Additionally, this implies that approximately USD 0.70 per kg of H_2 would be available for transport and storage, helping to bridge intermittencies from renewable hydrogen generation. Recent progress on this subject demonstrated the feasibility of hydrogen storage in salt caverns and via liquid organic hydrogen carriers at this price point.[26] The key advantage of H_{2}-based decarbonization approaches is the elimination of the need for CO_2 end-storage, which simplifies logistics. However, research consistently highlights that successful decarbonization of steel production depends heavily on access to low-cost iron ore and renewable hydrogen.[50]

7.3.3 Energy Efficiency and Life Cycle Assessment

The iron and steel sectors are one of the largest consumers of primary energy and one of the largest emitters of greenhouse gases in industry. In China, for example, emissions from the sector account for nearly one-fifth of national carbon emissions (Figure 7.9).

With growing pressure to decarbonize and reduce costs of energy consumption, global iron and steel sectors have made notable improvements, estimated at 0.7%

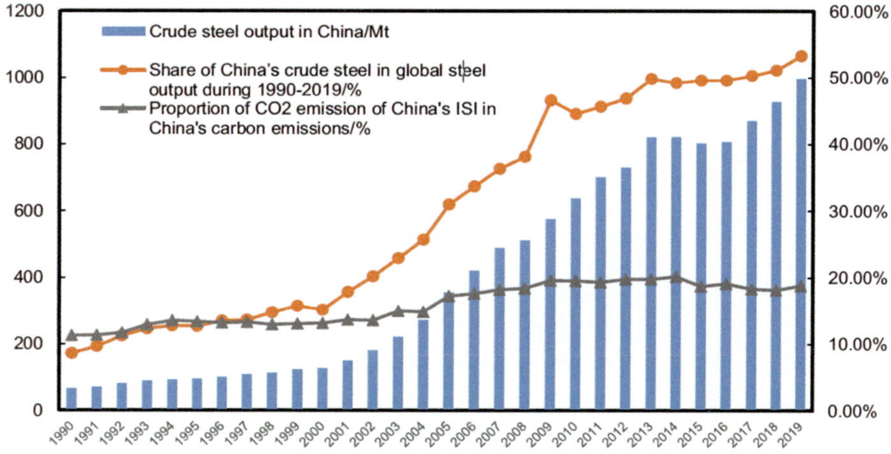

FIGURE 7.9 Crude steel production and associated CO2 emissions in proportion to China's carbon emissions.

(Reproduced with permission 7/29/2024.[51])

energy intensity reduction and 1.4% CO_2 intensity reduction per year between 2010 and 2016. These values and other environmental impacts associated with processes and specific products of the iron and steel sector are quantified through accounting of emissions. Referred to as Scope 1, 2, and 3 emissions, greenhouse gas emissions emitted directly from a process are Scope 1, while emissions related to energy consumption necessary for the process are Scope 2. Emissions related to the manufacturing of equipment and materials used in the process, as well as downstream effects of the process or product are referred to as Scope 3, and are much more challenging to completely capture. One method of accounting is life cycle assessment (LCA), which is a standardized approach for evaluating environmental impacts resulting from a product or process. LCA can be applied to account for all energy and material flows related to a base case and to alternative scenarios under consideration. This is especially useful when considering a diverse portfolio of options for decarbonizing mining processes or the production of a steel product.[52] When combined with technoeconomic analysis, policies and best available technologies can be compared across a set of metrics, including emissions reduction and impact of decisions on the sales price of products.

Numerous evaluations of the iron and steel sector indicate great potential for energy efficiency improvements and greenhouse gas emissions reductions through process redesign, renewable feedstock and energy source selection (Figure 7.10), and emerging technologies. For example, switching electricity consumption from grid sources to dedicated renewable sources would lower emissions for existing end uses of electricity, and would have a greater impact if primary BF-BOF production methods (i.e., integrated production routes) were replaced with scrap steel recycling through electrified EAF. As not all facilities have the land footprint or business case to purchase and operate dedicated renewables such as wind and solar power systems nearby, it is

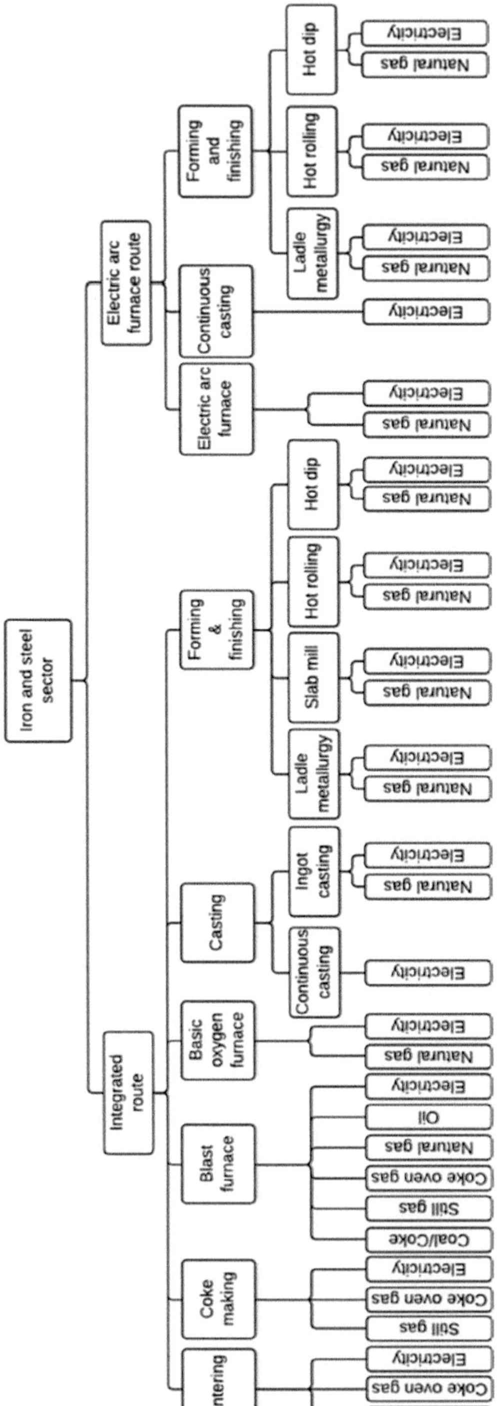

FIGURE 7.10 Steelmaking processes with energy demand and plausible energy sources.

(Reproduced with permission 7/29/2024.[53])

possible that facilities will run electrified systems on grid electricity. For many locations, grids remain too reliant on fossil fuels to generate low-carbon electricity.

Detailed scenario analyses evaluating mitigation strategies out to 2030 and 2050 have been developed at country and regional levels, supporting policy development.[53] The global warming potential (GWP), with a unit of kg of carbon dioxide equivalents, is a key impact of sustainability and is often presented for a target functional unit, for example, kg of hot-rolled steel coil (HRC).[54] Values characterizing blast furnaces range from 1.0 to 2.3 kg CO_2eq per kg HRC and cumulative energy demand of 15 to 25 MJ per kg HRC. Suer et al reviewed LCA in the metals and steel sector and concluded that increasing recycling of scrap steel is critical for improving the overall sector performance, but that primary steel production, which contributed the greatest emissions, will remain necessary to meet the growing demand for steel products.[54] In contrast to primary steel produced from blast furnaces, secondary steel produced in EAF has a GWP of 0.74-0.77 kg/kg HRC, while cumulative energy demand varies from 1.3 to 11 MJ/kg HRC.

Zang et al compared the adoption of natural gas DRI in place of blast furnace-basic oxygen furnace and suggested that 33% emissions reductions are possible.[47] The impact of displacing blast furnace-basic oxygen furnace with hydrogen-powered DRI greatly depends on the source of hydrogen and the source of energy powering hydrogen production. For example, a DRI coupled with SMR-sourced hydrogen can lower emissions if the system is coupled with carbon dioxide capture. The most dramatic emissions reductions (67%) are achieved if the hydrogen is generated with renewably powered electrolyzers, when considering only modifications to the reduction process. Additional emissions reductions are possible if the facility's heating and power requirements are also met with renewable power and renewable natural gas. Fischedick et al estimated the cumulative energy demand of hydrogen-based systems assuming a fluidized bed reactor with fine iron ores (Circored technology), quite different from the process assumed in Zang, making comparisons challenging.[55] Other approaches for hydrogen-based DRI-EAF have been evaluated for energy, cost, or impact on scrap consumption but are not comprehensively compared through LCA.[54]

Sustainability goes beyond carbon emissions and equivalent global warming potentials. However, much less data is available regarding the impact of decarbonization of iron and steelmaking on critical impact categories such as water stress, critical material consumption, cumulative or primary energy consumption, and gaseous, liquid, and solid pollution generation.[56,57] Still, attempts have been made to characterize "green" steel in several countries to set guidelines for domestic production and import requirements.

One major challenge of applying LCA to evaluate an industry is that LCA results can vary widely depending on methodologies but also basic assumptions, such as the purity of iron ore, size of facility, losses of product, or the source of electricity assumed. Furthermore, there is no true consensus on methodologies for allocating impacts associated with recycling to the recycled product or to the original primary product. In a recycle cutoff approach, the emissions from recycling are also allocated to the primary product and the recycled product has no associated emissions allocated to it, while in an avoided burden end-of-life approach, the recycled product carries emissions from the earlier life cycle stages of primary production.[58]

7.4 CLOSING REMARKS

The iron and steel sectors are coupled industries with a history rich in innovation and international partnership. Trends in these industrial sectors present opportunities for decarbonization through the use of hydrogen as a reducing agent and fuel. We highlight numerous investments in hydrogen by companies, government agencies, and research groups at precommercial and commercial stages. These demonstrations are critical for generating data on performance, educating stakeholders, and encouraging continued investment. As the environmental and economic implications of hydrogen technologies become clearer at the facility level, there remains a gap in knowledge related to the impact hydrogen use may have on vulnerabilities and challenges that characterize the iron and steel supply chain. Key industry challenges and interconnections noted in this chapter include dependency on volatile fuels and energy prices, limited flexibility and need for reliable productivity, historical supply chain development based on available domestic energy supply including renewable energy, complexity of supply chain optimization for material and energy transport and handling despite geographic constraints and heterogeneity, need for the reduction of primary (crude) steel production through increased recycling of scrap steel and use of EAF, challenges around retrofitting of existing infrastructure, variations in the definition of "green products" globally, and need for retirement and construction of brand-new facilities to lower greenhouse gas emissions. Stepping back and admiring the ingenuity of this sector, one clearly sees that the process of taking a mined ore and converting it into a portfolio of quality products with high fidelity reflects decades of experimentation and optimization. Alterations to any aspect of the supply chain must therefore be introduced within the context of practical near and future technical targets, as well as introduced with an eye to local energy and feedstock characteristics and availability. Hydrogen holds great promise for lowering environmental impacts in both the iron and steel sectors, but the ability of hydrogen solutions to meet cost targets and emissions criteria is sensitive to process designs, iron ore quality, and primary energy resources. Dedicated energy generation may be necessary when local utilities cannot provide necessary capacity. Changes to workforce will likely be critical for evaluation to gain the support of local communities and ensure regional benefits. This may require the development of training programs for a diverse pool of workers. Widespread adoption of decarbonization measures in the iron and steel sectors will require very large companies to signal confidence in new strategies, whether that be hydrogen, carbon capture, electrification, or another novel solution. Renewably generated hydrogen could decouple primary iron and steelmaking from coal, diesel, and natural gas industries worldwide while stimulating new international partnerships that will drive operations, performance, impact, and ultimately decarbonization.

ACRONYMS

ARA auxiliary reducing agent
BOF basic oxygen furnace
BF blast furnace

CF	capacity factor
CCF	capital charge factor
CCS	carbon dioxide capture and storage
CO_2e	carbon dioxide equivalence
CNY¥	Chinese Yuan
DRI	direct reduced iron
EAF	electric arc furnace
HBI	hot briquette iron
LCOP	levelized cost of production
MMT	million metric tons
NAICS	North American Industry Classification System
PEM	proton exchange membrane
SOEC	solid oxide electrolyzer cell
SMR	steam methane reforming
TOC	total overnight capital
USD	United States Dollar

NOTES

1. Country capacities: https://www.oecd.org/sti/ind/steelcapacity.htm.
2. Crude steel production: https://worldsteel.org/media/press-releases/2024/august-2024-crude-steel-production/.
3. MIDREX facility locations: https://www.midrex.com/about-midrex/midrex-plants-map/.
4. Electrolysis for iron: https://www.thechemicalengineer.com/features/electrochemistry-for-greener-steel/.
5. SOEC with DRI: www.hydrogen.energy.gov/docs/hydrogenprogramlibraries/pdfs/review23/ta052_brouwer_2023_o-pdf.pdf.
6. Taconite: https://www.dnr.state.mn.us/education/geology/digging/taconite.html.
7. NAICS code description: https://www.naics.com/naics-code-description/?code=212210.
8. Taconite Mining: https://libguides.mnhs.org/taconite.
9. China investments in Guinea region: https://asia.nikkei.com/Spotlight/Caixin/China-s-opportunities-and-risks-in-Africa-s-giant-iron-ore-field.
10. Cleveland-Cliffs facilities: https://www.clevelandcliffs.com/operations/steelmaking/iron-ore-facilities.
11. Taconite ore processing: https://www.epa.gov/stationary-sources-air-pollution/taconite-iron-ore-processing-national-emission-standards-hazardous.
12. China imports of iron ore: https://asia.nikkei.com/Spotlight/Caixin/China-s-opportunities-and-risks-in-Africa-s-giant-iron-ore-field.
13 Hydrogen at mines: https://www.enelgreenpower.com/stories/articles/2021/11/green-idrogen-mines.
14. Ammonia for explosives: https://ammoniaenergy.org/articles/the-mining-industry-a-driving-force-behind-green-ammonia/.
15. First Mode Trucks: https://firstmode.com/updates/life-of-first-zero-emission-haul-truck/.
16. nuGen trucks: https://southafrica.angloamerican.com/our-difference/futuresmart-mining/nugen.
17. Gibson Island project: https://www.hydrogeninsight.com/industrial/we-will-make-green-ammonia-but-it-will-be-too-expensive-for-us-to-use-for-fertiliser-production-chemicals-firm/2-1-1559448.

18. Enaex green ammonia project: https://www.enaex.com/cl/us/green-ammonia/.
19. Hydrogen in Europe: https://bellona.org/news/eu/2021-05-hydrogen-in-steel-production-what-is-happening-in-europe-part-two.
20. Kobe Steel demonstration: https://www.greencarcongress.com/2021/02/20210223-kobelco.html.
21. Trading Price Hot-Rolled Coil Steel: https://tradingeconomics.com/commodity/hrc-steel.
22. Trading Price Steel Rebar: https://tradingeconomics.com/commodity/steel.
23. Global price iron ore: https://fred.stlouisfed.org/series/PIORECRUSDM.
24. DRI pellet premium: www.fastmarkets.com/insights/brazilian-dr-iron-ore-pellet-premiums-expected-to-slip-in-q2/.
25. EAF Steel cost: www.frcglobal.com/blog/the-costs-of-producing-steel-with-electric-arc-furnaces#:~:text=In%20the%20US%2C%20steel%20scrap,16%2C000%20to%20%24%2067%2C500%20per%20load.
26. Coking coal prices: www.frcglobal.com/blog/the-costs-of-producing-steel-with-electric-arc-furnaces#:~:text=In%20the%20US%2C%20steel%20scrap,16%2C000%20to%20%24%2067%2C500%20per%20load.
27. US Energy Information Agency Natural Gas Prices: www.eia.gov/dnav/ng/NG_PRI_SUM_A_EPG0_PIN_DMCF_M.htm.
28. EAF energy consumption: http://heattreatconsortium.com/metals-advisor/electric-arc-furnace/electric-arc-furnace-energy-consumption/.
29. US Energy Information Agency Electricity prices: https://www.eia.gov/electricity/monthly/epm_table_grapher.php?t=epmt_5_03.
30. BOF: https://www.steel-technology.com/articles/oxygenfurnace#:~:text=Whilst%20a%20typical%20integrated%20(i.e.,of%20the%20initial%20capital%20outlay.
31. EAF capital: https://www.steelonthenet.com/capital-investment/eaf.html#:~:text=Typical%20capex%20investment%20costs&text=Note%20that%20cost%20per%20tonne,industry%20capital%20investment%20costs%20page.
32. DRI capital: https://thundersaidenergy.com/downloads/direct-reduced-iron-costs-and-projects/#:~:text=Leading%20DRI%20technologies%20include%20Midrex,%25%20below%20higher%2DCO2%20steel.
33. Federal Reserve Bank of St. Louis Employment NAICS 331110: https://fred.stlouisfed.org/series/IPUEN331110W010000000.
34. Mineral Commodity Summary Iron and Steel: https://pubs.usgs.gov/periodicals/mcs2024/mcs2024-iron-steel.pdf.
35. US Bureau of Labor Statistics NAICS 331100: https://www.bls.gov/oes/2023/may/naics4_331100.htm.

REFERENCES

1. Wrona, P., Różański, Z., Pach, G., Niewiadomski, A. P. & Veiga, J. P. Historical outline of iron mining and production in the area of present-day Poland. *Minerals* **11**, 1136 (2021).
2. Marchant, J. Iron in Egyptian relics came from space. *Nature* (2013). doi: 10.1038/NATURE.2013.13091
3. Eggert, G. *Medieval Technology and American History - In-Depth Articles - The Rise and Fall of Medieval Iron Technology*. https://www.engr.psu.edu/mtah/articles/rise_fall_medieval_technology.htm
4. Hauffe, K. L. von Bogdandy und H. J. Engell: The reduction of iron ores. Springer-Verlag, Berlin, und Verlag Stahleisen, Düsseldorf 1971, 576 Seiten. Preis: DM 188. *Berichte der Bunsengesellschaft für Phys. Chemie* **76**, 1285 (1972).

5. Geerdes, M., Toxopeus, H. & van der Vliet, C. *Modern Blast Furnace Ironmaking. An introduction* (2009).
6. Remus, R., Roudier, S., Aguado-Monsonet, M. A. & Delgado Sancho, L. Best Available Techniques (BAT) Reference Document for Iron and Steel Production: Industrial Emissions Directive 2010/75/EU: (Integrated Pollution Prevention and Control). 597 (2013) doi: 10.2791/98516
7. Wei, C. et al. Development of direct reduced iron in China: Challenges and pathways. *Engineering* (2024). doi: 10.1016/J.ENG.2024.04.025
8. Hasanbeigi, A., Arens, M. & Price, L. Alternative emerging ironmaking technologies for energy-efficiency and carbon dioxide emissions reduction: A technical review. *Renew. Sustain. Energy Rev.* **33**, 645–658 (2014).
9. Rukini, A., Rhamdhani, M. A., Brooks, G. A. & Van den Bulck, A. Metals production and metal oxides reduction using hydrogen: A review. *J. Sustain. Metall.* **8**, 1–24 (2022).
10. Ruth, M. et al. *The Technical and Economic Potential of the H2@Scale Hydrogen Concept within the United States*, 196 (2020).
11. Rosner, F. et al. Green steel: design and cost analysis of hydrogen-based direct iron reduction. *Energy Environ. Sci.* **16**, 4121–4134 (2023).
12. Sane, A., Buragino, G., Makwana, A. & He, X. *Enhancing Direct Reduced Iron (DRI) for Use in Electric Steelmaking*. Air Prod. Chem. Inc., Allentown, PA, USA, 14–18 (2020).
13. He, H., Guan, H., Zhu, X. & Lee, H. Assessment on the energy flow and carbon emissions of integrated steelmaking plants. *Energy Rep.* **3**, 29–36 (2017).
14. Noble, B. B., Konovalova, A., Moutarlier, L. J., Brogden, V. & Kempler, P. A. Electrochemical chlor-iron process for iron production from iron oxide and salt water. *Joule* **8**, 714–727 (2024).
15. Huijgen, W. J. J. & Comans, R. N. J. Carbonation of steel slag for CO_2 sequestration: Leaching of products and reaction mechanisms. *Environ. Sci. Technol.* **40**, 2790–2796 (2006).
16. Ma, A., Zheng, X., Li, S., Wang, Y. & Zhu, S. Zinc recovery from metallurgical slag and dust by coordination leaching in NH_3–CH_3COONH_4–H_2O system. *R. Soc. Open Sci.* **5** (2018).
17. Mastropasqua, L. et al. Molten carbonate fuel cells retrofits for CO_2 capture and enhanced energy production in the steel industry. *Int. J. Greenh. Gas Control* **88**, 195–208 (2019).
18. Dunstan, M. T., Donat, F., Bork, A. H., Grey, C. P. & Müller, C. R. CO_2 capture at medium to high temperature using solid oxide-based sorbents: Fundamental aspects, mechanistic insights, and recent advances. *Chem. Rev.* **121**, 12681–12745 (2021).
19. Yong, Z., Mata, V. & Rodrigues, A. E. Adsorption of carbon dioxide at high temperature—A review. *Sep. Purif. Technol.* **26**, 195–205 (2002).
20. Ge, B. et al. Innovative process integrating high temperature heat pump and direct air capture. *Appl. Energy* **355**, 122229 (2024).
21. Bhaskar, A., Assadi, M. & Somehsaraei, H. N. Can methane pyrolysis based hydrogen production lead to the decarbonisation of iron and steel industry? *Energy Convers. Manag. X* **10**, 100079 (2021).
22. Goswami, N., Ayyaswamy, A., Vishnugopi, B. S., Ciez, R. & Mukherjee, P. P. An integrated analytics of PEM electrolyzer and steel making. *ECS Meet. Abstr.* **243**, 1958 (2023).
23. Krüger, A., Andersson, J., Grönkvist, S. & Cornell, A. Integration of water electrolysis for fossil-free steel production. *Int. J. Hydrog. Energy* **45** (2020) Elsevier.
24. Korobeinikov, Y. et al. Reduction of iron-ore pellets using different gas mixtures and temperatures. *steel Res. Int.* **94**, 2300066 (2023).

25. Hurtado, C. et al. The melting behavior of hydrogen direct reduced iron in molten steel and slag: An integrated computational and experimental study. *Metals* **14**, 821 (2024).
26. Breunig, H. et al. Achieving gigawatt-scale green hydrogen production and seasonal storage at industrial locations across the U.S. *Nat. Commun.* (2024). doi: 10.21203/RS.3.RS-4326648/V1
27. Reznicek, E. P. et al. Techno-economic analysis of low-carbon hydrogen production pathways for decarbonizing steel and ammonia production. *Cell Press Sneak Peek* (2024). doi: 10.2139/ssrn.4785779
28. Zang, G. et al. Cost and life cycle analysis for deep CO_2 emissions reduction of steelmaking: Blast furnace-basic oxygen furnace and electric arc furnace technologies. *Int. J. Greenh. Gas Control* **128**, 103958 (2023).
29. Al Bari, S. *Iron Ore Mining in the US - Market Research Report (2014-2029).* https://www.ibisworld.com/united-states/market-research-reports/iron-ore-mining-industry/ (2024).
30. Khaustovich, V. *Iron & Steel Manufacturing in the US - Market Research Report (2014-2029).* https://www.ibisworld.com/united-states/market-research-reports/iron-steel-manufacturing-industry/ (2024).
31. Tiley, R. *Geophysical Technologies for Detecting Underground Coal Mine Voids An Interactive Forum.* https://www.fhwa.dot.gov/engineering/geotech/hazards/mine/workshops/ktwkshp/ky0302.pdf (2003).
32. EPA. *Taconite Ore Processing.* https://gaftp.epa.gov/ap42/ch11/s23/final/c11s23_feb1997.pdf (1997).
33. Patent Submissions - Solid State Hydrogen Storage - Australian Mines Limited (ASX:AUZ) - Listcorp.
34. *QNP Green Ammonia Project Feasibility Study - Australian Renewable Energy Agency (ARENA).* https://arena.gov.au/knowledge-bank/qnp-green-ammonia-project-feasibility-study/
35. Babich, A. Blast furnace injection for minimizing the coke rate and CO_2 emissions **48**, 728–741 (2021). doi: 10.1080/03019233.2021.1900037
36. Yilmaz, C. & Turek, T. Modeling and simulation of the use of direct reduced iron in a blast furnace to reduce carbon dioxide emissions. *J. Clean. Prod.* **164**, 1519–1530 (2017).
37. Chen, Y. & Zuo, H. Review of hydrogen-rich ironmaking technology in blast furnace **48**, 749–768 (2021). doi: 10.1080/03019233.2021.1909992
38. Barnes, R. S. Current state of iron and steel technology. *Ironmak. Steelmak.* **2**, 82–88 (1975).
39. Astier, J., Krug, J. C. & de Pressigny, Y. de L. Technico-economic potentialities of hydrogen utilization for steel production. *Int. J. Hydrogen Energy* **7**, 671–679 (1982).
40. Spreitzer, D. & Schenk, J. Reduction of iron oxides with hydrogen—A review. *Steel Res. Int.* **90**, 1900108 (2019).
41. Heidari, A., Niknahad, N., Iljana, M. & Fabritius, T. A review on the kinetics of iron ore reduction by hydrogen. *Materials (Basel, Switzerland)* **14**, (2021).
42. Pardo, N., Moya, J. A. & Vatopoulos, K. *Prospective Scenarios on Energy Efficiency and CO2 Emissions in the EU Iron & Steel Industry*. (2013). doi: 10.2790/64264
43. Agora Industry, Wuppertal Institute & Lund University. *Global Steel at a Crossroads: Why the Global Steel Sector Needs to Invest in Climate-neutral Technologies in the 2020s*. (2021).
44. Feng, Z. et al. Process optimization, energy consumption analysis, and environmental assessment of total CO_2 capture from syngas based on [NEMH][Ac] protic ionic liquid. *Int. J. Greenh. Gas Control* **131**, 104037 (2024).

45. Collard-Wexler, A. & De Loecker, J. Reallocation and technology: Evidence from the US steel industry. *Am. Econ. Rev.* **105**, 131–171 (2015).
46. Hermwille, L. et al. A climate club to decarbonize the global steel industry. *Nat. Clim. Chang. 126* **12**, 494–496 (2022).
47. Zang, G. et al. *Cost and Life Cycle Analysis for Deep CO_2 Emissions Reduction for Steel Making: Direct Reduced Iron Technologies*. (2023). doi: 10.1002/srin.202200297
48. Kim, J. et al. An integrative process of blast furnace and SOEC for hydrogen utilization: Techno-economic and environmental impact assessment. *Energy Convers. Manag.* **250**, 114922 (2021).
49. Benavides, K. et al. Mitigating emissions in the global steel industry: Representing CCS and hydrogen technologies in integrated assessment modeling. *Int. J. Greenh. Gas Control* **131**, 103963 (2024).
50. Xu, S., Wang, X., Jiang, Y., Yu, B. & Wei, Y. M. Optimum investment strategy for hydrogen-based steelmaking project coupled with multiple uncertainties. *J. Environ. Manage.* **356**, 120484 (2024).
51. Yue, Q. et al. Analysis of iron and steel production paths on the energy demand and carbon emission in China's iron and steel industry. *Environ. Dev. Sustain.* **25**, 4065–4085 (2023).
52. Koltun, P. & Klymenko, V. *Cradle-to-gate life cycle assessment of the production of separated mix of rare earth oxides based on Australian production route. Min. Miner. Depos.* **14**, 1–15 (2020).
53. Talaei, A., Ahiduzzaman, M., Davis, M., Gemechu, E. & Kumar, A. Potential for energy efficiency improvement and greenhouse gas mitigation in Canada's iron and steel industry. *Energy Effic.* **13**, 1213–1243 (2020).
54. Suer, J., Traverso, M. & Jäger, N. Review of life cycle assessments for steel and environmental analysis of future steel production scenarios. *Sustainability* **14**, 14131 (2022).
55. Fischedick, M., Marzinkowski, J., Winzer, P. & Weigel, M. Techno-economic evaluation of innovative steel production technologies. *J. Clean. Prod.* **84**, 563–580 (2014).
56. Chen, W. et al. Environmental sustainability challenges of China's steel production: Impact-oriented water, carbon and fossil energy footprints assessment. *Ecol. Indic.* **136**, 108660 (2022).
57. Liang, T. et al. Environmental impact evaluation of an iron and steel plant in China: Normalized data and direct/indirect contribution. *J. Clean. Prod.* **264**, 121697 (2020).
58. Broadbent, C. Steel's recyclability: Demonstrating the benefits of recycling steel to achieve a circular economy. *Int. J. Life Cycle Assess.* **21**, 1658–1665 (2016).

8 Decarbonization of Steel Industry

Chenn Q. Zhou, Orlando J. Ugarte, Tyamo Okosun, and Armin Silaen

8.1 INTRODUCTION

Steel is a critical material for manufacturing, infrastructure, and defense. Even as new composite materials are developed, there exists a continually increasing need for steel, with global consumption expected to reach approximately 2.45 Gt per year by 2050.[1] Part of this demand is to supply the construction of renewable power generation facilities, which have quickly become the most economical form of power generation and are necessary to reach carbon-neutrality goals set worldwide.

The greenhouse gas (GHG) emissions of iron and steelmaking are among the highest of any industry. Specifically, the iron and steel industry accounts for about 25% of direct industrial GHG emissions in the world, and for over 7% of global GHG emissions.[2] The energy demand for iron and steel industry processes is also high, reaching approximately 20% of the industrial energy use. In addition, steel production can generate a range of air pollutants such as sulfur dioxide, nitrous oxide, and volatile organic compounds, adversely impacting the health of steelworkers and their local communities.[3] The formation of air pollutants must be considered alongside GHG emissions as decarbonization goals for the steel industry are set. Significant efforts have been initiated to reduce CO_2 emissions worldwide, aiming to reduce GHG emissions such that average global temperature increase remains below 1.5°C as compared to pre-industrial levels.[4]

Advancements in steel manufacturing technologies over the past decade have enabled continuous improvements in steel quality as well as in increased energy efficiency and environmental stewardship. While many of these advancements have been driven by economics rather than environmental targets, they have nonetheless achieved significant reductions in emissions. One example is the rise of electric arc furnace (EAF) steelmaking in the United States (accounting for 70% of U.S. steel production today) and the corresponding displacement of the carbon-intensive blast furnace – basic oxygen furnace (BF-BOF) process.[5] This transition has enabled the U.S. steel industry to achieve very low per capita emissions compared to other producers worldwide. Although top steel producers are showing an active role in decarbonizing the steel industry, process improvement, innovation, and active implementation are largely needed in order to reach emission-reduction goals. In this chapter, an overview of current strategies and future trends is discussed, including

 DOI: 10.1201/9781003655442-11

limitations and challenges. The rise of advanced digital tools necessary for process analysis, optimization, and innovation is also discussed.

8.2 OVERVIEW OF STEEL INDUSTRY PROCESSES AND EMISSIONS

The iron and steelmaking sector is divided into two basic types of molten steel production. These include 1) the integrated steel mill, which combines a blast furnace (BF) with a basic oxygen furnace (BOF); and 2) the mini-mill, which uses EAFs. Both are shown in Figure 8.1. The two types of steel mills are distinguished by their raw material inputs. The integrated BF-BOF process consumes mostly agglomerated iron ore along with some scrap steel; the EAF process consumes mostly scrap steel as well as reduced iron, cast iron, and other iron-containing materials to produce raw steel. Both include casting, in which molten steel from BOF or EAF is solidified into long, semi-finished slabs, beams, billets, and blooms. These semi-finished products are then used to produce the finished steel products in rolling and finishing processes.

8.2.1 Integrated Mills

An integrated steel plant is one in which all the processes involved in steelmaking are done in one place. It uses a three-step process to produce molten steel, including coke production, ironmaking in blast furnaces, and basic oxygen furnace process. The fuel and carbon in the integrated mill process are mainly provided by coke, which is produced in coke ovens by heating coal without the presence of oxygen. The blast furnace process reacts the coke, iron ore, and limestone to produce molten pig iron. Alternative fuels such as pulverized coal, natural gas, syngas, hydrogen, or other low-carbon fuels can be used to partially replace coke in a blast furnace. The last step occurs in the BOF where oxygen, flux, and scrap steel are added to the pig iron to obtain the desired composition of steel. The fully integrated production is completed by forming and finishing operations. The blast furnace-basic oxygen furnace (BF-BOF) is the main route in the global iron and steel industry. However, the nature of the integrated mill process produces significant carbon emissions, especially at the coking and ironmaking stages.

8.2.2 Mini-Mills

A "mini-mill" is a smaller facility that typically produces molten steel in an EAF by applying electrical and chemical heating to melt scrap metal or direct reduced iron (DRI). The EAF includes electrodes that pass electrical current through the scrap steel, leading to the heating and melting of the scrap. The molten steel bath is then refined by injecting oxygen. The EAF process generally operates in a batch mode instead of a continuous mode as in the BF-BOF case. The process is fed with recycled steel scrap, small amounts of pig iron, and other alternative iron units. An alternative iron unit that is increasing in importance in recent years is DRI, where iron ore is chemically reduced at temperatures below the iron melting point in a shaft furnace similar to the top half of a blast furnace. DRI applies natural gas, syngas, and other blends of H_2/CO reducing gases to reduce virgin iron oxides in new

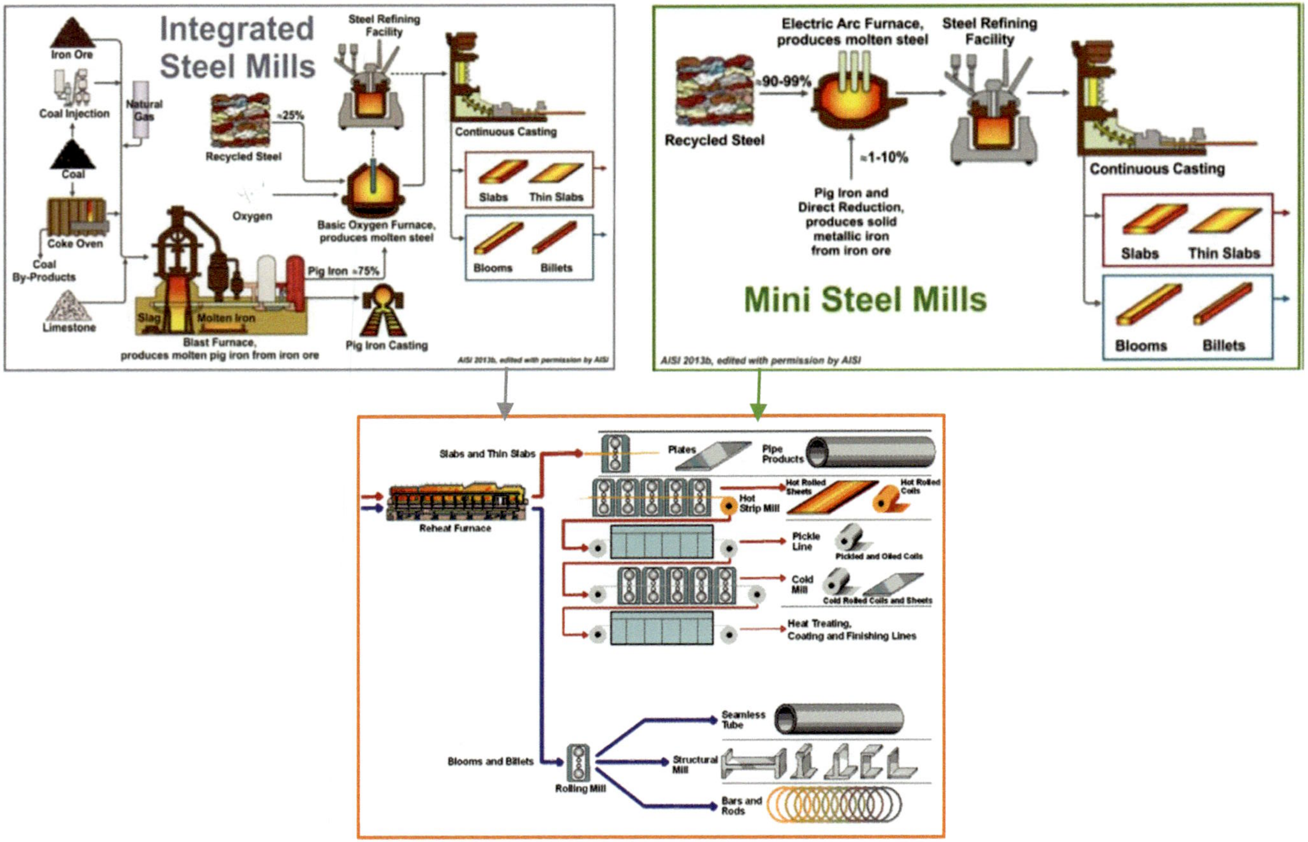

FIGURE 8.1 Process scheme of integrated mill and mini-mill process.[6]

solid-state iron and release fewer emissions per ton of product than a comparable blast furnace. Although mini-mills do not include coking and ironmaking processes (the main sources of emissions in integrated mills), the energy used to carry out the operation in mini-mills (electrical power and combustion of fossil fuels) results in GHG emissions generated both on and off-site. The emissions are further increased during refining operations as carbon is injected. When alternative iron units or pig iron are included as part of heat, the production of that iron-bearing material further increases the carbon emissions of the overall process.

8.2.3 Carbon Emissions in Integrated Mills and Mini-Mills

Integrated mills are the dominant technology for primary steelmaking globally. This accounts for over 71% of global crude steel production, representing approximately 1,335 million tons in 2024.[7] Considering typical U.S. operation parameters, the BF process generates roughly 1,514 kg CO_2/ton-HM. In addition, carbon emissions related to coke production, sintering, pelletizing, and BOF steelmaking increase the total emission in integrated mills to 2,225 kg CO_2 per ton-HM[8] (see Table 8.1). EAFs are mostly used to produce secondary steel from recycled steel scrap. The EAF is also the most important electrification alternative for the steelmaking industry. Studies show that the carbon footprint per ton of EAF steel can be as low as 842 kg CO_2 per ton of steel, depending on iron type (pig iron or scrap), electricity sources, and process efficiency.[9] EAFs operating on pure scrap can produce steel with an extremely low emissions profile. Due to supply limitations and product quality/purity requirements, the inclusion of some virgin iron material is often necessary, driving up emissions. Additionally, batch operation yields intermittent duty cycles, causing power quality problems for transmission and generation.[10]

As described earlier, EAF operation may also include melting of DRI. DRI composition is similar to pig iron and it can increase the efficiency of EAF process. The DRI-EAF process allows for higher electrification and can reduce carbon emissions if low-carbon feedstocks and electricity are used. Moreover, the carbon emissions in DRI generation can be further lowered if the reduction gas is replaced with higher H_2 blends. The BF-BOF process faces additional challenges in the use

TABLE 8.1
CO_2 Emissions Produced in Integrated Mill Process (Assumed Average Electricity Emissions of CO_2 460 kg/MWh in U.S.)[11]

Process	Kg of CO_2/Ton Hot Metal
Sintering	276
Pelletizing	46
Coke production	101
Blast furnace	1,514
Basic oxygen furnace	229
Total	**2,225**

TABLE 8.2
CO_2 Emissions Produced in Mini-mill Process (Assumed Electricity Carbon Intensity Is 460 kg/MWh)[13]

Production Method	EAF	DRI-EAF
Primary reactors [kg CO_2/ton]	0	1,048 (coal-based DRI), 522 (gas-based DRI)
Pre-treatment reactors [kg CO_2/ton]	420	307
Electricity intensity [KWh/ton]	918	1,298 (coal-based DRI), 1,231 (gas-based DRI)
Total [Kg CO_2/ton]	842	1,953 (coal-based DRI), 1,395 (gas-based DRI)

of high levels of H_2 in the reducing gas due to the endothermic nature of H_2 reduction and the corresponding energy requirements, given that all reduced material must be melted. The DRI-EAF route produces 1,953 kg CO_2 per ton of HM when it is coal-based and 1,395 kg CO_2 per ton of HM when the process is gas-based[12] (Table 8.2). This represents 12% and 37% less emissions than in traditional integrated mills, respectively.

8.3 CO_2 EMISSION-REDUCTION TECHNOLOGIES

Multiple strategies are applied to reduce carbon emissions in the steel industry, while new technologies and ideas at different levels of maturity are being added to this effort. The U.S. Department of Energy has set forth an industrial decarbonization roadmap, developed through nationwide collaboration with academia and industry.[14] This document points out the need for greenhouse emission reductions across industries to moderate the impacts of climate change. The roadmap establishes four technology pillars for carbon emission reductions: increasing the energy efficiency of the current processes, electrification of iron and steel processes, transition to low-carbon fuels, feedstocks, and energy sources, and implementing carbon capture strategies. These pillars are shown in Figure 8.2(a), and represent a portion of the U.S. strategy to reach a net-zero economy by 2050. They are being applied across the steel industry through investments by individual companies, collaborative industry organizations, and support from government-funded R&D projects. This effort aligns with steel decarbonization goals set by the International Energy Agency (IEA), which has proposed a net-zero emissions pathway by 2070. From May 2022 to December 2024, funded by a multiyear grant from the U.S. Department of Commerce and its National Institute of Standards & Technology, the Association and Iron and Steel Technology (AIST) developed a decarbonization roadmap on behalf of the U.S. steel industry. The roadmap presented the pathways for achieving decarbonization goals in the U.S. steel industry and prioritized research areas to address the technologies, infrastructure, and workforce needs that will decarbonize the iron and steel industry and advance steel manufacturing competitiveness across the steel industry value chain[15] (see Figure 8.2(b)). The four technology themes in the AIST roadmap are

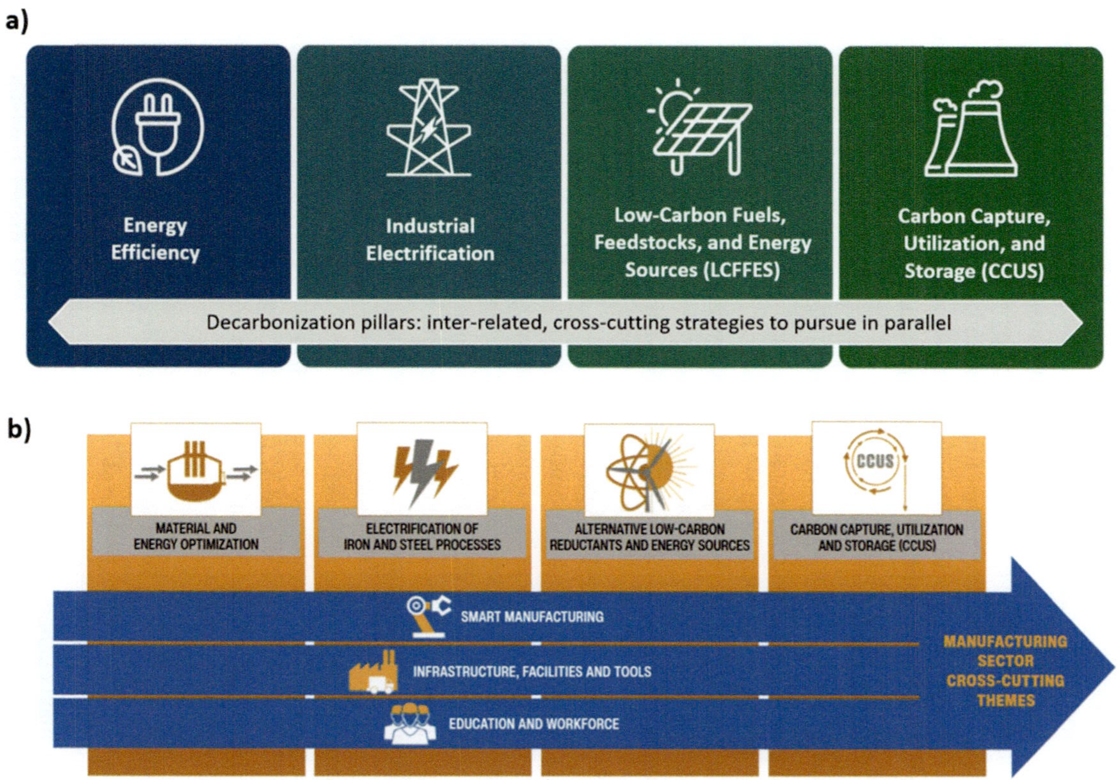

FIGURE 8.2 (a) Four industrial decarbonization pillars in DOE roadmap.[16] (b) The four technology themes and three cross-cutting themes for the AIST roadmap.[17]

aligned with those in the DOE's industrial decarbonization roadmap, which are discussed in more detail below.

8.3.1 Increasing Energy Efficiency

Energy efficiency is an especially important factor for U.S. iron and steel producers, as it impacts production, operational efficiency, and emissions. Reducing energy consumption has direct benefit on reducing carbon emissions. Blast furnace operation may be optimized to reduce energy consumption by decreasing coke consumption, directly resulting in lower carbon emissions, or optimization of hot metal silicon content, which can result in indirect reductions in emissions in downstream steelmaking. For example, each kilogram of coke replaced by an injected fuel such as natural gas reduces the corresponding CO_2 emissions by 25%. The EAF experiences significant energy losses through the walls, which require water-cooling systems to protect the furnace structure. Proper charging practices and adjustment of burner parameters in EAF operations may also reduce operation time and energy consumption. Reports indicate the possibility of increasing EAF efficiency by 24–33% through the application of proper material recovery practices and by optimizing energy transfer processes.[18] Similarly, current operations in iron and steel production can be revisited to increase energy efficiency. Here, some of these initiatives are discussed.

8.3.1.1 Set Initial Fabrication Close to the Final Product Conditions

Production efficiency can be significantly increased by setting the initial fabrication of a custom component as near as possible to the final size and shape of the product. For instance, by applying technologies such as twin roll and belt casting to obtain near-net shape products,[19] it is possible to reduce energy consumption by up to 3 GJ/ton of steel.[20] Near-net shape products minimize the need for additional operations such as machining and finishing of the surfaces, reducing scrap generation, waste material, costs, and carbon emissions.[21] Production technologies implemented to obtain near-net shapes could be applied to processes such as additive manufacturing, die casting, sand casting, investment casting, and injection molding, as well as linear and rotary friction welding. It should be noted that near-net shape production operations can affect scale formation and the ability to control product quality, surface quality, mechanical properties, and recyclability.

8.3.1.2 Applying Proper Steel Recycling Practices for Steelmaking

As mentioned earlier, EAF processes based on the melting of steel scrap can reduce carbon emissions significantly compared to BF-BOF and EAF-DRI routes.[22] However, the main challenge for utilizing scrap in EAF operations is the presence of undesired elements such as Cu, Ni, S, P, or Sn in the scrap material, which are very difficult to remove during the refining stage. These elements can lead to cracking, hot shortness, and hot deformation when the material is exposed to elevated temperatures. Although all steel can be recycled, the products that can be made from recycled material are limited to those that are not thin-gauged, as they represent lower value. For instance, Cu and Sn in steel scrap originating from tin-plate, wires, and motors in vehicles lead to surface cracking (surface hot shortness) during secondary cooling or thermo-mechanical treatment. Therefore, it is necessary to apply end-of-life processing of the steel scrap and to include product design for circularity across the broader economy. Moreover, sensor-based sorting technology of steel that involves camera-based optical recognition is under R&D to improve current sorting methods such as physical separation, magnetic separation, and shredding applied in scrap yards.

8.3.1.3 Producing Alloys with Less Carbon Emissions

Overall, it is possible to move the production trend to alloys that are less carbon intensive.[23] For instance, higher strength steels for auto-bodies[24] provide benefits in reducing vehicle weight and increasing fuel efficiency. Then, by reducing the weight of the product, it is possible to reduce the carbon footprint of steel in the production process and fuel consumption during product life. Similarly, alloys can increase the strength of structural steels for construction, which is necessary for affordable and sustainable housing. Electrical steels with high-silicon content[25] can be used in motor laminations and transformer cores, in addition to multiple applications such as wear-resistant and fatigue-resistant steels, durable bearings for wind turbines, corrosion-resistant coatings, or alloys for extreme environments. Each of these can extend structural life and enhance performance and energy efficiency. In some cases, the use of new materials is needed not only for energy efficiency but to implement new technologies such as those based on hydrogen. Namely, the "hydrogen economy" requires new hydrogen-cracking-resistant steels to store and transport hydrogen.

8.3.2 Electrification of Iron and Steel Processes

Electrification of iron and steelmaking processes aims to replace fossil fuel-based energy sources with renewable electricity, eliminating direct emissions. Some technologies that may contribute to the electrification of the steel industry are discussed in the following sections.

8.3.2.1 Electric Smelting Furnace Process

The electric smelting furnace (ESF) technology is a reduction process that is long-established in the nonferrous metals industry. ESFs for ironmaking normally employ DRI units, and the process is followed by the BOF process. ESF units have two designs: a spherical shell design and a rectangular design (six-in-line furnace). The two designs are extensively applied for smelting and processing of ferroalloys such as Ni, Cu, FeNi, FeTiO3, and platinum group metals. ESFs are large structures, that is, the round units can have diameters up to 20 m and heights up to 6 m, whereas rectangular smelters are usually 40 m x 15 m x 9 m. A main challenge in ESF operation is the high content of gangue oxides and varying levels of carbon in DRI/HBI units, which require an additional amount of lime, dolomitic lime, and doloma. The variations in the raw material supplied to the process impact the electrical and chemical energy demand, metallurgical conditions, and hot heel level in the furnace. Acidic fluxes are necessary to prevent reduction of the metal yield and basicity of the slag generated by the gangue in the DRI/HBI charge. Moreover, an increase in process slags as compared to standard operations when using steel scrap increases the electrical power demand and affects the lifetime of the refractory lining.[26] A key advantage of the use of ESFs is their reducing environment, as opposed to the oxidizing environment of an EAF. Slotted into an existing integrated steelmaking operation directly upstream of a BOF and replacing the blast furnace, an ESF could reduce carbon emissions significantly while maintaining the similar product quality and purity most easily provided by integrated steel mills. There are existing examples of process heating electrification for steelmaking in this vein, with significant government-industry collaboration. For instance, the NeoSmelt collaboration in Australia will include the largest ESF pilot plant. This plant would produce 30,000 to 40,000 tonnes of molten iron per year, starting in 2028.[27]

8.3.2.2 Applying Electrical Energy to Heating Operations

The steel industry could benefit from electrifying the reheat process as well as other downstream processes. Reheating furnaces have a key role in CO_2 emissions in downstream steel processing, as these use combustion to heat and homogenize temperatures in semi-finished cast products. Currently, CO_2 emissions of steel reheating furnaces are in the range of 63.2–76.0 kg CO_2/t steel, depending on steel quality, type of steel products, and available techniques.[28] Reheating furnaces are gas-fired in most cases as these have a lower energy cost. For instance, slab reheating furnaces, including walking beam and tunnel furnaces, operate mainly with natural gas. Electrical heating of the products via induction heating,[29] the use of electrically generated hot plasma, or direct/indirect resistance heating could be used to replace combustion in reheating furnaces. Induction heating technology can be also applied to melting furnaces, vacuum induction furnaces, and metal quenching, among other processes.

8.3.2.3 Use of Plasma Heating

Electrical power can also be applied to steel production through plasma technology. Plasma reactors can be adapted to DRI production, further electrifying the DRI-EAF process. Moreover, plasma torches could be implemented at the BF to supplement or provide the heating necessary for the hot blast. Plasma torches with an electrical efficiency of 85% and up to 2.4 MW are currently used to replace fuel burners.[30] The ULCOS project[31] studied the adaptation of 2 MW torches in the tuyeres of a large BF showing a power consumption of around 250 kWh/ton of hot metal. Furthermore, by using plasma torches to superheat air up to 1,800 °C in a BF, the coke consumption was reduced to physical process minimums, and coke was only required to provide the necessary carbon to the iron and to support the BF burden.[32] In this case, it might be possible to reduce BF CO_2 generation by up to 270 kg/ton, considering a production rate of 6,000 tons/day.

8.3.2.4 Electrolysis Reactors for Steelmaking Process

Electrolysis reactors for steelmaking are an emerging technology that can replace carbon (as in blast furnace) with electrons to produce iron electrolytically. This process has been demonstrated in laboratory and pilot scales by applying both low- and high-temperature electrolysis reactors.[33] An example of these electrolysis reactors is the aqueous electrolysis electrowinning of iron oxide ore that employs aqueous alkaline solutions. Here, iron ores are suspended in concentrated aqueous sodium hydroxide (NaOH) solutions with a cathode (e.g., graphite electrode) and an anode

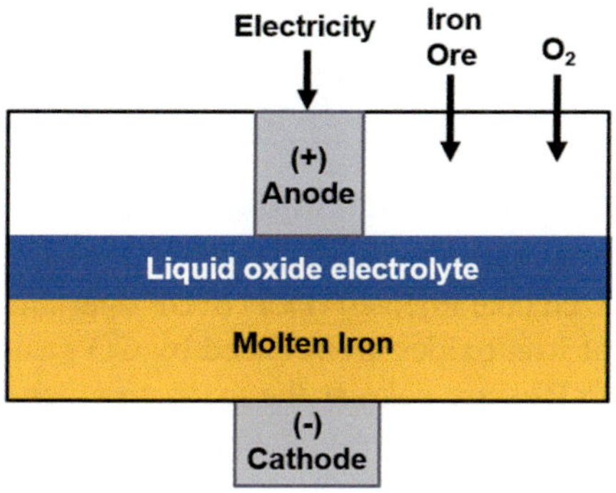

FIGURE 8.3 Boston metal process.[35]

(e.g., nickel) to extract iron metals by applying a constant current (Figure 8.3). This process has shown high efficiency at the laboratory scale: above 90% with respect to iron deposition.[34] High-temperature systems include processes such as molten oxide electrolysis, wherein iron oxide is melted and directly separated into liquid iron and oxygen gas via electrolysis. Scaling these systems to production levels needed for steel production, however, remains in progress.

8.3.3 Low-Carbon Fuels, Feedstocks, and Energy Sources

There is significant research in replacing fossil fuels with less carbon-intensive alternatives. In particular, hydrogen is one of the most promising alternatives to remove carbon usage altogether. However, replacing fuels with alternatives such as hydrogen requires a proper understanding of the process, as the alternative fuel or feedstock could impact other outputs of the operation. Some of the initiatives to apply low-carbon fuels in steel production are as follows.

8.3.3.1 Hydrogen DRI Process

Hydrogen can also lower carbon emissions in steel production by reducing iron ore in DRI process.[36] Namely, hydrogen-based DRI in combination with EAF operation reduces more than 60% of the carbon emissions in coal-based BF-BOF route, assuming hydrogen is generated by using a renewable energy process. The EAF operation can charge mixtures of DRI, HBI (hot briquetted iron, a more stable, premium form of DRI compacted at a temperature above 650°C), and scrap as material inputs, overcoming issues related to scrap availability. However, it is currently necessary to have carbon in the steelmaking operation as it drives the foaming of slag and the stirring of molten steel during refining due to the formation of CO bubbles. In addition, carbon provides additional heating as it reacts with available oxygen. On the contrary, H_2 reduction of iron oxides is an endothermic reaction. As a result, iron can revert back to iron oxide if temperatures decline enough, requiring an additional heating source in the reduction furnace. It has been determined thermodynamically that an optimal gas mixture for this process would be around 75% H2 and 25% CO.[37] Blends of CO and H_2 produce a maximum rate of carbon deposition at 500–600 °C temperatures, whereas there is no carbon deposition when the temperature is above 900 °C. Therefore, the reduction of iron ore by using H_2/CO blends is favored at high temperatures, whereas CO provides carbon on the DRI product.

8.3.3.2 Auxiliary Fuel Gas Injection in Blast Furnaces

In blast furnace operations, hydrogen, natural gas, and syngas can be used to replace part or all of the pulverized coal injected (PCI) in BF operations.[38,39] Typically, in BF processes, about 30% of iron oxides are reduced by CO gas, about 10% by H2 gas, and the rest by carbon.[40] BFs use coke as the main provider of energy and reducing gases. Although fuel gases can replace injected carbon in BF, these cannot replace coke as the gases do not provide the burden support necessary for the process operation. The amount of coke that is used in blast furnaces range between 300 and 450 kg/t hot metal, depending on the amount of auxiliary reducing agents injected. By replacing PCI and coke (to the maximum extent possible) with less carbon-intensive fuels, it is possible to reduce the emissions and, in some cases, reduce the energy consumption in molten iron production. Hydrogen is an attractive fuel gas that could be injected to reduce carbon emissions in BF. However, a limiting factor in the amount of hydrogen that can be supplied is the endothermic nature of hydrogen reactions and the impact hydrogen produces on bed permeability due to the reduction of coke consumption. The injection of H_2 in a BF was first tested by ThyssenKrupp in 2019. This has been followed by other efforts, including a test conducted by Cleveland-Cliffs.[41] ThyssenKrupp Steel also conducted a hydrogen ironmaking experiment in the BF using part of hydrogen instead of coal for smelting. In this trial, it reported a 20% reduction in CO_2 emissions.[42] Japanese Course50 project aims to replace some reductants in the BF with hydrogen, reducing the CO_2 emissions of the BF by 30%.[43]

8.3.3.3 Hydrogen in Reheat Furnace Operations

Another potential application for hydrogen in the steel industry is in the reheat furnace (RF) as a replacement for or blended with natural gas as fuel for combustion. Hydrogen leads to faster flames and higher adiabatic temperatures than natural gas, and injection parameters need to be adjusted properly to control the reheating performance. Partial replacement of natural gas with hydrogen can also change the zoning of the RF. Moreover, the high moisture produced by hydrogen combustion may impact the quality of the product and the refractory life. Hydrogen can also be used to partially provide the energy required by the RF, having the rest of the energy supplied by induction heating in hybrid reheating furnaces. Hydrogen-based RF tests at the industry scale have been started. Linde Gas and Ovako have demonstrated H_2/Oxy-flue combustion in a 650kW batch pit furnace without any negative impact on the steel quality.[44] With funding support from DOE's Industrial Efficiency and Decarbonization Office (IEDO) in 2024, PNW CIVS is leading a nationwide collaboration effort to develop and demonstrate an industry-scale H_2-fired large continuous reheating furnace in Gerdau's Monroe plant in the United States.[45] The project aims to 1) demonstrate H_2-firing in an industry-scale reheating furnace guided by computational fluid dynamics (CFD) simulations and laboratory experiments, and supported by life cycle assessment (LCA) and technoeconomic analysis (TEA) to evaluate energy, economic, and emissions benefits, and 2) establish roadmap strategies and operation scheme for zero-CO_2 emission production in the full furnace with CFD, LCA, and TEA modeling and a state-of-the-art Integrated Virtual Reheating Furnace (IVRF) to ensure production at existing quality standards.

8.3.3.4 Using Waste Streams for Biocoal Injection in BF and EAF Operations

The steel industry can also benefit from alternative carbon reductants. This includes replacing natural gas with gasified waste or using biocoal injection in BF and EAF. For instance, plastic injection has been applied in the tuyeres of BF[46] and biomass is used in EAF.[47] The possibility of injecting charcoal and multiple blends of biomass materials, such as charcoal, rice husk, sugarcane, and eucalyptus, in blast furnaces has been studied experimentally.[48] De Assis et al.[49] concluded that blends showed better combustion rates than using pure materials, and that the combustion performance for multiple blends also depended on the injection rate. Namely, the 25% eucalyptus bark/charcoal blend showed the best combustion performance for injection rates of 50 kg/ton hot metal, whereas 25% sugarcane bagasse and charcoal presented better combustion rate for 100 kg/ton hot metal injection rates. Biomass utilization in the slag foaming of EAF has also been considered. For instance, the wettability of different carbon carriers, such as coke, natural graphite, petrol coke, charcoal, and synthetic graphite, was studied when they interacted with the EAF slag.[50] Among these materials, natural graphite showed the most favorable conditions to increase the volume of slag foaming, whereas charcoal led to the lowest foaming of the slag.

8.3.4 Carbon Capture, Utilization, and Storage

Among the various technologies currently available to reduce carbon emissions in industry, carbon capture, utilization, and storage (CCUS) is one of the less invasive methods to reach significant emission reductions. Namely, CCUS technologies can be applied to existing furnaces and equipment without interfering with the main processes. IEA forecasts that 21% of the world's crude steel production will include carbon capture and storage (CCS) systems by 2026, which translates to 506 Mt CO/year.[51] CCS technologies include capturing CO_2 from large emissions sources (point-source capture) or removing CO_2 from the atmosphere (direct air capture). The cost of CCS implementation varies with the process type, CO_2 transport, capture technology, and storage location. For instance, cost of carbon capture can increase three to six times when it is implemented in cement or power generation facilities as compared to its implementation in natural gas processing plants.[52]

8.3.4.1 CCUS Technology in Integrated Mills

In integrated steel mills, most of the carbon emissions (about 70%) come from flue gas produced in the hot stoves of blast furnaces and in the power plant. The coke oven gas (COG) and BOF gas, in turn, produce approximately 9% and 7%, respectively. CCUS in BF lowers the usage of coke and coal by re-using off-gas, where CO_2 is first removed from the off-gas to obtain the reducing CO and H_2 gases. The ultralow CO_2 steelmaking blast furnace process (ULCOS-BF) replaces hot blast with oxygen and recycles CO and H_2 from the top gas to inject them into the lower shaft and hearth tuyeres, resulting in higher chemical energy efficiency. ULCOS-BF also includes a full CO_2 capture and storage process and utilizes low-purity O_2 to obtain the reducing gases from pulverized coal. The project aims at reducing CO_2 emissions in the BF by at least 50%.[53]

8.3.4.2 CCUS Methods in DRI Production

DRI production can also benefit from CCUS technologies. Currently, coal and natural gas are the two main fuels used in DRI production. CCUS implemented in DRI natural gas-based processes can lower CO_2 emissions to 0.77–0.92 tons of CO_2 per ton of steel. The range depends on the type of electricity used.[54] Natural gas is used in the most known DRI technology processes such as MIDREX and HYL III. Moreover, post-combustion CCUS systems are able to remove CO_2 from the flue gas, reducing up to 89% of the CO_2 that would otherwise be released into the atmosphere.[55] DRI and HIsarna are alternatives to the BF-BOF steelmaking route, and these generate increased CO_2 concentrations (up to 90%), which make them ideal for CCUS implementation.[56]

8.3.4.3 Carbon Capture Technologies in Steel Production

Among the carbon capture (CC) technologies available for the steelmaking processes, four main technologies can be identified: absorption, adsorption, membranes, and cryogenic. The absorption processes separate the gases by using the difference in affinity of substances in the gas. The substances are absorbed through chemical reactions in liquid state and then separated in the following step by changing the pressure (or temperature) of the liquid. The adsorption processes separate the gas based on one of the following effects: steric (molecular size and shape), kinetic (diffusion rates), or equilibrium (chemical interactions). The membrane process separates the gas based on filters. This process requires compression, which can make the operation expensive. Currently, membrane separation for CO gas is not at a commercial scale and it is not applicable to most steelmaking gases.[57] In the cryogenic separation process, the gas temperature is lowered to liquefy the gas and separate the CO_2. This technology, however, is energy intensive as it needs large power for refrigeration. Also, the feedstock gas needs to be pre-treated and dehydrated so that hydrate formation and CO_2 freezing in the equipment are prevented.[58]

8.4 CASE STUDIES: USING DIGITALIZATION AS AN ENABLING TECHNOLOGY FOR DECARBONIZATION

Digitalization is becoming an enabling technology for decarbonization. It uses advanced digital tools to create new or modify existing products, services, and operations, playing a key role in smart manufacturing (also called Industry 4.0 or Digital Transformation). Advanced digital technologies include multi-physics CFD modeling and simulation, virtual reality (VR) and augmented reality (AR) visualization, machine learning, and more. These technologies can be integrated to create physics-based and data-driven digital twins or virtual processes used by researchers, engineers, and operators for 1) smarter and quicker decision-making; 2) process or product design, troubleshooting, and optimization for increasing productivity, energy efficiency, and product quality as well as reducing pollutant emissions; 3) real-time operation control; 4) low-risk and timely emerging technology research, development, demonstration, and deployment; and 5) more effective training for workforce and safety.

Many iron and steelmaking processes involve high temperature and harsh environments where taking sensor measurements is difficult, expensive, and, in some cases, impossible. In the past 20+ years, Purdue University Northwest (PNW) researchers have closely worked with the steel industry to use advanced digital technologies to provide fundamental insights and practical guidance. In 2009, PNW established the Center for Innovation through Visualization and Simulation (CIVS). In 2016, a nationwide, industry-led Steel Manufacturing Simulation and Visualization Consortium (SMSVC), housed at CIVS, was launched with the mission of developing and implementing innovative technical solutions through the integration of simulation and visualization technologies to ensure a competitive advantage for U.S. steel manufacturing. CIVS and SMSVC have developed a well-recognized integrated and application-driven approach for integrating CFD with advanced VR and AR visualization, machine learning, and high-performance computing to develop digital twins and interactive virtual training tools for various iron and steelmaking processes.

PNW CIVS is a multidisciplinary center that integrates advanced digital technologies to solve real-world problems, generating significant economic and educational impact. The SMSVC is a nationwide, industry-led consortium currently with 15-member companies including steel producers and suppliers. The recent focus areas at CIVS and the SMSVC include energy efficiency, decarbonization, digitalization, workforce development, and safety, with applications to blast furnace, DRI, EAF, reheating furnace, ladle, continuous casting, and other iron and steelmaking processes. Since 2022, CIVS has been leading three multimillion-dollar projects and participating in several other large projects funded by DOE for decarbonizing the steel industry. In all of these projects, advanced digital technologies are used to enable the development and demonstration of new decarbonization technologies. CIVS and the SMSVC have developed and applied a number of state-of-the-art high-fidelity multi-physics CFD models. These models can 1) provide fundamental insights into complex process flow patterns, chemical reactions, and heat transfer, as well as stress and fatigue to determine mechanical failure; 2) develop practical knowledge of the impacts of operating conditions on process performance; and 3) function akin to virtual sensors to generate large databases of flow properties, which can be used by machine learning. CIVS and the SMSVC also used advanced VR and AR for 1) visualizing scientific data for better understanding and communicating complex CFD simulation results, 2) developing 3D physics-based and data-driven interactive 3D virtual simulators or digital twins for process control and workforce training with real-world scenarios. They also used machine learning to develop reduced order models (ROMs) based on a high-fidelity CFD simulation database, which dramatically reduced computational time and made it possible for real-time or almost real-time monitoring and optimization. The following examples will illustrate this digitalization work.

8.4.1 Integrated Virtual Blast Furnace

As a key component in integrated steel mills, blast furnaces (BFs) are extremely complex, energy-intensive processes with many dynamic and physical-chemical phenomena, operating at more than 1500°C. The operability and observability of the process are also rather limited. Consuming about 65% of energy used in the entire mill, the BF represents a significant opportunity for improving energy efficiency and reducing CO_2 emissions. Due to the process complexity and difficulties in attaining sensor measurements, modeling and simulation are promising tools for further improvements.

In the past two decades, PNW has worked with the steel industry to develop state-of-the-art CFD models and virtual blast furnaces for simulating and visualizing the complex physical phenomena in industrial-scale blast furnaces.[59,60] The CFD models include a hearth model, a tuyere model, a raceway model, and a shaft model, integrated in functionality to simulate the entire process. Tuyere simulation is handled with commercial CFD solvers, while in the raceway region, an iterative approach between commercial solvers and in-house models is used to calculate the shape and volume of the raceway. This involves the use of a cold flow model that tracks the momentum-based interaction of granular and gas phases to generate a coke distribution. This distribution is then supplied to calculate the reactions and heat transfer processes inside the raceway region. The phenomena within the blast furnace shaft are modeled using an in-house CFD code. The process is assumed to be in a quasi-steady state and uses as inputs the bosh gas temperature, bosh gas velocity, and bosh gas composition profiles generated in the raceway simulation. The shaft model solves transport equations for the gas and burden phases, including reduction reactions, solution loss, water gas shift, direct reduction, and flux decomposition reactions. The cohesive zone (CZ) position is calculated using isotherms in the burden temperature at the softening and liquidus temperatures of the iron ore.

The integrated BF model has been used to study the impact of multiple fuel injection scenarios on CO_2 emissions. These scenarios include injection of natural gas at various rates and preheated temperatures, injection of syngas at various temperatures, and injection of hydrogen at various injection rates.[61,62,63] Results show that natural gas injection reduces CO_2 emissions by up to 48 kg/thm, when the fuel injection increases from 95 to 130 kg/thm. CO_2 further declines up to 65 kg/thm when natural gas (NG) is preheated by 300 K. This represents a 4.3% CO_2 reduction in the BF process from relatively minor modifications to operation.

Modeling syngas injection predicted a larger impact on the reduction of CO_2 than natural gas. Results at similar injection rates and temperatures show that syngas reduces 7.5 times more CO_2 than natural gas. Syngas enhances CO_2 reduction due to its low-carbon content (28%) and similar coke replacement ratio as natural gas. The CO_2 reduction achieved by syngas injection increases with the injection temperature. Results also show that H_2 injection requires careful operational adjustments for effective use in the blast furnace. Modeling results indicate that the largest CO_2 reductions could be obtained when H_2 is co-injected with natural gas. Specifically, by co-injecting 95 kg of natural gas along with 15 kg of H_2, CO_2 emissions are reduced by 82 kg/thm, which is a 5.4% reduction when compared to gas-only injection operation.[64] It should be mentioned that the CO_2 impact shown in this study did not account for the CO_2 emissions associated with fuel production and cokemaking.

These models have allowed PNW researchers to collaborate with industry to investigate the impacts of specific design and operational changes on furnace efficiency and troubleshooting. Furthermore, advanced 3D visualization technologies were applied in concert with these models to develop a virtual blast furnace for interactive training. These have resulted in millions in cost savings and significant downtime reduction. Figure 8.4 shows sample results of BF CFD simulation and visualization. The following are examples of outcomes:

- Troubleshot a lance failure issue at the Stelco Hamilton Works E blast furnace, resulting in $8.5 million per year in cost savings and downtime avoidance of 50%.
- Simulated design and operation modifications for the Cleveland-Cliffs (formerly AK Steel) Dearborn Works BF, which increased coal burnout by 30% and had the potential to increase productivity by 2.5%.
- Identified the cause of insufficient cooling for tuyere failures in a U. S. Steel blast furnace.

Built upon over 20 years of success, the CIVS team is leading a national effort to develop next-generation physics-based and data-driven, interactive, and integrated virtual blast furnace (IVBF), supported by the U.S. Department of Energy (DOE). This IVBF is intended for use in both offline and real-time scenarios by integrating CFD simulation, machine learning, reduced order dynamic modeling, and advanced visualization technologies. The IVBF is a software suite with multiple modules, providing a near-digital-twin tool for blast furnace operators and engineers to create smarter and more efficient production environment. The IVBF will allow operators and engineers to conduct real-time analysis to monitor BF performance and internal states, perform rapid off-line modeling to quickly adapt to changing conditions and

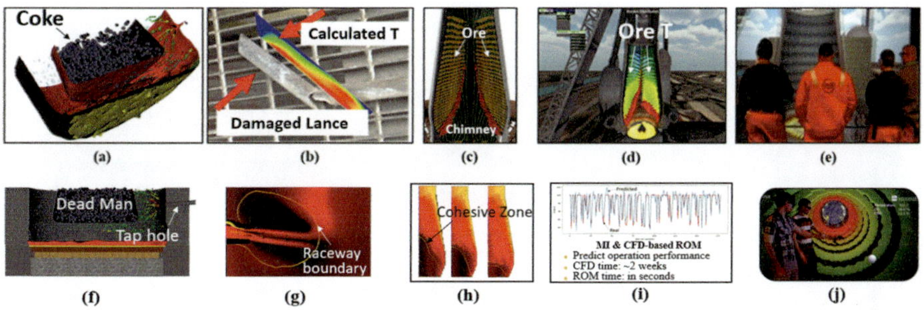

FIGURE 8.4 Blast furnace CFD simulation and visualization results: (a) hearth flow and inner profiles, (b) calculated lance surface temperature vs a damaged lance, (c) burden structures, cohesive zone, and gas flows, (d) virtual BF, and (e) virtual training, (f) hearth erosion profile and velocities, (g) raceway shape and gas temperatures for waste plastic injection, (h) H_2 injection impacts on cohesive zones and gas temperatures, (i) MI & CFD-based ROM, (j) CFD soft sensor measurement in a virtual blast furnace.

investigate "what-if" scenarios, and evaluate new technologies based on the results of high-fidelity analysis. When complete, the IVBF will provide the U.S. steel industry with a unique and powerful tool for improving BF operation, stability, and energy efficiency, as detailed in the feedback structure in Figure 8.5.

In the IVBF, a neural network-based ROM has been trained on a large database of CFD simulation results. This ROM can accelerate the speed at which blast furnace operators and engineers can obtain predictions of key parameters in near real time.[65] The rapid-prediction module uses reduced order models based on both neural network and XGBoost algorithms (depending on whether the prediction is outside or inside of the training database range) to provide predicted outputs based on user-generated inputs. Results from the ROMs compare favorably to existing validated CFD models, with R^2 values above 0.95 for most variables, and predictions generated within fractions of a second from user input, significantly faster than week-long run-times of a traditional CFD modeling approach.

The IVBF also includes a system to monitor the casting rate of the molten iron released from the blast furnace in real time.[66] This remains a significant challenge in blast furnace operations, and while methods of accurately estimating average furnace production over periods of 24 hours exist, direct measurement of the instantaneous flow rate of material leaving the furnace would significantly benefit furnace operators in establishing the stability of the process. An image-based method to estimate iron casting rate based on a live video of the hot metal jet has been developed. The image is recorded and processed with an image processing algorithm to extract both the velocity and the size of the jet, which are then used to estimate the casting rate. During trials of the system at a fully operational blast furnace, the system has performed effectively in the harsh casthouse environment and successfully predicted the instantaneous casting rate across multiple casts.

Real-time analytics based on existing sensor data from the blast furnace have also been developed. The IVBF includes a forecasting model for blast furnace hot metal

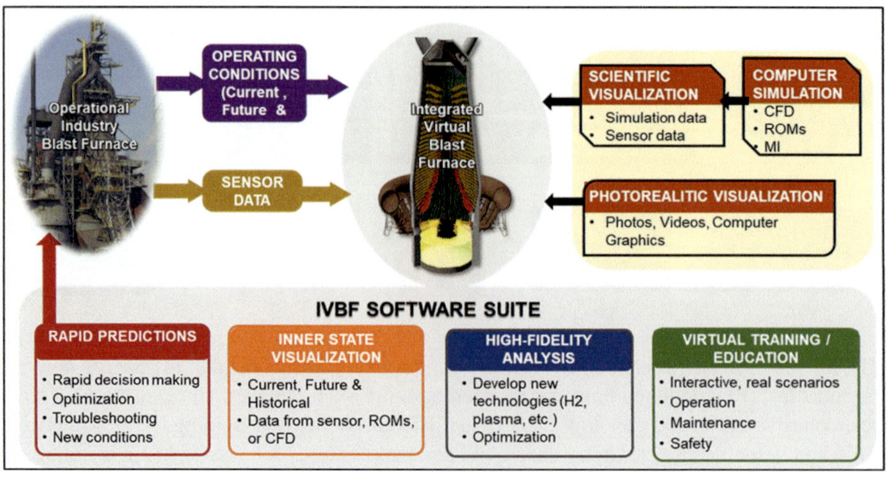

FIGURE 8.5 Schematic of the IVBF and its application on BF operation.

silicon content, which can influence downstream carbon emissions in the BOF process. A better understanding of the hot metal silicon content prior to shipping to the BOF will allow downstream operators to better control their processes, saving energy and reducing emissions for integrated steelmaking. A generalized data processing scheme was developed to accept the diverse range of potential furnace data parameters recorded, and a machine learning model using the XGBoost algorithm was employed to predict the cast-average silicon content for the upcoming cast using a rolling input of the past shift's recorded data with an accuracy of 91% when compared to real-world data.[67]

8.4.2 CFD Modeling of Hydrogen-Based DRI and Operation Scaling

As part of a DOE project, CIVS is developing a CFD model of the DRI process for an existing HYL Energiron ZR furnace, with CO and H_2 reduction as well as steam methane reforming being considered. In support of an effort led by Carnegie Mellon University (CMU), this model will utilize kinetics developed by CMUs experimental testing to simulate a 100% H2 atmosphere for low-carbon DRI processing.[68] The CFD model of the DRI reactor is being developed using ANSYS Fluent, where the furnace is modeled as a 3D steady-state process, and interactions between the ore and gas phases are tracked with an Eulerian multiphase model. The iron ore reduction reactions are handled using the unreacted shrinking core model (URC), where three distinct boundaries inside of the pellet are assumed as hematite (Fe_2O_3) reduces to metallic iron (Fe). These account for both hydrogen and carbon monoxide reductions. The model includes reaction resistances from species transport across the gas film, the diffusion of reducing gas inside of the pellet, and the chemical kinetics of the reactions. The final goal for this model is a complete CFD twin of the DRI furnace that is able to predict production and metallization based on prescribed gas inlet conditions (see Figure 8.6).

8.4.3 Comprehensive Modelling of EAF Operations

U.S. steelmaking is a world leader in the use of EAFs, employing 50–60% electrical energy to melt and refine recycled steel scrap and metallics into liquid steel. The remainder of the energy is supplied by burners and chemical reactions. EAF scope one emissions are roughly 100 kg of CO_2 per ton of steel, of which around 45% are due to the carbon injection needed to reduce yield losses, ensure electrical energy efficiency, and obtain the desired steel quality.

The EAF operation includes complex chemical-physical phenomena such as combustion, arc heating, phase change, chemical reactions, and heat transfer. The multiple (and concurrent) phenomena in the EAF process make proper understanding and accurate prediction of EAF operation difficult, which prevents further improvements and process efficiency. CFD can provide significant insight into the EAF operation. Since 2016, the researchers at PNW CIVS have closely worked with the SMSVC members to develop and implement comprehensive EAF CFD models to simulate the multi-stages of the entire EAF process. The EAF CFD models use a step-by-step and integrated methodology. The CFD approach includes a series of models

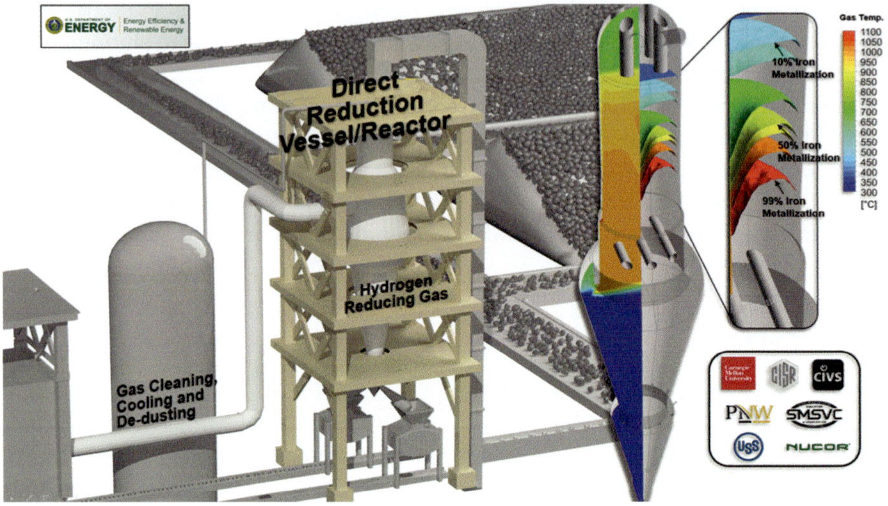

FIGURE 8.6 DRI process overview and simulation contours of gas temperature with metallization iso-surfaces.

developed specifically to address the complex phenomena of preheating, melting, and refining stages of EAF operations. The CFD models include the coherent jet model, electrical arc model (both AC and DC modes), scrap melting model, oxidation model, and slag foaming model. These models are then integrated to simulate actual EAF operations. Figure 8.7 shows the EAF models and examples of simulation and visualization of EAFs.

The CFD models have been validated against data provided by industrial partners from SMSVC. In particular, simulation of a complete AC tap-to-tap operation showed 10% difference in melting rate, whereas the electrode position after bore-in differed by less than 5%. Similar results were obtained when applying the scrap melting model to a complete DC EAF operation. The validated model has been used to determine the impact of burner parameters during preheating. Results showed that by increasing the burner power by 25%, the scrap temperature by 300 K, and the scrap porosity by 6%, it is possible to increase the melting percentage by 63% during EAF preheating stage. Similarly, the melting model has been applied to determine the impact of scrap layering on melting performance of scrap/HBI material. Simulation of a real two-bucket EAF operation concluded that by increasing the number of layers from 4 to 10 it is possible to increase the melting rate by 12% during melting of the 1st bucket. The CFD platform has also been used to determine optimal oxygen injection on molten bath mixing during refining. The mixing time controls the efficiency of the process as it determines the time needed for the injected oxygen to disperse throughout the molten steel. Results showed that it is possible to reduce the mixing time by 6.7% by increasing the total oxygen injection by 25%. Also, it was concluded that non-uniform injection rates can increase mixing efficiency by up to 10%.

These modeling capabilities formed a solid foundation for developing integrated virtual EAF simulators as part of a project funded by the U.S. DOE.[69] In this project,

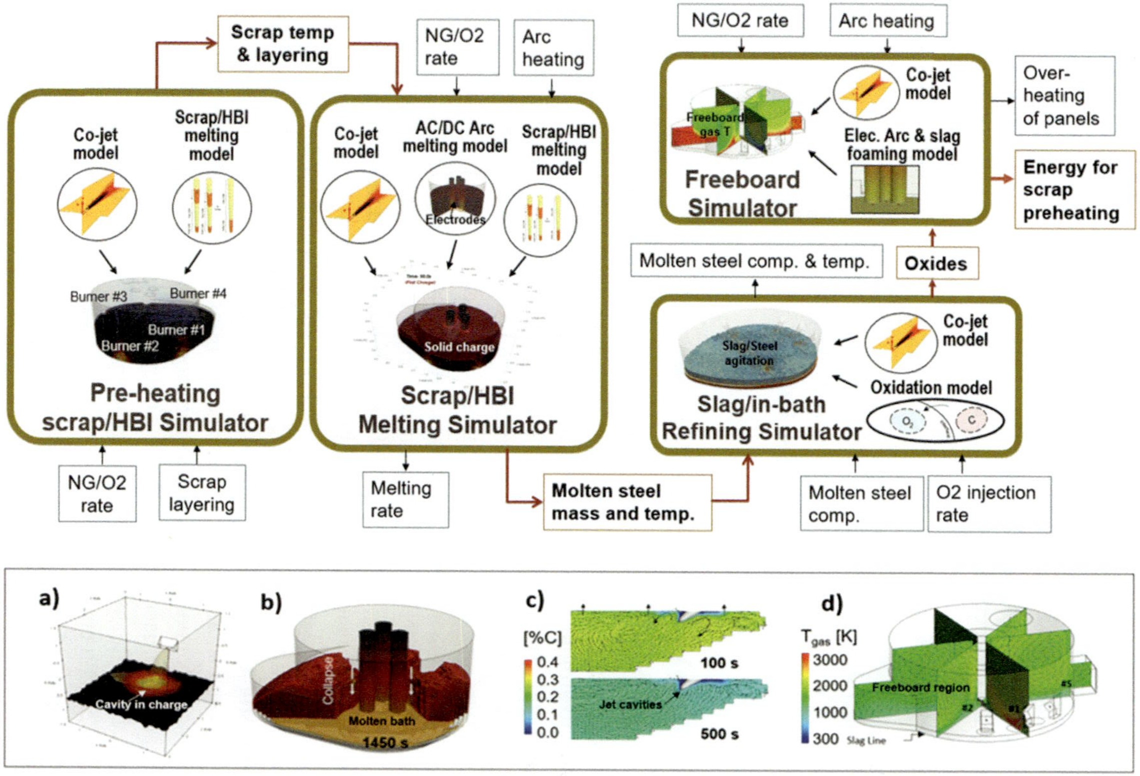

FIGURE 8.7 Top: Step-by-step and integrated approach of EAF CFD modeling. Bottom: (a) EAF scrap melting under gas burners, (b) scrap collapse during EAF melting, (c) carbon content during EAF refining, and (d) freeboard gas temperature in an EAF.

CIVS, in collaboration with Carnegie Mellon University and SMSVC member companies Nucor and Linde, aims to investigate the use of alternative injectants during EAF refining. Namely, carbon and oxygen injection are replaced by natural gas, CO, and CO_2 to perform refining operations. In EAF processes, 45% of scope one emissions are due to the carbon injection needed to reduce yield losses, ensure electrical energy efficiency, and obtain the desired steel quality. Replacing these carbon additions with alternative reductants would directly eliminate 3.6 million tons of CO_2 per year from the 155 EAFs in the United States, with the added benefit of enabling a pathway to net-zero carbon EAF steelmaking. Electro-magnetic stirring (EMS) will also be simulated to reduce the need for O_2 injection, helping avoid steel bath over-oxidation and excess FeO formation. The project goals are to demonstrate stable alternative injection of reducing gases, soft oxidants, and slag foaming agents during EAF refining, to advance technical knowledge and develop guidelines for alternative injection EAF operation with EMS, to optimize EAF operation by adjusting electrical and chemical energy use, and to positively impact underserved communities through workforce development and virtual training.

8.4.4 Integrated Virtual Reheating Furnace

Reheating furnace (RF) is a key and highly energy-intensive process in hot rolling mills, which primarily use NG to reheat semi-finished products such as blooms, slabs, or billets to around 1,200 °C. There are more than 80 RFs in the United States, which burn NG and generate roughly 200 kg CO_2/ton of steel. Alternative fuels such as H_2 could be used to reduce or eliminate CO_2 emissions; however, this introduces additional variables and environmental changes, requiring in-depth knowledge and guidance to ensure efficient production and good product quality.

In the past 20 years, PNW has collaborated with industry and national laboratories to develop several existing CFD RF models, which have been used to investigate the impacts of RF operational changes on slab quality, scale growth, and furnace efficiency, and to gain fundamental insights and practical guidance for design, troubleshooting, and optimization for reheating furnaces. Recently, PNW has begun conducting investigations of industrial H_2-fired RF operation, indicating the potential to eliminate CO_2 emissions under controlled conditions. These capabilities and experiences pave a solid foundation for further developing and implementing CFD modeling of reheating furnaces using H_2 or other low-carbon fuels, as well as for developing an IVRF to make the steel industry hydrogen-ready in the pathways toward reduced carbon emissions.

Under the support of DOE, CIVS is leading a nationwide team, including ANL, ORNL, and SMSVC member companies ArcelorMittal, Cleveland-Cliffs, Gerdau, and Linde, to develop and demonstrate an Industrial Hydrogen-Fired Steel Reheating Furnace.[70] In this project, a unique, integrated approach employing high-fidelity computational modeling, machine learning, LCA & TEA, advanced visualization, detailed lab experimentation, and industrial field demonstration is used to 1) provide knowledge and guidance for incrementally implementing H_2-firing in industrial RFs at various conditions for CO_2 reduction, and 2) develop a state-of-the-art multifunctional IVRF platform as an effective tool for developing strategies and evaluate impacts to advance the deployment of H_2-fired RF technology in steel industry (see Figure 8.8).

In this project, CFD simulations are used to investigate heating rates, scale formation, thermal stresses and uniformity, NOx formation, and moisture generation prior to and in parallel with the field demonstration. Lab testing and plant data serve to validate CFD models and characterize H_2 burners. The IVRF platform will include modules for virtual demonstration, roadmapping and optimization, control, and virtual training for workforce and safety. A field demonstration will be conducted of single-zone H_2-firing in an industry RF with various $H_2/CH_4/O_2$ fuel blends, showing the feasibility of incremental conversion to H_2 without influencing quality and productivity. The Virtual Demonstration Module in the IVRF integrates CFD, LCA, and TEA, and visualization to show impacts on the RF from H_2-firing in partial zones and all zones. This supplementary demonstration will advance technical knowledge and provide scenarios to form a precedent for full conversion to H_2 and strategies to be implemented in response to ongoing market changes. Prior to the onset of viable market conditions for full H_2-fired conversions, this will build confidence that the industry need not wait for a future tipping point for economic H_2 use and may follow a plan of increasing usage at the steadily moving breakeven point.

8.4.5 Simulation of Ladle Furnaces and Process Optimization

The U.S. steel industry is an energy-intensive yet critical component of the domestic and global economy. With over 5% of produced steel ending up rejected and reprocessed as scrap,[71] inefficiencies and production faults represent an enormous energy and financial burden. A critical stage to prevent product rejection is the refining ladle process. However, steel refining in a ladle is a nonlinear dynamic and stochastic process. Furthermore, refining ladle is a necessary component in the continuous casting process.[72] Improper homogenization or cleaning of steel in the refining ladle can lead to costly downtime of the casting process as well as defective products that must be reprocessed upstream. To address these issues, numerical simulation is needed to model ladle operation, with HPC being necessary due to the computational requirements to simulate details accurately. Furthermore, reduced order modeling using simulation and industry data allows for real-time process control improvements. Reduction of product rejection due to improper refining and improvement of processing time can save an estimated $106 million annually.

PNW in collaboration with the steel industry has developed a comprehensive, multiphase CFD model to simulate the ladle process for over 20 years (Figure 8.9). The ladle CFD models can simulate argon gas stirring, EMS, or combination of argon gas stirring (through top lance or bottom plug) and EMS. The Population Balance Method (PBM) is applied to model the billions of inclusion particles (which are at the micrometer scale) in the molten steel as a separate continuous phase in which the inclusions exist. Inclusion growth through aggregation is modeled based on statistical particle density in a cell volume. These models have been validated and extensively applied to actual ladle operations to investigate parametric effects on temperature distribution, slag eye size, flow characteristics, wall shear stress, and

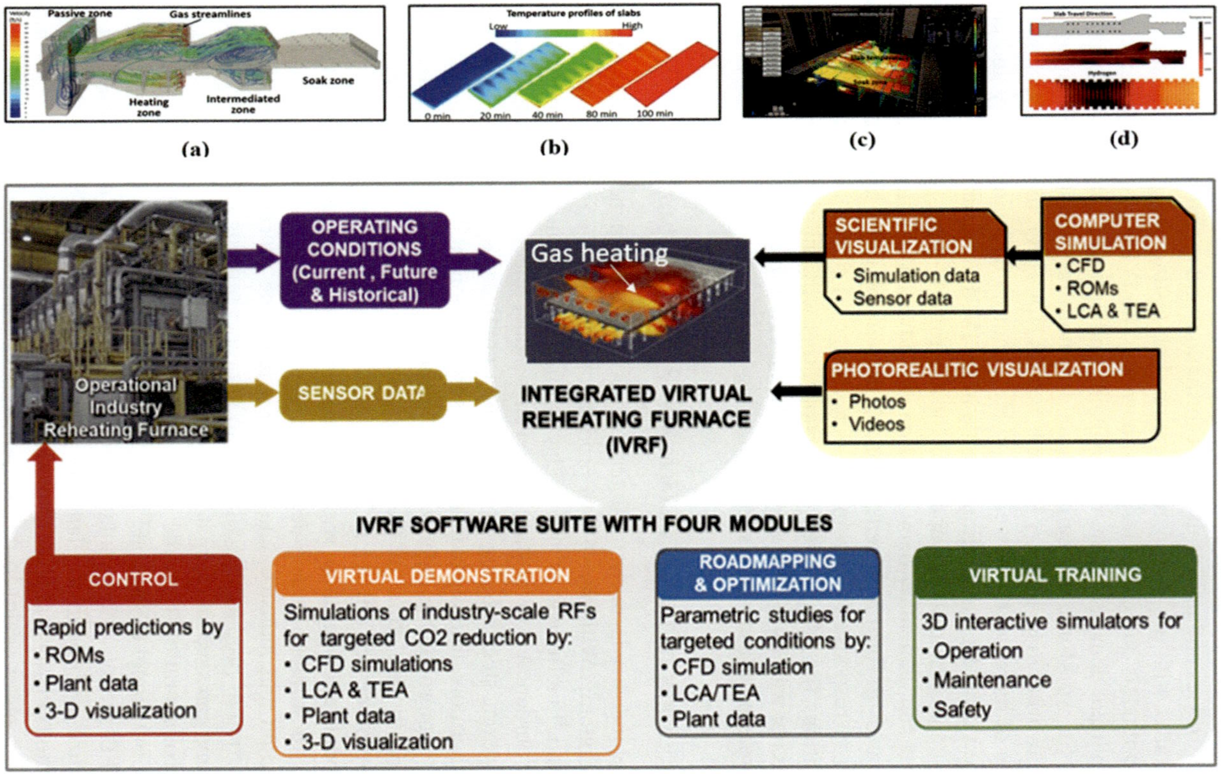

FIGURE 8.8 CFD simulation and visualization of reheating furnace (RF) at CIVS (a) Flow inside an RF, (b) slab temperatures, (c) a virtual RF for control and training, (d) H2-fired furnace, and (e) IVRF software suite, including control, virtual demonstration, roadmapping and optimization, and virtual training modules.

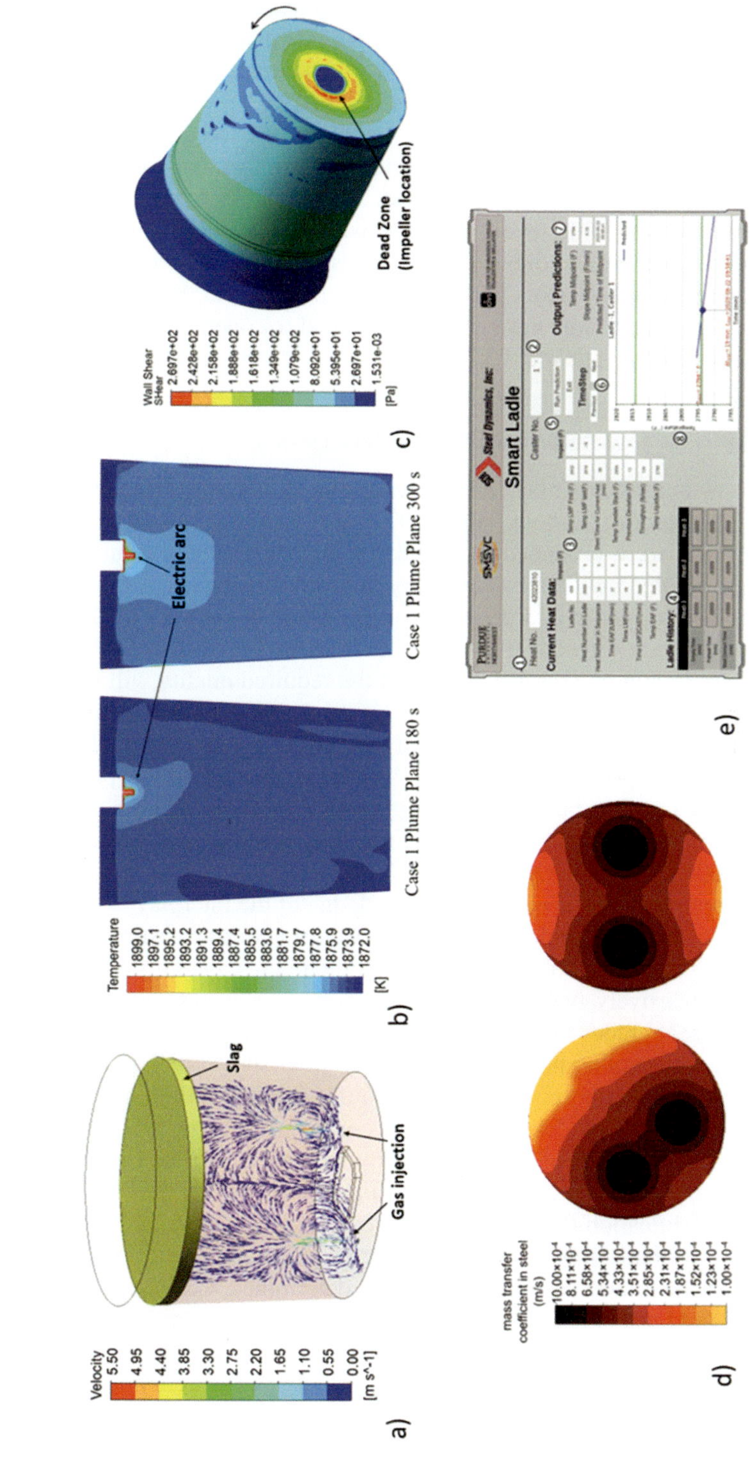

FIGURE 8.9 (a) Multiphase flow in the industrial-scale ladle, (b) Temperature distribution during ladle arc heating, (c) Wall shear stress, (d) sulfur mass transfer coefficient, (e) "smart ladle" digital tool.

inclusions, under different operation conditions. For instance, a study performed in a real ladle scenario showed that an increase in the flow rate of 100% leads to a reduction in the temperature difference in the ladle of 17.4%, whereas the slag eye-opening increases by 116.4% and the wall shear stress increases 49.9%.[73]

CIVS & SMSVC also developed a Smart Ladle, an AI tool, to quantify and respond to the variables that affect steel casting temperature, so that consistent casting quality and optimal productivity are reached. Namely, a deep-learning network was used to develop quantifiable relationships between the casting temperature and various factors during the ladle refining process. These enable predictions of casting temperature and precise adjustments to steel temperature prior to the ladle reaching the casting stage. The smart ladle has been implemented for operators' use in plants of SMSVC member companies.

8.4.6 Modeling of Continuous Casting Processes

The continuous casting process plays a significant role in ensuring structural integrity of the final steel product. About 98% of 92.4 million short tons of crude steel produced in the U.S. were produced via continuous casters, where liquid steel is solidified by water-cooled mold and water spray to form semi-finished slabs, billets, or blooms. Continuous casting is a delicate process because the flow of the liquid steel and the cooling rate directly affect the slab/billet/bloom surface and internal quality. Any slab/billet/bloom that does not meet the required quality will be downgraded or scrapped (and remelted). Re-melting of steel can constitute a significant loss of revenue and energy use. An EAF-based facility consumes 1.5 MMBtu electricity per short ton of tapped steel on average,[74] meaning each ton of lost production has a high direct energy cost besides other losses such as additional material consumptions. In addition, re-melting of steel can result in both production delays and the emission of 400–800 kg of CO_2.

Due to the complexity of the process, the knowledge of the interplay between the physical, chemical, and metallurgical phenomena occurring during casting is incomplete. It is critical to develop a robust solidification modeling capability for the steel industry in comprehensively understanding and optimizing casting operation, and conditions for different grades of steel in order to reduce casting defect rates, where determining secondary cooling heat transfer coefficient of spray cooling in continuous casting is extremely important for steel manufacturing. Multiple efforts have been made to use numerical modeling to simulate continuous casting process.[75,76,77,78] Since 2016, CIVS & SMSVC has used a step-by-step and integrated approach to develop and apply a number of CFD models for simulating primary cooling, secondary cooling, and solidification through the whole caster (Figure 8.10).[79,80,81,82]

These models have been applied to industry-scale casters. In particular, the CFD simulations of the integrated casting model have determined the impact of argon injection on flow field and surface standing wave, the impact of nozzle clogging on solidification, and chemical segregation. Moreover, secondary cooling models have been applied to optimize cooling intensity in terms of both suitable cooling (avoiding slab defects due to overcooling or undercooling) and energy savings. Here, the models aim to optimize the nozzle spray flow rate and the distance between spray

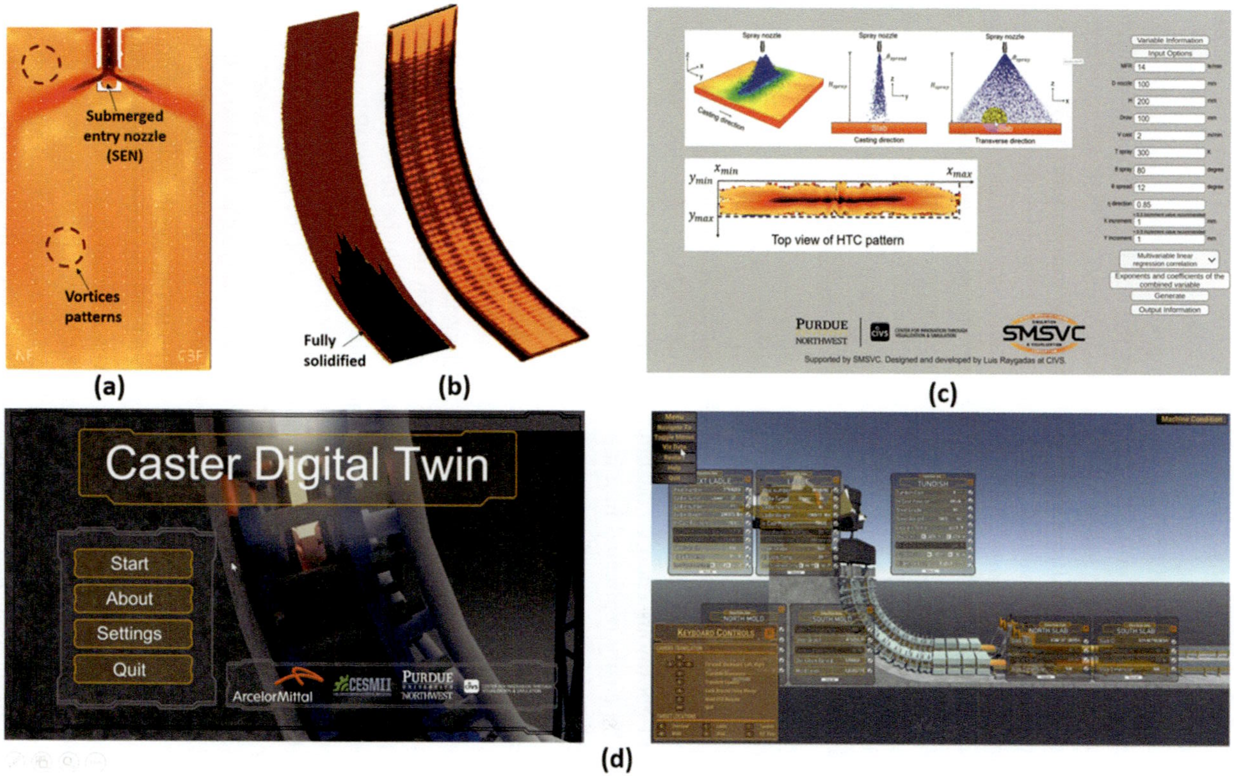

FIGURE 8.10 (a) Simulated velocities in primary cooling, (b) iso-surface shell and surface temperature of secondary cooling, (c) CFD results-based HTC calculator for secondary cooling process, (d) digital twin caster showing main-menu (left), and process data (right).

nozzles, developing more accurate and reliable heat transfer coefficient correlation. Combining with finite element analysis (FEA) simulations allows the investigation of slab thermal stress distribution along the caster strand.

CIVS and SMSVC companies have also made efforts to develop a digital twin solution for continuous casting, which integrates sensor data such as the gap and alignment data into the caster digital twin model for better visualization. This is combined with the solidification profile obtained from the secondary cooling computer model. The secondary cooling model was previously tuned for high accuracy utilizing an array of strain gages, linear variable differential transformers (LVDTs), two color pyrometers, and line scanning pyrometers. The combination of the roll gap information and the solidification information in the digital twin enables better real-time decisions to be made by operators and engineers to operate only when the caster is within specification, and select casting speeds aimed at optimized solidification time in the casting strand. Proper interpretation of the data provided by measurement tools such as these can help to decrease defects and prevent further time spent in data collection and analysis.[83]

8.5 MAJOR CHALLENGES AND FUTURE TRENDS

Adaptation of the technologies described in the four pillars of steel decarbonization requires significant RD&D effort. Namely, the steel industry demands large-scale production, and enormous energy consumption and includes much harsher conditions than can be easily replicated in a laboratory or even at a pilot scale. Some of these challenges and future trends are summarized below.

Needs for additional equipment to change plant layouts: Challenges with replacing BF with DRI – EAF production routes at integrated steel mills include the integration of ladle furnaces (LF) – EAF systems in existing BOF steel plants, as well as the integration of DRI production into the integrated metallurgical plant. Ideally, the location of the EAF should be close to the DRI production site, so that the EAF operation could benefit from hot DRI supply. However, providing space at the plant site for both the EAF and DRI production facilities may create difficulties with the logistics at the production site.

Impacts of batch operation and tap weight of EAF on other processes: An issue in EAF operations is that EAF typically has a smaller batch size than the BOF, which may require more heating and refining time in secondary metallurgy. Also, the tap weight of EAF has an impact on the pacing of continuous casting processes. Namely, a smaller EAF tap weight may require shorter tap-to-tap times and high-power EAFs. The tap-to-tap time will further affect the pace of ladle processing and the speed of the continuous casting, which also depends on the geometry and size of the cast. This leads to adding more chemical and temperature analysis. Overall, to replace a single blast furnace it is required approximately four EAFs and proper-sized DRI shaft furnaces in the production line.

Needs for RD&D of materials in waste heat and gas recovery processes: Application of waste heat and gas recovery (WHR) to steel production processes requires further RD&D. Specifically, RD&D could help determine phase-stable materials and functional surfaces, as well as obtaining embrittlement-resistant materials to prevent the effect of material aging. WHR also requires technological advancement of materials for application in harsh environments. WHR systems can be implemented for oxy-fired RFs, whereas coke gas, blast furnace gas, and furnace gas can be recycled back into the process or they can be used to produce hot water, steam, and electricity. However, commercialization of these technologies faces economic viability. RD&D is also needed to reduce the initial investment cost for waste heat to power (WHP) systems, such as organic Rankine cycle and supercritical CO_2 power cycles.

Challenges in electrification of steel production processes: Several technical challenges may arise in the electrification of steel production. For instance, corrosive gases included in the production environment could result in frequent failures of the equipment used for electrical heating. Also, difficulties in modifying steel processes may increase depending on the type of steel product. Namely, modifying fuel-fired burners to work with induction heating may only be feasible for thin slabs or billets at present. In this case, production plants require significant redesign to apply electrical heating in the process, as the steel needs to reach temperatures above 1,100 °C. Another area that requires further attention is the increase in electricity demand associated with electrification of the steel industry. RD&D could investigate the optimal ways to provide the capacity needed in industrial zones or clusters, as these require high-voltage electricity transmission infrastructure to deliver electricity for steel production.[84]

Process scalability to meet steel production demand: Scalability of the new technologies proposed for steel production is perhaps the most challenging part of the steel decarbonization pathway. For instance, DRI-H_2 is a technically proven method that can allow for nearly emission-free production of steel. The DRI-H_2 technology, however, is restricted by cost and availability of hydrogen, the limitations in iron-ore quality for post-reduction processing, and the process competitiveness of hydrogen with natural gas.

Similarly, scaling-up of electro-technologies in the steel industry may involve high capital costs. Although some of the technologies under consideration show good performance in lab-scale and small-scale applications, it does not necessarily scale up with the same efficiency. For instance, systems that can process a million metric-ton of steel per year using electrolysis of iron ore are not yet as economical as traditional systems. These systems require large-scale testing and process optimization to improve the operational efficiency, so that costs can be reduced.[85] Also, RD&D must focus on modifying the design of furnaces to apply resistance heating in batch and continuous modes when these operate at large production scales.

Moreover, the technical viability of iron electrolysis has been demonstrated in laboratory settings, and results show that it can lower the electricity consumption needed to synthesize hydrogen.[86] Direct electrolysis of iron is named a transformative technology in the long term, as it can potentially decarbonize the steel industry if no-carbon electricity is used. RD&D of electrolysis processes includes molten oxides at high temperatures (1,600 °C) and aqueous electrowinning at low temperatures (110 °C). However, the yield level and process scalability of electrolysis of iron ore are not at a commercial scale yet, and active RD&D is being performed.

Commercialization of CCS technologies: CCS could decarbonize different processes of iron and steel production, such as top-gas recycling in blast furnaces, DRI, oxygen-rich smelt reduction, bath smelting reduction, HIsarna, and direct smelting reduction.[87] These pathways vary greatly in their commercialization status, with blast furnace CCS being at the pilot stage, DRI CCS in the development stage, and smelting reduction CCS in the pilot stage. The top-gas recirculation blast furnace process with CCS could also reduce coal inputs and increase the percentage of CO_2 in the exhaust gas, which would also lower the cost of carbon capture.[88] Also, CCS could be combined with oxy-fuel combustion in reheat, sintering, or pelletizing furnaces, though this combination has not yet been demonstrated.

Challenges in digitalization: A major challenge is the lack of real-time process optimization and control strategies, methods, and systems for data collection. Non-intrusive characterization tools, both spatial and temporal, are currently lacking. This contributes to data acquisition challenges, as does the complexity of process equipment. In situ parameters that cannot be measured may lead to model errors. Knowledge and technology for effectively measuring these parameters are currently limited. In addition, many operators rely on (and are comfortable with) rules of thumb for decision-making rather than real-time data. CFD simulations can function like soft sensors, which can provide data to generate machine learning-based reduced order models (ROMs) for fast predictions. However, validation of CFD models and ROMs is critical to ensure their reliability, which requires data from laboratory-scale and commercial plant scales.

Creating a smart manufacturing enterprise is a "large" problem with many complex elements. The challenge can be so large it is difficult to target the phases and starting points. For example, while creating a "streamlined integrated plant model" is highly desirable, it represents an enormous undertaking with many components and uncertainties. In many cases, processes are still treated as individual systems that do not lend well to integration. For example, understanding and controlling the stability of material flows, from raw material inputs to various processes and finally in finished materials, is an important but daunting technical problem as it spans multiple parts of the production process.

Scaling is another issue for advanced modeling, simulation, and visualization (MSV) using high-performance computing. For example, scalable high-performance computing (HPC) multi-physics models and simulation codes are lacking and could be used for a number of process units. An example is using scalable HPC to model the entire blast furnace, which would allow for effective start-up diagnostics and modeling of other important operating parameters.

There is inherent risk (i.e., issues of trust and reliability) involved in the adoption of model-based systems, especially if large operational changes and capital investment are required. In addition, adoption of smart manufacturing techniques may require installation of high-technology equipment, which can be problematic in a plant setting where harsh environments require very robust systems.

Critically needed workforce development: The current workforce in the steel industry is aging and will need a fresh influx of operators, engineers, and scientists. Furthermore, decarbonization and digitalization require higher skilled workforce. Students often have a limited understanding of the steel industry and the many types of opportunities it provides. College curricula and technical training may not necessarily target the skills needed for working in the industry or prepare students for steel industry careers. A comprehensive training, curriculum, and outreach program is needed to build a new steel industry workforce. The approach is to develop real-world interactive problem scenarios to introduce students to and instruct students in the skills and concepts of the steel industry. Virtual training tools are one way to enhance operator maintenance and overall industry performance. New training tools would combine all aspects of operations (i.e., safety, maintenance, reliability) and contribute to workforce development. Such tools are also cross-cutting and could be applied across steel companies to replicate benefits.

8.6 SUMMARY

The steel industry has played a key role in industrialization, economic growth, and improving living standards. As the world moves towards sustainability, iron and steelmaking will be crucial in achieving global decarbonization goals. The push to decarbonize the steel sector has sparked the next wave of technological innovation. Four technology pillars have been identified. However, achieving decarbonization will require multiple, different technologies working together to create a more sustainable future for the industry. The technologies discussed in this chapter are promising but also face significant technical and economic challenges to scale commercially and deliver the anticipated impact. Steelmakers are making substantial investments in the transition to net-zero production. To succeed in this transformation, it is crucial to utilize advanced digital tools for optimizing current processes, developing new decarbonization technologies, and building the skilled workforce needed to meet the net-zero goals.

NOTES

1 International Energy Agency, Iron and Steel Technology Roadmap, October 2020.
2 International Energy Agency, Iron and Steel Technology Roadmap, October 2020.
3 Hasanbeigi, A., et al. Air Pollution from Global Steel Industry: An International Benchmarking of Criteria Air Pollutants Intensities, August 2022.
4 International Energy Agency, Net Zero Emissions by 2050: A Roadmap for the Global Energy Sector, May 2021.
5 Department of Energy, Industrial Decarbonization Roadmap, September 2022.
6 Steelmaking Flowlines, AISI, 2013.
7 Worldsteel Association (2024).
8 Fan and Friedmann, *Joule* 5, 829–862, 2021.
9 Fan and Friedmann, *Joule* 5, 829–862, 2021.
10 Seker et al. *Elektron. Elektrotech.* 23, 25–33.

11 Fan and Friedmann, *Joule* 5, 829–862, 2021.
12 Fan and Friedmann, *Joule* 5, 829–862, 2021.
13 Fan and Friedmann, *Joule* 5, 829–862, 2021.
14 Industrial Decarbonization Roadmap, Department of Energy, 2022.
15 Roadmap for Iron and Steel Manufacturing: Revolutionizing U.S. Global Leadership for a Sustainable Industrial Supply Chain, 2024.
16 Industrial Decarbonization Roadmap, Department of Energy, 2022.
17 Roadmap for Iron and Steel Manufacturing: Revolutionizing U.S. Global Leadership for a Sustainable Industrial Supply Chain, 2024.
18 Fruehan, R.J. and Paxton, H., "U.S. DOE, Office of Industrial Technology, Report on Minimum Energy for Steel Manufacturing," 2000.
19 Slater, C. & Davis, C. Metallurgical and Materials Transactions B, 2532, 51B, 2020.
20 Ge, S. et al. *ISIJ Int.* 52, 2109–2122, 2012.
21 Near net shape manufacturing, Aerospace Technology Institute, 2021.
22 Fan and Friedmann, *Joule* 5, 829–862, 2021.
23 Slater, C. & Davis, C. *Metallur. Mater. Trans. B* 2532, 51B, 2020.
24 https://www.ssab.com/en-us/brands-and-products/docol/ultra-high-strength-steel.
25 Liang, R. et al. *J. Magn. Magn. Mater.* 441, 511–16, 2017.
26 Kirowitz, J. Electric melting furnaces for green steel transformation of integrated steel plants -requirements, challenges, and solutions from a refractory perspective, EOSC, Aachen, Germany October 2022.
27 https://www.bhp.com/news/media-centre/releases/2024/12/bluescope-bhp-and-rio-tinto-select-wa-for-australias-largest-ironmaking-electric-smelting-furnace
28 Steinboeck, A. et al. *J. Process Control* 21(2), 279–292, 2011.
29 https://www.clevelandcliffs.com/sustainability/steel-as-a-sustainable-material/dept-of-energy-butler-works-project.
30 Westinghouse Plasma Corporation. Westinghouse Plasma Torches for Foundry & Ironmaking Applications. Calgary, Canada, 2014.
31 J. Borlée, "European Steel Industry Main Pathways Toward the Smart, Low Carbon Industry of the Future," ESTEP Workshop, February 2018.
32 M. Sukhram, "The Use of Plasma Torches in Blast Furnace Ironmaking," ECIC2016, Linz, Austria.
33 Siderwin concept, Climate Action Report, May 2019, ArcelorMittal.
34 https://dmse.mit.edu/news/making-steel-with-electricity/.
35 https://dmse.mit.edu/news/making-steel-with-electricity/.
36 https://www.energy.gov/nepa/articles/cx-028861-scaling-hydrogen-direct-reduced-iron-pathways-decarbonize-iron-and.
37 https://www.midrex.com/tech-article/impact-of-hydrogen-dri-on-eaf-steelmaking/.
38 Nielson, S. et al. (2021). *Processes* 9(8), 1447.
39 Okosun, T. et al. (2022). *JOM* 74(4), 1521–1532.
40 Wei, Y.J. Technology prospect for achieving the long term goal of reducing CO2 emission from iron and steel. World Met. Guide 2020.
41 https://www.businesswire.com/news/home/20240126665523/en/.
42 Gao, Y.M. *Metall. Manag.* 2020, 20, 4–14.
43 Tang, J., et al. *Int. J. Miner. Metall. Mater.* 2020, 27, 713–723.
44 https://fuelcellsworks.com/news/ovako-first-in-the-world-to-heat-steel-using-hydrogen/.
45 https://www.pnw.edu/civs/2024/07/16/project-team-and-doe-representatives-visit-gerdaus-monroe-steel-plant-for-h2-demonstration-in-reheating-furnace/.
46 Kim, Dongsu, et al. *J. Hazard. Mater.* 94.3 (2002), 213–222.
47 Echterhof, Thomas. *Metals* 11.2 (2021): 222.
48 Pohlmann, J. G., et al. *J. Cleaner Prod.* 129 (2016): 1–11.

49 Campos de Assis, C. et al. *Ironmaking & Steelmaking* 47.3 (2020): 284–289.
50 Sahajwalla, V. et al. *AISTech 2005 Proceedings*. Charlotte, NC, USA, 9–12 May 2005.
51 https://www.globalccsinstitute.com/#_ftn5.
52 https://www.iea.org/commentaries/is-carbon-capture-too-expensive.
53 Alexandra Hirsch et al., ULCOS top gas recycling blast furnace, April 2012 Stahl und Eisen 132(4): 31–40.
54 https://www.ispatguru.com/natural-gas-and-its-usage-in-iron-and-steel-industry/.
55 https://climatescience.org/advanced-post-combustion-carbon-capture.
56 https://www.globalccsinstitute.com/news-media/insights/ccs-a-necessary-technology-for-decarbonising-the-steel-sector/.
57 Jason Collis et al., *Front. Energy Res.*, 19 May 2021 doi:10.3389/fenrg.2021.642162.
58 Jebin James et al., *Front. Chem. Eng.*, 2023.
59 Okosun, T. et al., *Steel Res. Int.*, 90 (2019).
60 Zhou, C. et al. *JOM*, 68 (2016).
61 Okosun, T. et al. (2020). *Processes*, 8(7), 771.
62 Nielson, S. et al. (2021). *Processes*, 9(8), 1447.
63 Okosun, T. et al. (2022). *JOM*, 74(4), 1521–1532.
64 Ugarte, O. et al. (2023). *Steel Res. Int.*, 94(12), 2300188.
65 Calix, R. *Dynamics* 3 (4), 636–655.
66 Shang, W. *JOM*, Vol. 75, No. 7, 2023.
67 Raza, M. O., et al. *Materials* 18 (2025), https://doi.org/10.3390/ma18030632.
68 https://www.energy.gov/nepa/articles/cx-028861-scaling-hydrogen-direct-reduced-iron-pathways-decarbonize-iron-and.
69 https://www.pnw.edu/civs/2024/12/02/civs-lands-7-million-doe-iedo-grant-for-electric-arc-furnace/.
70 https://www.pnw.edu/civs/2024/07/16/project-team-and-doe-representatives-visit-gerdaus-monroe-steel-plant-for-h2-demonstration-in-reheating-furnace/.
71 USGS, Mineral Commodity Summaries 2019, 2019.
72 USGS, Mineral Commodity Summaries 2019, 2019.
73 Guo, X. etal. *Steel Res. Int.*, 94(12), p. 2300195.
74 Association of Iron and Steel Technology, "2019 AIST Electric Arc Furnace Roundup," Iron & Steel Technology, pp. 128–151, 2019.
75 Thomas, B.G., *Steel Res. Int.*, 89(1) 2018.
76 Thomas, B.G., et al. *Metall. Mater. Trans. B*, 45 (2014), 22–35.
77 Meng, Y.A., and Thomas, B.G., *Metall. Mater. Trans. B*, 34(5) 2014, 685–705.
78 Thomas, B.G., et al. *ISIJ Int.*, 41(10) 2001, 1262–1271.
79 Meng, D., et al. *Proceedings of the AMSE 2024 International Mechanical Engineering Congress and Exposition*, paper: IMECE2024-145700, 2024.
80 Thapa, S., et al., *Materials*, 16(3) 2023, 1042.
81 Aniusuba, V., et al. *Energies*, 14(21) 2021, 7339.
82 Ma, H., et al. *Front. Mater.*, 7, 2020, Article 577265.
83 Toth, K. et al. Iron & Steel Technology Magazine, December 2022. 1–7.
84 Chris Bataille, Low and Zero Emissions in the Steel and Cement Industries: Barriers, Technologies, and Policies, OECD Green Growth and Sustainable Development Forum, Paris, November 26–27, 2019).
85 Kiran Thirumaran et al., *ACEEE Summer Study on Energy Efficiency in Industry*, Portland, Oregon, 2019.
86 ArcelorMittal, Climate Action Report 1, May 2019.
87 Hasanbeigi, Ali et al. Emerging Energy Efficiency and Carbon Dioxide Emissions-Reduction Technologies for the Iron and Steel Industry, Lawrence Berkeley National Laboratory, BNL-6106E, January 2013.
88 Anne Carpenter, CO2 Abatement in the Iron and Steel Industry, IEA Clean Coal Centre.

Part IV

Strategies for Decarbonizing Construction Materials

9 Heidelberg Materials Carbon Capture and Storage Demonstration Projects

Heidelberg Materials

9.1 INTRODUCTION

Carbon capture is a decades-old concept that has historically been implemented in industrial sectors such as petroleum and coal-fired power generation. The application of carbon capture and storage (CCS) to cement kilns is more recent. Kiln applications entail significant differences and complexities compared to earlier use cases, including differences in flue gas makeup and process characteristics.

Nevertheless, it is critical for the cement industry to decarbonize. Most estimates identify the cement/concrete industry as being responsible for 5–8% of global anthropogenic greenhouse gas emissions, and using net zero carbon concrete has become a goal for many building professionals. Reducing CO_2 emissions from cement production is an important step toward achieving net-zero carbon concrete. Currently, carbon-neutral concrete calculations must rely on the purchase of offset credits. Capturing and storing CO_2, on the other hand, allows for direct greenhouse gas (GHG) reduction benefits to be realized on the product and hence avoid the requirement to use offsets. According to the International Energy Agency (IEA), carbon capture utilization and storage is virtually the only technology solution for deep CO_2 emissions reductions from cement production.[1]

The flue gas from cement production has a higher CO_2 concentration than traditional combustion flue gases due to chemical conversion of the main raw material, limestone ($CaCO_3$). Typical flue gas composition is 60% nitrogen (N), 17% water (H_2O), 13% carbon dioxide (CO_2), and 10% oxygen (O_2). Of the 13% CO_2, approximately 66% results from process emissions (derived from the fundamental chemical process of converting limestone to lime, also referred to as calcination) and approximately 33% results from combustion emissions (related to the fuel required to heat the kiln) (Figure 9.1).

The general manufacturing process for ordinary portland cement requires sources of calcium, silica, iron, and alumina. Calcium, the largest component by mass, is provided primarily by limestone. The process begins by crushing limestone and blending other sources of silica (typically from sand), as well as alumina and iron

DOI: 10.1201/9781003655442-13

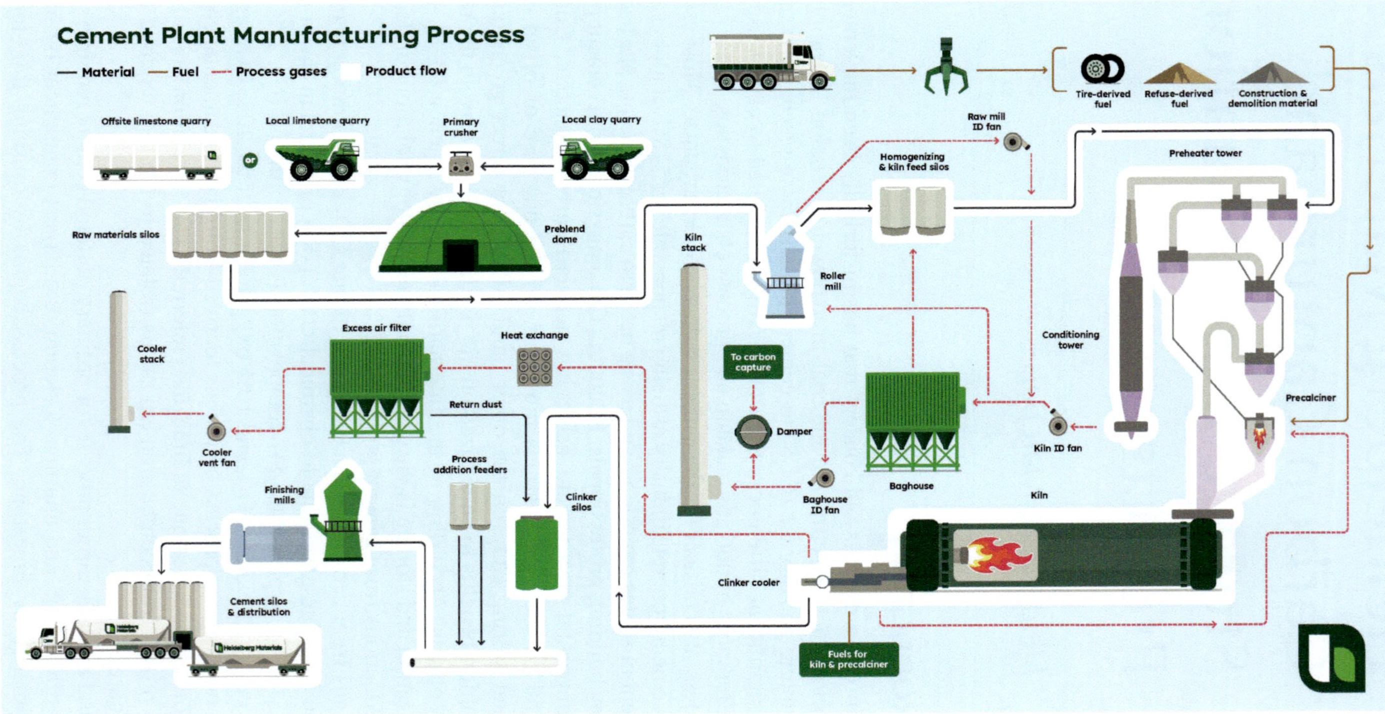

FIGURE 9.1 Typical cement plant manufacturing process and potential location of carbon capture tie-in equipment for post-combustion technologies.

(typically from clay). The crushed material is transported to a raw material storage area and homogenized. The desired raw mix of crushed material and any additional components required for the type of cement is prepared using metering devices. Roller grinding mills or ball mills grind and further dry the raw mixture to a fine powder referred to as raw meal, after which it is conveyed to silos for further homogenization. The pyro-processing of the raw meal at approximately 1,450 °C is carried out in the kiln system. The types of pyro-processing lines, in order of efficiency, are wet process, long-dry, and preheater/precalciner systems. A chemical conversion process known as sintering creates a new intermediate product referred to as clinker. After pyro-processing, the clinker is cooled and stored in silos, domes, or other similar structures. From there the clinker is conveyed to the finish mill system, consisting of ball mills or roller mills, where it is blended and interground with gypsum, limestone, and other additives to very fine powder referred to as cement. The finished cement is stored in separate silos by type.

9.2 TYPES OF CARBON CAPTURE TECHNOLOGY

CCS technology is not a simple "add-on" to existing equipment. Carbon capture facilities include multiple structures and are often of such significant size and complexity that they are located on sites neighboring the existing plant. Additionally, there is no simple or single mechanism for capturing CO_2. Currently, a variety of technologies for carbon capture are commercially available:

- *Chemical absorption* is achieved by a reaction between CO_2 and a chemical solvent. Chemical absorption using amine, a chemical compound derivative of ammonia, is the most commonly used CO_2 separation technique in the cement industry.
- *Physical separation*. Methods can be based on adsorption, absorption, cryogenic separation, dehydration, and compression. These technologies are currently being evaluated for a number of different cement production applications.
- *Oxyfuel separation* is achieved by the combustion of a fuel using nearly pure oxygen. The result is a flue gas containing predominantly CO_2 and water vapor. Separation is then achieved via condensation resulting in a near pure CO_2 stream.
- *Membrane separation* is achieved by filtering using polymeric or inorganic barriers. Membranes for flue gas treatment are currently only in the development stage.
- *Calcium looping*. In a reactor, lime (CaO) is used as a sorbent to capture CO_2 from a gas stream to form calcium carbonate ($CaCO_3$), which is then sent to a second reactor where it is regenerated, resulting in lime and a pure stream of CO_2.
- *Chemical looping*. Another two-reactor technology, this method uses particles of metal in the first reactor to form a metal oxide; in the second reactor the oxide reacts with fuel, producing energy and a concentrated stream of CO_2.

- *Direct separation.* Captures process emissions from cement production by indirectly heating the limestone using a calciner prior to introduction into the kiln. This method allows the CO_2 to be removed directly from limestone, before it would otherwise combine with combustion gases.

9.3 AMINE-BASED CARBON CAPTURE IN THE CEMENT INDUSTRY

Amine-based carbon capture uses a succession of cooling and heating cycles combined with an amine solvent to first remove CO_2 from flue gas and then capture the pure CO_2, making it suitable for permanent storage. Amine-based carbon capture has been used in the petroleum sector since 1996 and in the coal-fired power industry since 2014. Cement plant applications were pioneered at Heidelberg Materials' cement production facility in Brevik, Norway, where a small-scale trial was completed prior to a current project that entails scaling up the trial, with the goal of capturing up to 400,000 metric tons of CO_2 per year.

Amines themselves are created by replacing one or more hydrogen atoms in ammonia with an alkyl or aryl group. Multiple amines are possible, with a small number (<10) being common for CO_2 capture. Challenges in amine-based carbon capture technologies have revolved around the development of solvents and their subsequent commercial availability. Currently available amine solvents are grouped into two generations, the first of which includes monoethanolamine (MEA), which is the most widely used solvent for CO_2 absorption. Second-generation solvents include blends that have been improved to achieve, for example, lower regeneration temperatures.

The first component in an amine-based carbon capture facility is a motorized damper installed at the base of the plant's exhaust stack. This damper routes the kiln's flue gas to the CCS equipment. The diverted gas is cooled to 30–40 °C in a direct contact cooler (cooled gas being required for the proper functioning of the amine). This also cleans out some of the unwanted compositions in the flue gas. Then the gas is sent to the absorber, a vertical tower into which the amines are sprayed, and while falling react with the CO_2 in the flue gas. The amine solvent, hereafter called simply "amine," in solution with water, binds with and absorbs the CO_2 from the gas.

After binding, the remaining flue gas containing less than 1% CO_2 is released into the atmosphere. Liquid amine accumulates in the bottom of the absorber, bound with CO_2, to become a CO_{2}-rich amine solution. This amine CO_2 solution is routed to a desorber, where it is heated to about 120 °C, releasing water and the CO_2 from the amine. "Lean" amine, or amine that does not contain any carbon dioxide, is returned to the absorber for re-use. The remaining CO_{2}-rich solution is cooled upon leaving the desorber, resulting in a solution of water and pure CO_2. The pure CO_2 is dehydrated and compressed in a compression facility. This involves increasing pressure over several stages. Once the CO_2 gas is again cooled, it adopts a liquid state before the pressure is reduced. Liquified CO_2 is then ready for transportation and permanent storage.

9.4 CARBON TRANSPORT AND STORAGE

Most CO_2 is transported via pipeline or ship, although truck and rail transport are viable in some situations. Pipelines offer a well-established transport method, with a substantial amount of existing infrastructure. Transport via ship offers flexibility, compared to pipelines, and in recent years, capture hubs have been developed to offer a way of moving and consolidating CO_2 until a more extensive network of pipelines can be established.

Permanent CO_2 storage occurs underground, at depths and locations that are typically thousands of meters below the surface and far below potable water layers and oil and gas reservoirs. Geologically, storage reservoirs must consist of porous rock (which allows injected CO_2 to flow and fill pore space) overlaid by impermeable cap-rock layers (which prevent upward migration of the CO_2). Deep saline formations, in which the water-permeable rocks are saturated with salt water, represent suitable conditions and are often used for CO_2 storage. These saline formations not only offer large capacity but are geologically common both on- and offshore. Depleted oil and gas reservoirs, where porous rock previously trapped crude oil, are another suitable type of formation. According to the IEA,[2] geological analysis suggests that the world has ample CO_2 storage capacity, with most estimates placing it between 8,000 Gt and 55,000 Gt.

Stringent Measurement, Monitoring, and Verification (MMV) protocols govern CO_2 storage operations. Approved MMV plans are required prior to the deployment of any subsurface CO_2 injection and storage.

9.5 HEIDELBERG MATERIALS CASE STUDIES

The following case studies represent research and pilot/demonstration projects being conducted by Heidelberg Materials. Heidelberg Materials is one of the world's largest integrated manufacturers of building materials and sustainable solutions, with locations in approximately 50 countries across the world. It is a leading supplier of cement, supplementary cementitious materials, aggregates, ready-mixed concrete, and asphalt. The company has committed to cumulatively capturing and storing 10 million metric tons of CO_2 by 2030. Its Brevik CCS plant is under construction, and seven other large CCS projects are in various stages of development around the world.

9.5.1 Case Study: Carbon Capture and Storage (CCS) at the Brevik, Norway Cement Plant

A carbon capture and storage (CCS) facility at Heidelberg Materials' Norcem cement plant in Brevik, Norway, is on track to be the world's first site to capture carbon emissions from clinker production at an industrial scale. This demonstration project will provide learnings and model processes not only for the cement industry but for other sectors as well. It is also part of a larger national effort in Norway to be a leader in CCS.

The annual volume of CO_2 captured is expected to be approximately 400,000 metric tons per year, which represents 50% of the plant's total emissions. The capture plant will predominantly operate on steam generated from waste heat recovered from the cement plant flue gas, and excess low-grade heat from the cement plant will be used for the amine heat exchanger. This will eliminate the need for additional fuel to achieve the carbon capture at the Norcem Brevik plant.

The CCS technology is being built under the project name "Brevik CCS" by Heidelberg Materials Sement Norge, the only cement producer in Norway. Brevik CCS is part of the Norwegian government's Longship program, which aims to demonstrate the capture, transport, and safe storage of CO_2 from industrial sources. The CCS technology is expected to be operational in Q1 2025.

Heidelberg Materials employees conceived of the carbon capture project as early as the mid-2000s. Norcem Brevik represented an excellent location for cost-effective carbon capture, conditioning, and ship transportation because of its adjacency to a deep water port. The project matured into a desktop study, and in 2011 Heidelberg Materials received support from the ECRA (European Cement Research Academy) for further studies. From 2013 to 2017, Heidelberg Materials received funding from the CLIMIT program (Norway's national program for research, development, and demonstration of CCS) to test four different technologies on actual flue gas at the cement plant in Brevik. In 2016, the Norwegian CCS Demonstration Project (NDC) was launched and Heidelberg Materials received support from Gassnova and the Ministry of Petroleum and Energy to carry out concept and front-end engineering design (FEED) studies. (Gassnova, established by Norwegian authorities to promote technology development and competence-building around CCS, is the Norwegian government's closest adviser in this field.) In December 2020, the Norwegian government gave final approval and support for the construction of the world's first full-scale carbon capture plant at a cement plant. Of Heidelberg Materials' eight large CCS projects in development around the world, CCS at the Brevik location is furthest along in its construction.

9.5.1.1 CCS Design at the Heidelberg Materials Norcem Cement Plant in Brevik, Norway

CCS at the Brevik plant is an amine-based, post-combustion CO_2 removal system. While amine-based CO_2 removal has been used in the petroleum sector and the coal-fired power industry for many years, the tests conducted at Brevik were the first use of the system in the cement industry. Providers selected as a result of the FEED study were Aker Solutions for the design and delivery of CO_2 capture, conditioning, intermediate storage, and loading, and Norsk Energi for the design and delivery of waste heat recovery from the existing cement production process (Figure 9.2).

9.5.1.1.1 Carbon Capture, Conditioning, Intermediate Storage, and Loading

Technology for CO_2 capture and conditioning includes Aker Solutions' advanced carbon capture technology (ACC™). ACC uses a proprietary ACC S26 solvent and includes an emission control system and anti-mist design, intended to minimize emissions of amine and amine degradation products. The ACC emission control technology was tested using Aker Solutions' mobile test unit (MTU) on actual flue

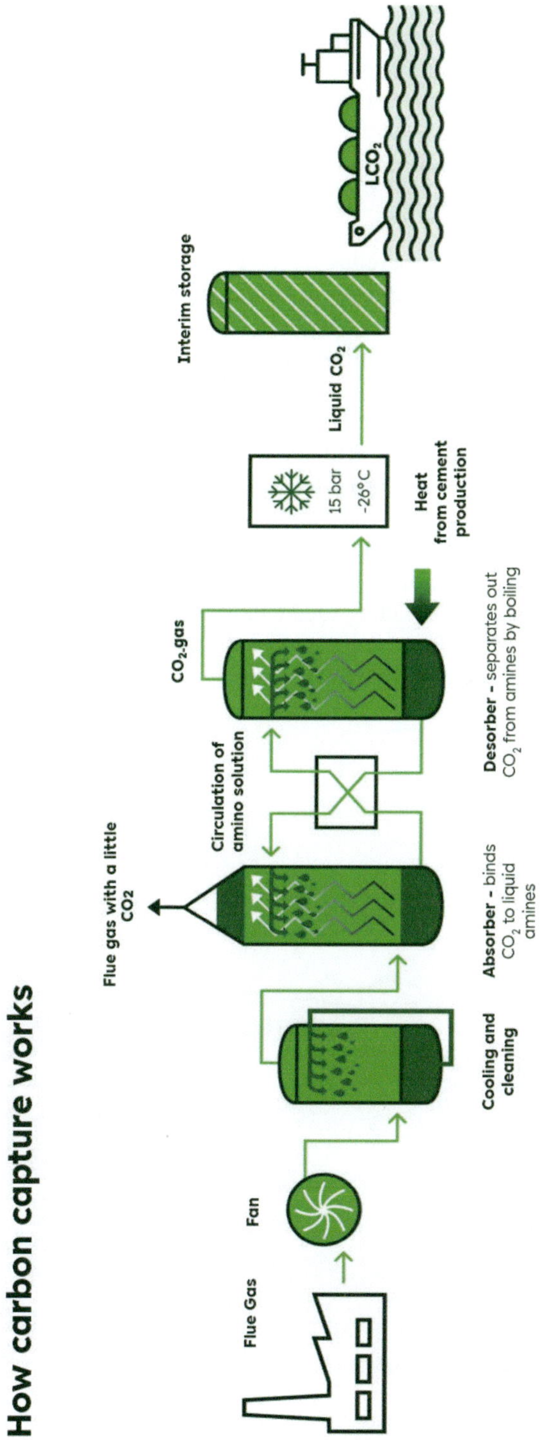

FIGURE 9.2 Brevik, Norway carbon capture process showing amine technology.

gas from the Brevik plant and determined to be viable. The MTU has successfully tested carbon capture from various flue gases for more than 20,000 operating hours at different industrial plants in Norway, UK, and USA. Testing on actual flue gas in Brevik was a strict requirement from Heidelberg Materials to secure the correct amine composition.

Flue gas pretreatment will be accomplished using a direct contact cooler (DCC), which will condense water from the flue gas. This process water will be used to supplement the water supply for gas suspension absorption (which requires spray water and diluting water for lime), reducing the plant's consumption of potable water. There is also potential for the liquid stream of water from the DCC to wash out pollutants from the cement plant flue gas.

Other than the cooling water returns and boiler blowdown (i.e., water removed from the boiler to keep dissolved solids within given parameters), the only continuous liquid stream from the CO_2 capture plant is from the DCC. An effluent water treatment plant (EWTP) will treat effluent water from the DCC. The EWTP will consist of ultrafiltration units for particulate removal and an activated carbon bed for the removal of dissolved mercury and dioxins. Filtered emissions from the DCC to the sea will be mixed with cooling water (seawater) before discharge (Figure 9.3).

A novel technology developed for CO_2 capture at Norcem Brevik is Aker Solutions' CO_2 compressor with integrated heat recovery. The technology will result in a significant reduction in thermal energy required to operate the capture plant and should provide findings that are of interest industry-wide, especially for applications

FIGURE 9.3 Successful installation of the absorber in the summer of 2023. The absorber is a key part of the Brevik CCS facility.

where no heat (steam) source is available for the capture plant and new heating plant/boiler installation is required. Currently, all full-scale amine-based carbon capture units used in the industry require a greater amount of energy to operate than is available using waste heat, unless a combined heat and power plant is incorporated to optimize the system and meet the dual need for steam and electricity.

The CCS system's net production rate of liquid CO_2 (i.e., liquid CO_2 routed to the storage tanks minus any displaced and evaporated CO_2 gas) is equal to the CO_2 capture rate of 55 metric tons per hour during normal operation. Normal operation is defined as 7,400 hours of cement plant operation in an average year, which is adequate to maintain a flue gas flow rate and CO_2 content sufficient for capturing 55 ton/h of CO_2. With 7,400 hours of operation, there will be 1,400 hours of outage in an average year, with 500 planned (3 weeks of winter repair) and 900 unplanned. Outage times for the cement plant will coincide with outages of the carbon capture and conditioning equipment, providing ample opportunities to perform maintenance on that equipment. Therefore, it is expected the carbon capture equipment will have high availability for maximum CO_2 capture (Figure 9.4).

The Norcem Brevik CCS project will include Aker Solutions' proprietary technology for heat integration in the CO_2 liquefaction and inert stripping processes.

Because the main connection points between the existing manufacturing plant and the new CCS plant are within the flue gas system (which contains both waste heat and CO_2), the new plant is being installed parallel to the existing flue gas system (Figure 9.5).

FIGURE 9.4 CO_2 tanks at the Heidelberg Materials cement plant in Brevik, Norway.

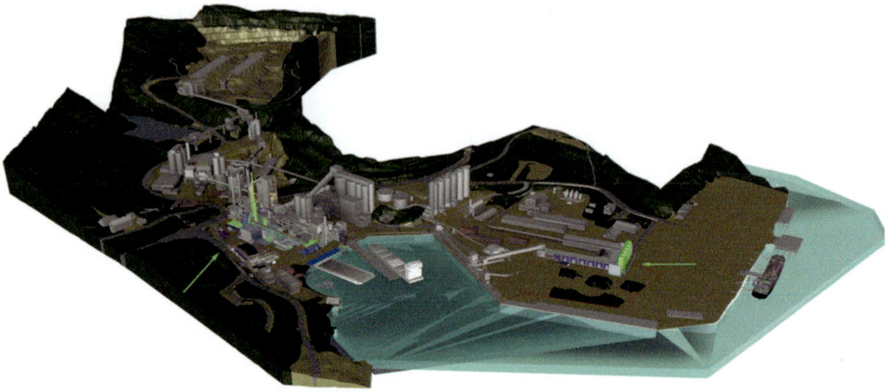

FIGURE 9.5 Brevik cement plant rendering. Arrows and shading indicate CCS equipment.

9.5.1.1.2 Waste Heat Recovery

Waste heat from cement production will be recovered by converting it into steam at the quality and pressure required for utilization in the CO_2 capture process. The waste heat recovery system will include an auxiliary electrode steam boiler for reclaimer service and to provide partial backup for fluctuations in the waste heat recovery. Waste heat recovery boilers were designed based on testing at Norcem Brevik on actual flue gas, with the main design data being flue gas velocity for achieving self-cleaning, and the heat transfer coefficient confirming the total heat recovery potential.

The technology designed for waste heat recovery (shell boilers) has not previously been operated with the same type of cement off-gas that is found at Norcem Brevik. Norsk Energi brought deep experience in engineering and commissioning shell boilers and water-tube boilers for waste heat recovery from off-gases with high dust content, whose typical applications have been at brown-field smelting plants. Boilers at smelting plants must handle process fluctuations, rapid energy peaks, and high dust content; the shell boilers used at these plants have a unique design that ensures a self-cleaning effect for heat transfer surfaces and results in low fouling and high boiler efficiency.

At Norcem Brevik, approximately 60% of the heat demand in the CO_2 capture process is covered by the waste heat recovery boilers. Steam may be produced at a very low cost when applying shell boilers. Furthermore, the steam-from-power ratio is high (COP = 6–7) and significantly higher than the COP of an industrial heat pump (COP = 3–4). Proof-of-concept from Norcem Brevik waste heat recovery and steam production using shell boilers should pave the way for cost-effective CO_2 capture at other industrial plants with waste heat sources, especially cement plants.

9.5.1.1.3 Operational Changes and Equipment Modifications at the Cement Plant

Various equipment and process modifications were required at the existing plant to accommodate the incorporation of the CCS and waste heat recovery systems. One of the main operational changes was a reduction in cooling temperature for the flue gas in the gas conditioning towers, which in turn necessitated the addition of a variable cooling service (water spray). Upgrades to dampers, dust-handling equipment, and fans were among the other changes required.

9.5.1.2 Carbon Transport and Storage

Transport and storage will be provided by Northern Lights, a partnership owned in equal shares by TotalEnergies, Equinor, and Shell. Northern Lights was established as a demonstration project in 2024 to serve industrial companies with the transport and sequestration stages of CCS and is part of the Norwegian government's initiative, "Longship."

CO_2 from the Norcem Brevik plant and Hafslund Oslo Celsio (formerly Fortum Oslo Varme) will be transported in a liquefied state at –26 °C by one or two ships over a distance of approximately 700 km. (Hafslund Oslo Celsio, part of the Longship project, produces heat from local sources of surplus energy and distributes it to commercial buildings and homes in the Oslo Metropolitan Area.) Shipped CO_2 will arrive at an onshore terminal in Øygarden, Hordaland, which is equipped with buffer storage, a pump, and a heater. A 110 km pipeline, 12 inches in diameter, will route the CO_2 to a single injection well, with permanent storage taking place at a depth of approximately 2,600 meters below the seabed on the Norwegian continental shelf (Figure 9.6).

The total cost for the Norcem Brevik CCS plant is estimated to be €400 million, with 85% subsidized by the government of Norway.

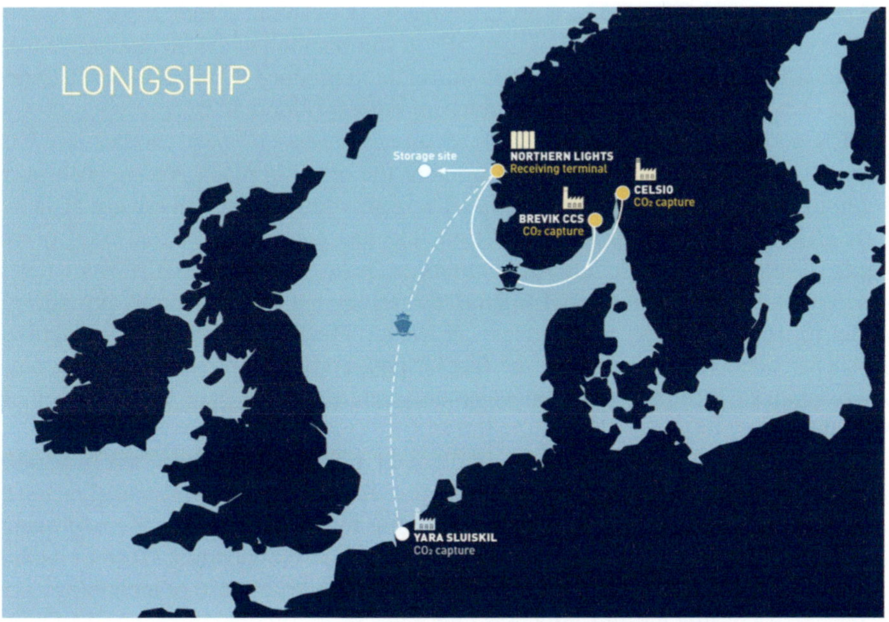

FIGURE 9.6 Map of transport route for Brevik, Norway captured carbon.

9.5.2 Case Study: Carbon Capture and Storage at Heidelberg Materials Edmonton, Canada

Development of North America's first industrial full-scale CCS solution in the cement industry is taking place at Heidelberg Materials' cement plant in Edmonton, Alberta, Canada. The project goal is to capture approximately 1,000,000 metric tons of CO_2 each year. The design involves transporting captured emissions via pipeline to be permanently and safely stored, and addresses Scope 1, 2, and 3 emission targets. Scope 1 emissions will be reduced at the production level, Scope 2 emissions will be reduced as a result of the plant creating its own energy, and Scope 3 emissions will be offset by the capture of biogenic carbon from the recently installed alternative fuel system.

A feasibility study supported by the organization Emissions Reduction Alberta was successfully completed and Heidelberg Materials launched a two-stage competitive Front End Engineering Design (FEED) process in the summer of 2022. The plan is on target for an investment decision in late 2024, which will support the facility going into service projected between the summers of 2027 and 2028. The Edmonton plant's CCS process is expected to capture 95% of the CO_2 generated from the cement plant along with the combined heat and power facility, which provides the energy for the new CCS facility and the existing cement plant.

The existing Heidelberg Materials Edmonton cement plant operates a four-stage preheater/precalciner kiln and produces 1 million metric tons of cement per year. A mile-long, 12-foot-diameter flue gas pipe will connect the main plant to the CCS operation. The new CCS design consists of four major systems that are integrated to capture the CO_2. These systems include the flue gas duct, carbon capture plant, combined heat and power plant (CHP), and heat rejection plant. The capture plant's capacity to treat the flue gas from the kiln as well as the CHP, in combination with kiln alternative fuels that contain biogenic carbon, will allow this plant to produce the world's first net-zero carbon cement at full scale without purchasing offsets (Figure 9.7).

CCS at the Edmonton plant will be an amine-based, post-combustion CO_2 removal system that is located on a site near the existing plant. Amine absorption technology was selected because it has been proven to work effectively at large-scale CO_2 capture from post-combustion flue gas generated by coal-fired power plants. Canadian power plant applications of post-combustion CCS were determined to have similar flue gas characteristics to cement plants, thereby providing a model for the Edmonton project team. However, Heidelberg Materials recognized risks associated with developing the largest post-combustion carbon capture plant downstream of a cement plant, most notably related to the particulates in cement plant flue gas.

The company chose, therefore, to implement a detailed pilot program in parallel with the FEED process. During the pilot period, Heidelberg Materials and the chosen technology providers will gather data pertaining to their technology's performance against the Edmonton cement plant's specific flue gas composition. Testing against "real world" cement flue gas conditions will provide a true sense of technology performance. As of this writing, FEED is ongoing. After completion of the FEED, one proponent will be selected for the scope of work.

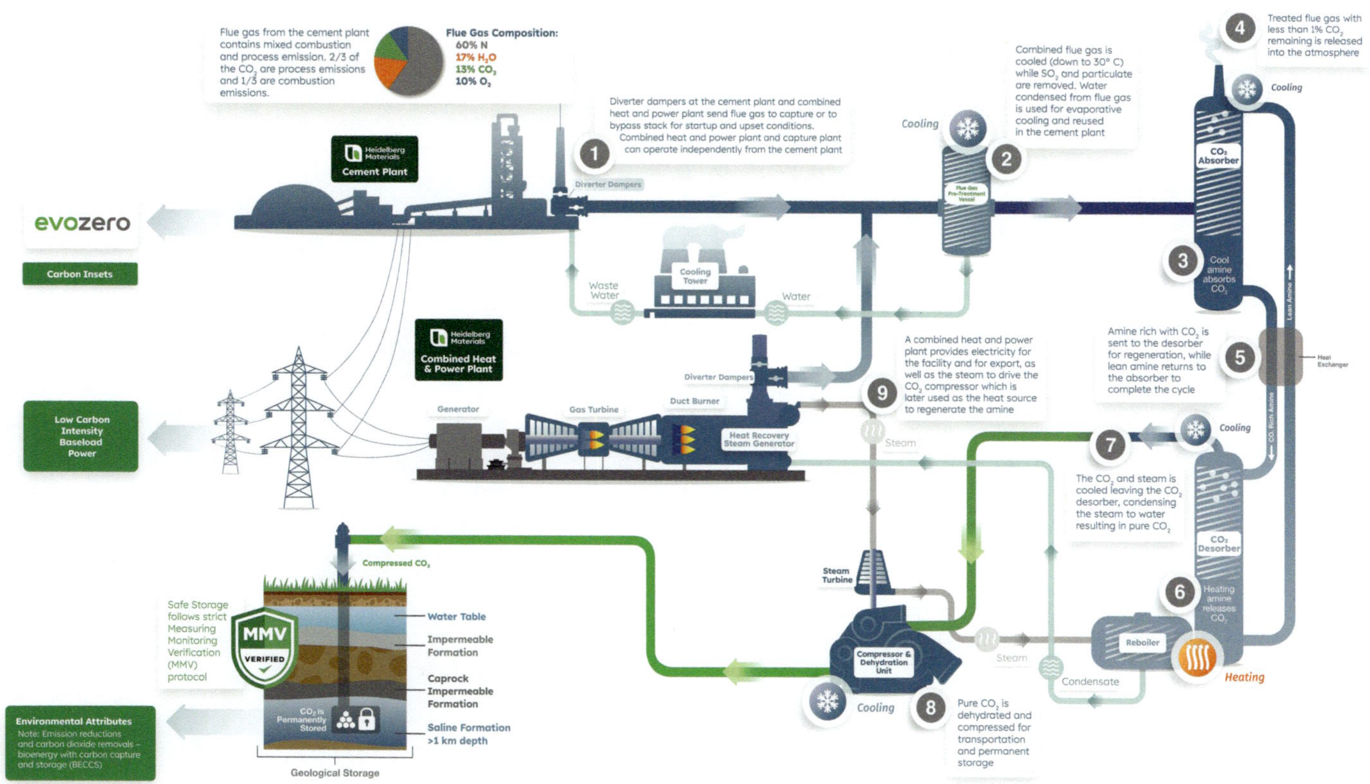

© 2023, Heidelberg Materials

FIGURE 9.7 Process overview specific to Heidelberg Materials Edmonton, Alberta, Canada CCS.

9.5.2.1 Heidelberg Materials Edmonton CCS Technologies

A Mitsubishi Heavy Industries (MHI) pilot plant was leased by Heidelberg Materials and installed between the fabric filter baghouse and exhaust stack of the existing Edmonton plant (see Figure 9.8). (Baghouses are so-called because of the dust-capturing bags that reside within the structure, or "house." In accordance with environmental permit requirements, flue gases are routed via ductwork into the baghouse, where the bags serve as high-efficiency filters for flue gas particulate matter.)

The MHI pilot plant has been operational since July 2023 with a capture rate of 300 kg/day. It was the second of its kind to be installed at a cement plant by MHI (see Figure 9.9).

A second pilot plant, owned by Heidelberg Materials, began installation at the Edmonton cement plant in January 2024. The Technip/Kanfa-designed modular pilot plant with a capacity of 3,000 kg/day will be operating with a Shell/Cansolv amine system until full-scale plant start-up. This pilot plant includes pretreatment plants to test the effects of pretreatment on the flue gas composition. The first pretreatment is a DeNOx unit to remove nitrogen oxides and the second pretreatment plant is Wet Electrostatic Precipitator (WESP) to remove particulate from the flue gas. The pilot plant and pretreatment plants were designed to run in various operating modes during the pilot period, testing the need for pretreatment of flue gas and helping inform larger project decisions for the full-scale CCS plant contracts (Figure 9.10).

The pilot program results support selection of the winning proponent, with that proponent continuing to operate their amine within the Heidelberg Materials-owned pilot plant in order to continue data acquisition and serve as a testing platform until the full-scale facility is complete and commissioned. The pilot plant results will also help inform other Heidelberg Materials cement plants for use in future projects.

FIGURE 9.8 Conceptual rendering of Heidelberg Materials Edmonton, Alberta, Canada CCS facility.

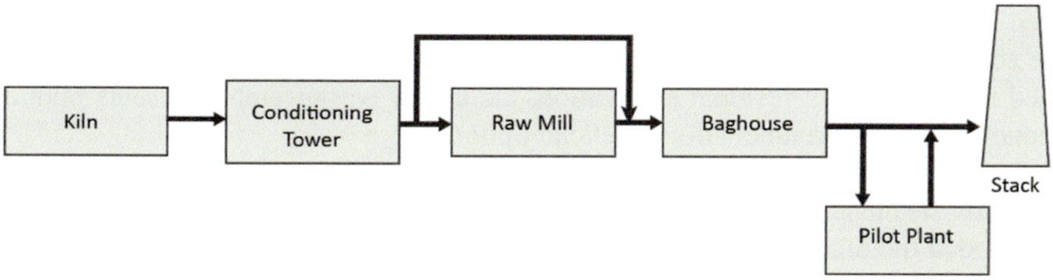

FIGURE 9.9 Simplified process flow diagram for the pilot plants at the Edmonton, Alberta, Canada CCS facility.

FIGURE 9.10 Mitsubishi Heavy Industries (MHI) pilot plant installed at Heidelberg Edmonton, Alberta, Canada cement plant.

9.5.2.2 Complementary Technologies

Seventy-five thousand liters of water per hour will be extracted as part of the CCS process. Whereas in traditional cement production, such water would have been exhausted through the stack as vapor, the Edmonton facility will use 90% of the wastewater to provide cooling for the capture process. The remaining 10% will be used in residual applications such as in cement product finish grinding mills and in the cement process gas conditioning tower. Since fresh municipal water is currently used for this type of residual application, the use of wastewater represents both an economic and environmentally beneficial option.

To optimize the design of the Edmonton plant's CCS system and meet the dual need for steam and electricity, a combined heat and power (CHP) plant will be installed at the plant. The United States Environmental Protection Agency (USEPA) has stated that conventional, separate facilities for power and heat generation waste nearly two-thirds of the energy generated.[3] CHP technology eliminates that inefficiency, taking advantage of on-site power production to route excess heat to applications within the facility such as producing steam or heating water, reducing facility demand on the grid. CHP is not yet common in the cement industry, despite the fact that recovering carbon dioxide from a flue gas stream is energy-intensive. The Edmonton CHP plant is therefore expected to yield significant advances in implementing the technology more broadly for the cement sector.

At Edmonton, in addition to generating the electricity needed for the carbon capture process and for other power requirements across the facility, the CHP plant will generate the steam needed to drive the facility's CO_2 compressor and regenerate the amine in the absorber. The CHP plant will produce a total of 124 MW from two combustion turbines to meet the capture and cement plant's requirements, with excess power generated being sold into the Alberta electricity grid. It is estimated that approximately 70MW of low-emission electricity will be exported to this local grid. Inclusion of a CHP plant in the Edmonton project is also what will allow Heidelberg Materials to mitigate Scope 2 electricity emissions, which is an integral part of producing carbon-free products.

While Edmonton's CCS process will capture 95% of the CO_2 present in kiln exhaust, energy consumption to power the CCS process represents another emissions source. Therefore, to achieve maximum emission reductions, the Edmonton plant's CCS equipment will also remove CO_2 from the CHP exhaust, achieving the same 95% removal rate.

The CCS and CHP plants are designed to operate at various modes, with two main modes, addressing the fact that there is a higher level of intermittency of operation for a cement plant compared to thermal power plants and energy infrastructure (which have been the target for previous large-scale post-combustion carbon capture plants). The two main modes are:

1. Design mode: The CCS plant will operate at design capacity with the cement kiln on and one or both CHP turbine units (CHP A and CHP B) in operation.
2. Electrical or power production mode: The CCS plant will operate with the cement kiln offline and one or both combustion turbine units in operation. The CHP plant must provide sufficient steam to the capture plant to capture and compress CO_2 generated from the CHP plant only.

The power production mode will allow the start-up and operation of the capture plant to be independent of the cement plant. It will also permit the power generation and

CO_2 capture systems to operate during both planned and unplanned outages of the cement plant. The power production mode will increase the utilization of the CCS plant and lead to fewer process interruptions. It is expected that this concept can be applied to various industrial sectors by combining several flue gas sources to obtain the greater benefits of scale and maximum utilization of the CCS plant, making CCS economically viable.

When the kiln is in operation, up to 4,085 metric tons of CO_2 per day will be captured; when only the CHP is operating, that number is 1,600 metric tons per day, representing an increased capture percentage of 98%.

The detailed matrix of different operating scenarios is shown in Figures 9.11–9.13. Figures 9.12 and 9.13 show the variation of the flue gas and CO_2 mass flow entering the CCS plant. To operate in all listed scenarios, the CCS facility must be able to function at a significant turndown (i.e., with a significant ratio of maximum capacity to minimum capacity). During the design process, close attention was paid to the design criteria of each major piece of equipment to ensure this turndown was feasible.

9.5.2.3 CCS Design at the Heidelberg Materials Edmonton Cement Plant

Each heat recovery steam generator (HRSG) is designed to provide sufficient steam to capture 95% of the CO_2 produced from the cement and CHP plants if one of the

FIGURE 9.11 3D rendering and construction of the second Edmonton pilot plant owned by Heidelberg Materials and designed by Technip/Kanfa.

Plant Operation Mode	CASE 1	CASE 2	CASE 3	CASE 4	CASE 5	CASE 6	CASE 7	CASE 8
Cement Plant	ON	ON	ON	OFF	OFF	OFF	ON	OFF
CHP A	ON	ON	OFF	ON	ON	OFF	ON	ON
CHP B	ON	OFF	ON	ON	OFF	ON	ON	ON
CO2 Capture Plant	ON	ON	ON	ON	ON	ON	OFF	OFF

FIGURE 9.12 Matrix of different operating scenarios for the Edmonton, Alberta, Canada CCS facility.

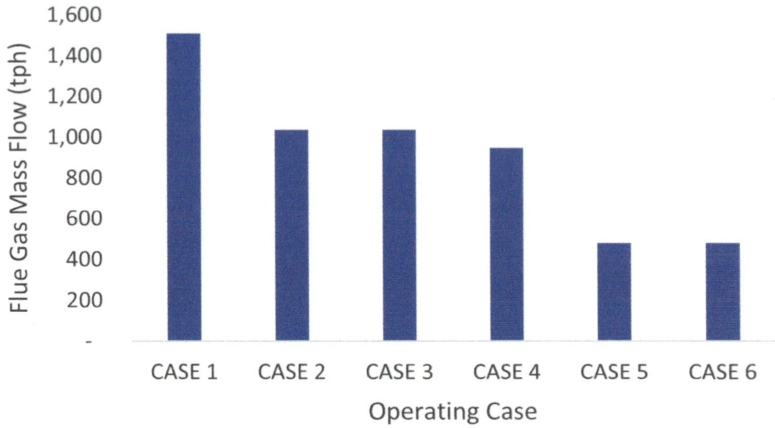

FIGURE 9.13 Flue gas mass flow inlet to the Edmonton, Alberta, Canada CCS plant at different operating modes.

combustion turbines and HRSG are not in service. Each HRSG will have two bypass stacks (one hot and one cold). The hot stack will be required during a simple cycle mode, providing emergency power, and the cold stack will be used to facilitate the start-up of the CCS plant.

Flue gas pretreatment will be required upstream of the CO_2 absorption process to condition flue gas to a suitable condition for CO_2 separation. An evaluation was performed on whether the two-flue gas sources should be pretreated together or separately. The justification was based on the contaminants in the two-flue gas streams, the level of treatment required, and the water discharge management. In the case of Heidelberg Materials Edmonton, the flue gas from the kiln has a much higher contaminant level than that of the CHP flue gas, and this leads to each flue gas stream being pretreated in a separate tower before combined and entering the CO_2 separation process. The CO_2 absorption process will be a single train consisting of an absorber and a regenerator. This is more cost-effective than two separate trains, and it was ensured that the turndown operation could be met, and that the absorber size was within the modulization capability.

Heidelberg Materials Edmonton selected a steam-driven CO_2 compressor, and the exhaust steam from the turbine will be fed to the amine reboiler for amine regeneration. This design requires intensive heat interaction, and a dump condenser will be added to manage the exceeded steam during the turndown operation.

Integrating a CCS system into a plant of this scale necessitates significant cooling capacity. The estimated heat rejection duty of the Heidelberg Materials Edmonton project is as high as 250 MW. Figure 9.14 illustrates the distribution of this heat rejection load across the different stages of the CCS process. Approximately 40% of the cooling requirement will be to cool the flue gas from the cement plant (kiln quench water cooler) and the CHP plant (CHP quench water cooler) down to the desired temperature for the amine absorption process. (The amine absorption process is exothermic, thus favoring lower temperatures.) During flue gas cooling,

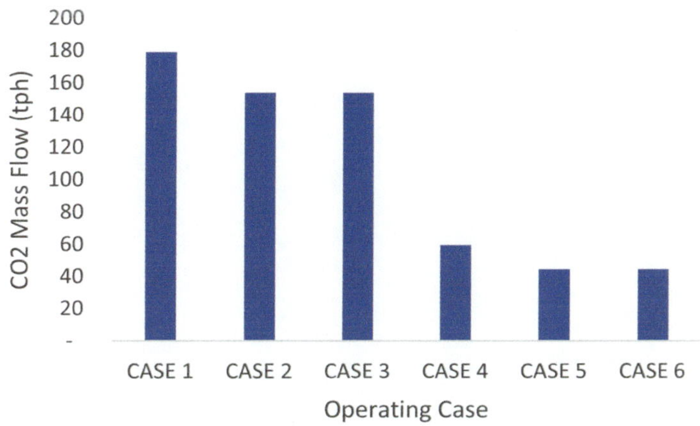

FIGURE 9.14 CO_2 mass flow inlet to the Edmonton, Alberta, Canada CCS plant at different operating modes.

75 m^3/hr of flue gas wastewater will be condensed, requiring careful management. Approximately half of the total cooling load is assigned to the amine process, which includes absorber wash, amine, and overhead condenser cooling. The remaining cooling duty is required to dissipate the heat generated during the CO_2 compression.

With its location in the city of Edmonton, the Heidelberg Materials plant faces the challenge of limited water resources for evaporative cooling. Efficient management of the flue gas condensate will be imperative, and its utilization for evaporative cooling will be vital to ensure zero liquid discharge within the plant. Therefore, a hybrid heat rejection system was chosen, strategically balancing cooling water loads, cooling temperatures, and water availability. The hybrid cooling system is an integration of the dry cooling system (the air-cooled heat exchanger (ACHE)), and the wet cooling system (the wet surface air cooler (WSAC)). WSAC will be installed downstream of ACHE to take advantage of the lower temperature that can be achieved. Flue gas water condensate will be treated to maintain the total suspended solids (TSS), total dissolved solids (TDS), chlorides, sulfates, etc., to the acceptable level by the WSAC and used as WSAC makeup water.

Each operating mode of the CHP and CCS has distinct cooling requirements and water availability as illustrated in Figure 9.15. When the cement plant is shut down, the kiln flue gas quencher will go offline, as well, leaving only the CHP flue gas quencher to produce a limited amount of CO_2 and effluent water. Consequently, the heat rejection and water treatment system must be capable of operating in a turndown state to ensure the continuous functionality of the CO_2 capture system (Figure 9.16).

Besides the higher heat rejection load and the water management for variable operating modes, Edmonton's range of climatic conditions was another challenge in the heat rejection system design. The ambient temperature can be as high as 40 °C in summer and as low as –45 °C in winter. A 40% glycol water solution was selected as a cooling medium to avoid winterization requirements for the hybrid cooling system.

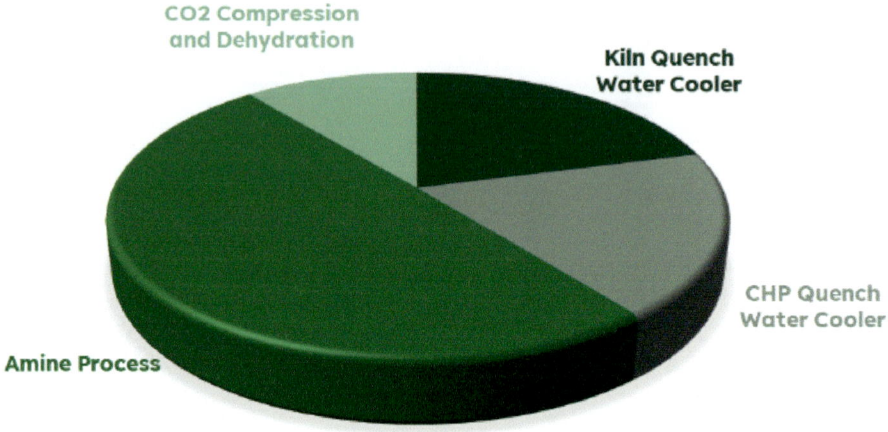

FIGURE 9.15 Distribution of heat rejection load across the different stages of the CCS process.

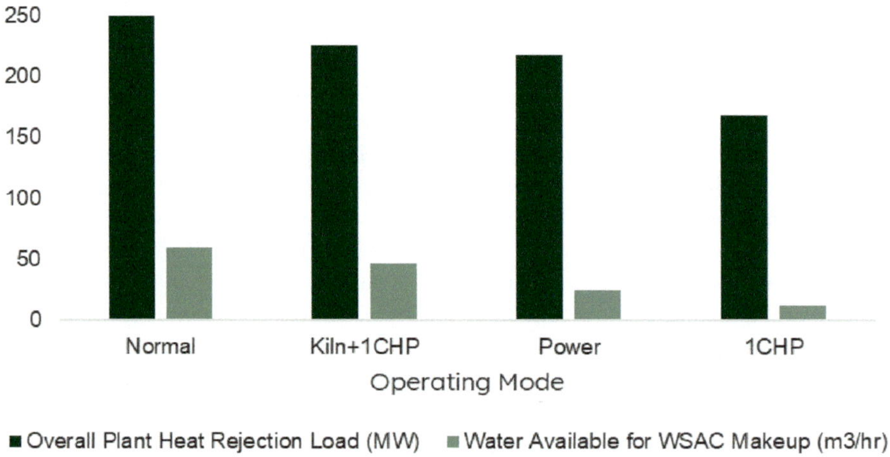

FIGURE 9.16 Heat rejection load and water availability at different operating modes.

Heidelberg Materials Edmonton selected the 85th percentile of the annual ambient temperature to design the heat rejection system. In contrast, the conventional approach would be to design the heat rejection system based on 0.4% of the highest ambient temperature month. Designing the heat rejection system based on 85th percentile ambient temperature reduces the heat rejection plant sizing. However, during summer (15% of the year) the heat rejection will not be able to supply the cooling medium at the desired temperature. This can be resolved by adjusting the CO_2 capture and compression process. With increased energy consumption by 1%, the CCS plant can maintain the CO_2 capture rate with the higher cooling medium temperature. Furthermore, to take advantage of the cold season in Edmonton, the ACHE and

WSAC fans and cooling medium feed pumps will be equipped with variable speed controllers. This will also benefit when the plant operates in turn-down modes.

9.5.2.4 Carbon Transport and Storage

An underground pipeline will transport compressed CO_2 25–35km from the CCS site to the Wabamun Carbon Hub. Proximity to this Hub, which is a geological fit for carbon storage, was one critical factor in selecting the Edmonton location for CCS implementation, since having access to carbon storage is a determinant of whether such a project is feasible for a given location. Storage will follow strict MMV protocols.

The large CO_2 pipeline is designed with branches at its end in order to utilize a number of injection wells. The CO_2 will be injected to a depth of more than 2 km into the Hub's saline aquifer. There are four key contributors to the successful storage of CO_2 at the Wabamun Hub. A layer of impermeable bedrock caps the hub, creating a barrier that prevents leakage. The pores of a saline aquifer trap CO_2 via capillary action. Over time, the CO_2 will dissolve in the brine solution. Finally, CO_2 will chemically interact with the brine solution and minerals in the rock formation to mineralize over time.

The cost of the Heidelberg Materials Edmonton project is estimated at nearly $1.65 billion Canadian dollars.

In addition to capturing 95% of the CO_2 from the cement plant and CHP, environmental benefits associated with the project will include improved air emissions, reduced freshwater consumption, selling excess clean-electricity to the grid, and selling net-zero carbon cement for carbon-neutral concrete for regional markets and beyond (Figure 9.17).

9.5.3 Case Study: Carbon Capture and Storage at Heidelberg Materials Geseke, Germany

With European CCS projects relying on overwater transport and undersea storage of CO_2, the aim of Heidelberg Materials' GeZero (Geseke Zero Emissions) project in Geseke, North-Rhine Westphalia, Germany, is to demonstrate the first full-chain, full-scale CCS project (from source to sink) in a remote inland location. The site is approximately 280 km from the sea, where key offshore CO_2 storage sites are under development.

Because cement plants are located in proximity to their primary raw material (i.e., limestone deposits), the inland location of the Geseke cement plant is representative of the majority of cement plants in the EU. Full-scale CCS developments at such locations have so far been hindered by the distance (and therefore transportation costs) to major end-storage developments, which today are occurring in the North Sea. GeZero's concept of transporting CO_2 via railway is Germany's first large-scale CO_2 end-to-end logistic value chain based on railway transportation, and it is being independently developed with resources of Heidelberg Materials. GeZero is also the first CCS project in Germany to have this logistical setup and scale. Given these factors, the project is expected to be a model for other inland industrial sites. It is

FIGURE 9.17 Aerial view of Heidelberg Materials cement plant in Geseke, Germany.

designed to achieve CO_2 capture of 700,000 metric tons per year – equivalent to about 3.5 per cent of the annual emissions of the German cement industry.

While amine-based chemical absorption technology currently sees the widest use in industrial applications, the Geseke plant will use oxyfuel separation for carbon capture. It will be the first plant to have an oxyfuel second-generation kiln (OxyFuel II). This beyond-state-of-the-art kiln will require installation of an air separation unit (ASU) to produce the oxygen that is required for combustion. As the facility also captures emissions from biomass that substitutes for fossil fuels, Heidelberg Materials Geseke will be one of the first European cement plants able to produce fully decarbonized cement and clinker.

GeZero is in its early stages. A FEED study began in 2024 and construction is expected to begin in 2026. €191m, granted by the EU Innovation Fund, supports the project, with Heidelberg Materials covering the remaining amount. Commissioning is projected to occur in 2029.

9.5.3.1 Development of Oxyfuel Technology

Not as long-established as amine technology, oxyfuel separation has nevertheless proven successful in several large prototype and pre-demonstration projects in coal-based power generation plants and cement plants. Oxyfuel technology uses pure oxygen instead of ambient air for combustion in the kiln. This enables a higher concentration of CO_2 in the exhaust gas as it is not being diluted with large amounts of nitrogen from the air, leading to more efficient carbon capture.

Collaborative efforts of the ECRA and ThyssenKrupp Industrial Solutions underpin the Heidelberg Materials Geseke plant's choice to use oxyfuel technology with cement kiln processes. From 2019 to 2023, Accelerating Carbon Capture using Oxyfuel Technology in Cement Production (AC²OCem) invested in pilot-scale experiments and analytical studies to further develop oxyfuel technology for cement plants and reduce the time to market of the oxyfuel technology in the cement sector. AC²OCem was co-financed by the ACT (Accelerating CCS Technologies) European Development Program. Heidelberg Materials was a participant in the AC²OCem program.

One objective of AC²OCem was to investigate both first-generation oxyfuel technology, used to retrofit existing kiln lines, and second-generation oxyfuel technology, which can be implemented in new-build cement plants. Research into first-generation oxyfuel technology focused on optimizing the oxyfuel calciner operation and advancing the kiln burner technology for combusting 100% alternative fuels. The program advanced innovations in second-generation oxyfuel technology, which avoids recirculating flue gas. AC²OCem's pilot-scale facility for replicating cement kiln conditions featured an oxyfuel kiln burner that allowed up to 100% oxygen combustion and over-stoichiometric combustion (i.e., a situation in which some of the oxygen used for the combustion reaction remains unreacted). Second-generation oxyfuel technology is more efficient, producing smaller gas volumes that in turn require less equipment and processing. Excess oxygen from the kiln is used in downstream, calciner processes. Second-generation oxyfuel technology is expected to reduce the cost of CCS for cement plants.

More recently, Heidelberg Materials and other industrial partners launched a €120m joint venture pilot oxyfuel project. Titled Cement Innovation for Climate (C14C), this pilot will test a rotary kiln with a 450 metric ton per day clinker capacity using a proprietary ThyssenKrupp oxyfuel calciner. The C14C pilot system is expected to be commissioned in early 2025. Findings from the pilot will be used to reduce operational risks at the Heidelberg Materials Geseke plant, which are associated with employing highly innovative methods that have low technology readiness levels (TRLs). For example, findings may provide data for material and mix selection.

For the Geseke site, it was determined based on existing plant equipment that the pure oxyfuel technology would achieve the highest CO_2 capture rate economically, ecologically, and technically.

9.5.3.2 CCS Design at the Heidelberg Materials Geseke Cement Plant

Major components of the Heidelberg Materials Geseke CCS system include an ASU for oxygen generation, an oxyfuel kiln, comprehensive gas cleaning installations, a carbon processing unit (CPU) for purification, and a liquefaction unit. All carbon capture components will be accommodated on the existing plant site. Electrical energy will be provided exclusively by renewable energies, with some power generation on the premises. A rail station with storage and loading equipment for liquid CO_2 will also be on-site, since rail transport will be required until pipeline infrastructure is available (Figure 9.18).

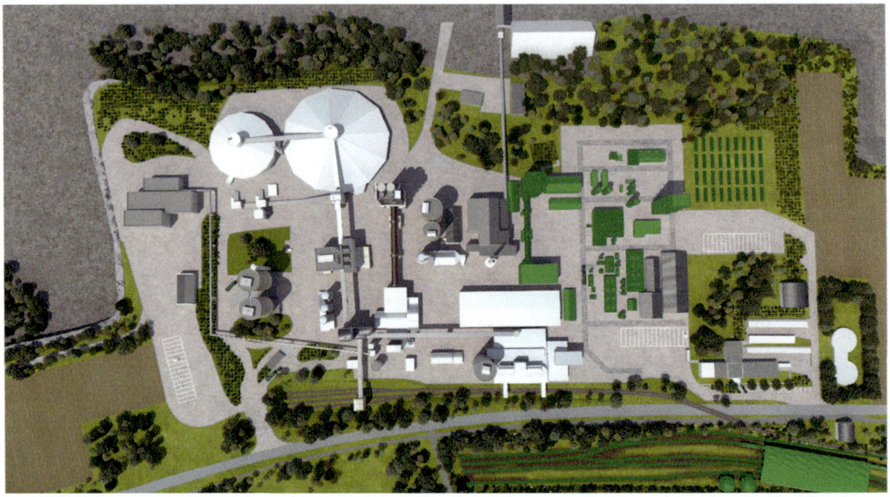

FIGURE 9.18 Plan rendering of Heidelberg Materials Geseke, Germany CCS components and rail station.

Heidelberg Materials Geseke's ASU will deliver the pure oxygen to the kiln that is required for the oxyfuel separation method. The design of the Heidelberg Materials Geseke ASU was dependent on the clinker capacity of the plant's new kiln, as well as on the planned fuel mix.

The plant's second-generation oxyfuel kiln is expected to be installed in 2025. In a process adapted for the Heidelberg Materials Geseke project, the oxyfuel kiln will create a highly concentrated CO_2 stream, expected at 90% (dry), which will then be fed to the CPU. The adapted process largely eliminates the need for the recirculation of process gas to maintain the hydraulic properties of a standard preheater tower.

Cryogenic purification and liquefaction systems were chosen because they achieve a CO_2 compaction level that meets transport and storage specifications. The size and design of the CPU were influenced by the off-gas volume requiring treatment, as well as by the required gas purity, which is 99% (a purity that ensures high-quality CO_2 for storage). The liquefaction unit will be configured for reliable storage and railcar loading.

To optimize the heat and energy balance of the new plant as much as possible, an Organic Rankine Cycle (ORC) power plant will be installed. Power will be generated using waste heat from various sources and by post-combustion of the waste gas stream from the CPU. A new photovoltaic plant on the premises will cover a small portion of the electricity demand (Figure 9.19).

9.5.3.3 Carbon Transport and Storage

Captured CO_2 in its liquid form (22 bar, −35.5 °C) will be transported using pressurized railcars, currently planned to consist of a maximum train size of 36 cars organized into two strands of 18 each. The train transport solution is expected to bridge the gap until a pipeline infrastructure is developed. Transport infrastructure

FIGURE 9.19 Orthographic rendering of Heidelberg Materials Geseke, Germany CCS components.

developed for the Geseke project will also accommodate emissions from surrounding small- and medium-sized industries, thus extending GeZero's carbon-sequestration benefit beyond the scope of a single manufacturer.

Trains departing Heidelberg Materials Geseke will be destined for the CO_2 hub "CO_2nnectNow," located at the HES Wilhelmshaven Tank Terminal. The Wilhelmshaven terminal is situated on a deep draft quay and allows access to industry and rail infrastructure. CO_2nnectNow is under application for Project of Common Interest (PCI) status. (PCI status improves access to financing for projects that promote shared infrastructure interests in the European region.) From the hub, transportation via pipeline will carry CO_2 to its storage site in the North Sea (Figure 9.20).

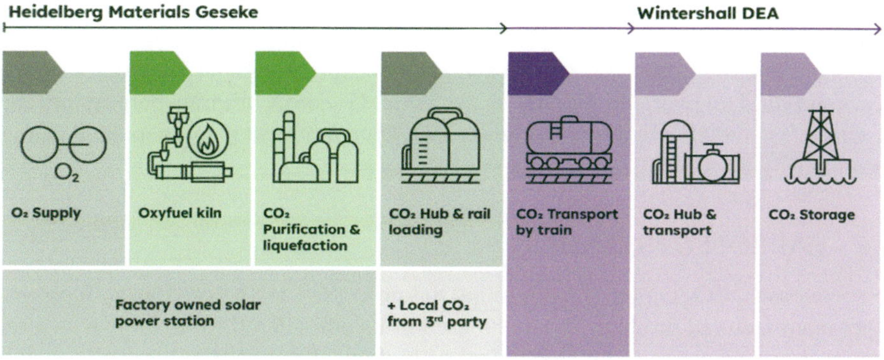

FIGURE 9.20 The GeZero process covers the entire carbon capture, storage, and utilization value chain.

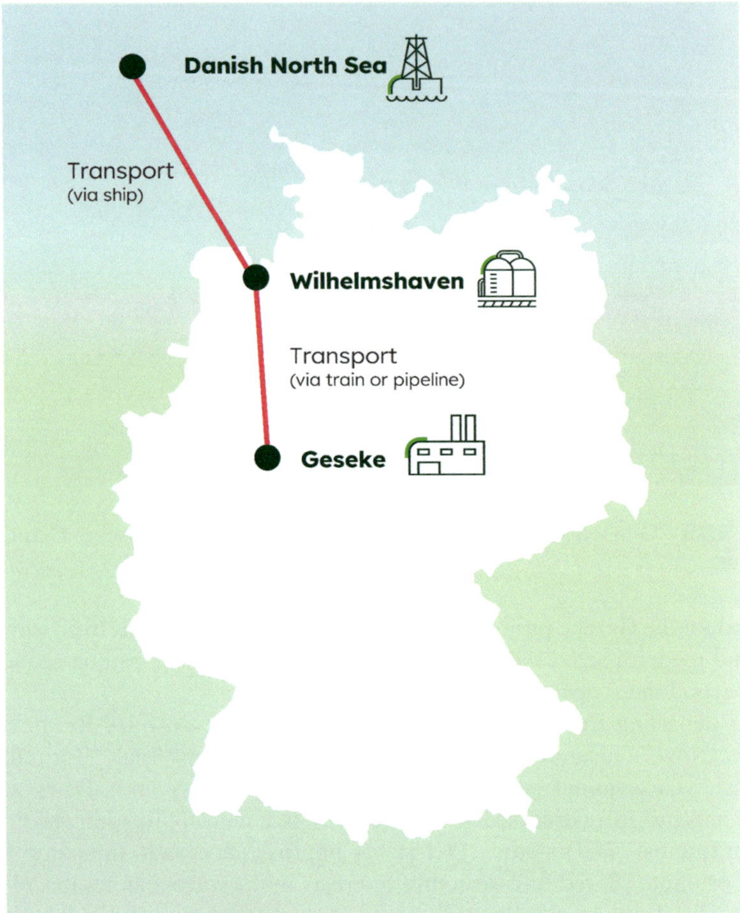

FIGURE 9.21 Map of transport for GeZero CCS.

GeZero will become Germany's first working logistical connection for transporting inland emissions and will remain a model concept throughout the project's initial years and until an overland pipeline is available. GeZero's rail transport system represents a fast and effective way to convey CO_2 from a cement plant to its final storage location (Figure 9.21).

9.6 CARBON UTILIZATION

In some cases, CO_2 captured from industrial processes can be used as an input into other manufactured products. It can be used in its non-altered form, known as direct use, or it can be converted via chemical or biological means. Captured CO_2 can be used in fertilizer production, food and beverage production, water treatment, and as an addition to greenhouse environments to increase crop yields. Uses under development include using CO_2 after a reaction with green hydrogen as fuel (e.g., using CO_2 to create energy for chemical production) and in building materials (e.g., using concentrated CO_2 to carbonate concrete products).

The Office of Fossil Energy and Carbon Management's Carbon Conversion program is exploring a mineralization pathway for CO_2. CO_2 can be converted using alkaline reactants to produce inorganic minerals such as synthetic aggregates. Aggregates and other building-material applications represent an effective long-term storage option for CO_2, as well as a potentially large market. These applications, which involve CO_2 mineralization during curing, are considered to be a permanent sequestration strategy because they combine CO_2 gas with calcium oxides and other materials to form a solid. When applied to calcium-rich materials, this process would produce $CaCO_3$ (limestone), essentially reversing the calcination that occurs in a cement kiln.

A large-scale carbon capture and liquefaction plant in Lengfurt, Germany, is expected to capture CO_2 from cement production that has a level of purity suitable for use as a raw material in both the food and chemical industries. Heidelberg Materials and Linde, a global industrial gases and engineering company, have established a joint venture to build and operate this large-scale carbon capture and utilization (CCU) facility at the existing Heidelberg Materials Lengfurt plant. The joint venture goes under the name "Capture-to-Use" (CAP2U).

The companies' funding of the project will be supplemented by around €15 million from the "Decarbonisation of Industry" funding program on behalf of the German Federal Ministry for Economic Affairs and Climate Action (BMWK). A groundbreaking ceremony for the CCU facility was held in June of 2024, and it is scheduled to go into operation as early as 2025. The planned volume of purified and liquefied CO_2 is around 70,000 metric tons per year. The majority of the CO_2 generated will be marketed by Linde, with applications in the chemical and food industry including, for example, use in carbonated water.

9.7 NEXT STEPS IN DECARBONIZING THE CEMENT INDUSTRY

CCS is expected to be a strong focus of future efforts to reduce emissions associated with cement production. To support the technology's continued development and implementation, investment in infrastructure to support geologic sequestration is critical. Having access to dependable carbon storage is a determinant of whether or not a CCS project is feasible for a given facility/location. Pipelines and Class VI wells (which are used to inject CO_2 into deep rock formations) are the key components of this infrastructure and permitting and provision for these elements of storage will be critical for future success. Ongoing federal and state government support in the evaluation, development, and implementation of different CCS technologies through funding of front-end engineering and design (FEED) studies, demonstration projects, and deployment at scale is an essential component of the future success of CCS for the cement industry.

REFERENCES

1 IEA. "About CCUS – Analysis." IEA, Apr. 2021, www.iea.org/reports/about-ccus
2 Malischek, Raimund, and Samantha McCulloch. "The World Has Vast Capacity to Store CO_2: Net Zero Means We'll Need It – Analysis." IEA, 1 Apr. 2021, www.iea.org/commentaries/the-world-has-vast-capacity-to-store-co2-net-zero-means-we-ll-need-it
3 U.S. EPA, Combined Heat and Power: Frequently Asked Questions, updated April 2022. https://www.epa.gov/sites/default/files/2015-07/documents/combined_heat_and_power_frequently_asked_questions.pdf

10 CO_2 Utilization in Cement and Concrete

Wenyu Liao, Kangyi Cai, Yuyan Huang, Nannan Zhang, Gao Deng, Muhammad Haris Javed, and Hongyan Ma

10.1 INTRODUCTION

The cement and concrete industry is one of the largest contributors to global CO_2 emissions, accounting for nearly 10% of the world's total anthropogenic emissions, primarily due to the production of cement [1]. Cement manufacturing is energy-intensive and emits significant amounts of CO_2, both due to the combustion of fossil fuels and the chemical reactions (e.g., decomposition of limestone) during clinker production [2]. As global efforts to mitigate climate change intensify, developing sustainable methods to reduce and repurpose CO_2 emissions has become a critical priority. CO_2 utilization in cement and concrete represents a promising pathway for reducing the industry's carbon footprint while generating value-added products.

CO_2 utilization aligns with circular economy principles, transforming waste emissions and solid waste (or synthetic) materials into value-added commodity materials, specifically cement and concrete products in this article. This approach has several environmental and economic benefits: it reduces the overall carbon intensity (or embodiment) of cement and concrete, sequesters CO_2 in durable forms (e.g., $CaCO_3$) for permanent storage, and may enhance certain material properties such as strength and durability. Moreover, CO_2 utilization technologies support the transition to low-carbon infrastructure, addressing the growing demand for sustainable construction materials (e.g., the U.S. Department of Transportation's Low-Carbon Transportation Materials Grants Program [3]) and compliance with stricter environmental regulations (e.g., the Federal Buy Clean Initiative [4]). By fostering innovation in materials science and industrial processes, CO_2 utilization contributes to climate change mitigation and promotes resilience in the built environment.

The state of the art in CO_2 utilization for cement and concrete spans a range of technologies and applications. Key advancements include but are not limited to: (1) *Carbonation Curing*: This technique involves curing precast concrete products with CO_2 to accelerate the formation of calcium carbonates ($CaCO_3$), which can enhance strength and durability while permanently storing CO_2; (2) *Enhanced Mineralization*: Captured CO_2 is reacted with industrial or mining byproducts, such as steel slag, fly ash, and recycled concrete, to produce carbonates-based alternative cementitious binders (i.e., non-portland-cement binders), supplementary cementitious materials

DOI: 10.1201/9781003655442-14

(SCMs), functional fillers, or aggregates in concrete; (3) *Direct Injection*: CO_2 is injected during the mixing phase of concrete production, forming carbonates that impact material properties in certain manners; and (4) *Bio-mediated Carbonation*: Using bio-molecules to mediate carbonation or harnessing microorganisms to precipitate $CaCO_3$ offers a novel, biologically inspired method for CO_2 sequestration – even using atmospheric CO_2 directly – in concrete.

For future implementation, these technologies will require advancements in CO_2 capture, temporary storage, and transportation to enhance their economic and technical feasibility. Despite significant progress, challenges persist in scaling these technologies, ensuring consistent material performance, and meeting regulatory requirements (e.g., ASTM standards and design codes). This chapter reviews and evaluates several promising technologies that have been extensively studied, while also briefly discussing other relevant or emerging approaches, aiming to stimulate mindset shifts and technology advancements in this field.

10.2 CARBONATION CURING AND HARDENING

"Carbonation curing and hardening" refers to a curing process of prefabricated cement-based or solid waste-based materials in the aqueous/moist CO_2-rich environment. Chemical reactions occur between the CO_2 and the alkaline-earth-metal-bearing phases [5], like the Ca-bearing components in Portland cement – alite (C_3S) and belite (β-C_2S) – as well as their hydration products of calcium hydroxide (CH) and calcium silicate hydrate (C-S-H) [6–8]. Different minerals and solid wastes can be introduced into such materials as sole binders, alternative SCMs/fillers, and/or aggregates, for example, reactive magnesia [9], blast furnace slag and steel slags from basic oxygen furnace (BOF), electric arc furnace (EAF), ladle furnace [10–12], coal ash [13–17], and cement kiln dust (CKD) [18], along with fines and aggregates from recycled concrete [19], in which the Ca/Mg-bearing phases can also be carbonated. The hardening process in these binder systems is achieved through the synergy of hydration and carbonation, which can also be exclusively accomplished by carbonation reaction. For example, various calcium silicates, including wollastonite (CS), rankinite (M-C_3S_2), and calcio-olivine (γ-C_2S), possess non-hydraulic but strong carbonation capability [20–22]. Other alternative cement systems with noticeable carbonation hardening capacities have also been proposed, such as magnesium slag (a byproduct from magnesium manufacturing) [23] and iron-silica system (combination of metallic iron powder from the shot-blasting process of steel members with fly ash, limestone, and metakaolin) [24].

10.2.1 Mechanisms and Factors

The essence of carbonation curing/hardening is the generation of carbonates (with or without hydrate water molecules, see Figure 10.1) in the aqueous/moist environment, which involves two steps – CO_2 dissolution and carbonate precipitation [5, 25, 26]. Since the alkaline-earth-metal-bearing phases are not the sole components in the binder materials, carbonate precipitate is always accompanied by other chemical changes, like the formation of Ca-modified silica gels from the (non-)hydraulic calcium silicates

FIGURE 10.1 Schematic diagram of the carbonation curing process. M^{2+} denotes divalent metal cations such as Ca^{2+} and Mg^{2+}; and $MCO_3 \cdot xH_2O$ means that the formed carbonates can be anhydrous carbonates (e.g., $CaCO_3$) or carbonate hydrates (e.g., $MgCO_3 \cdot 3H_2O$).

[20, 21]. A decelerating growth trend of carbonation evolution after an initial surge is widely observed, reflecting three stages of carbonation curing distinguished by the controlling factors of carbonation rate [27–29]. (a) *Boundary-control stage*: CO_2 from outside dissolves into the aqueous solution surrounding the binder particles, combines with the dissolved metal ions, and then precipitates at the particle surface, whose reaction rate is determined by the aqueous-solid phase boundary area. (b) *Diffusion-control stage*: Carbonates and other reaction products gradually accumulate and encapsulate the binder particles as carbonation goes on. CO_3^{2-} and HCO_3^- ions must penetrate the carbonation layer to maintain the reaction, while the immigration of metal ions from the binder particles is also gradually hindered. The carbonation rate at this stage is limited by the ion diffusion rate at the carbonation layer. (c) *Stationary stage*: As the carbonation layer densifies and thickens, the reaction rate reaches a stationary plateau eventually. The metric, degree of carbonation (DOC), is widely utilized to quantify the carbonation process, while the amount of CO_2 sequestrated in the unit quantity of the materials is commonly measured through thermogravimetric analysis as an index of carbonation capacity [30, 31]. Empirical and numerical kinetic models have been developed to describe the carbonation evolution [32].

Effects of different influential factors on carbonation have been broadly investigated, which can be classified into two categories – material properties and curing conditions.

Different materials exhibit varying carbonation reactivities. For example, the reactivity of γ-C_2S varies under different doping dosages of barium [31]. Among calcium silicate minerals that have been studied for carbonation curing and hardening, γ-C_2S is found to be the most reactive one, followed by its counterparts – C_3S_2, CS, β-C_2S, and C_3S – under an identical curing environment [20]. Moreover, chemical additives can regulate the carbonation reactions, for example, chitosan facilitates calcium ion dissolution [33] and amino acid promotes specific polymorph (e.g., vaterite) formation of $CaCO_3$ [34]. Apart from the compositional effects, the physical characteristics of materials also play a significant role. Finer binder particles imply more rapid and in-depth carbonation but require more energy-intensive mechanical grinding [35]. Additionally, influences derived from mixture proportioning parameters are expected. For example, a low water-to-binder ratio (near 0.20) is commonly adopted to realize high CO_2 uptakes for calcium silicates [20, 21, 36, 37] while maintaining high mechanical performance.

The condition factors of carbonation curing/hardening include CO_2 purity or concentration, pressure, temperature, humidity, gas fluidity, and more. Among them, CO_2 purity/concentration is a critical factor as suggested by the far stronger carbonation process at the 100% CO_2 environment compared to the atmospheric CO_2 level (0.04% CO_2 purity) [38]. Maintaining pressure at several to tens of folds higher than air pressure is substantiated to be able to promote CO_2 diffusion through the carbonation layer [27]. Besides, the temperature factor has a complex effect on the carbonation reaction, as higher temperatures can reduce CO_2 dissolution and accelerate evaporation, while also increasing molecular kinetic energy and promoting ion migration in the pore solution as well as the CO_2 diffusivity [5, 39]. Striking a balance between low equilibrium CO_2 concentration and fast reaction kinetics is crucial for improving reaction; and then, this balance must be integrated with the influence on the carbonation curing process to achieve a positive impact on the final performance of carbonation-cured concrete [5, 39]. A slight increase in temperature from 20°C to 50°C was found to enhance the strength of carbonation-cured cement paste. [7]. In contrast, humidity and gas fluidity are taken as indirect influential factors considering that they mainly affect water and CO_2 supply rather than the reaction itself. Adequate moisture/water with sufficient CO_2 supply is the prerequisite for facilitating the carbonation reaction [21, 27]. Moreover, the duration factor plays a predominant role in carbonation, affecting the carbonation depth and overall CO_2 uptake. The major portion of carbonates is generated within the first few hours [20], while extended curing duration (prolonged to several days or even weeks) allows the decayed carbonation to continue and leads to better engineering performance [9].

In practice, all the above-mentioned factors need to be balanced under a comprehensive consideration of performance requirements, efficiency, and cost. This balance requires extensive testing due to the wide compositional variability of feedstock and the complex interplay of a range of influential factors during curing, which necessitates standardized protocols that are still lacking [39–41]. Emerging data-driven artificial intelligence (AI) offers another promising solution, in which machine learning models trained on immense curated datasets have the potential to predict optimal conditions for maximizing CO_2 uptake or product quality, while adaptive algorithms enable real-time adjustments in dynamic systems. Early studies demonstrate AI's potential to significantly reduce experimental burdens, in either optimizing carbon sequestration processes [42, 43] or predicting the performance of conventional concrete [44, 45]. However, further research is still needed to refine these models, and the AI model for carbonation curing and its integration with the final performance prediction of carbonation-cured concrete remain unexplored. Furthermore, scaling these tools requires efforts to build robust standardized databases and establish cross-industry frameworks to ensure reproducibility and broader adoption.

10.2.2 Characteristics and Performances

Different products with diverse crystallinities can be generated during the carbonation curing process, with their composition and polymorph determined by the materials used, preparation approach, and curing conditions [36]. Taking non-hydraulic calcium silicates, for example, various calcium carbonates can be found, including calcite (cubic shape), aragonite (clavate shape), and vaterite (spheric shape) [20, 27, 46]. "Sheet-like" amorphous Ca-modified silica gels possessing a noticeable amount of Q^3 and Q^4 silicate species – according to the results from Fourier-transform infrared (FT-IR) spectroscopy [27] and ^{29}Si nuclear magnetic resonance (NMR) spectroscopy [47]) – are also discovered, implying the existence of highly connected silicate networks. The whole carbonated matrix can be viewed as a carbonate skeleton embedded with unreacted calcium silicate particles and Ca-modified silica gels [20]. The formed carbonation products gradually fill up the micropores, decreasing the general porosity, while nanopores in the silica gels increase [27]. The Ca-modified silica gels have a comparable elastic modulus to the C-S-H gel in hydrated Portland cement per nano-indentation test results [48, 49].

Carbonation-cured cement-based materials have much faster strength development than their hydration counterparts. After one day of CO_2 curing, carbonated β-C_2S and CS have a compressive strength equal to 80 MPa and 20 MPa, respectively [20]. A surprising strength improvement at high temperatures (400–500°C) has also been observed on carbonated γ-C_2S, attributed to the carbonate polymorph transition, implying its extraordinary fire-resistant performance [50]. In general, the compressive strength of carbonation-cured cement-based materials can range from 10 MPa to 100 MPa, with γ-C_2S and Portland cement pastes at the high end. Additionally, when properly controlled, carbonation-induced volume expansion can be utilized to reduce permeability and thus improve the material's resistance to environmental degradation [37].

10.2.3 Current and Future Applications

As a promising sustainable technique, carbonation curing/hardening has been adopted for manufacturing precast concrete components, such as lightweight building bricks, blocks, and even reinforced structural members [51, 52]. Some representative companies that exist or existed in the market include Novacem, Solidia, CarbonBuilt, and Carbicrete. Components in thin-walled shapes with low to medium compaction densities are preferable, as carbonation curing, being a self-suppression process, is inherently limited to the superficial zone. Nevertheless, the novel idea of combining carbonation curing with additive manufacturing may provide a good solution to overcome this inherent limitation [53]. The layer-by-layer casting approach of 3D-printed concrete (thin layer but requires fast strength gain for supporting next layer) fits right in the characteristics of carbonation curing (superficial but rapid reaction). The application of carbonation curing to non-thin-walled products may only help densify the surface layer toward improved durability without remarkable CO_2 uptake for permanent storage [51].

10.3 CO_2 UTILIZATION IN SUPPLEMENTARY CEMENTITIOUS MATERIALS

CO_2 mineralization by alkaline solid waste powder has been widely studied in the field of environmental engineering [54–56]. However, it is a waste of resources if such mineralization processes end up with disposal or landfill. If designed scientifically, the CO_2-mineralized solid wastes can be used as alternative, multi-phase supplementary cementitious materials (SCMs) to substitute cement, enabling remarkable CO_2 sequestration and decarbonizing cement and concrete production. This strategy is discussed in this section, focusing on several feedstock materials – recycled concrete fines, steel slag, and coal combustion fly ash, among others.

10.3.1 Recycled Concrete Fines

Recycled concrete fines (RCFs) are fine particles (< 0.15 mm) derived from the crushing and processing of demolished concrete [5, 57, 58]. They consist of a blend of un-hydrated cement grains, hydration products, aggregates, and minor impurities [59, 60]. In practice, the composition and quality of RCFs differ from the source concrete, degree of hydration and deterioration, and crushing and processing techniques. The cement paste portion may include residual phases such as C-S-H, CH, un-hydrated C_3S, and C_2S [59, 60]. Additionally, impurities like chlorides, sulfates, and carbonates need attention on the characterization and preprocessing to optimize their performance in cementitious systems [61–63]. If used correctly, RCFs can potentially contribute to enhancing the binding properties of cementitious systems due to the calcium-rich phases when synergized with clinker and benefiting the mechanical and durability properties [64, 65]. This can improve early-age strength and potentially reduce reliance on clinker, which is the most energy- and carbon-intensive component of Portland cement [66–68]. Additionally, their fine particle size and high surface area of RCFs can refine the microstructure of concrete, improving impermeability and enhancing durability, such as resistance to chloride attack, which is critical for structures exposed to aggressive environments [69–72].

The high-calcium composition of RCFs also implies that they have the potential for carbon sequestration, realized through carbonation of the calcium-rich phases that convert such phases into calcium carbonate and reactive alumina/silica. These products can be re-integrated into concrete as carbon-negative SCMs or fillers, depending on their reactivity [73]. In other words, the logic behind using RCFs for CO_2 capture lies in their ability to act as both a carbon sink and a reactive SCM [74]. When exposed to CO_2, the calcium-rich phases in RCFs may undergo the following representative chemical reactions [75, 76]:

$$Ca_3SiO_5 + 3CO_2 + xH_2O \rightarrow 3CaCO_3 + SiO_2\,xH_2O \tag{10.1}$$

$$Ca_2SiO_4 + 2CO_2 + xH_2O \rightarrow 2CaCO_3 + SiO_2\,xH_2O \tag{10.2}$$

$$Ca(OH)_2 + 2CO_2 + H_2O \rightarrow CaCO_3 + 2H_2O \tag{10.3}$$

$$C - S - H + CO_2 \rightarrow CaCO_3 + SiO_2 + H_2O \tag{10.4}$$

The reaction kinetics of carbon mineralization in RCFs are primarily governed by the interaction between calcium-rich phases and CO_2 gas or dissolved CO_2 in aqueous systems [77–79]. The carbonation process proceeds in three main stages: diffusion of CO_2, chemical reaction, and product formation [79–81]. First, CO_2 gas or dissolved CO_2 diffuses through the liquid medium to reach the reactive phases in RCFs. Then, CO_2 reacts with calcium ions released from the dissolution of portlandite or other calcium-rich phases, forming $CaCO_3$ [79, 80]. This step is typically rapid and exothermic [64, 82]. However, the reaction could be slowed down by the formed dense $CaCO_3$ layer on the surface of RCF particles thus limiting further diffusion of CO_2 to unreacted phases [82, 83]. Figure 10.2 illustrates the total $CaCO_3$ content in RCFs as a function of carbonation duration. On average, RCFs can capture CO_2 at 10–20% by their mass, translating to approximately 0.2–0.4 kg of CO_2 per kg of RCFs, depending on their reactivity and processing conditions [64, 82]. The overall CO_2 uptake is influenced by several factors, including the composition (particularly the content of calcium-rich phases), the fineness of the RCF particles, the carbonation environment (e.g., CO_2 concentration, gas pressure, carbonation duration, temperature, and humidity), carbonation methods (dry, semi-dry, and wet) and the effectiveness of preprocessing techniques such as grinding or thermal activation [74, 88, 89]. Proper optimization of these factors can maximize the CO_2 sequestration potential of RCFs, making them an effective tool for carbon capture and utilization.

After being added into cement-based materials as a multi-phase SCM, the $CaCO_3$ in carbonated RCFs participates in the nucleation and growth of cement hydration products [5, 75, 76]. The presence of fine carbonated particles provides additional nucleation sites for C-S-H, accelerating the hydration of clinker phases [90, 91]. Furthermore, the silica-alumina gels formed during carbonation can react with calcium hydroxide produced during hydration, contributing to secondary C-S-H formation through pozzolanic activity [5, 74]. This synergistic interaction can enhance the strength and durability of the matrix [5, 90, 91]. However, the extent of these benefits depends on the degree of carbonation, particle size, and composition of the RCFs [82, 83]. Over-carbonation can lead to excessive $CaCO_3$ formation, which may cover the reactive phases and reduce the overall reactivity of the material, while

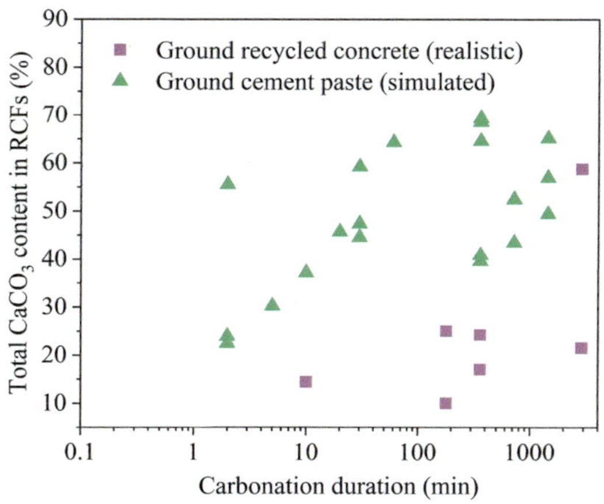

FIGURE 10.2 Total $CaCO_3$ content as a function of carbonation duration [61, 84–87].

under-carbonation means a waste of the carbon storage capacity of the calcium-rich resource [64, 82]. Therefore, balancing carbon uptake and reactivity is critical in using RCFs to produce carbon-negative SCMs. In addition, advanced techniques such as thermal activation and chemical treatments have made it possible to optimize the reactivity of heavily carbonated RCFs, thus optimizing the carbonation process [74, 88, 89].

Carbonation alters the particle size distribution of RCFs, normally creating finer particles that augment the surface area [75, 76]. Thus, the incorporation of carbonated RCFs into cement-based materials, on the one hand, will negatively impact workability (e.g., flowability), thereby increasing water demand for the mixture [90]. This adverse effect can be modified by using superplasticizers or optimizing the particle size distribution of the blend [92, 93]. On the other hand, the incorporation of carbonated RCFs tends to improve the compressive strength of cement-based materials if dispersed well and proportioned properly, attributed to the enhanced packing density and reduced porosity [19, 94–96]. Figure 10.3 illustrates the strength evolution (up to 90 days) of cement paste and ultra-high-performance concrete with varying dosages of RCF and carbonated RCF (CRCF). As discovered, RCF and CRCF enhance the early-age strength of normal cement paste, while the long-term strength is either comparable or lower than the control. In contrast, the incorporation of RCF and CRCF in ultra-high-performance concrete (UHPC) slurry reduces strength overall; however, CRCF consistently outperforms uncarbonated RCFs in both short-term and long-term strength. To effectively achieve the desired strength improvement in the applications of carbonated RCFs, striking a balance between factors (e.g., carbonation degree, dispersion quality, and cement replacement ratio) is always essential [58, 97, 98]. Moreover, the denser microstructure resulting from the incorporation of carbonated RCF may also improve the durability of cement-based materials [61, 64]. For example, reduced porosity limits the attack of ions like chlorides and sulfates and enhances resistance to chemical attack and freeze-thaw cycles [64, 90, 91]. At the concrete scale, the presence of fine calcium carbonate and silica-alumina gels in carbonated RCF also contributes to improved interfacial transition zones (ITZ) between bulk paste and aggregate particles [81, 99, 100]. This change tends to enhance the mechanical performance and durability of concrete [101, 102]. These positive effects make carbonated RCF a valuable resource for the production of sustainable cement and concrete [103].

10.3.2 Steel Slag

Steel slag is a byproduct generated during the steelmaking process, with 190–280 million tons produced globally every year [104]. The chemical compositions and mineralogical characteristics of steel slags are highly complicated, depending significantly on the raw materials used and the steelmaking process composed of many complex processes [105, 106]. Generally, there are three types of steel slags produced during the steel manufacturing process, namely, the BOF slag, the EAF slag, and the ladle metallurgy furnace (LMF) slag [107]. BOF slag and EAF slag are generated during the initial refining of steel using distinct production methods, thereby exhibiting similar chemical compositions [108]. In contrast, LMF slag is a byproduct

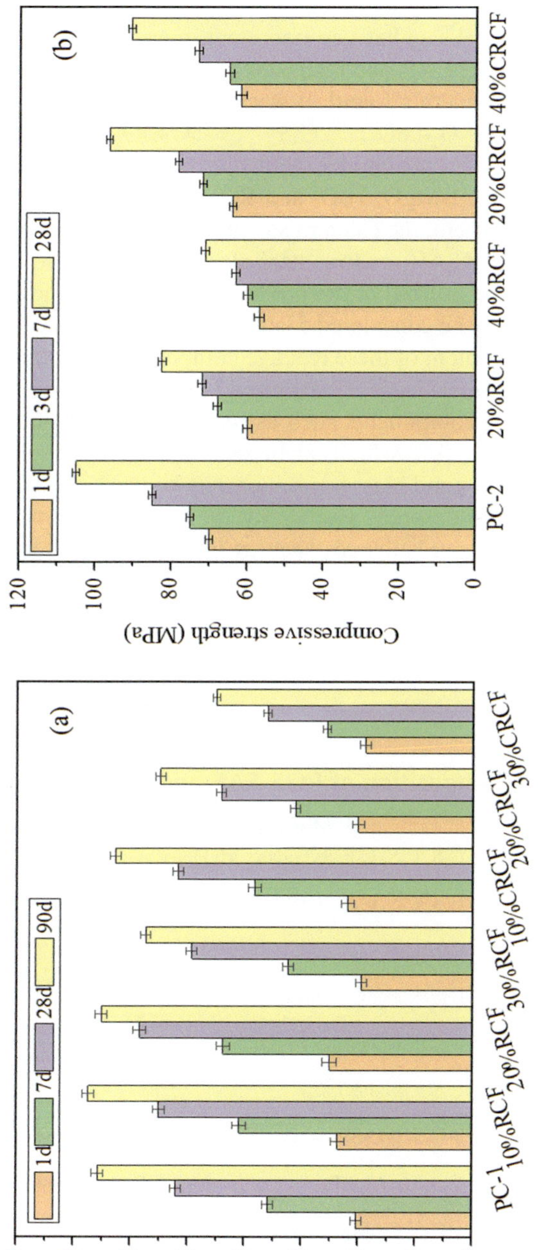

FIGURE 10.3 Compressive strength with different replacement ratios of RCFs (uncarbonated and carbonated) in (a) cement paste and (b) ultra-high-performance concrete (UHPC) slurry [68, 83, 88].

obtained through a secondary metallurgy process, where the molten steel from BOF and EAF undergoes further refinement to reduce the content of sulfur and phosphorous and achieve the desired metallurgical composition [105, 109].

Mineral properties of steel slag have an important impact on its usage. The mineralogical composition of BOF, EAF, and LMF slags has been summarized, as shown in Table 10.1. The primary minerals found in steel slags include β-C_2S ($2CaO{\cdot}SiO_2$), γ-C_2S, C_4AF ($4CaO{\cdot}Al_2O_3{\cdot}FeO_3$), C_3S ($3CaO{\cdot}SiO_2$), C_3A($3CaO{\cdot}Al_2O_3$), $C_{12}A_7$($12CaO{\cdot}7Al_2O_3$), C_2F($2CaO{\cdot}Fe_2O_3$), gehlenite ($2CaO{\cdot}Al_2O_3{\cdot}SiO_2$), merwinite ($3CaO{\cdot}MgO{\cdot}2SiO_2$), f-CaO, f-MgO, FeO, and RO phase (a solid solution of CaO-FeO-MnO-MgO). Among them, the calcium silicates (C_2S and C_3S) and calcium aluminates (C_3A, $C_{12}A_7$, and C_4AF) endow steel slag with hydraulic and cementing properties, allowing it to be utilized in the production of cement and concrete [112]. However, steel slag can only be considered a latent hydraulic material (which does not manifest itself) since the content of these reactive compounds is much lower than that in Portland cement due to their large crystal size [113]. Besides, the Fe in steel slag exists in the form of inert components such as ferrite phases, including RO phase, FeO, and C_2F ($Ca_2Fe_2O_5$) [5, 114], which can be separated from the steel slag by magnetic separation for the sinter and blast furnace or may be used in steelmaking [115].

Therefore, despite their widespread availability, the utilization of most steel slags as SCMs remains limited due to challenges such as low hydration reactivity, potential volume instability, and poor grindability [104, 116]. However, with high carbonation reactivity, steel slag is considered one of the most suitable materials for CO_2

TABLE 10.1
Mineralogical Phases Composition of BOF, EAF, and LMF Slags [110, 111]

Slag Type	Mineralogical Phases Composition
BOF	C_2S, C_3S, C_2F, RO phase, MgO, f-CaO
	C_2S, C_3S, C_2F, $2FeO{\cdot}SiO_2$, $2MgO{\cdot}SiO_2$, $C_{12}A_7$, MgO, f-CaO
	C_2S, C_3S, C_2F, Fe_3O_4, RO phase, $C_{12}A_7$, f-CaO
	C_2S, C_2F, $2CaO{\cdot}P_2O_5$, f-CaO
	β-C_2S, RO phase, MgO
	C_2S, C_3S, FeO, C_2F
EAF	γ-C_2S, β-C_2S, $Ca_7Mg(SiO_4)_4$, $Mg_2Si_2O_6$, $Ca_3Mg(SiO_4)_2$, $CaSiO_3$, MgO
	C_2S, $Ca_2Al_2SiO_7$, Fe_2O_3, $Ca_{14}Mg_2(SiO_4)_8$, $MgFe_2O_4$, Mn_3O_4, MnO_2
	γ-C_2S, $Ca_3Mg(SiO_4)_2$, CFMS, RO phase
	C_2S, C_3S, $4CaO{\cdot}Al_2O_3{\cdot}Fe_2O_3$, $Ca_2Al_2SiO_7$, FeO, Fe_3O_4, MgO, SiO_2
LMF	C_2S, $CaO{\cdot}Al_2O_3{\cdot}2SiO_2$, CaS, Al_2O_3 $C_{12}A_7$, MgO, β-C_2S, γ-C_2S, $Ca_2Al_2SiO_7$

mineralization [117, 118]. Moreover, it has been demonstrated that carbonation can diminish the negative effects of steel slags as SCMs [119], including the improvement of volume stability resulting from the consumption of free-CaO and free-MgO [120, 121], and the increase in hydration activity due to the formation of $CaCO_3$ and amorphous alumina/silica gel [110]. When blended with cement, besides serving as a well-known filler or nucleus in cement [122], the precipitated $CaCO_3$ can engage in a reaction with the available alumina, which may originate from either the cement clinkers or the carbonated slags themselves, resulting in the formation of hemi- and mono-carbonate and contributing to the overall performance of the cement [122]. Moreover, the amorphous silica/alumina gel demonstrates reactivity as a pozzolan in cementitious materials to increase the formation of C-S-H and carboaluminate (hemicarboaluminate, Hc, and monocarboaluminate, Mc), which proves effective in refining porosity and enhancing strength of cementitious materials. In this context, the carbonation of steel slags can significantly improve the hydration behavior and microstructure of blended cement when carbonated steel slags are applied as SCMs, thereby enhancing the mechanical and durability performance of cement and concrete. Generally, the compressive strength and volume stability of the blended cement with carbonated steel slag can be improved correspondingly. Improvements in different properties of cement blended with carbonated steel slags are summarized in Table 10.2.

In general, steel slags can mineralize CO_2 in the range of 1–46% by weight of dry slags via the direct carbonation pathway, with carbonation efficiency varying based on reaction conditions [110, 125, 128, 129]. When used to replace 10–40% cement, the blended cement-based materials can achieve strengths of 32–65 MPa [117, 121, 123–127] – which can easily meet the technical requirements of ASTM C1157 [130] though not necessarily exceed the strength of the corresponding plain cement counterpart. However, the incorporation of carbonated steel slag normally has negative effects on workability (or water demand), so superplasticizers are normally needed at a relatively large dosage to compensate for this effect [123, 126].

10.3.3 Coal Combustion Fly Ash

Coal combustion fly ash (FA) from coal combustion is a fine, powdered substance resulting from the inorganic residue left behind during the burning of pulverized coal in power plants. FA can be categorized into low-calcium FA (LFA), usually with less than 10% CaO, and high-calcium FA (HFA) depending on its calcium content [131]. In the construction industry, LFA is widely utilized as an SCM in concrete, owing to its ability to improve workability, boost long-term strength, and enhance durability [132]. On the other hand, the use of HFA is often less favored as it contains free-CaO, which may compromise the soundness and durability of concrete by causing potential volume expansion [133].

Following the development of accelerated carbonation technologies for CO_2 mineralization in solid wastes, the carbonation treatment of coal ashes, especially HFA, has gained strong interest, as it not only efficiently captures CO_2 but also removes free-CaO, enabling safe use in construction materials [133–135]. Siriruang et al. [133] and Pei et al. [136] investigated the expansion behavior of cement paste incorporating FA, both with and without carbonation treatment. The findings revealed that

TABLE 10.2
Changes in the Properties of Cement Blended with Carbonated Steel Slag

Slag Type	Carbonation Process	Substitution Ratio (%)	Properties Improvements	Ref.
BOF	High-gravity carbonation for 3 h at 60 °C	10	Shortened setting time; accelerated early mechanical strength	[123]
BOF	Gas-solid carbonation for 72 h at 25 °C	15–45	Higher compressive strength; denser microstructure; reduced leaching of V and Cr	[124]
NA	Aqueous carbonation for 0.5–10 h at 60°C	0–60	Improved volume stability; filled microcracks or pores; enhanced 28-day strength	[120]
BOF	Aqueous carbonation for 2 h at 40 °C	10–30	Decreased 28-day compressive strength	[125]
NA	Gas-solid carbonation for 30 min at 20 °C	30	9.7% increase in compressive strength; improved volume stability	[117]
NA	Bio carbonation for 24 h at 25 °C	30	Enhanced stability and reactivity; significant increase in flexural and compressive strength (up to 86.8% and 87.1%, respectively)	[121]
BOF	Aqueous carbonation for 0–1.5 h at 40 °C	15	Shortened setting time; increased compressive strength at early ages (1, 3, and 7 days); negatively affected workability; improved soundness	[126]
BOF	Gas-solid carbonation for 5 min–72 h at 20 °C	20–40	Improved pore structure; increased compressive strength	[127]

carbonation treatment of FA markedly enhanced the stability of cement pastes, attributed to the reduction of free-CaO following the carbonation reaction. Chen et al. [137] employed carbonation curing and carbonated fly ash (CFA) to enhance the performance of mortar samples, reporting improvements in freeze-thaw resistance and sulfate attack resistance. These enhancements were attributed to the consumption of free-CaO and calcium hydroxide during carbonation curing. Furthermore, they observed that carbonation treatment of FA could lead to an increase in particle size and a reduction in specific surface area, attributed to reaction products adhering to the surface [134], which was proposed to decrease the reactive contact area and influence pozzolanic reactivity [138]. The pozzolanic reactivity of specific materials depends on their chemical composition or mineralogical properties, particularly active phases, as well as physical characteristics such as surface area, particle size, and surface morphology [139–141]. Carbonation treatment significantly modifies the surface properties of FA, thereby inevitably affecting ion dissolution and reaction behaviors.

Furthermore, the carbonation treatment of FA modifies the phase distribution, especially in the aluminate hydration products within CFA-blended cementitious materials. The $CaCO_3$ present in CFA as a result of carbonation treatment can interact with aluminate phases to produce monocarboaluminate (Mc) and hemicarboaluminate (Hc), thereby stabilizing ettringite [142–144]. This alteration in the phase assemblage can substantially influence the micro- and macro-scale mechanical properties, as well as the transport characteristics of concrete, thereby affecting the durability and longevity of cement-based materials [145, 146].

Table 10.3 provides a summary of the maximum CO_2 sequestration capacities of various coal FAs under diverse operating conditions. Generally, coal FAs can mineralize CO_2 in the range of 0.2–24.9% by dry ash weight, with HFA at the high end. When used to replace 10–30% cement, the blended cement-based materials can achieve strengths of 22–38 MPa (Table 10.4).

10.3.4 Other Alkaline Materials

Municipal Solid Waste Incineration (MSWI) ashes are byproducts of burning municipal solid waste for power generation. These ashes include bottom ash (solid residue of incineration, relatively coarse) and fly ash (fine particles collected from scrubbers and bag houses). It is estimated that hundreds of millions of tons of MSWI ashes have already been generated and ponded/landfilled over the past decades [157]. MSWI ashes are complex materials composed of a mix of oxides such as SiO_2, Al_2O_3, CaO, and Fe_2O_3, along with trace amounts of heavy metals [158, 159], with bottom ash normally containing <20% CaO while fly ash containing >40% CaO. These features make them reactive and suitable for CO_2 mineralization via either gas-solid carbonation or aqueous carbonation [160]. In addition, some surface activation techniques, like grinding, steaming, and ultrasonication, can further enhance the carbon uptake reactivity and capacity of the ashes [161]. MSWI fly ash can uptake over 25% CO_2 in less than an hour via aqueous carbonation, with calcite as the dominant reaction product, while semi-dry, gas-solid carbonation typically takes more than 10 hours, mainly forming vaterite [160]. Another study reported that MSWI fly ash can sequester CO_2 at 16.3–21.4% by weight of ash, influenced by the dosage of water used in the carbonation process [162]. Ultrasound treatment was found to improve CO_2 capture efficiency, improving carbonation from 14.1% to 18.8% by enhancing mass transfer and strengthening the gas-liquid-solid reactions [163]. The resultant carbonated MSWI ashes could demonstrate enhanced pozzolanic properties, making them promising candidates as (alternative) SCMs in concrete production [164]. Carbonation of MSWI ash can also help reduce the leaching of heavy metals by converting them into less soluble carbonate forms, which are more stable and less prone to environmental contamination [165, 166].

The reaction mechanisms of carbonated MSWI ashes in cement-based materials are not different from those of carbonated RCF, steel slag, and coal fly ash. Potential benefits could be attributed to the nano-/sub-micro-calcium carbonate and activated aluminosilicate phases in the carbonated ashes, which improve the microstructure

TABLE 10.3
CO_2 Sequestration Compacity for Various Coal FAs under Different Reaction Conditions

Reaction Type	Composition	Reaction Conditions	Efficiency (%)	CO_2 Compacity (g CO_2/kg FA)	Ref.
Gas-solid	6.48% CaO; 15.33% MgO	300 °C; 1 bar; 15% CO_2	—	2.6	[147]
	26.16% CaO; 3.65% MgO	23 °C; 2 bar; 99.99% CO_2	27	—	[135]
	6.74% CaO; 2.22% MgO	25 °C; 10 bar; 99.99% CO_2	35.53	26.3	[148]
	28.42% CaO; 1.91% MgO	600 °C; 1 bar; 15% CO_2	28.74	60	[149]
	35.27% CaO; 3.21% MgO	160 °C; 6 bar; 99.99% CO_2	48.14	117.7	[150]
Aqueous	25.84% CaO; 2.17% MgO	40 °C; 80 bar; 99.99% CO_2; L/S: 30	24.2	54.9	[151]
	24.8% CaO; 13% MgO	80 °C; 30 bar; 99.99% CO_2; L/S: 0.3	13.6	27.04	[152]
	24.3% CaO; 0.9% MgO	55 °C; 1 bar; 40% CO_2; L/S: 10	86	102.9	[153]
	6.74% CaO; 2.22% MgO	25 °C; 4 bar; 99.99% CO_2; L/S: 15	57.87	50.3	[148]
	62.8% CaO; 0.83% MgO	57.3 °C; 1 bar; 99% CO_2; L/S: 18.9; High-gravity	77.2	249.3	[154]
	3.35% CaO; 3.45% MgO	30 °C; 1 bar; 100% CO_2; L/S: 15	83.5	32	[155]
	33.06% CaO; 0.95% MgO	60 °C; 1 bar; 15% CO_2; L/S: 10	78.2	120	[156]

and enhance the properties of cement-based materials [8, 167]. Such benefits have been verified in past studies [160]. Replacing cement with carbonated MSWI fly ash (up to 50%) increases compressive strength from 12.6 MPa (raw ash) to 22.8 MPa (conventionally carbonated) and 32.2 MPa (ultrasonically carbonated), due to void filling by fine particles and enhanced hydration from nano-$CaCO_3$ [168]. Additionally, the inclusion of up to 50% carbonated municipal solid waste incineration fly ash significantly improves compressive strength by over 25% compared to the reference mortar [169]. Aqueous carbonated fly ash has been found to enhance compressive strength at all ages (1, 3, 7, 28, 56, and 90 days) across replacement levels of 10% to 30%, whereas dry carbonated fly ash increases strength at lower replacement levels but, at 30%, reduces early-age strength (up to 28 days) before improving it at 56 and 90 days [160].

Apart from the complex solid wastes discussed above, CO_2-mineralized SCMs could also be produced using relatively simple feedstocks. For example, reactive magnesia or magnesium-rich rock minerals could be carbonated into nesquehonite, dypingite, or other hydrated polymorphs of magnesium carbonate – sometimes accompanied by amorphous silica; and wollastonite could also be easily carbonated, forming calcium carbonate and amorphous silica. These CO_2-mineralized blended are also suitable to be used as SCMs, enhancing certain properties of cement-based materials [170, 171].

10.4 CARBONATED ARTIFICIAL AGGREGATE

Every year, the global construction industry consumes aggregate at a scale of 50 billion metric tons (Gt). This massive consumption includes aggregate used in concrete, asphalt, road base, and various other applications, with the majority of sources directly from natural deposits. From an environmental perspective, identifying alternative aggregates, especially by utilizing solid wastes as a natural aggregate replacement in the construction sector, is of critical importance [172, 173]. Extensive research has been conducted to pre-process or pre-treat various waste materials for use as raw materials in the production of artificial aggregates. Given the vast market demand for aggregate and the alkaline nature of solid wastes, utilizing carbonation to process solid waste into carbonated artificial aggregates (CAAs) has garnered significant attention because of the Gt-scale potential of permanent CO_2 storage in such CAAs.

Carbonated (or CO_2-cured) recycled concrete aggregate (RCA) is perhaps the most extensively studied form of CAA [174, 175]. However, carbonated RCA is out of the scope of the present review, since the overall CO_2 uptake in carbonated RCA is usually limited due to the relatively low content of cement paste (the only phase in RCA capable of reacting with CO_2). To achieve high CO_2 uptake in CAAs, an ideal process should begin with solid waste powders, proceed with granulation techniques, and culminate in carbonation [176]. A typical process is shown in Figure 10.4.

Granulation is a size enlargement process by consolidating fine moisturized particles into larger granules. This technique has been widely studied in many fields, including pharmaceuticals, the food-processing industry, and agricultural fertilizers, as well as the production of CAAs. During granulation, CO_2 can be injected into the granulator (if sealed) to facilitate the layer-by-layer enlargement of the particle through carbonation hardening. After granulation, a dedicated hardening process (e.g., accelerated carbonation) is required to solidify the fresh granules and enhance the mechanical strength of the granules to be used as aggregate.

Accelerated carbonation is normally used to turn freshly prepared pellets into a solid compound of aggregates or enhance partially hardened aggregate. Under a high CO_2 concentration, the strength development of the pellet can be accelerated to a certain extent. For instance, after four days of carbonation curing of fresh pellets, the strength of CAAs derived from basic oxygen furnace slag (BOFS) could increase by 3.2 times as compared to normal cured aggregates [177].

TABLE 10.4
Compressive Strength of Blended Cement-Based Materials Incorporated Carbonated Fly Ash

Types of FAs	Substitution Ratio (%)	Compressive Strength 28d (MPa)	Ref.
Petroleum coke fly ash	10	31.9	[136]
	15	21.52	
HFA	10	37.5	[134]
	20	38.2	
	30	38.4	
LFA	30	35.1	
HFA	20	45.0	[133]

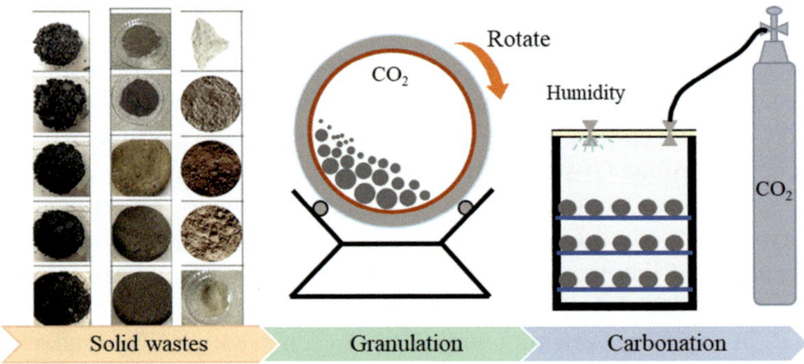

FIGURE 10.4 Flowchart of CAA production.

10.4.1 Mechanical and Physical Properties

Figure 10.5 summarizes the relationship between the crushing strength (i.e., aggregate crushing value) and CO_2 uptake of CAAs based on previous studies utilizing various raw materials [172, 177–185]. The CO_2 uptake of CAAs in these studies ranged from 3% to 17%, with crushing strength varying between 1.5 MPa and 8.5 MPa compared to approximately 9.8 MPa for natural limestone aggregate [186]. It can be seen that an increase in CO_2 uptake leads to an increase in crushing strength. At the same extent of CO_2 uptake, CAAs produced with steel slag can have relatively high crushing strength while the counterparts made from sintered red mud show poorer resistance to crushing.

CAAs demonstrate higher densities compared to uncarbonated artificial aggregates. This is mainly due to the fact that the CAAs fix CO_2, which in turn fills the

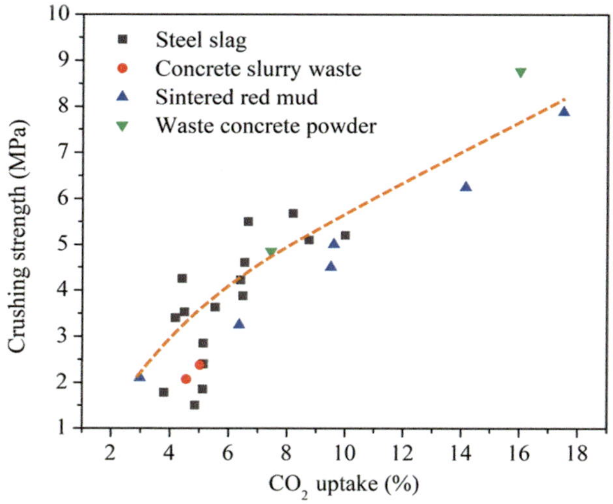

FIGURE 10.5 Relationship between CO_2 uptake and crushing strength of CAAs.

voids in the system and hence leads to an increase in density. In general, it is inevitable that a higher CO_2 uptake leads to a greater packing density within the CAAs. The bulk density of CAA is mainly related to the density of the raw materials, alongside the positive effect of carbonation. As shown in the literature [185], the density of CAAs made from BOFS can be relatively high, followed by CAAs made from waste concrete powder and slurry waste in descending order. Normally aggregates with a bulk density below 1200 kg/m^3 are classified as lightweight aggregates. Most CAAs fall into the category of lightweight aggregates, with the exception of CAAs made with BOFS that can exceed 1200 kg/m^3.

Water absorption stands out as one of the critical physical properties of CAAs since it affects the mix proportion design and fresh properties of concrete. This property is linked to the microstructure of raw materials and CAAs. Due to inherent structural differences among various raw materials, no exact linear correlation can be observed between different raw materials and CO_2 uptake. Nevertheless, for the same raw material, water absorption can show a tendency to decrease as the CO_2 uptake increases due to the fact that higher CO_2 uptake results in a more compact microstructure, impeding water infiltration (see Figure 10.6). Previous studies, as summarized in Figure 10.5, indicated that the water absorption of the CAAs was generally low, mostly below 20% [116, 172, 180–182, 185, 186], except for CAAs made from low-quality feedstocks like sintered red mud.

10.4.2 Properties and Characteristics of Concrete with CAAs

Although CAAs can be used in various applications, their use in making structural concrete offers the highest market value. Therefore, understanding how the incorporation of CAAs impacts the mechanical performance of structural concrete is

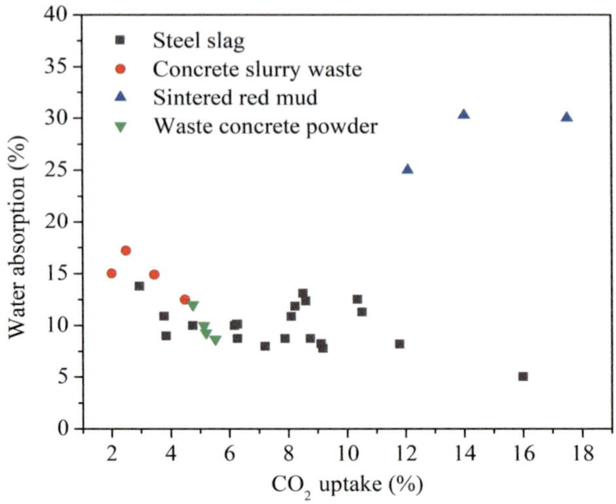

FIGURE 10.6 Relationship between CO_2 uptake and water absorption of CAAs.

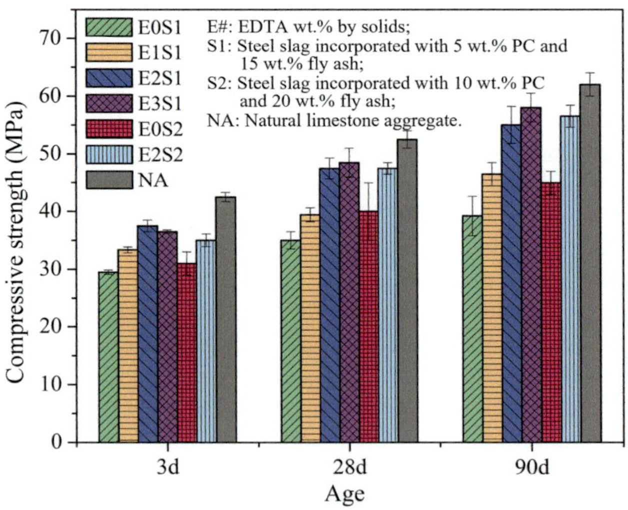

FIGURE 10.7 Compressive strength of concrete cast with carbonated steel slag aggregates.

critically important. Previous studies have indicated that the mechanical properties of concrete made with CAAs could exhibit comparable compressive strength to concrete made with natural aggregates. As shown in Figure 10.7 (reproduced using data in [186]), the 28d and 90d strengths of concrete made with CAAs reached 48 MPa and 59 MPa, respectively, which were just less than 10% lower than the natural aggregate concrete. The CAAs used in this study were produced from the carbonation of blended feedstock consisting of steel slag, Portland cement, and fly ash, facilitated by ethylenediamine tetra-acetic acid (EDTA).

10.5 OTHER CO_2 UTILIZATION TECHNOLOGIES

In addition to the extensively studied technologies discussed in the above sections, there are a number of other carbon utilization strategies in concrete, including (1) using water-dissolved or supercritical CO_2 to enhance carbonation curing and hardening, (2) using carbonated water as mixing water for concrete, (3) injecting CO_2 into concrete during mixing, and (4) using CO_2-derived solid carbon or organic chemicals in concrete. These four strategies are discussed, respectively, in the subsequent paragraphs as follows.

The efficiency of carbonation curing and hardening is often constrained by the slow diffusion kinetics of CO_2 and its limited penetration depth in prefabricated construction products [187, 188]. To improve CO_2 uptake and penetration depth, one approach is to dissolve CO_2 into a solvent (e.g., water) and then use it – instead of gaseous CO_2 – to cure the prefabricated products [189]. Another approach is to cure prefabricated concrete using supercritical CO_2 – carbon dioxide held above its critical temperature and pressure (see Figure 10.8) – which is found to be a more effective carbonation medium than CO_2 gas or water solution under standard conditions, due to its superior diffusivity and reactivity [190]. In CO_2 gas-based carbonation, the diffusion of CO_2 into the cement paste pores governs the reaction rate, and the carbonation front advances inward at a rate proportional to the square root of time based on the analytic solution of Fick's first law of diffusion [191]. Such processes can span from hours to years. While under supercritical conditions, the carbonation reaction can be accelerated significantly since the solubility of CO_2 in water increases by nearly a hundred-fold as pressure rises from 0.1 MPa to 20 MPa [192–194]. Moreover, the ease with which supercritical CO_2 penetrates into the micropores increases the reaction volume and enhances the carbonation efficiency. Unlike gas-based carbonation, supercritical CO_2 carbonation is not governed by CO_2 diffusion but rather by the chemical reaction kinetics of $CaCO_3$ precipitation, dictated by the detachment of calcium ions from the calcium-rich phases. This process can be completed within

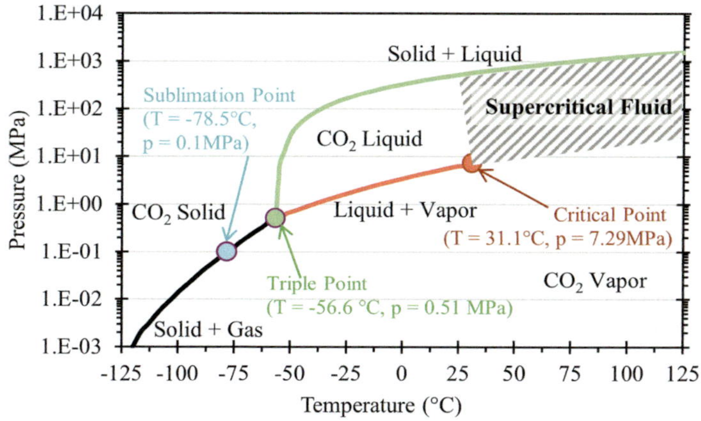

FIGURE 10.8 Phase diagram of CO_2 showing the stability fields of the solid, liquid, and vapor states, including the critical point after which a supercritical fluid exists.

minutes to hours [195]. Notably, increasing pressure and temperature beyond the critical point does not significantly enhance carbonation kinetics [196]. When supercritical CO_2 is used to cure Portland cement concrete, the system rapidly attains high concentrations of CO_3^{2-} and Ca^{2+} during the initial stages, driving a swift and extensive initial precipitation of $CaCO_3$ on the portlandite's surface. As the reservoir of readily available portlandite diminishes, further carbonation proceeds through the decalcification of the C-S-H gel, leading to $CaCO_3$ precipitation between the fibrils of the C-S-H outer product [192, 196]. Supercritical CO_2 curing has been proven to reduce porosity, improve compressive strength, and enhance durability [197, 198]. Decreased total pore volume is attributed to $CaCO_3$ deposition; however, supercritical conditions favor the development of a higher volume of gel pores compared to natural carbonation [196]. Supercritical CO_2 curing has also proven effective for enhancing the strength and durability of concrete incorporating industrial wastes, such as concrete sludge waste [199, 200]. These advancements highlight the transformative potential of supercritical CO_2 curing. However, practical applications still face challenges, including the scalability of the technology and the energy costs associated with maintaining supercritical conditions [201].

By replacing conventional mixing water of concrete with an aqueous CO_2 solution (i.e., carbonated mixing water), the CO_2 dissolved in water can more efficiently react with calcium ions throughout the cement matrix [202, 203]. This reaction forms $CaCO_3$ nanoparticles, which can serve as nucleation sites for the formation of C-S-H and stimulate early-stage hydration of cement. The CO_2 uptake capacity of such carbonated mixing water can be improved by adding a small dosage of quick lime, cement, or calcium-rich solid waste (e.g., steel slag and high-calcium coal ash) powder into the water to significantly increase the carbonate and bicarbonate ions, as well as $CaCO_3$ nanoparticles (suspended), in the mixing water before concrete making [204–206]. Based on the same principle, concrete wash water can be beneficiated with CO_2, transforming it into a better alternative to conventional mixing water [202]. By leveraging the suspended solids of hydrating cement and other hydration products in the concrete wash water, the introduction of CO_2 promotes in-situ $CaCO_3$ formation and increases CO_2 uptake compared to normal water. More specifically, calcium hydroxide converts to calcium carbonate; C-S-H partially transforms into calcium carbonate and a decalcified silicate hydrate gel; and ettringite and other sulfoaluminate phases are converted into calcium carbonate, calcium sulfate hydrates (gypsum and hemihydrate), and alumina gel [207]. Through these reactions, the wash water is "beneficiated" to generate valuable mineral fillers and reactive aluminosilicate phases, refining the microstructure and benefiting the performance of concrete.

Injecting CO_2 into the concrete during the mixing stage has also been implemented as a carbon utilization strategy. The injected CO_2 can react with the calcium in the fresh concrete to form nano-calcite, which is sometimes believed to enhance hydration efficiency by accelerating the formation of calcium silicate hydrate (C-S-H) and resulting in greater strength (e.g., an 8–9% increase of early-age strength) [208, 209]. CarbonCure, perhaps the only commercial company promoting this technique, reported that a 1-minute exposure to CO_2 gas during the mixing process improved the heat of hydration over 24 hours by 8% [210], consistent with the extent of early-age strength increase. This effect is comparable to that achieved by a 1%

addition of nano-carbonate powder, which enhances compressive strength by approximately 8% [211, 212]. However, the CO_2 uptake capacity of this approach is relatively small, typically sequestering only 0.1–0.5% by weight of cement in the concrete, to avoid the potential negative effects of over-carbonation on long-term strength development (due to the wrapping of cement particles by a dense calcite layer) and on steel reinforcement (due to the reduced pH value of the pore solution) [213]. This shortcoming constrains the overall impact of this strategy on CO_2 utilization/storage, though it may save some cement to reduce the overall carbon footprint of concrete.

The fast development of electrochemical, thermochemical, and biological methodologies has also enabled another possibility of CO_2 utilization in concrete, that is, converting CO_2 into carbon-based products and using such products as functional ingredients of concrete. First, CO_2 can be converted into solid carbon, such as carbon black, carbon nanotubes, and carbon fibers, through electrochemical reduction or thermochemical reactions [214]. These solid carbon products can be incorporated into concrete, where they provide multiple benefits. Carbon black, with its high surface area and electrical conductivity, can improve concrete's durability and conductivity [215, 216]. Carbon fibers and nanotubes can enhance compressive and tensile strength and crack resistance, acting as reinforcements within the cementitious material's matrix [217]. Second, CO_2 can be converted into small organic molecules (e.g., citric acid, oxalic acid, and acetic acid) and polymeric molecules [218–220], which may serve as chemical admixtures in concrete [221]. Some of the organic acid molecules can also be proportioned with certain alkaline solid wastes to make alternative binders, including acid-base cements (e.g., ferrous oxalate cement [222]) and alkali-activated materials (e.g., alkali acetate-activated blast furnace slag [223]). Moreover, CO_2 can be converted into certain minerals via bio-mediated processes, which can be leveraged either to produce reactive fillers or additives for concrete or as a bio-cementation mechanism [224, 225]. However, it is still very challenging to manage the biological processes efficiently and ensure consistent mineralization or precipitation in large-scale applications.

10.6 DISCUSSION

We envision that carbon utilization technologies utilizing specific feedstocks can advance rapidly. However, for future scaling and implementation, other critical factors must be considered as well, including feedstock variability, supply chain logistics, techno-economic feasibility, lifecycle analysis, and policy frameworks.

Feedstock variation poses a significant challenge for carbon utilization in cement and concrete. The composition of materials such as CO_2 sources and industrial byproducts can vary widely depending on their origin and processing methods. Figure 10.9 maps the distribution of composition across various types and sources of feedstocks, illustrating their wide compositional variation. This variability can affect the reactivity, compatibility, and performance of the final products. Addressing these challenges requires robust quality control measures, adaptable processing technologies, and a comprehensive understanding of feedstock characteristics to ensure consistent performance. This challenge may be addressed through comprehensive

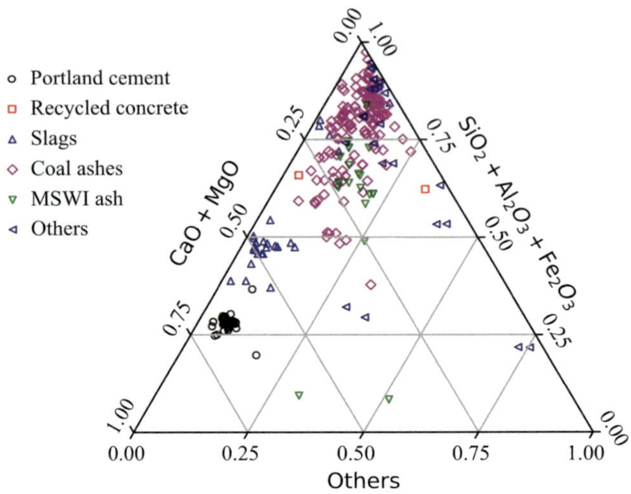

FIGURE 10.9 Wide compositional variation of various potential feedstocks.

sampling and characterization, database curation, and the application of AI. Achieving this requires a collaborative alliance to standardize sampling and characterization methods, as well as data structures that unify how relevant data should be recorded and reported for public access and use.

The development of a reliable supply chain for CO_2 utilization technologies involves logistical complexities in capturing, transporting, and integrating CO_2 into production processes, as well as the logistics of industrial solid wastes. Given that the market value of cement and concrete is in the range of \$50–150 per metric ton, long-distance transportation of CO_2 and solid waste feedstock is economically infeasible. Techno-economic analysis (TEA) is a case-by-case need to balance cost efficiency with scalability. Establishing regional hubs that combine CO_2 and solid waste sources, processing facilities, and end-use markets may help optimize supply chains and reduce costs, making these technologies more commercially viable.

Life cycle assessment (LCA) is critical for evaluating the environmental benefits of carbon utilization technologies. However, challenges arise in accurately modeling the full life cycle, including CO_2 capture, transportation, and incorporation into cement and concrete. Variability in data, assumptions, and boundary conditions can lead to inconsistent results, complicating the comparison of technologies. Standardized LCA methodologies and comprehensive datasets are needed to provide reliable insights and guide sustainable development. These are also needed to support the development of Environmental Product Declarations (EPDs) that are being accepted by the cement and concrete industry and federal agencies to produce and select "low-carbon" materials.

Policy and regulatory frameworks play a crucial role in fostering carbon utilization technologies, but inconsistent or insufficient regulations can hinder progress [226]. Challenges include the lack of clear guidelines for measuring, reporting, and verifying (MRV) carbon sequestration in construction materials, as well as limited financial incentives to promote adoption. To facilitate large-scale deployment as soon as possible to curb climate change, policymakers will need to establish supportive regulations, including carbon pricing, tax credits, and funding for research and development, to encourage investment and market adoption of these innovative technologies.

10.7 CONCLUSION

The integration of CO_2 utilization into cement and concrete production has the potential to revolutionize the construction industry by reducing emissions, enhancing material performance, and contributing to global sustainability goals. Various strategies can be implemented to promote such CO_2 utilization, including (1) using CO_2 in the production of concrete feedstock materials (e.g., water, supplementary cementitious materials, alternative binders, aggregate, and admixtures); (2) injecting CO_2 into concrete during mixing; and (3) using CO_2 to cure prefabricated concrete products. However, technical, economic, and political challenges still exist, hindering large-scale applications of relevant innovative technologies in practice. Continued innovation and collaboration across academia, industry, and government will be essential to overcome existing barriers and unlock the full potential of CO_2 utilization in cement and concrete.

REFERENCES

[1] F. Belaïd, "How does concrete and cement industry transformation contribute to mitigating climate change challenges?," *Resources, Conservation & Recycling Advances*, vol. 15, p. 200084, 2022.

[2] Z. Li, X. Zhou, H. Ma, and D. Hou, *Advanced Concrete Technology*, 2nd Edition. Hoboken, NJ: Wiley, 2022.

[3] *Low-Carbon Transportation Materials Grants Program*. Available: https://www.fhwa.dot.gov/lowcarbon/

[4] *Federal Buy Clean Initiative*. Available: https://www.sustainability.gov/buyclean/

[5] M. Zajac, I. Maruyama, A. Iizuka, and J. Skibsted, "Enforced carbonation of cementitious materials," *Cement and Concrete Research*, vol. 174, p. 107285, 2023.

[6] J. F. Young, R. L. Berger, and J. Breese, "Accelerated curing of compacted calcium silicate mortars on exposure to CO_2," *Journal of the American Ceramic Society*, vol. 57, no. 9, pp. 394–397, 1974.

[7] B. Lu, S. Drissi, J. Liu, X. Hu, B. Song, and C. Shi, "Effect of temperature on CO_2 curing, compressive strength and microstructure of cement paste," *Cement and Concrete Research*, vol. 157, p. 106827, 2022.

[8] M. Zajac, L. Irbe, F. Bullerjahn, H. Hilbig, and M. Ben Haha, "Mechanisms of carbonation hydration hardening in Portland cements," *Cement and Concrete Research*, vol. 152, p. 106687, 2022.

[9] R. I. Khan, S. Siddique, and W. Ashraf, "Effects of magnesia in semi-hydraulic and non-hydraulic calcium silicate binders during carbonation curing," *Construction and Building Materials*, vol. 338, p. 127628, 2022.

[10] Z. Liu, H. Zeng, and F. Wang, "Development of high performance carbonatable concrete for steel slag valorization," *Construction and Building Materials*, vol. 291, p. 123317, 2021.

[11] M. Mahoutian, Z. Ghouleh, and Y. Shao, "Carbon dioxide activated ladle slag binder," *Construction and Building Materials*, vol. 66, pp. 214–221, 2014.
[12] X. Cheng et al., "Effect of carbon dioxide mineralization curing on mechanical properties and microstructure of Portland cement–steel slag–granulated blast furnace slag ternary paste," *Construction and Building Materials*, vol. 431, p. 136553, 2024.
[13] B. Song, S. Liu, X. Hu, K. Ouyang, G. Li, and C. Shi, "Compressive strength, water and chloride transport properties of early CO_2-cured Portland cement-fly ash-slag ternary mortars," *Cement and Concrete Composites*, vol. 134, p. 104786, 2022.
[14] Z. Wei et al., "Clinkering-free cementation by fly ash carbonation," *Journal of CO_2 Utilization*, vol. 23, pp. 117–127, 2018.
[15] P. Zhang, W. Liao, A. Kumar, Q. Zhang, and H. Ma, "Characterization of sugarcane bagasse ash as a potential supplementary cementitious material: Comparison with coal combustion fly ash," *Journal of Cleaner Production*, vol. 277, p. 123834, 2020.
[16] Q. Jin, W. Liao, X. Ni, and H. Ma, "Low-grade fly ash in portland cement blends: A decoupling approach to evaluate reactivity and hydration effects," *Cement*, vol. 18, p. 100119, 2024.
[17] W. Liao, H. Ma, H. Sun, Y. Huang, and Y. Wang, "Potential large-volume beneficial use of low-grade fly ash in magnesia-phosphate cement based materials," *Fuel*, vol. 209, pp. 490–497, 2017.
[18] W. O. Alimi, S. K. Adekunle, S. Ahmad, and A. O. Amao, "Carbon dioxide sequestration characteristics of concrete mixtures incorporating high-volume cement kiln dust," *Case Studies in Construction Materials*, vol. 17, p. e01414, 2022.
[19] B. Zhan, C. S. Poon, and C. J. Shi, "Materials characteristics affecting CO_2 curing of concrete blocks containing recycled aggregates," *Cement and Concrete Composites*, vol. 67, pp. 50–59, 2016.
[20] Y. Mu, Z. Liu, and F. Wang, "Comparative study on the carbonation-activated calcium silicates as sustainable binders: Reactivity, mechanical performance, and microstructure," *ACS Sustainable Chemistry & Engineering*, vol. 7, no. 7, pp. 7058–7070, 2019.
[21] J. M. Bukowski and R. L. Berger, "Reactivity and strength development of CO_2 activated non-hydraulic calcium silicates," *Cement and Concrete Research*, vol. 9, no. 1, pp. 57–68, 1979.
[22] W. Ashraf and J. Olek, "Carbonation behavior of hydraulic and non-hydraulic calcium silicates: Potential of utilizing low-lime calcium silicates in cement-based materials," *Journal of Materials Science*, vol. 51, no. 13, pp. 6173–6191, 2016.
[23] L. Mo, Y. Hao, Y. Liu, F. Wang, and M. Deng, "Preparation of calcium carbonate binders via CO_2 activation of magnesium slag," *Cement and Concrete Research*, vol. 121, pp. 81–90, 2019.
[24] S. Das, B. Souliman, D. Stone, and N. Neithalath, "Synthesis and properties of a novel structural binder utilizing the chemistry of iron carbonation," *ACS Applied Materials & Interfaces*, vol. 6, no. 11, pp. 8295–8304, 2014.
[25] W. Ashraf and J. Olek, "Carbonation activated binders from pure calcium silicates: Reaction kinetics and performance controlling factors," *Cement and Concrete Composites*, vol. 93, pp. 85–98, 2018.
[26] X. Hu, P. He, and C. Shi, "Carbonate binders: Historic developments and perspectives," *Cement and Concrete Research*, vol. 175, p. 107352, 2024.
[27] S. Zhao, Z. Liu, F. Wang, S. Hu, and C. Liu, "Effect of extended carbonation curing on the properties of γ-C2S compacts and its implications on the multi-step reaction mechanism," *ACS Sustainable Chemistry & Engineering*, vol. 9, no. 19, pp. 6673–6684, 2021.

[28] X. Liu, P. Feng, Y. Cai, X. Yu, C. Yu, and Q. Ran, "Carbonation behavior of calcium silicate hydrate (C-S-H): Its potential for CO_2 capture," *Chemical Engineering Journal*, vol. 431, p. 134243, 2022.
[29] X. Liu, P. Feng, Y. Cai, X. Yu, and Q. Liu, "Carbonation behaviors of calcium silicate hydrate (C-S-H): Effects of aluminum," *Construction and Building Materials*, vol. 325, p. 126825, 2022.
[30] W. Ashraf, J. Olek, and S. Sahu, "Phase evolution and strength development during carbonation of low-lime calcium silicate cement (CSC)," *Construction and Building Materials*, vol. 210, pp. 473–482, 2019.
[31] Y. Mu, Z. Liu, F. Wang, and X. Huang, "Effect of barium doping on carbonation behavior of γ-C_2S," *Journal of CO_2 Utilization*, vol. 27, pp. 405–413, 2018.
[32] X. You, X. Hu, P. He, J. Liu, and C. Shi, "A review on the modelling of carbonation of hardened and fresh cement-based materials," *Cement and Concrete Composites*, vol. 125, p. 104315, 2022.
[33] S. Zhao, Z. Liu, Y. Mu, F. Wang, and Y. He, "Effect of chitosan on the carbonation behavior of γ-C_2S," *Cement and Concrete Composites*, vol. 111, p. 103637, 2020.
[34] R. I. Khan, W. Ashraf, and J. Olek, "Amino acids as performance-controlling additives in carbonation-activated cementitious materials," *Cement and Concrete Research*, vol. 147, 2021.
[35] W. J. J. Huijgen and R. N. J. Comans, "*Carbon Dioxide Sequestration by Mineral Carbonation: Literature Review update 2003–2004*," Energy Research Centre of the Netherlands 2005, Available: https://publications.ecn.nl/E/2005/ECN-C-05-022
[36] J. Chang, Y. Fang, and X. Shang, "The role of β-C_2S and γ-C_2S in carbon capture and strength development," *Materials and Structures*, vol. 49, no. 10, pp. 4417–4424, 2016.
[37] Y. Mu, Z. Liu, F. Wang, and X. Huang, "Carbonation characteristics of γ-dicalcium silicate for low-carbon building material," *Construction and Building Materials*, vol. 177, pp. 322–331, 2018.
[38] B. Zhan, C. Poon, and C. Shi, "CO_2 curing for improving the properties of concrete blocks containing recycled aggregates," *Cement and Concrete Composites*, vol. 42, pp. 1–8, 2013.
[39] Z. Liu and W. Meng, "Fundamental understanding of carbonation curing and durability of carbonation-cured cement-based composites: A review," *Journal of CO_2 Utilization*, vol. 44, p. 101428, 2021.
[40] O. Ahmed, S. Ahmad, and S. K. Adekunle, "Carbon dioxide sequestration in cementitious materials: A review of techniques, material performance, and environmental impact," *Journal of CO_2 Utilization*, vol. 83, p. 102812, 2024.
[41] A. Sanna, M. Uibu, G. Caramanna, R. Kuusik, and M. M. Maroto-Valer, "A review of mineral carbonation technologies to sequester CO_2," *Chemical Society Reviews*, vol. 43, no. 23, pp. 8049–8080, 2014.
[42] Y. Liu, J. Zhang, S. Zhang, A. A. Zhang, J. Peng, and Q. Yuan, "Machine learning-guided optimization of coarse aggregate mix proportion based on CO_2 intensity index," *Journal of CO_2 Utilization*, vol. 85, p. 102862, 2024.
[43] M. Ma, M. Chen, T. Zhang, M. Zhang, and X. Cui, "Research on carbonation percentage of carbonated recycled concrete fine aggregate: Experimental investigation and machine learning prediction," *Journal of Sustainable Cement-Based Materials*, pp. 1–22, 2025.
[44] P. Hosseini, "21 - Integration of new technologies with sustainable concrete materials and structures," in *Sustainable Concrete Materials and Structures*, A. Ashour, X. Wang, and B. Han, Eds.: Woodhead Publishing, 2024, pp. 641–667.

[45] V. Q. Tran, H. V. T. Mai, Q. T. To, and M. H. Nguyen, "Machine learning approach in investigating carbonation depth of concrete containing Fly ash," *Structural Concrete*, vol. 24, no. 2, pp. 2145–2169, 2023.

[46] W. Ashraf and J. Olek, "Elucidating the accelerated carbonation products of calcium silicates using multi-technique approach," *Journal of CO_2 Utilization*, vol. 23, pp. 61–74, 2018.

[47] W. Ashraf, J. Olek, and J. Jain, "Microscopic features of non-hydraulic calcium silicate cement paste and mortar," *Cement and Concrete Research*, vol. 100, pp. 361–372, 2017.

[48] W. Ashraf, J. Olek, and N. Tian, "Multiscale characterization of carbonated wollastonite paste and application of homogenization schemes to predict its effective elastic modulus," *Cement and Concrete Composites*, vol. 72, pp. 284–298, 2016.

[49] W. Ashraf and J. Olek, *Nanomechanical Characterization of the Carbonation Process in Calcium Silicate Hydrate*. CRC Press, 2018.

[50] Y. Mu, Z. Liu, and F. Wang, "Study on heat resistance of carbonated γ-C_2S binder: Strength, phase and microstructure evolution," *Construction and Building Materials*, vol. 329, p. 127049, 2022.

[51] X. Xian, D. Zhang, Y. Shao, and S. Zhang, "Evaluation of corrosion resistance of precast reinforced concrete subjected to early-age ambient pressure carbonation curing by accelerated impressed current method," *Journal of Sustainable Cement-Based Materials*, vol. 12, pp. 1–17, 2022.

[52] X. Xian and Y. Shao, "Microstructure of cement paste subject to ambient pressure carbonation curing," *Construction and Building Materials*, vol. 296, p. 123652, 2021.

[53] K. Zhong, Z. Liu, and F. Wang, "Development of CO_2 curable 3D printing materials," *Additive Manufacturing*, vol. 65, p. 103442, 2023.

[54] W. Liu et al., "CO_2 mineral carbonation using industrial solid wastes: A review of recent developments," *Chemical Engineering Journal*, vol. 416, p. 129093, 2021.

[55] S.-Y. Pan et al., "CO_2 mineralization and utilization by alkaline solid wastes for potential carbon reduction," *Nature Sustainability*, vol. 3, no. 5, pp. 399–405, 2020.

[56] I. Walker, R. Bell, and K. Rippy, "Mineralization of alkaline waste for CCUS," *npj Materials Sustainability*, vol. 2, no. 1, 2024.

[57] B. Cantero, M. Bravo, J. de Brito, I. F. Sáez del Bosque, and C. Medina, "Mechanical behaviour of structural concrete with ground recycled concrete cement and mixed recycled aggregate," *Journal of Cleaner Production*, vol. 275, p. 122913, 2020.

[58] A. Gholizadeh-Vayghan, A. Bellinkx, R. Snellings, B. Vandoren, and M. Quaghebeur, "The effects of carbonation conditions on the physical and microstructural properties of recycled concrete coarse aggregates," *Construction and Building Materials*, vol. 257, p. 119486, 2020.

[59] C. Sun, Q. Chen, J. Xiao, and W. Liu, "Utilization of waste concrete recycling materials in self-compacting concrete," *Resources, Conservation and Recycling*, vol. 161, p. 104930, 2020.

[60] L. Likes, A. Markandeya, M. M. Haider, D. Bollinger, J. S. McCloy, and S. Nassiri, "Recycled concrete and brick powders as supplements to Portland cement for more sustainable concrete," *Journal of Cleaner Production*, vol. 364, p. 132651, 2022.

[61] Y. Jiang, Z. Gu, Z. Ma, Y. Gao, P. Shen, and C. S. Poon, "Carbonation behaviors of recycled concrete fines containing supplementary cementitious materials: Compositions, microstructures, and mechanisms," *ACS Sustainable Chemistry & Engineering*, vol. 12, no. 16, pp. 6252–6267, 2024.

[62] G.-H. Cai, Y.-F. Zhou, J.-S. Li, L.-J. Han, and C. S. Poon, "Deep insight into mechanical behavior and microstructure mechanism of quicklime-activated ground granulated blast-furnace slag pastes," *Cement and Concrete Composites*, vol. 134, p. 104767, 2022.

[63] P. Ma, W. Liao, Y. Zhuo, H. Ma, Y. Zhu, and G. Chen, "Characterization of alkali-silica reaction (ASR) products and C-S-H using SWIR spectroscopy for nondestructive detection of ASR," *Construction and Building Materials*, vol. 416, p. 135207, 2024.
[64] Z. Ma, Y. Jiang, P. Shen, and C. S. Poon, "14 - High-volume recycled concrete fines blended cements," in *High-Volume Mineral Admixtures in Cementitious Binders*, D. C. W. Tsang and X. Zhu, Eds.: Woodhead Publishing, 2025, pp. 275–292.
[65] A. Morandeau, M. Thiéry, and P. Dangla, "Investigation of the carbonation mechanism of CH and C-S-H in terms of kinetics, microstructure changes and moisture properties," *Cement and Concrete Research*, vol. 56, pp. 153–170, 2014.
[66] V. Horsakulthai, "Effect of recycled concrete powder on strength, electrical resistivity, and water absorption of self-compacting mortars," *Case Studies in Construction Materials*, vol. 15, p. e00725, 2021.
[67] Y. Jiang, B. Li, S. Liu, J. He, and A. G. Hernandez, "Role of recycled concrete powder as sand replacement in the properties of cement mortar," *Journal of Cleaner Production*, vol. 371, p. 133424, 2022.
[68] H. Mehdizadeh, T.-C. Ling, X. Cheng, and K. H. Mo, "Effect of particle size and CO_2 treatment of waste cement powder on properties of cement paste," *Canadian Journal of Civil Engineering*, vol. 48, no. 5, pp. 522–531, 2020.
[69] G. S. dos Reis, B. G. Cazacliu, R. Artoni, and J.-M. Torrenti, "Effect of the accelerated carbonation treatment on the recycled sand physicochemical characteristics through the rolling carbonation process," *Journal of CO_2 Utilization*, vol. 39, p. 101181, 2020.
[70] C. Liang, N. Lu, H. Ma, Z. Ma, and Z. Duan, "Carbonation behavior of recycled concrete with CO_2-curing recycled aggregate under various environments," *Journal of CO_2 Utilization*, vol. 39, p. 101185, 2020.
[71] X. Fang, D. Xuan, P. Shen, and C. S. Poon, "Fast enhancement of recycled fine aggregates properties by wet carbonation," *Journal of Cleaner Production*, vol. 313, p. 127867, 2021.
[72] R. Cai *et al.*, "Skin effect of chloride ingress in marine concrete: A review on the convection zone," *Construction and Building Materials*, vol. 262, p. 120566, 2020.
[73] M. R. M. Saade, A. Yahia, and B. Amor, "Is crushed concrete carbonation significant enough to be considered as a carbon mitigation strategy?," *Environmental Research Letters*, vol. 17, no. 10, p. 104049, 2022.
[74] M. Zajac, J. Skocek, P. Durdzinski, F. Bullerjahn, J. Skibsted, and M. Ben Haha, "Effect of carbonated cement paste on composite cement hydration and performance," *Cement and Concrete Research*, vol. 134, p. 106090, 2020.
[75] J. Xiao, C. Wang, T. Ding, and A. Akbarnezhad, "A recycled aggregate concrete high-rise building: Structural performance and embodied carbon footprint," *Journal of Cleaner Production*, vol. 199, pp. 868–881, 2018.
[76] M. G. Beltrán, A. Barbudo, F. Agrela, A. P. Galvín, and J. R. Jiménez, "Effect of cement addition on the properties of recycled concretes to reach control concretes strengths," *Journal of Cleaner Production*, vol. 79, pp. 124–133, 2014.
[77] C. Liang, B. Pan, Z. Ma, Z. He, and Z. Duan, "Utilization of CO_2 curing to enhance the properties of recycled aggregate and prepared concrete: A review," *Cement and Concrete Composites*, vol. 105, p. 103446, 2020.
[78] S. Liu et al., "A comparison of liquid-solid and gas-solid accelerated carbonation for enhancement of recycled concrete aggregate," *Cement and Concrete Composites*, vol. 118, p. 103988, 2021.
[79] L. Li, D. Xuan, A. O. Sojobi, S. Liu, and C. S. Poon, "Efficiencies of carbonation and nano silica treatment methods in enhancing the performance of recycled aggregate concrete," *Construction and Building Materials*, vol. 308, p. 125080, 2021.

[80] Y. Pu et al., "Accelerated carbonation technology for enhanced treatment of recycled concrete aggregates: A state-of-the-art review," *Construction and Building Materials*, vol. 282, p. 122671, 2021.
[81] T. Zhang, M. Chen, Y. Wang, and M. Zhang, "Roles of carbonated recycled fines and aggregates in hydration, microstructure and mechanical properties of concrete: A critical review," *Cement and Concrete Composites*, vol. 138, p. 104994, 2023.
[82] Q. Zhou, W. Wang, and T. Noguchi, "Recyclable calcium carbonate-based concrete: Utilizing calcium carbonate to bond recycled concrete fines through an in-situ heterogeneous dual-precipitation approach," *Cement and Concrete Research*, vol. 186, p. 107679, 2024.
[83] T. Zhang, J. Cui, M. Chen, X. Ouyang, and Z. Yan, "Effect of carbonated recycled concrete fines on early-age engineering properties, hydration and microstructure of cement paste under high substitution level and low water-to-binder ratio," *Journal of Building Engineering*, vol. 96, p. 110376, 2024.
[84] Z. Ma et al., "Revealing the connection between carbonation regimes and early pozzolanic reactivity of recycled concrete powder: Impact of composition and microstructure," *Cement and Concrete Research*, vol. 186, p. 107697, 2024.
[85] Z. Ma et al., "Utilizing waste cement for carbon dioxide sequestration and capture: The role of water content on the growth of calcium carbonate," *ACS Sustainable Chemistry & Engineering*, vol. 12, no. 6, pp. 2273–2288, 2024.
[86] Y. Jiang, L. Peng, Z. Ma, J.-X. Lu, P. Shen, and C. S. Poon, "Enhancing the treatment efficiency of recycled concrete fines with aqueous carbonation," *Cement and Concrete Research*, vol. 174, p. 107338, 2023.
[87] P. Shen et al., "Phase assemblance evolution during wet carbonation of recycled concrete fines," *Cement and Concrete Research*, vol. 154, p. 106733, 2022.
[88] B. Lu, C. Shi, J. Zhang, and J. Wang, "Effects of carbonated hardened cement paste powder on hydration and microstructure of Portland cement," *Construction and Building Materials*, vol. 186, pp. 699–708, 2018.
[89] B. Zhan, C. S. Poon, Q. Liu, S. Kou, and C. Shi, "Experimental study on CO_2 curing for enhancement of recycled aggregate properties," *Construction and Building Materials*, vol. 67, pp. 3–7, 2014.
[90] L. Peng et al., "Mechanism underlying early hydration kinetics of carbonated recycled concrete fines-ordinary portland cement (CRCF-OPC) paste," *Cement and Concrete Composites*, vol. 144, p. 105275, 2023.
[91] Y. Zhao et al., "Highly reactive carbonated recycled concrete fines prepared via mechanochemical carbonation: Influence on the early performance of cement composites," *Cement and Concrete Composites*, vol. 152, p. 105636, 2024.
[92] C. Medina, W. Zhu, T. Howind, M. I. Sánchez de Rojas, and M. Frías, "Influence of mixed recycled aggregate on the physical – mechanical properties of recycled concrete," *Journal of Cleaner Production*, vol. 68, pp. 216–225, 2014.
[93] M. V. A. Florea, Z. Ning, and H. J. H. Brouwers, "Activation of liberated concrete fines and their application in mortars," *Construction and Building Materials*, vol. 50, pp. 1–12, 2014.
[94] D. Xuan, B. Zhan, and C. S. Poon, "Development of a new generation of eco-friendly concrete blocks by accelerated mineral carbonation," *Journal of Cleaner Production*, vol. 133, pp. 1235–1241, 2016.
[95] L. Evangelista, M. Guedes, J. de Brito, A. C. Ferro, and M. F. Pereira, "Physical, chemical and mineralogical properties of fine recycled aggregates made from concrete waste," *Construction and Building Materials*, vol. 86, pp. 178–188, 2015.

[96] R. Zaharieva, F. Buyle-Bodin, F. Skoczylas, and E. Wirquin, "Assessment of the surface permeation properties of recycled aggregate concrete," *Cement and Concrete Composites*, vol. 25, no. 2, pp. 223–232, 2003.
[97] D. Pedro, J. de Brito, and L. Evangelista, "Structural concrete with simultaneous incorporation of fine and coarse recycled concrete aggregates: Mechanical, durability and long-term properties," *Construction and Building Materials*, vol. 154, pp. 294–309, 2017.
[98] D. Zhang, Z. Ghouleh, and Y. Shao, "Review on carbonation curing of cement-based materials," *Journal of CO_2 Utilization*, vol. 21, pp. 119–131, 2017.
[99] N. Li and C. Unluer, "Enhancement of the wet carbonation of artificial recycled concrete aggregates in seawater," *Cement and Concrete Research*, vol. 175, p. 107387, 2024.
[100] J. Li and E.-H. Yang, "Macroscopic and microstructural properties of engineered cementitious composites incorporating recycled concrete fines," *Cement and Concrete Composites*, vol. 78, pp. 33–42, 2017.
[101] J. Skocek, M. Zajac, and M. Ben Haha, "Carbon capture and utilization by mineralization of cement pastes derived from recycled concrete," *Scientific Reports*, vol. 10, no. 1, p. 5614, 2020.
[102] M. Zajac, J. Skocek, J. Skibsted, and M. Ben Haha, "CO_2 mineralization of demolished concrete wastes into a supplementary cementitious material – a new CCU approach for the cement industry," *RILEM Technical Letters*, vol. 6, pp. 53–60, 2021.
[103] P. J. M. Monteiro, S. A. Miller, and A. Horvath, "Towards sustainable concrete," *Nature Materials*, vol. 16, no. 7, pp. 698–699, 2017.
[104] P. Liu, L. Mo, and Z. Zhang, "Effects of carbonation degree on the hydration reactivity of steel slag in cement-based materials," *Construction and Building Materials*, vol. 370, p. 130653, 2023.
[105] S. Zhang, Z. Ghouleh, A. Mucci, O. Bahn, R. Provençal, and Y. Shao, "Production of cleaner high-strength cementing material using steel slag under elevated-temperature carbonation," *Journal of Cleaner Production*, vol. 342, p. 130948, 2022.
[106] P. Nielsen, M. A. Boone, L. Horckmans, R. Snellings, and M. Quaghebeur, "Accelerated carbonation of steel slag monoliths at low CO_2 pressure – microstructure and strength development," *Journal of CO_2 Utilization*, vol. 36, pp. 124–134, 2020.
[107] Y. Jiang, T.-C. Ling, C. Shi, and S.-Y. Pan, "Characteristics of steel slags and their use in cement and concrete—A review," *Resources, Conservation and Recycling*, vol. 136, pp. 187–197, 2018.
[108] L. Li, T.-C. Ling, and S.-Y. Pan, "Environmental benefit assessment of steel slag utilization and carbonation: A systematic review," *Science of The Total Environment*, vol. 806, p. 150280, 2022.
[109] B. Xu and Y. Yi, "Treatment of ladle furnace slag by carbonation: Carbon dioxide sequestration, heavy metal immobilization, and strength enhancement," *Chemosphere*, vol. 287, p. 132274, 2022.
[110] Y. Zhang, L. Yu, K. Cui, H. Wang, and T. Fu, "Carbon capture and storage technology by steel-making slags: Recent progress and future challenges," *Chemical Engineering Journal*, vol. 455, 2023.
[111] J. O'Connor et al., "Production, characterisation, utilisation, and beneficial soil application of steel slag: A review," *Journal of Hazardous Materials*, vol. 419, p. 126478, 2021.
[112] W. Gao et al., "Comprehensive utilization of steel slag: A review," *Powder Technology*, vol. 422, p. 118449, 2023.

[113] A. C. P. Martins et al., "Steel slags in cement-based composites: An ultimate review on characterization, applications and performance," *Construction and Building Materials*, vol. 291, p. 123265, 2021.

[114] Z. Chen, R. Li, X. Zheng, and J. Liu, "Carbon sequestration of steel slag and carbonation for activating RO phase," *Cement and Concrete Research*, vol. 139, p. 106271, 2021.

[115] J. Guo, Y. Bao, and M. Wang, "Steel slag in China: Treatment, recycling, and management," *Waste Management*, vol. 78, pp. 318–330, 2018.

[116] L. Mo, S. Yang, B. Huang, L. Xu, S. Feng, and M. Deng, "Preparation, microstructure and property of carbonated artificial steel slag aggregate used in concrete," *Cement and Concrete Composites*, vol. 113, 2020.

[117] P. Liu, J. Zhong, M. Zhang, L. Mo, and M. Deng, "Effect of CO_2 treatment on the microstructure and properties of steel slag supplementary cementitous materials," *Construction and Building Materials*, vol. 309, p. 125171, 2021.

[118] S. Srivastava, M. Cerutti, H. Nguyen, V. Carvelli, P. Kinnunen, and M. Illikainen, "Carbonated steel slags as supplementary cementitious materials: Reaction kinetics and phase evolution," *Cement and Concrete Composites*, vol. 142, p. 105213, 2023.

[119] S. Srivastava, R. Snellings, and P. Cool, "Clinker-free carbonate-bonded (CFCB) products prepared by accelerated carbonation of steel furnace slags: A parametric overview of the process development," *Construction and Building Materials*, vol. 303, p. 124556, 2021.

[120] Z. Chen, R. Li, and J. Liu, "Preparation and properties of carbonated steel slag used in cement cementitious materials," *Construction and Building Materials*, vol. 283, p. 122667, 2021.

[121] X. Zhang, C. Qian, Z. Ma, and F. Li, "Study on preparation of supplementary cementitious material using microbial CO_2 fixation of steel slag powder," *Construction and Building Materials*, vol. 326, p. 126864, 2022.

[122] Y. Rui, C. Qian, X. Zhang, and Z. Ma, "Different carbon treatments for steel slag powder and their subsequent effects on properties of cement-based materials," *Journal of Cleaner Production*, vol. 362, p. 132407, 2022.

[123] K.-W. Chen, S.-Y. Pan, C.-T. Chen, Y.-H. Chen, and P.-C. Chiang, "High-gravity carbonation of basic oxygen furnace slag for CO_2 fixation and utilization in blended cement," *Journal of Cleaner Production*, vol. 124, pp. 350–360, 2016.

[124] G. Liu, K. Schollbach, P. Li, and H. J. H. Brouwers, "Valorization of converter steel slag into eco-friendly ultra-high performance concrete by ambient CO_2 pre-treatment," *Construction and Building Materials*, vol. 280, p. 122580, 2021.

[125] M. Ma, H. Mehdizadeh, M.-Z. Guo, and T.-C. Ling, "Effect of direct carbonation routes of basic oxygen furnace slag (BOFS) on strength and hydration of blended cement paste," *Construction and Building Materials*, vol. 304, p. 124628, 2021.

[126] Y. Li, H. Mehdizadeh, K. H. Mo, and T.-C. Ling, "Co-utilization of aqueous carbonated basic oxygen furnace slag (BOFS) and carbonated filtrate in cement pastes considering reaction duration effect," *Cement and Concrete Composites*, vol. 138, p. 104988, 2023.

[127] L. Li, T. Chen, and X. Gao, "Effects of superimposed carbonation synergy on BOFS cement-based materials," *Cement and Concrete Composites*, p. 105008, 2023.

[128] X. Huang, J. Zhang, and L. Zhang, "Accelerated carbonation of steel slag: A review of methods, mechanisms and influencing factors," *Construction and Building Materials*, vol. 411, 2024.

[129] N. Zhang, G. Deng, W. Liao, H. Ma, and C. Hu, "Aqueous carbonation of steel slags: A comparative study on mechanisms," *Cement and Concrete Composites*, vol. 155, p. 105838, 2025.

[130] ASTM C1157-08a. *Standard Performance Specification for Hydraulic Cement.*
[131] G. Deng, Y. He, L. Lu, F. Wang, and S. Hu, "Comparison between fly ash and slag slurry in various alkaline environments: Dissolution, migration, and coordination state of aluminum," *ACS Sustainable Chemistry & Engineering*, vol. 9, p. 12109–12119, 2021.
[132] G. Deng, Y. He, L. Lu, and S. Hu, "Evolution of aluminate hydrate phases in fly ash-cement system under the sulfate conditions," *Construction and Building Materials*, vol. 252, 2020.
[133] C. Siriruang, P. Toochinda, P. Julnipitawong, and S. Tangtermsirikul, "CO_2 capture using fly ash from coal fired power plant and applications of CO_2-captured fly ash as a mineral admixture for concrete," *Journal of Environmental Management*, vol. 170, pp. 70–78, April 1 2016.
[134] T. Chen, M. Bai, and X. Gao, "Carbonation curing of cement mortars incorporating carbonated fly ash for performance improvement and CO_2 sequestration," *Journal of CO_2 Utilization*, vol. 51, 2021.
[135] G. Deng et al., "Pozzolanic reactivity of carbonated high-calcium fly ash: A mechanism study," *Construction and Building Materials*, vol. 446, 2024.
[136] S.-L. Pei, S.-Y. Pan, X. Gao, Y.-K. Fang, and P.-C. Chiang, "Efficacy of carbonated petroleum coke fly ash as supplementary cementitious materials in cement mortars," *Journal of Cleaner Production*, vol. 180, pp. 689–697, 2018.
[137] A. Su, T. Chen, X. Gao, Q. Li, and L. Qin, "Effect of carbonation curing on durability of cement mortar incorporating carbonated fly ash subjected to freeze-thaw and sulfate attack," *Construction and Building Materials*, vol. 341, 2022.
[138] I. Mehdipour and K. H. Khayat, "Effect of particle-size distribution and specific surface area of different binder systems on packing density and flow characteristics of cement paste," *Cement and Concrete Composites*, vol. 78, pp. 120–131, 2017.
[139] Y. Liu, Q. Yuan, Y. Tan, and M. J. Garba, "Comparative analysis and modification of evaluation methods for the pozzolanic reactivity of natural volcanic powders," *Journal of Building Engineering*, vol. 78, 2023.
[140] J. Yoon, K. Jafari, R. Tokpatayeva, S. Peethamparan, J. Olek, and F. Rajabipour, "Characterization and quantification of the pozzolanic reactivity of natural and non-conventional pozzolans," *Cement and Concrete Composites*, vol. 133, 2022.
[141] J. Skibsted and R. Snellings, "Reactivity of supplementary cementitious materials (SCMs) in cement blends," *Cement and Concrete Research*, vol. 124, 2019.
[142] T. Matschei, B. Lothenbach, and F. P. Glasser, "The role of calcium carbonate in cement hydration," *Cement and Concrete Research*, vol. 37, no. 4, pp. 551–558, 2007.
[143] B. Lothenbach and M. Zajac, "Application of thermodynamic modelling to hydrated cements," *Cement and Concrete Research*, vol. 123, 2019.
[144] B. Lothenbach, G. Le Saout, E. Gallucci, and K. Scrivener, "Influence of limestone on the hydration of Portland cements," *Cement and Concrete Research*, vol. 38, no. 6, pp. 848–860, 2008.
[145] C. Hu, Y. Ruan, S. Yao, F. Wang, Y. He, and Y. Gao, "Insight into the evolution of the elastic properties of calcium-silicate-hydrate (C-S-H) gel," *Cement and Concrete Composites*, vol. 104, 2019.
[146] D. Hou et al., "Insights on the molecular structure evolution for tricalcium silicate and slag composite: From 29Si and 27Al NMR to molecular dynamics," *Composites Part B: Engineering*, vol. 202, 2020.
[147] A. Coppola, F. Scala, and M. Azadi, "Direct dry carbonation of mining and industrial wastes in a fluidized bed for offsetting carbon emissions," *Processes*, vol. 10, no. 3, p. 582, 2022.

[148] R. R. T. Dananjayan, P. Kandasamy, and R. Andimuthu, "Direct mineral carbonation of coal fly ash for CO_2 sequestration," *Journal of Cleaner Production*, vol. 112, pp. 4173–4182, 2016.
[149] W. Liu et al., "CO_2 sequestration by direct gas–solid carbonation of fly ash with steam addition," *Journal of Cleaner Production*, vol. 178, pp. 98–107, 2018.
[150] A. Ćwik, I. Casanova, K. Rausis, N. Koukouzas, and K. Zarębska, "Carbonation of high-calcium fly ashes and its potential for carbon dioxide removal in coal fired power plants," *Journal of Cleaner Production*, vol. 202, pp. 1026–1034, 2018.
[151] Q. Yuan, G. Yang, Y. Zhang, T. Wang, J. Wang, and C. E. Romero, "Supercritical CO_2 coupled with mechanical force to enhance carbonation of fly ash and heavy metal solidification," *Fuel*, vol. 315, p. 123154, 2022.
[152] N. L. Ukwattage, P. G. Ranjith, M. Yellishetty, H. H. Bui, and T. Xu, "A laboratory-scale study of the aqueous mineral carbonation of coal fly ash for CO_2 sequestration," *Journal of Cleaner Production*, vol. 103, pp. 665–674, 2015.
[153] L. Ji et al., "Feasibility and mechanism of an amine-looping process for efficient CO_2 mineralization using alkaline ashes," *Chemical Engineering Journal*, vol. 430, p. 133118, 2022.
[154] T.-L. Chen, S.-L. Pei, and P.-C. Chiang, "Integrated leaching–carbonation kinetic model on CO_2 mineralization of alkaline solid wastes in a high-gravity rotating packed bed," *Reaction Chemistry & Engineering*, vol. 5, no. 10, pp. 1929–1938, 2020.
[155] A. Ebrahimi, M. Saffari, D. Milani, A. Montoya, M. Valix, and A. Abbas, "Sustainable transformation of fly ash industrial waste into a construction cement blend via CO_2 carbonation," *Journal of Cleaner Production*, vol. 156, pp. 660–669, 2017.
[156] E. Miao et al., "Kinetic analysis on CO_2 sequestration from flue gas through direct aqueous mineral carbonation of circulating fluidized bed combustion fly ash," *Fuel*, vol. 342, p. 127851, 2023.
[157] J. Li, "Municipal solid waste incineration ash-incorporated concrete: One step towards environmental justice," *Buildings*, vol. 11, no. 11, 2021.
[158] A. P. Bayuseno, "Mineral phases in raw and processed municipal waste incineration residues," Ruhr-Universität Bochum, Universitätsbibliothek, 2006.
[159] Q. Xue, Y. Ji, Z. Ma, Z. Zhang, and Z. Xu, "Solidification mechanism of Pb and Cd in S2−-enriched alkali-activated municipal solid waste incineration fly ash," *Materials*, vol. 16, no. 10, 2023.
[160] P. Ren, T.-C. Ling, and K. H. Mo, "CO_2 pretreatment of municipal solid waste incineration fly ash and its feasible use as supplementary cementitious material," *Journal of Hazardous Materials*, vol. 424, p. 127457, 2022.
[161] S. Arickx, T. Van Gerven, and C. Vandecasteele, "Accelerated carbonation for treatment of MSWI bottom ash," *Journal of Hazardous Materials*, vol. 137, no. 1, pp. 235–243, 2006.
[162] J.-G. Jiang, X.-J. Du, M.-Z. Chen, and C. Zhang, "Continuous CO_2 capture and MSWI fly ash stabilization, utilizing novel dynamic equipment," *Environmental Pollution*, vol. 157, no. 11, pp. 2933–2938, 2009.
[163] J. Chen et al., "Study on the accelerated carbonation of MSWI fly ash under ultrasonic excitation: CO_2 capture, heavy metals solidification, mechanism and geochemical modelling," *Chemical Engineering Journal*, vol. 450, p. 138418, 2022.
[164] X. Shen, H. He, C. He, B. Li, W. Luo, and P. Ren, "Low-carbon blended cement containing wet carbonated municipal solid waste incineration fly ash and mechanically activated coal fly ash," *Case Studies in Construction Materials*, vol. 21, p. e03671, 2024.

[165] T. Van Gerven, E. Van Keer, S. Arickx, M. Jaspers, G. Wauters, and C. Vandecasteele, "Carbonation of MSWI-bottom ash to decrease heavy metal leaching, in view of recycling," *Waste Management*, vol. 25, no. 3, pp. 291–300, 2005.

[166] E. Rendek, G. Ducom, and P. Germain, "Carbon dioxide sequestration in municipal solid waste incinerator (MSWI) bottom ash," *Journal of Hazardous Materials*, vol. 128, no. 1, pp. 73–79, 2006.

[167] S. Monkman, Y. Sargam, and L. Raki, "Comparing the effects of in-situ nano-calcite development and ex-situ nano-calcite addition on cement hydration," *Construction and Building Materials*, vol. 321, p. 126369, 2022.

[168] J. Chen *et al.*, "Resource utilization of ultrasonic carbonated MSWI fly ash as cement aggregates: Compressive strength, heavy metal immobilization, and environmental-economic analysis," *Chemical Engineering Journal*, vol. 472, p. 144860, 2023.

[169] O. López-Zaldívar, P. L. Mayor-Lobo, F. Fernández-Martínez, and F. Hernández-Olivares, "Improved cement mortars by addition of carbonated fly ash from solid waste incinerators," *Materiales de Construcción*, vol. 65, no. 319, p. e062, 2015.

[170] A. Leemann, F. Winnefeld, B. Münch, and F. Läng, "Carbonated wollastonite – An effective supplementary cementitious material?," *Journal of Microscopy*, vol. 286, no. 2, pp. 120–125, 2022.

[171] S. A. Walling and J. L. Provis, "Magnesia-based cements: A journey of 150 years, and cements for the future?," *Chemical Reviews*, vol. 116, no. 7, pp. 4170–4204, 2016.

[172] S. Deng, P. Ren, Y. Jiang, X. Shao, and T.-C. Ling, "Use of CO_2-active BOFS binder in the production of artificial aggregates with waste concrete powder," *Resources, Conservation and Recycling*, vol. 182, 2022.

[173] W. C. Tang, P. C. Ryan, H. Z. Cui, and W. Liao, "Properties of self-compacting concrete with recycled coarse aggregate," *Advances in Materials Science and Engineering*, vol. 2016, pp. 1–11, 2016.

[174] A. K. Shuvo, P. K. Sarker, and F. U. A. Shaikh, "Efficacy of various accelerated carbonation techniques to improve recycled concrete aggregates: A comprehensive review," *Journal of Building Engineering*, vol. 95, no. 15, p. 110257, 2024.

[175] V. W. Tam, A. Butera, K. N. Le, and W. Li, "Utilising CO_2 technologies for recycled aggregate concrete: A critical review," *Construction and Building Materials*, vol. 250, p. 118903, 2020.

[176] P. Ren, T.-C. Ling, and K. H. Mo, "Recent advances in artificial aggregate production," *Journal of Cleaner Production*, vol. 291, 2021.

[177] Y. Jiang and T.-C. Ling, "Production of artificial aggregates from steel-making slag: Influences of accelerated carbonation during granulation and/or post-curing," *Journal of CO_2 Utilization*, vol. 36, pp. 135–144, 2020.

[178] M. Xu, Q. Zhao, L. Mo, and B. Chen, "Preparation of carbon-negative artificial lightweight aggregates by carbonating sintered red mud (SRM): CO_2 sequestration, microstructure and performance," *Journal of Cleaner Production*, vol. 469, p. 143207, 2024.

[179] Z. X. Chen, N. T. Zhang, S. R. Yan, J. Fish, and S. H. Chu, "CO_2 mineralization into waste-valorized lightweight artificial aggregate," *Construction and Building Materials*, vol. 409, p. 133861, 2023.

[180] M. Xu, Y. Zhang, S. Yang, L. Mo, and P. Liu, "Effects of internal CO_2 curing provided by biochar on the carbonation and properties of steel slag-based artificial lightweight aggregates (SALAs)," *Cement and Concrete Composites*, vol. 142, p. 105197, 2023.

[181] A. Wang, P. Ren, Q. Zeng, and T.-C. Ling, "Performance investigation and optimization of the granulation-CO_2 concentration for the production of high-strength BOFS aggregates," *Journal of CO_2 Utilization*, vol. 64, p. 102160, 2022.

[182] H. Liu and Q. Li, "Preparation of artificial aggregates from concrete slurry waste and waste brick masonry powder: CO_2 uptake and performance evaluation," *Construction and Building Materials*, vol. 382, p. 131356, 2023.
[183] Y. X. Chen, G. Liu, K. Schollbach, and H. J. H. Brouwers, "Development of cement-free bio-based cold-bonded lightweight aggregates (BCBLWAs) using steel slag and miscanthus powder via CO_2 curing," *Journal of Cleaner Production*, vol. 322, p. 129105, 2021.
[184] X. Zhang, C. Qian, and D. Xie, "Preparation of artificial aggregate using waste concrete powder and CO_2 fixed by microorganisms," *Clean Technologies and Environmental Policy*, vol. 24, no. 5, pp. 1453–1467, 2022.
[185] M. Xu and L. Mo, "Towards sustainable artificial aggregate production using industrial waste and CO_2: A comprehensive review," *Journal of Building Engineering*, vol. 97, 2024.
[186] S. Yang, L. Mo, and M. Deng, "Effects of ethylenediamine tetra-acetic acid (EDTA) on the accelerated carbonation and properties of artificial steel slag aggregates," *Cement and Concrete Composites*, vol. 118, p. 103948, 2021.
[187] X. Fu, A. Guerini, D. Zampini, and A. F. Rotta Loria, "Storing CO_2 while strengthening concrete by carbonating its cement in suspension," *Communications Materials*, vol. 5, no. 1, p. 109, 2024.
[188] T. Y. Lo, W. Liao, C. K. Wong, and W. Tang, "Evaluation of carbonation resistance of paint coated concrete for buildings," *Construction and Building Materials*, vol. 107, pp. 299–306, 2016.
[189] S. H. Han, Y. Jun, and J. H. Kim, "The use of monoethanolamine CO_2 solvent for the CO_2 curing of cement-based materials," *Sustainable Materials and Technologies*, vol. 35, p. e00559, 2023.
[190] S. A. Angus, B. Armstrong, K. M. de Reuck, *International Thermodynamic Tables of the Fluid State*. New York: Pergamon Press: Oxford, 1973.
[191] H. Klopfer, "The carbonation of external concrete and the control of it," *Bautenschutz Bausanier*, vol. 1, no. 3, pp. 86–97, 1978.
[192] C. A. García-González et al., "Modification of composition and microstructure of Portland cement pastes as a result of natural and supercritical carbonation procedures," *Industrial & Engineering Chemistry Research*, vol. 45, no. 14, pp. 4985–4992, 2006.
[193] A. D. Dukes, "Measuring the Henry's law constant for carbon dioxide and water with uv-visible absorption spectroscopy," *Analytical Sciences*, vol. 36, no. 8, pp. 971–975, 2020.
[194] M. Poulain et al., "Experimental measurements of carbon dioxide solubility in Na–Ca–K–Cl solutions at high temperatures and pressures up to 20 MPa," *Journal of Chemical & Engineering Data*, vol. 64, no. 6, pp. 2497–2503, 2019.
[195] V. F. Zhuravlev, I. G. Lesokhin, and R. G. Tempelman, "Kinetics of reactions in the formation of aluminates and the contribution of mineralizers to the process," *Journal of Applied Chemistry of the USSR*, vol. 21, no. 9, p. 887, 1948.
[196] K. Vance, G. Falzone, I. Pignatelli, M. Bauchy, M. Balonis, and G. Sant, "Direct carbonation of $Ca(OH)_2$ using liquid and supercritical CO_2: Implications for carbon-neutral cementation," *Industrial & Engineering Chemistry Research*, vol. 54, no. 36, pp. 8908–8918, 2015.
[197] Y. Liu et al., "Experimental Investigation of How Supercritical CO_2 Changes the Chemistry and Microstructure of Cements," in *SPE Annual Technical Conference and Exhibition*, 2024, vol. SPE Annual Technical Conference and Exhibition, New Orleans, Louisiana, USA.

[198] S. F. Santos, R. Schmidt, A. E. F. S. Almeida, G. H. D. Tonoli, and H. Savastano, "Supercritical carbonation treatment on extruded fibre–cement reinforced with vegetable fibres," *Cement and Concrete Composites*, vol. 56, pp. 84–94, 2015.

[199] M.-S. Kim, S.-R. Sim, and D.-W. Ryu, "Supercritical CO_2 curing of resource-recycling secondary cement products containing concrete sludge waste as main materials," *Materials*, vol. 15, no. 13, p. 4581, 2022.

[200] Y. Gu, X. Wang, Z. Xu, S. Wei, Q. Yuan, and Y. Zhang, "Effect of supercritical CO_2 carbonation parameters on heavy metal solidification and environmental stability of coal fly ash," *Journal of Material Cycles and Waste Management*, vol. 26, no. 5, pp. 3047–3058, 2024.

[201] W. Gao et al., "Industrial carbon dioxide capture and utilization: State of the art and future challenges," *Chemical Society Reviews*, vol. 49, no. 23, pp. 8584–8686, 2020.

[202] D. Suescum-Morales, J. M. Fernández-Rodríguez, and J. R. Jiménez, "Use of carbonated water to improve the mechanical properties and reduce the carbon footprint of cement-based materials with recycled aggregates," *Journal of CO_2 Utilization*, vol. 57, p. 101886, 2022.

[203] J. Kwasny, P. A. M. Basheer, and M. Russell, "CO_2 Sequestration in Cement-based Materials During Mixing Process Using Carbonated Water and Gaseous CO2," Purdue University Libraries Scholarly Publishing Services.

[204] X. Qian, J. Wang, Y. Fang, and L. Wang, "Carbon dioxide as an admixture for better performance of OPC-based concrete," *Journal of CO_2 Utilization*, vol. 25, pp. 31–38, 2018.

[205] B. Tan, "Effect of in-situ formed nano-calcite particles on hydration and properties of cement-based materials," M.S. in Civil Engineering, Department of Civil, Architectural and Environmental Engineering, Missouri University of Science and Technology, 2022.

[206] X. Qian et al., "Nanosilica in-situ produced with sodium silicate as a performance enhancing additive for concretes," *Cement and Concrete Composites*, vol. 142, p. 105198, 2023.

[207] B. Šavija and M. Luković, "Carbonation of cement paste: Understanding, challenges, and opportunities," *Construction and Building Materials*, vol. 117, pp. 285–301, 2016.

[208] CarbonCure Technologies Inc., "Ready mix technology trial results: Technical note," Dartmouth, NS, Canada.

[209] CarbonCure Technologies Inc. (January 26). *CarbonCure Precast: Frequently Asked Questions for Structural Engineers*. Available: https://www.carboncure.com/wp-content/uploads/2023/05/CC_FAQ-Precast-Engineers_FINAL.pdf

[210] S. Monkman, P. A. Kenward, G. Dipple, M. Macdonald, and M. Raudsepp, "Activation of cement hydration with carbon dioxide," *Journal of Sustainable Cement-Based Materials*, vol. 7, no. 3, pp. 160–181, 2018.

[211] X. Liu, L. Chen, A. Liu, and X. Wang, "Effect of nano-$CaCO_3$ on properties of cement paste," *Energy Procedia*, vol. 16, pp. 991–996, 2012.

[212] G. Land and D. Stephan, "Controlling cement hydration with nanoparticles," *Cement and Concrete Composites*, vol. 57, pp. 64–67, 2015.

[213] M. Thomas, "Impact of CO_2 utilization in fresh concrete on corrosion of steel reinforcement," CarbonCure Technologies Inc., Dartmouth, NS, Canada 2019.

[214] X. Cheng, M. Wu, J. Li, W. Wang, N. Mitsuzaki, and Z. Chen, "Conversion of carbon dioxide into solid carbon materials – A mini review," *Catalysis Science & Technology*, vol. 13, no. 13, pp. 3891–3900, 2023.

[215] S. Monkman, M. Macdonald, and L. Sutter, "Beneficiation of concrete wash water with carbon dioxide," *Materials and Structures*, vol. 54, no. 2, 2021.

[216] M. Li, V. Lin, J. Lynch, and V. C. Li, "Multifunctional carbon black engineered cementitious composites for the protection of critical infrastructure," Springer Netherlands, 2012, pp. 99–106.

[217] M. Ramezani, A. Dehghani, and M. M. Sherif, "Carbon nanotube reinforced cementitious composites: A comprehensive review," *Construction and Building Materials*, vol. 315, p. 125100, 2022.

[218] K. R. Choi, Y.-J. Ahn, and S. Y. Lee, "Bacterial conversion of CO_2 to organic compounds," *Journal of CO_2 Utilization*, vol. 58, p. 101929, 2022.

[219] I. Omae, "Recent developments in carbon dioxide utilization for the production of organic chemicals," *Coordination Chemistry Reviews*, vol. 256, no. 13–14, pp. 1384–1405, 2012.

[220] E. Schuler, M. Demetriou, N. R. Shiju, and G. J. M. Gruter, "Towards sustainable oxalic acid from CO_2 and biomass," *ChemSusChem*, vol. 14, no. 18, pp. 3636–3664, 2021.

[221] Y. Fang et al., "Densifying hydration products of alite by a bio-inspired admixture," *Materials & Design*, vol. 225, p. 111490, 2023.

[222] Z. Luo et al., "Preparation and characterization of ferrous oxalate cement—A novel acid-base cement," *Journal of the American Ceramic Society*, vol. 104, no. 2, pp. 1120–1131, 2021.

[223] Y. Huang, A. T. M. Marsh, Z. Yue, S. Krishnan, S. Adu-Amankwah, and S. A. Bernal, "Hybrid organic–inorganic blast furnace slag binders activated with alkali acetates," *ACS Omega*, vol. 9, no. 33, pp. 35888–35905, 2024.

[224] D. M. Iqbal, L. S. Wong, and S. Y. Kong, "Bio-cementation in construction materials: A review," *Materials*, vol. 14, no. 9, p. 2175, 2021.

[225] M. Khanjani, D. J. Westenberg, A. Kumar, and H. Ma, "Tuning polymorphs and morphology of microbially induced calcium carbonate: Controlling factors and underlying mechanisms," *ACS Omega*, vol. 6, no. 18, pp. 11988–12003, 2021.

[226] M. Fikru, T. Shen, J. Brodmann, and H. Ma, "A strategic insight into the market for carbon management capacity," *Sustainable Futures*, vol. 8, p. 100374, 2024.

11 Low-Carbon Materials and Practices for Asphalt Pavement Decarbonization

Sepehr Mohammadi and Zhanping You

11.1 INTRODUCTION

The transportation industry, as a major contributor to global greenhouse gas (GHG) emissions and energy consumption, has emerged as a critical focus area for reducing environmental impacts. Efforts to lower the carbon footprint across various aspects of transportation, including the use of low-carbon materials in asphalt pavements, are increasingly recognized as critical for achieving sustainability goals. In recent years, the asphalt industry has significantly increased its emphasis on utilizing green additives and low-carbon materials to improve sustainability and minimize environmental impact and carbon footprint associated with paving materials and operations. To effectively address these environmental challenges, it is first necessary to understand the various stages of the asphalt pavement life cycle including production, construction, maintenance, use, end-of-life, and their respective contributions to greenhouse gas emissions. This understanding paves the way for the adoption of innovative materials and sustainable practices that can significantly reduce the carbon footprint across each life cycle stage for asphalt pavements. National Asphalt Pavement Association (NAPA) reported that reducing carbon footprint of asphalt pavements requires a holistic understanding of GHG emissions across the whole pavement life cycle. Integrating low-carbon materials with technologies that enhance pavement longevity presents a significant opportunity to lower embodied carbon while boosting overall performance (De Pascale et al., 2025; Liu et al., 2024; Rangelov et al., 2021; Shacat et al., 2024).

According to ISO21930, for asphalt pavements, as shown in Figure 11.1, the primary life cycle stages include production (A1–A3), construction (A4, A5), use (B1, B6, B7), maintenance and rehabilitation (B2–B5), and end-of-life (C1–C4) in a cradle-to-grave scope. In a cradle-to-gate scope, A1 stage includes the material supply. In other words, the primary focus of A1 life cycle stage is identifying the sources of GHG emissions resulting from the extraction and processing of raw materials that are the main ingredients of a typical asphalt mix design, such as asphalt binder and aggregates. Next, the raw materials are transported to the asphalt plant for mix

DOI: 10.1201/9781003655442-15

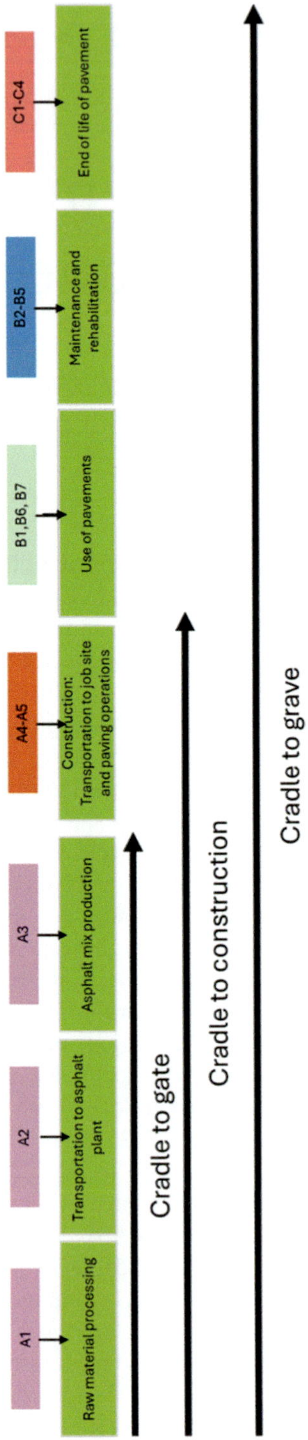

FIGURE 11.1 Main life cycle stages for asphalt pavements (Shacat et al., 2024).

production, which encompasses the A2 stage. The final stage in the cradle-to-gate process, which focuses on asphalt mix production, is A3. On the other hand, the gate-to-construction stages of A4 and A5 include the transportation of the produced asphalt mix from asphalt plant to construction site and paving operations for installing the pavement, respectively (ISO 21930, 2017; Shacat et al., 2024).

The use of pavements, or the B stage, starts with B1, which is for the emissions released during the use of a roadway as a primary result of the fuel combustion of vehicles driving on the pavement. Additionally, B2, B3, and B5 stages mainly encompass the GHG emissions resulting from the main activities for the maintenance, rehabilitation, and reconstruction of the roads, respectively. According to ISO21930, the B4 stage primarily focuses on the emissions from the replacement activities and is not applicable at the product level. As the last two stages of the use phase, B6 and B7 mainly include the emissions associated with the operational energy and water use, respectively. Finally, the end-of-life stages start with C1, which primarily encompasses the activities that are conducted for removal or reclamation of a pavement. On the other hand, in C2 stage, the reclaimed asphalt is transported from a construction site to a waste processing facility. In a waste processing facility, the C3 and C4 stages are related to emissions released during the waste processing and waste disposal activities, respectively (ISO 21930, 2017; Shacat et al., 2024).

The structure of this chapter is designed to provide a comprehensive review of sustainability strategies across key stages of the asphalt pavement life cycle. While the entire life cycle is essential to achieving sustainability, this chapter primarily focuses on the cradle-to-gate stages, encompassing material extraction, production, and transportation; the gate-to-construction stage, which includes the transportation of asphalt mixtures to construction site and construction activities; and the phases of maintenance, rehabilitation, and reconstruction, which are critical for extending pavement life and minimizing long-term environmental impacts. Furthermore, subsequent sections will delve into innovative technologies and practices to enhance sustainability at each stage. For the cradle-to-gate stages, the discussion includes reclaimed and recycled materials, bio-inspired solutions, and energy-efficient asphalt production methods. For the gate-to-construction stages, it covers optimized transportation and efficient equipment usage. Strategies for sustainable maintenance and rehabilitation are also reviewed, providing a holistic perspective on reducing the carbon footprint of asphalt pavements.

11.2 SOURCES OF EMISSIONS IN PRODUCTION STAGES

Cradle-to-gate stages encompass the production stages of asphalt mixtures including processing of raw materials (A1), transportation of raw materials to asphalt plant (A2), and mix production (A3). Examples of A1 stage include the carbon emissions associated with the mining, extraction, and production of aggregates. Another significant source of GHG emissions in the A1 stage is the production of asphalt binder, which involves processes such as crude oil extraction, transportation to refineries, and the refining operations required to convert crude oil into asphalt binder. In addition to processing of the main ingredients, upstream GHG emissions associated with any recycled or reclaimed materials, as well as the additives or modifiers that are incorporated into asphalt pavement, are also included in the A1 impacts. It should be noted that the impacts associated with recycled materials such as reclaimed asphalt pavement (RAP) or recycled asphalt shingles (RAS) are assessed using a cut-off approach. Cut-off approach assumes that the recycled materials enter the current product system without any life cycle burdens, and their system boundaries are typically defined at the point where these materials reach the main processing or storage location. For instance, operations such as milling and initial transportation of RAP are not included as a part of A1 emissions. Instead, only the energy and emissions from transportation of RAP from the central collection or processing facility to the asphalt plant, and the associated processes such as crushing and screening, are accounted for (Mukherjee, 2016; Shacat et al., 2024).

A1 impacts are determined solely by the mix design, making them deterministic. In other words, A1 impacts are not stochastic, meaning that each contractor can control the A1 impacts associated with their mix designs. For example, the blending contents of asphalt binder, aggregates, RAP, or any additives/modifiers are determined based on a specific mix design of a contractor or asphalt mix producer, as well as the agency standards and specifications. This highlights the crucial role of using recycled materials with lower environmental impacts, compared to virgin materials in reducing the carbon footprint of the A1 stage. For example, if a contractor does not fully utilize the maximum allowable percentage of RAP in their mix design that is allowed by agency, they will miss out on the maximum potential reductions in GWP that could be achieved by incorporating RAP into its allowable limit. This highlights the importance of fully utilizing the allowable RAP content or any other recycled material to maximize sustainability benefits, particularly in terms of lowering the environmental impact as measured by GWP in A1 stage (Bhat & Mukherjee, 2019; Miller et al., 2024a, 2024b).

As mentioned earlier, asphalt binder plays a significant role in the A1 stage emissions, contributing 94% of the GHG emissions in raw material extraction and processing and 53% of the total cradle-to-gate (A1-A3) emissions (Shacat et al., 2024). As a result, understanding the environmental impacts of asphalt binder production is crucial for developing strategies to reduce GWP in asphalt mix designs. Reports have indicated that carbon footprint of asphalt binder is primarily driven by crude oil extraction methods, with the highest impacts associated with oil sands and tertiary extraction processes. On the other hand, at the refinery stage, electricity consumption is the largest contributor to emissions, followed by on-site thermal energy generation and direct emissions. Terminal operations also contribute up to 20% of potential environmental impacts, mainly due to thermal energy use and the transportation of asphalt to terminals. These findings are based on an industry-average assessment of asphalt binders, including those with additives such as styrene butadiene styrene (SBS), GTR, and polyphosphoric acid (PPA), as well as binders without additives. The scope of the assessment for the study included raw material sourcing, transportation, refining, terminal operations, and final blending processes, focusing on activities directly associated with asphalt production (Wildnauer et al., 2019).

For A2 stage, the examples of emissions include transportation of the liquid asphalt binder from the refinery to the asphalt plant, or the transportation of aggregates from the production facilities, utilizing various modes of transportation such as trucks, trains, inland barges, or ocean vessels. A2 emissions are entirely dependent on the distance traveled, the mode of transportation, and the availability of the aggregates and other mixed ingredients at the local level.

On the other hand, A3 emissions are derived from the operations of an asphalt plant, primarily resulting from the environmental impacts of fuel and electricity use during asphalt mix production. Examples include operations of on-site machinery and equipment, drying aggregates by burners, heating asphalt binder in storage tanks, storing asphalt mixtures in silos and asphalt binders in tanks, processing RAP and RAS, and additional activities required for incorporating additives or modifiers into asphalt mixtures. Electricity and energy flow values at an asphalt plant can vary depending on its geographical location and are influenced by daily fluctuations in temperature and humidity caused by the day-night cycle (Mukherjee, 2016; Shacat et al., 2024).

In a baseline scenario defined by NAPA and based on Mukherjee's study, in a typical asphalt mixture that contains 95% aggregates and 5% asphalt binder, when the transportation distances between the asphalt plant and aggregate sources and asphalt binder refinery are 21.5% and 3.9%, respectively, total cradle-to-gate GHG emissions amount to 53.7 kg CO_2 equivalent per ton of mix. This value is distributed across raw material extraction and processing (A1, 57%), transportation of materials (A2, 5%), and asphalt mix production (A3, 38%). Moreover, asphalt binder has been identified as the primary contributor to emissions in the A1 stage, accounting for 94% of GHG emissions and 53% of the total cradle-to-gate emissions (A1–A3), whereas aggregates contribute only 6% in A1. Even though the impact of transportation emissions may seem minor in cradle-to-gate stages, they can become significant when aggregates are acquired from long distances. In summary, NAPA's report emphasizes the critical role of asphalt binder in reducing GWP, the potential influence of transportation distances, and the need to enhance production efficiency to minimize environmental impacts (Miller et al., 2024a, 2024b; Shacat et al., 2024). Different low-carbon materials and practices, alongside related case studies for production stages, will be reviewed in the upcoming sections.

11.3 SOURCES OF EMISSIONS IN CONSTRUCTION, MAINTENANCE, AND REHAB STAGES

GHG emissions during the construction stage of asphalt pavements are primarily associated with two activities: the fuel consumed during the transportation of materials to the construction site (A4) and the fuel used in paving operations (A5). Compared to the cradle-to-gate emissions (A1–A3), the emissions from construction activities are relatively low, typically contributing around 10% or less of the overall environmental footprint. Carbon emissions in paving operations primarily stem from diesel combustion in paving machinery such as pavers and rollers. Key factors influencing the emissions in gate-to-construction stages include the number and type of equipment being used in the paving train, engine sizes, type of fuel used, energy efficiency, mechanical condition of the equipment, paving width and thickness, type of the mixture being paved, and hours of operation. Although GHG emissions for

A4–A5 stages are smaller compared to mix production, improving hauling and trucking activities and energy efficiency in construction equipment can help to further reduce the environmental impacts of these stages (Cass & Mukherjee, 2011; Peng et al., 2017; Gatiganti et al., 2024; Shacat et al., 2024).

Other studies that also included the construction stages of the flexible pavements have indicated that material production, which includes raw material extraction and resource supply in the cradle-to-gate stages (A1–A3), accounts for the majority of impacts in most environmental categories, largely driven by the significant contribution of asphalt binder as explained in previous sections. Thus, the environmental impacts of the construction stages (A4–A5) are comparatively lower, highlighting the substantial influence of material production on the overall life cycle impacts of asphalt pavements (Gulotta et al., 2019).

On the other hand, GHG emissions during the maintenance and rehab activities arise from multiple sources, which include material production and pavement construction (similar to A1–A5 stages, but in smaller scale and with lower quantities of materials), temporary infrastructure preparation for traffic management, and emissions from traffic mainly caused by work zone congestion. Additionally, emissions that occur during pavement removal activities, such as milling operations, can be a topic of investigation for the end-of-life stages. Generally, the main activities for maintenance, rehab, and reconstruction include chip seals, slurry seals, fog seals, cape seals, hot mix asphalt (HMA) patching, mill and overlay, microsurfacing, using stress-absorbing membrane interlayers (SAMIs), cold in-place recycling (CIR), hot in-place recycling (HIR), and full-depth reclamation (FDR). Various practices for reducing emissions across construction, maintenance, and rehab stages, along with relevant case studies, will be discussed in the upcoming sections (Shacat et al., 2024; You et al., 2018).

11.4 DECARBONIZATION STRATEGIES IN PRODUCTION STAGES

The first and most critical strategy for decarbonization in the cradle-to-gate stages, particularly in A1 stage, is the use of recycled materials with lower environmental impacts to replace virgin materials. Among these, RAP and bio-based materials stand out as the most significant contributors to reducing greenhouse gas emissions and resource depletion. Another example includes the incorporation of GTR into asphalt pavements. Additionally, for other production stages, such as A3, technologies including WMA provide significant benefits in terms of emissions reduction. While Table 11.1 presents the key materials examined in this chapter along with a summary of their performance characteristics and GWP reduction benefits, subsequent discussions will explore how each material contributes to sustainability and carbon footprint reduction.

The recycled materials not only lower the carbon footprint but also promote circular economy principles by reusing and repurposing resources. For example, the use of RAP affects the overall GWP of asphalt mixtures by replacing both virgin binders and aggregates. In this case, although the majority of the replaced material is aggregate, the primary GWP reduction comes from the substitution of virgin binder, which is the biggest contributor to A1's carbon footprint. NAPA reported that as each

TABLE 11.1

A Summary of the Emissions Reduction Effects and Performance Benefits of Sustainable Materials

Material	GHG Emission Reduction Effects	Performance Benefits
RAP	Lower A1 GWP intensity compared to virgin materials of aggregates and asphalt binder 5.92 kg of CO_2 eq. reduction per short ton in A1 stage	Higher rutting resistance
GTR	Lower A1 GWP intensity when 8% GTR is compared to virgin asphalt binder	Higher rutting, cracking, and moisture resistance, reduced noise levels
Bio-oil	Lower mixing and compaction temperatures and reduced emissions in A3 stage	Better workability and improved low-temperature cracking resistance
WMA	9–16% reduction in A3 emissions by reducing the temperatures in the range of 30 °F and 50 °F	Improved workability

Source: Miller et al. (2024b) and Shacat et al. (2024).

percentage of RAP is incorporated into asphalt mixtures, there is 5.92 kg of CO_2 equivalent reduction per short ton in A1 emissions (Miller et al., 2024a, 2024b; Shacat et al., 2024).

NAPA's latest report highlights the significant differences in the A1 GWP intensity factors (kg CO_2 equivalent per short ton of each ingredient) for materials used in asphalt mixtures. For upstream materials, neat asphalt binder has a GWP intensity of 573.06, while aggregates range between 1.761 and 3.42, and RAP is only 0.710. These values emphasize the critical role of incorporating recycled materials, such as RAP, in reducing the carbon footprint of asphalt mixtures (Miller et al., 2024b).

When RAP is incorporated as a secondary material into a typical asphalt mix design, the energy required for the additional processing is also accounted for in the A1 stage (Schoenauer et al., 2024). Additionally, while RAP offers significant environmental benefits in the A1 stage by reducing the need for virgin materials, its incorporation has also raised potential concerns in the A3 stage. As an example, to prevent further aging of RAP binder, millings are not heated directly at asphalt plants. Instead, virgin aggregates are overheated before their addition and mixing with the RAP in the drum. As a result, this additional heating process may lead to increased fuel and energy consumption. However, based on actual large production runs, it was reported that although overheating virgin aggregates to incorporate RAP consumes additional energy, this is often counterbalanced by the reduced need to heat virgin aggregates, since a portion of the aggregate mass is replaced by RAP. As a result, the overall energy demand and carbon footprint may remain unchanged (Aurangzeb et al., 2014). Additionally, it was reported that despite higher consumption of fuel and emissions by RAP in A3 stage, the additional carbon footprint is offset by the energy savings resulting from replacing virgin materials in the A1 stage, specifically by replacing virgin asphalt binder (Schoenauer et al., 2024).

Additionally, ensuring the consistent quality and performance of the final asphalt mix when higher blending contents of RAP are incorporated, particularly in terms of aging, can present additional challenges. In this context, bio-based rejuvenators play a crucial role. Bio-based rejuvenators offer renewable and environmentally sustainable alternatives to traditional petroleum-based additives, making them a responsible choice for sustainable pavement construction. Key benefits include their biodegradability, lower carbon footprint, and reduced dependency on finite fossil fuel sources. These materials are often derived from renewable feedstocks like plants and biomass, and they are often non-toxic and safer for both workers and the environment. Popular examples include waste vegetable oils such as waste cooking oils, waste oils derived from wood, and food waste bio-oils (Abdalla et al., 2022; Cavalli et al., 2024; Ji et al., 2017; Zhang et al., 2018). Bio-based rejuvenators not only enhance the performance and durability of asphalt mixtures with high blending contents of RAP but also can further contribute to reducing cradle-to-gate GHG emissions. NAPA reported that since asphalt mixtures are rarely combusted at the end of their life cycle, the biogenic carbon in asphalt additives, especially bio-based rejuvenators, is considered to remain permanently sequestered (Shacat et al., 2024). In terms of performance, bio-rejuvenators derived from waste wood have proven to enhance the performance of aged asphalt binders by reducing stiffness and increasing viscous components. The high concentration of light compounds in bio-oil balances the chemical composition of aged asphalt binder. Effectively restoring its rutting behavior as well as the resistance against fatigue and cold-weather cracking (Zhang et al., 2018).

Similarly, rejuvenators derived from waste cooking oils have been identified as excellent candidates for restoring the properties of aged asphalt binders, particularly in terms of viscosity. Studies have shown that adding these bio-based rejuvenators can achieve viscosities comparable to those of conventional asphalt binders when no RAP is utilized. Consequently, they address the issue of higher mixing and compaction temperatures associated with increased RAP utilization, which can lead to higher emissions during A3 stage, which is mix production at asphalt plant (Ji et al., 2017). At the same time, it is recommended to utilize the most efficient methods of extraction for bio-based rejuvenators such as pyrolysis and hydrothermal liquefaction, especially in cases when mass production is intended, to ensure the reduction of carbon footprint during the A1 stage for processing of the raw materials (Abdalla et al., 2022).

In the same way, the application of bio-based rejuvenators has also promoted the utilization of RAS by addressing many negative aspects of utilization, including aged asphalt binder, lack of interactions between RAS particles and asphalt binder, and poor performance. Case studies of bio-rejuvenators derived from swine manure have demonstrated that these bio-based materials can mitigate the stiffening effects associated with higher RAS content, enabling lower mixing and compaction temperatures. As a result, this adjustment facilitates the use of higher percentages of shingles in asphalt mixtures. Additionally, the incorporation of bio-rejuvenators has proved to enhance the low-temperature performance of RAS-modified asphalt, particularly in terms of cold-weather cracking resistance (Oldham et al., 2015).

Bio-oil-modified asphalts, also known as bio-asphalts, offer dual benefits in enhancing sustainability across multiple stages. Most of the asphalt binders currently used in asphalt pavements are petroleum residues that are extracted from crude oil. However, petroleum is increasingly scarce as a non-renewable resource, resulting in a short supply of petroleum asphalt. On the other hand, with the recent continual improvement in oil refinery technology, the quality and quantity of asphalt have downgraded gradually. However, fossil fuels are considered the main source of greenhouse gas emissions and global warming, and asphalt plants are considered as one of the most significant sources of pollution. Many of the researchers in the asphalt industry have been making an effort to reduce the use of petroleum-based asphalt binders in an environmentally friendly manner. One of the approaches to reduce the use of petroleum asphalt is to find alternative materials that can partially or fully replace petroleum-based asphalt (Mohammadi & Solatifar, 2023; Su et al., 2018).

Bio-asphalts are gaining interest because petroleum is a finite resource. Researchers are exploring sustainable alternatives to reduce reliance on virgin petroleum-based asphalt binders in pavement construction. Similar to bio-rejuvenators, the key raw material in the production of bio-asphalt is bio-oil, derived from a variety of biomass sources. Some of the most common sources include waste cooking oil, waste wood, waste engine oil, animal manure, corn stover, and soybean, and fast pyrolysis has been identified as one of the most effective ways of producing bio-oils (Fini et al., 2011; Lei et al., 2018; Li et al., 2019; Mohammadi & Solatifar, 2023; Su et al., 2018; Tokede et al., 2020; Wang et al., 2018; Yang & Suciptan, 2016; Yang et al., 2014; Zhang et al., 2021).

Bio-oils have the potential to replace a significant portion of conventional asphalt binder and, in some cases, even serve as a complete substitute. Thus, the most significant positive impact of using bio-oils is the reduction in the carbon footprint associated with the production of petroleum-based binders. Case studies of bio-asphalt utilizing bio-oils from lignin and swine manure have demonstrated that the cradle-to-gate environmental footprint of bio-asphalt is lower than that of conventional petroleum-based asphalt (Samieadel et al., 2018; Tokede et al., 2020). At the same time, the softening and viscosity-reducing effects of bio-oils have been identified as another aspect of their sustainability. For instance, studies focusing on bio-asphalt derived from waste wood and waste engine oil have indicated that bio-oils can reduce the mixing and compaction temperatures, thus reducing the emissions during asphalt production, which is A3 stage (Mohammadi & Solatifar, 2023; Su et al., 2018; Zhang et al., 2021).

The softening effects of bio-oils could also become beneficial to reduce the emissions during the construction of polymer-modified asphalt. For instance, Asphalt Institute reported that the carbon emissions for SBS-modified asphalt binder with a blending content of 3.5% is 20% higher than conventional asphalt. The higher carbon footprint of SBS-modified asphalt binder also results in a 13% increase in emissions across the cradle-to-gate stages of A1–A3 (Shacat et al., 2024; Wildnauer et al., 2019). On the other hand, when compared to conventional asphalt, polymer-modified asphalt requires higher mixing and compaction temperatures, which could lead to higher energy consumption and GHG emissions during mix production at asphalt plant (A3 stage). However, the findings from the case studies focusing on bio-asphalt alongside polymer-modified asphalt have indicated that composite modification of asphalt binders with both polymers and bio-oils can not only result in a desirable high-temperature performance but also eliminate some of the limitations of the application of polymers due to their high viscosity. For example, bio-modification of SBS-modified asphalt led to a reduction of its viscosity, which is beneficial for construction. It was concluded that the addition of bio-oil derived from waste wood and the increase in its content reduced the mixing and compaction temperatures of SBS-modified asphalt binder. Furthermore, when the bio-oil content was more than 10%, the mixing and compaction temperature of the SBS-modified bio-binder was the same as the base binder. Furthermore, the addition of bio-oils to SBS-modified asphalt led to less temperature sensitivity and enhanced rutting performance when compared to conventional asphalt (Zhang et al., 2017).

Ground tire rubber (GTR) is another recycled material that can offer a sustainable solution for asphalt modification by enabling the reutilization of waste tires in large quantities. This approach not only can reduce the landfilling of waste tires but also enhance the performance of asphalt mixtures. As mentioned earlier, NAPA's report provides valuable insights into the environmental impacts of asphalt mixture ingredients. For an 8% GTR-modified binder, the A1 GWP intensity factor is 558.98 kg CO_2 equivalent per short ton. This is significantly lower than the GWP intensity of neat asphalt binder at 573.06 kg CO_2 equivalent, highlighting the potential of GTR as a sustainable modifier to reduce the reliance on conventional materials (Miller et al., 2024b).

Furthermore, various case studies on the incorporation of waste tires into asphalt, utilizing different forms and methods, have shown promising outcomes in reducing carbon emissions while also enhancing performance. An example of this could be seen in case studies where recycled rubber pellets were incorporated into asphalt in Michigan. The case study demonstrated that even though achieving the desired air void percentage requires more asphalt binder, the higher density of rubber pellet-modified asphalt reduces the total amount of mixtures needed, thereby lowering the environmental impact across the cradle-to-gate stages (Xin et al., 2024).

In another case study, where about 2,270 waste tires were recycled for the reconstruction of the road's surface course, it was concluded that GTR-modified asphalt mix demonstrated better overall performance than the polymer-modified HMA in terms of low-temperature cracking, rutting, and moisture resistance. This study included the dry process for incorporating GTR with a dosage of 10% by total weight of the asphalt binder and performed a comparison to the scenario when SBS polymers were utilized instead of waste tires. As shown in Figure 11.2, the GTR particles were directly mixed with aggregates and then asphalt binders, and the incorporation process included feeding rubber particles into the drying drum, where they were blended with heated aggregates and asphalt to produce the hot mix asphalt (Jin et al., 2023).

In terms of performance, the disc-shaped compact tension (DCT) test results showed that the addition of GTR increased the cold-weather fracture energy between 13 and 16% when compared to polymer-modified HMA. Additionally, in terms of the performance at the asphalt binder level, dynamic shear rheometer (DSR) and asphalt binder cracking device (ABCD) test outcomes showed that the addition of GTR can enhance cold-weather, fatigue, and rutting performance. Consequently, the case study showed that GTR-modified asphalt is a reliable option for utilization on high-traffic volume roads in wet and freezing climates like Michigan. At the same time, NAPA's report also indicated that A1 GWP intensity factor for SBS-modified

FIGURE 11.2 Asphalt plant used to incorporate rubber into asphalt.

asphalt is higher than GTR-modified asphalt. This suggests that GTR not only offers potential performance benefits but may also help to reduce the carbon footprint compared to polymer-modified asphalt (Jin et al., 2023; Miller et al., 2024b).

In another case study aimed at increasing waste tire recycling, both laboratory testing and field demonstrations were conducted to assess the effectiveness of using both GTR and tire fibers for road construction. As shown in Figure 11.3, the study utilized 2% fabric fiber and 12% GTR by total weight of the asphalt binder, and the modified asphalt was added to the mixing drum by feeding system for injection. The field demonstration involved constructing three 0.8 km sections: one for controlling conventional asphalt mixture, one for GTR-modified asphalt mixture, and one for a combination of both GTR and tire fibers. Similar to other case studies, laboratory testing concluded the superior performance of GTR, as the results showed that GTR-modified asphalt had higher crack resistance, anti-rutting properties, and anti-stripping capabilities. For instance, the GTR-modified asphalt's Hamburg wheel tracking device test (HWDT) showed 177.9% improvement based on the failure pass numbers, compared to conventional asphalt. Furthermore, the other recycling material from waste tires, which was fiber, further improved the performance in terms of rutting, fatigue, and cracking. Additionally, field demonstrations and testing showed that recycling waste tires in asphalt led to noise reduction as high as 2 dB outside the vehicles and 1 dB inside the vehicles. The case study's second-year assessments also reported that no cracking was identified in any of the sections that utilized GTR and fiber (Jin et al., 2024). Figure 11.2 shows a typical asphalt plant that was utilized for

FIGURE 11.3 Rubber feeding system used to produce rubber-modified asphalt.

incorporating rubber into asphalt. Figures 11.3 and 11.4 also demonstrate the rubber feeding systems and aggregate bins, respectively.

Warm mix asphalt (WMA) technology also offers significant benefits in terms of carbon reduction, specifically in A3 stage by lowering the mixing and compaction temperatures when compared to HMA. The primary goal of WMA is to reduce energy and fuel consumption during the asphalt production process while

FIGURE 11.4 Aggregate bins in the asphalt plant.

maintaining the required performance criteria and workability. Unlike HMA, which is typically produced at temperatures in the ranges between 150 °C and 190 °C, WMA mixtures can be produced and installed at relatively lower temperatures, ranging from 100 °C to 140 °C. The carbon reduction benefits from WMA primarily stem from lower burner fuel consumption and decreased stack emissions (Shacat et al., 2024; Sukhija & Saboo, 2021). Given the increasing demand from population growth and traffic, reduction of emissions during the use phase of asphalt pavements remains a challenge. Therefore, Mazumder et al. reported that applying technologies like WMA is an essential step toward lowering the carbon footprint, particularly during the cradle-to-gate stages (Mazumder et al., 2016).

In a case study from China, a separate analysis for the production stage of WMA and HMA showed that there was a 46.7% reduction in the GWP when WMA technology was used. The study also indicated that despite the raw material processing stage of WMA additives being slightly higher than HMA, the overall benefits of reducing the mixing and compaction temperatures, and lower energy consumption and emissions during mix production stage outweigh the initial differences (Ma et al., 2019). To further highlight the benefits of WMA, for an asphalt plant using natural gas, NAPA reported that decreasing the mix temperature by about 30°F and 50°F results in a 9–16% reduction in plant-level (A3) emissions. This corresponds to an overall cradle-to-gate (A1–A3) emissions reduction of 3.6–6.0% (Shacat et al., 2024).

Finally, in addition to material and mix design considerations, another key factor in the production stage, particularly A3 stage, is the drying process of aggregates. In the drying process, the choice of heating energy, which can be coal, diesel, heavy oil, or natural gas, usually varies based on the location and the asphalt plant operations and, thus, can significantly influence the emissions. A case study from China revealed that the emissions from natural gas as heating energy source are significantly lower when compared to heavy oil and coal. This is mainly because the carbon emissions produced by natural gas are much smaller than other types of fuel used for the same heat output. As a result, natural gas, as a cleaner energy source, is a key focus for energy-saving and emissions reduction efforts during production stages. NAPA also reported that plant operation emissions (A3) for a mix produced with natural gas are 26% lower compared to a mix using diesel fuel as the burner, which also results in 12% reduction in cradle-to-gate emissions (A1-A3) (Shacat et al., 2024; Wang et al., 2024).

11.5 DECARBONIZATION STRATEGIES IN CONSTRUCTION, MAINTENANCE, AND REHAB STAGES

To reduce the carbon footprint during the paving operations, optimizing the construction equipment's efficiency has been identified as a crucial factor. Generally, key factors include engine power, energy efficiency, and mechanical condition, as well as the machinery's model and service life. Peng et al. reported that older models and extended service life can lead to increased fuel consumption and emissions. This highlights the need for upgrading the equipment and maintaining optimal mechanical performance (Peng et al., 2017).

In a case study involving 11 field projects, including the MnROAD experiment sections and the FHWA Climate Challenge project in Alabama, it was revealed that GWP from paving operations (A5) ranged between 1,030 and 1,590 kg CO_2e per lane-mile. Despite the different nature of the projects, such as each project incorporating different additives and modification methods such as dry rubber, wet rubber, dry plastic, wet plastic, and fibers, variations in A5 GWP results were reported to be relatively narrow. Additionally, transportation of asphalt mixtures (A4) was a more significant contributor to environmental impacts in construction stages than paving operations (A5). It was also reported that eliminating paving equipment idling at construction sites could reduce GHG emissions by 8% to 21%, according to different scenarios (Gatiganti et al., 2024). Another case study for construction of airfield runways also revealed that the environmental impacts associated with the production and combustion of fuel used during the construction of the runway account for less than 2% of total cradle-to-construction (A1–A5) stages (Yang & Al-Qadi, 2017).

Minimizing the GHG emissions from maintenance activities is crucial for improving the efficiency of each road's life cycle. Enhancing preventive maintenance can reduce the frequency of corrective repairs, leading to lower overall emissions. Thus, key strategies for reducing the carbon footprint of maintenance and rehab activities include timely preventive treatments, efficient supply chain management, and adoption of low-carbon manufacturing technologies (Ma et al., 2021). The data regarding carbon emissions can further support the decision-making by guiding the selection between different maintenance, rehabilitation, and reconstruction practices, especially when road agencies have to decide between two comparable treatment options (FHWA, 2023a).

In a case study by the Federal Highway Administration (FHWA) and Arizona Department of Transportation (ADOT), environmental considerations, particularly in terms of carbon footprint were incorporated into the treatment selection process, between mill and filling (M&F) and HIR. The scope of the study included key life cycle stages such as extraction, production, transportation, and construction, and the declared unit for the comparison of the two treatment options was selected as square yard of pavement being constructed. The outcome of the case study and GWP results revealed that the total kg of CO_2 equivalent per square yard of Hot in Place Recycling (HIPR) across two projects were 7.55 kg and 8.47. On the other hand, these numbers were reported to be 9.29 and 9.61 for M&F treatment, which points out that HIPR has a lower impact than M&F treatment. Additionally, the case study shows that although HIPR has a higher carbon footprint during the construction stage, which is mainly due to its energy-intensive heating and scarifying process, due to the higher volume of asphalt mixture required in M&F treatment, recycling methods like HIPR have a lower overall impact (FHWA, 2023a).

Finally, other key strategies that can lower the carbon footprint of maintenance and rehab stages include accelerated paving operations to avoid work zone congestions, scheduling the operations outside peak traffic hours, and designing perpetual pavements that will require less frequent maintenance. As an example, the perpetual pavement design approach offers a significant opportunity to reduce GHG emissions during the pavement design process. This method eliminates the need for structural pavement repairs by optimizing the design of each layer in the pavement structure to resist various distresses. By focusing on efficient layer optimization in this methodology, perpetual pavements limit necessary maintenance to periodic maintenance and rehab activities, such as mill, and overlay rather than activities that can have a bigger environmental impact such as full-depth reclamation. As a result, the frequency and scope of repairs are significantly reduced, and both life cycle costs and environmental impacts are lowered at the same time. Additionally, three of the recent case studies in Texas, Iowa, and Columbia have demonstrated that when compared to conventional asphalt and concrete pavements, implementation of perpetual pavement can reduce the environmental impacts and total costs of the projects, specifically due to less usage of materials over the life cycle, which was 50 years (FHWA, 2023b; Polo-Mendoza et al., 2023; Shacat et al., 2024; Walubita et al., 2022).

In summary, while the A1–A3 stages of asphalt pavements have been well-documented, many aspects of the construction and maintenance stages remain underexplored and can be a topic of investigation for future research. Additionally, evaluating how variations in project-specific factors such as project scale, contractor practices, or equipment usage may impact the environmental impacts of these stages, particularly in terms of carbon footprint and GWP, can be another direction for the research in this area. Furthermore, it is essential to develop a comprehensive yet practical data collection methodology that can be utilized by contractors. This methodology should systematically capture key project-specific variables, enabling straightforward application across diverse projects and regions. By analyzing these factors and their contributions to environmental impacts, this approach can provide actionable insights and establish consistent frameworks for evaluating and reducing GWP during construction and maintenance activities.

11.6 CONCLUSIONS AND THE ROAD FORWARD

This chapter aimed to evaluate the environmental impacts of asphalt pavement construction, maintenance, and rehabilitation, with a particular focus on carbon emissions. Through a detailed analysis of material production, construction, and maintenance practices, this work aimed to identify key factors influencing carbon emissions and suggest strategies for reduction. The following conclusions summarize the main findings related to the critical stages and practices that can effectively mitigate carbon emissions:

- When considering the production, construction, and maintenance stages, the most significant emissions in the asphalt life cycle come from the production stages, particularly extraction and processing of raw materials. These stages should be the primary focus of carbon reduction strategies, as they account for an important portion of carbon emissions.
- Since A1 impacts are deterministic and depend on mix designs, the choice of ingredients for asphalt mix design can play a vital role in production emissions. Reducing the amount of virgin materials, specifically virgin asphalt binder, by using recycled materials such as RAP and exploring alternative asphalt binders such as bio-asphalt can significantly lower emissions.
- Technologies such as WMA, as well as the utilization of bio-oils and bio-asphalt, offer significant benefits in terms of lowering the emissions during mixing and compaction processes and plant operations in A3 stage.
- The type of energy used during plant operations (A3 stage) can significantly affect emissions. The use of natural gas has proved to be a more environmentally friendly option for heating and drying processes compared to diesel or heavy oils.
- Recycled materials, such as GTR, offer both sustainability and performance benefits compared to virgin materials.
- The main efforts for carbon reduction in A4 and A5 stages should focus on equipment's efficiency, engine power, energy efficiency, mechanical condition, and the machinery's model and service life. Assessing how project-specific factors, such as project size, contractor practices, or equipment use, influence the environmental impacts of these stages should be the direction for future studies.
- Key strategies for reducing the carbon footprint of maintenance and rehab activities include timely preventive treatments, efficient supply chain management, accelerated paving operation to avoid work zone congestions, scheduling operations outside peak traffic hours, and designing perpetual pavements that will require less frequent maintenance.

Looking forward, several key areas require further research and development to further aid in enhancing the sustainability and lowering the carbon footprint of asphalt pavements. Key examples include:

- Standardized assessment methodologies for carbon reduction and development of testing protocols that can quickly establish connections between performance of asphalt pavements and GWP.
- Future research should focus on assessing how project-specific factors such as project scale, contractor practices, and equipment usage can affect carbon footprint and GWP in construction stages. Additionally, developing a standardized and practical data collection methodology for contractors is essential to systematically capture key project variables, which will also enable consistent evaluation and application across diverse regions.
- The long-term field performance of bio-based materials, specifically bio-asphalt, needs to be under further investigation to determine their overall performance and environmental benefits.
- Advancements in artificial intelligence (AI) and machine learning present new opportunities for optimizing pavement sustainability by improving material selection, predictive maintenance, and energy-efficient construction practices. Addressing these areas will be critical in advancing the next generation of sustainable pavement solutions.

REFERENCES

Abdalla, A., Faheem, A. F., & Walters, E. (2022). Life cycle assessment of eco-friendly asphalt pavement involving multi-recycled materials: A comparative study. *Journal of Cleaner Production*, 362. https://doi.org/10.1016/j.jclepro.2022.132471

Aurangzeb, Q., Al-Qadi, I. L., Ozer, H., & Yang, R. (2014). Hybrid life cycle assessment for asphalt mixtures with high RAP content. *Resources, Conservation and Recycling*, 83, 77–86. https://doi.org/10.1016/j.resconrec.2013.12.004

Bhat, C. G., & Mukherjee, A. (2019). Sensitivity of life-cycle assessment outcomes to parameter uncertainty: Implications for material procurement decision-making. *Transportation Research Record*, 2673(3), 106–114. https://doi.org/10.1177/0361198119832874

Cass, D., & Mukherjee, A. (2011). Calculation of greenhouse gas emissions for highway construction operations by using a hybrid life-cycle assessment approach: Case study for pavement operations. *Journal of Construction Engineering and Management*, 137(11), 1015–1025. https://doi.org/10.1061/(asce)co.1943-7862.0000349

Cavalli, M. C., Wu, W., & Poulikakos, L. (2024). Bio-based rejuvenators in asphalt pavements: A comprehensive review and analytical study. *Journal of Road Engineering*, 4(3), 282–291. KeAi Publishing Communications Ltd. https://doi.org/10.1016/j.jreng.2024.04.007

De Pascale, B., Tataranni, P., Indacoechea-Vega, I., Rodriguez-Hernandez, J., Lantieri, C., & Bonoli, A. (2025). Enhancing road performance and sustainability: A study on recycled porous warm mix asphalt. *Science of the Total Environment*, 960. https://doi.org/10.1016/j.scitotenv.2025.178370

Fini, E. H., Kalberer, E. W., Shahbazi, A., Basti, M., You, Z., Ozer, H., & Aurangzeb, Q. (2011). Chemical characterization of biobinder from swine manure: Sustainable modifier for asphalt binder. *Journal of Materials in Civil Engineering*, 23(11), 1506–1513. https://doi.org/10.1061/(asce)mt.1943-5533.0000237

Gatiganti, S.C., Bowers, B.F., & Tran, N.H. (2024). Life cycle assessment of asphalt concrete construction stage and impacts of construction equipment idling. In: Flintsch, G.W., Amarh, E.A., Harvey, J., Al-Qadi, I.L., Ozer, H., & Lo Presti, D. (eds.), *Pavement, Roadway, and Bridge Life Cycle Assessment 2024. ISPRB LCA 2024. RILEM Bookseries*, vol. 51. Springer, Cham. https://doi.org/10.1007/978-3-031-61585-6_6

Gulotta, T. M., Mistretta, M., & Praticò, F. G. (2019). A life cycle scenario analysis of different pavement technologies for urban roads. *Science of the Total Environment*, 673, 585–593. https://doi.org/10.1016/j.scitotenv.2019.04.046

ISO 21930. (2017). Sustainability in buildings and civil engineering works – Core rules for environmental product declarations of construction products and services. International Organization for Standardization, Geneva, Switzerland. https://www.iso.org/standard/61694.html.

Ji, J., Yao, H., Suo, Z., You, Z., Li, H., Xu, S., & Sun, L. (2017). Effectiveness of vegetable oils as rejuvenators for aged asphalt binders. *Journal of Materials in Civil Engineering*, 29(3). https://doi.org/10.1061/(asce)mt.1943-5533.0001769

Jin, D., Boateng, K. A., Ge, D., Che, T., Yin, L., Harrall, W., & You, Z. (2023). A case study of the comparison between rubberized and polymer modified asphalt on heavy traffic pavement in wet and freeze environment. *Case Studies in Construction Materials*, 18. https://doi.org/10.1016/j.cscm.2023.e01847

Jin, D., Mohammadi, S., Xin, K., Yin, L., & You, Z. (2024). Laboratory performance and field demonstration of asphalt overlay with recycled rubber and tire fabric fiber. *Construction and Building Materials*, 438. https://doi.org/10.1016/j.conbuildmat.2024.136941

Lei, Y., Wang, H., Fini, E. H., You, Z., Yang, X., Gao, J., Dong, S., & Jiang, G. (2018). Evaluation of the effect of bio-oil on the high-temperature performance of rubber modified asphalt. *Construction and Building Materials*, 191, 692–701. https://doi.org/10.1016/j.conbuildmat.2018.10.064

Li, J., Zhang, F., Liu, Y., Muhammad, Y., Su, Z., Meng, F., & Chen, X. (2019). Preparation and properties of soybean bio-asphalt/SBS modified petroleum asphalt. *Construction and Building Materials*, 201, 268–277. https://doi.org/10.1016/j.conbuildmat.2018.12.206

Liu, X., Deng, L., Zhang, H., & Yu, J. (2024). Quantitative study on carbon emissions of modified recycled asphalt mixture based on life cycle assessment method. *Developments in the Built Environment*, 20. https://doi.org/10.1016/j.dibe.2024.100584

Ma, F., Dong, W., Fu, Z., Wang, R., Huang, Y., & Liu, J. (2021). Life cycle assessment of greenhouse gas emissions from asphalt pavement maintenance: A case study in China. *Journal of Cleaner Production*, 288. https://doi.org/10.1016/j.jclepro.2020.125595

Ma, H., Zhang, Z., Zhao, X., & Wu, S. (2019). A comparative life cycle assessment (LCA) of warm mix asphalt (WMA) and hot mix asphalt (HMA) pavement: A case study in China. *Advances in Civil Engineering*, 2019. https://doi.org/10.1155/2019/9391857

Mazumder, M., Sriraman, V., Kim, H. H., & Lee, S. J. (2016). Quantifying the environmental burdens of the hot mix asphalt (HMA) pavements and the production of warm mix asphalt (WMA). *International Journal of Pavement Research and Technology*, 9(3), 190–201. https://doi.org/10.1016/j.ijprt.2016.06.001

Miller, L., Ciavola, B., & Mukherjee, A. (2024a). EPD Benchmark for Asphalt Mixtures SIP-108 February 27, 2024 Prepared for NAPA by WAP Sustainability.

Miller, L., Ciavola, B., & Mukherjee, A. (2024b). EPD Benchmark for National Asphalt Pavement Association version 2.0: Including Round 2 of Data Collection.

Mohammadi, S., & Solatifar, N. (2023). Optimized preparation and high-temperature performance of composite-modified asphalt binder with CR and WEO. *Journal of Materials in Civil Engineering*, 35(2). https://doi.org/10.1061/(ASCE)MT.1943-5533.0004568

Mukherjee, A. (2016). *Life Cycle Assessment of Asphalt Mixtures in Support of an Environmental Product Declaration*.

Oldham, D. J., Fini, E. H., & Chailleux, E. (2015). Application of a bio-binder as a rejuvenator for wet processed asphalt shingles in pavement construction. *Construction and Building Materials*, 86, 75–84. https://doi.org/10.1016/j.conbuildmat.2015.03.085

Peng, B., Fan, X., Wang, X., & Li, W. (2017). Key steps of carbon emission and low-carbon measures in the construction of bituminous pavement. *International Journal of Pavement Research and Technology*, 10(6), 476–487. https://doi.org/10.1016/j.ijprt.2017.03.002

Polo-Mendoza, R., Mora, O., Duque, J., Turbay, E., Martinez-Arguelles, G., Fuentes, L., Guerrero, O., & Perez, S. (2023). Environmental and economic feasibility of implementing perpetual pavements (PPs) against conventional pavements: A case study of Barranquilla city, Colombia. *Case Studies in Construction Materials*, 18. https://doi.org/10.1016/j.cscm.2023.e02112

Rangelov, M., Dylla, H., & Sivaneswaran, N. (2021). Life-cycle assessment of asphalt pavements with recycled post-consumer polyethylene. *Transportation Research Record*, 2675(12), 1393–1407). SAGE Publications Ltd. https://doi.org/10.1177/03611981211033862

Samieadel, A., Schimmel, K. & Fini, E.H. Comparative life cycle assessment (LCA) of bio-modified binder and conventional asphalt binder. *Clean Technologies and Environmental Policy*, 20, 191–200 (2018). https://doi.org/10.1007/s10098-017-1467-1

Schoenauer, P., Gruber, M.R., Hofko, B. (2024). Assessing the ecological impact of different asphalt mix designs following a cradle-to-gate approach. In Flintsch, G.W., Amarh, E.A., Harvey, J., Al-Qadi, I.L., Ozer, H., Lo Presti, D. (eds.), *Pavement, Roadway, and Bridge Life Cycle Assessment 2024. ISPRB LCA 2024. RILEM Bookseries*, vol. 51. Springer, Cham. https://doi.org/10.1007/978-3-031-61585-6_2.

Shacat, J., Willis, R., & Ciavola, B. (2024). *The Carbon Footprint of Asphalt Pavements: A Reference Document for Decarbonization.*

Su, N., Xiao, F., Wang, J., Cong, L., & Amirkhanian, S. (2018). Productions and applications of bio-asphalts – A review. *Construction and Building Materials*, 183, 578–591). Elsevier Ltd. https://doi.org/10.1016/j.conbuildmat.2018.06.118

Sukhija, M., & Saboo, N. (2021). A comprehensive review of warm mix asphalt mixtures-laboratory to field. *Construction and Building Materials*, 274. Elsevier Ltd. https://doi.org/10.1016/j.conbuildmat.2020.121781

Tokede, O. O., Whittaker, A., Mankaa, R., & Traverso, M. (2020). Life cycle assessment of asphalt variants in infrastructures: The case of lignin in Australian road pavements. *Structures*, 25, 190–199. https://doi.org/10.1016/j.istruc.2020.02.026

U.S. Department of Transportation, Federal Highway Administration (FHWA). Using Life Cycle Assessment (LCA) to Inform the Pavement Treatment Selection Process (FHWA-HIF-23-014). (2023a).

U.S. Department of Transportation, Federal Highway Administration (FHWA). Improved Asphalt Pavement Sustainability Through Perpetual Pavement Design (FHWA-HIF-19-080). (2023b).

Walubita, L. F., Martinez-Arguelles, G., Polo-Mendoza, R., Ick-Lee, S., & Fuentes, L. (2022). Comparative environmental assessment of rigid, flexible, and perpetual pavements: A case study of Texas. *Sustainability (Switzerland)*, 14(16). https://doi.org/10.3390/su14169983

Wang, C., Xue, L., Xie, W., You, Z., & Yang, X. (2018). Laboratory investigation on chemical and rheological properties of bio-asphalt binders incorporating waste cooking oil. *Construction and Building Materials*, 167, 348–358. https://doi.org/10.1016/j.conbuildmat.2018.02.038

Wang, P., Li, K., & Mao, J. (2024). Research on the carbon neutralization of hot mix asphalt pavement based on life cycle assessment theory. *Science of the Total Environment*, 954. https://doi.org/10.1016/j.scitotenv.2024.176691

Wildnauer, M., Mulholland, E., & Liddie, J. (2019). *Life Cycle Assessment of Asphalt Binder (Version March 14, 2019)*, Conducted for the Asphalt Institute by thinkstep AG.

Xin, K., Jin, D., Handler, R.M., & You, Z. (2024). Life cycle assessment of rubber modified asphalt pavement in Michigan: A case study. In Flintsch, G.W., Amarh, E.A., Harvey, J., Al-Qadi, I.L., Ozer, H., & Lo Presti, D. (eds.), *Pavement, Roadway, and Bridge Life Cycle Assessment 2024. ISPRB LCA 2024. RILEM Bookseries*, vol. 51. Springer, Cham. https://doi.org/10.1007/978-3-031-61585-6_22

Yang, R., & Al-Qadi, I. L. (2017). Development of a life-cycle assessment tool to quantify the environmental impacts of airport pavement construction. *Transportation Research Record*, 2603(1), 89–97. https://doi.org/10.3141/2603-09

Yang, S. H., & Suciptan, T. (2016). Rheological behavior of Japanese cedar-based biobinder as partial replacement for bituminous binder. *Construction and Building Materials*, 114, 127–133. https://doi.org/10.1016/j.conbuildmat.2016.03.100

Yang, X., You, Z., Dai, Q., & Mills-Beale, J. (2014). Mechanical performance of asphalt mixtures modified by bio-oils derived from waste wood resources. *Construction and Building Materials*, 51, 424–431. https://doi.org/10.1016/j.conbuildmat.2013.11.017

You, Z., Gilbertson, C., & Van Dam, T. (2018). *Identifying Best Practices in Pavement Design, Materials, Construction, and Maintenance in Wet-Freeze Climates Similar to Michigan.*

Zhang, R., Wang, H., Gao, J., You, Z., & Yang, X. (2017). High temperature performance of SBS modified bio-asphalt. *Construction and Building Materials*, 144, 99–105. https://doi.org/10.1016/j.conbuildmat.2017.03.103

Zhang, R., You, Z., Ji, J., Shi, Q., & Suo, Z. (2021). A review of characteristics of bio-oils and their utilization as additives of asphalts. *Molecules*, 26(16). MDPI AG. https://doi.org/10.3390/molecules26165049

Zhang, R., You, Z., Wang, H., Chen, X., Si, C., & Peng, C. (2018). Using bio-based rejuvenator derived from waste wood to recycle old asphalt. *Construction and Building Materials*, 189, 568–575. https://doi.org/10.1016/j.conbuildmat.2018.08.201

12 Thermal Energy Storage Materials in Buildings

Michael J. Adams, Jason Woods, Nadia Zaleski, and Akanksha K. Menon

12.1 INTRODUCTION

Global energy consumption and associated greenhouse gas emissions have continued to increase with population and economic growth [1], and this has contributed significantly to climate change. In 2017, the global average temperature reached 1°C above pre-industrial levels (1850–1900), and it is predicted to increase at a rate of 0.2°C per decade. Ambitious mitigation efforts are urgently needed to limit the rise in global average temperature below 1.5°C [2]. Of these emissions, a large fraction can be attributed to operations from residential and commercial buildings such as space heating, cooling, and ventilation. According to the International Energy Agency (IEA), building operations account for approximately 30% of global final energy consumption and 26% of global energy-related emissions, with an annual increase in energy demand of 1% over the past decade [3]. Reducing energy consumption and transitioning to renewable energy technologies in the residential and commercial building sectors will thus be important for achieving global net-zero emissions goals in the upcoming decades [4].

To this end, efforts include the construction of new net-zero buildings, renovation of existing buildings, shifting behavioral practices, and a push toward electrification [5]. Decarbonization is broader than electrification, and these efforts must be accompanied by a clean electric grid to maximize their impact. Furthermore, electrification can reduce operational emissions from buildings, but refrigerants used in air conditioners and heat pumps are potent greenhouse gases. Still, IEA found that switching to heat pumps with modern refrigerants can reduce emissions from 20% with fossil-fuel-intensive electricity up to 80% with clean electricity [6]. In a global comparison of building energy use and decarbonization strategies for 32 countries conducted by Camarasa et al., the United States (65 kUS$/capita and ~10,000 kWh/capita), North Western European region (36 kUS$/capita and ~5000 kWh/capita), and China were found to have the highest GDP and final energy consumption, with these numbers trending upwards with population and economic growth [7]. That work also discussed strategies for decarbonization, such as electrification of buildings by up to 50% by 2050, and annual renovation rates of 2.4%. Another study by Ang et al. explored carbon reduction strategies for existing buildings in eight cities around the world, and they highlighted the importance of retrofitting old buildings to meet decarbonization goals [8].

DOI: 10.1201/9781003655442-16

The decarbonization goals for the United States and Europe are important since they make up a large fraction of total global emissions [3]. The European Union has constructed the *Energy Performance of Buildings Directive*, which declares goals for renovating buildings among countries in the European Union to make them more sustainable. Using data from the EEA Greenhouse Gas Inventory from 2023, it was estimated that 85% of buildings in this group were built before 2000, of which 75% are characterized as having "poor performance" [9, 10]. A few specific initiatives from the directive include reducing greenhouse gas (GHG) emissions from the buildings sector by 60% by 2030, a set of building energy performance codes and standards, and for new buildings to be zero-emission. Other programs such as the *Renewable Energy Directive, Energy Efficiency Directive*, and *The Clean Energy for All Europeans Package* have been implemented to support building renovation efforts. However, one main challenge in the past has been limited public funds to support these efforts, so the EU has published the *2021 Annual Sustainable Growth Strategy* and *Renewed Sustainable Finance Strategy* to help direct financing for decarbonization projects [11].

The United States has developed similar goals for achieving decarbonization by 2050 with the most notable being the *Buildings Decarbonization Blueprint* – this includes specific actions the federal government can take to help reduce GHG emissions, increase energy efficiency, and transform the grid. A few specific goals include reducing GHG emissions from US buildings by 65% in 2030 and 90% by 2050 (below 2005 levels), deploying the Bipartisan Infrastructure Law and Inflation Reduction Act to provide financial incentives, and developing a set of building codes such as "electricity-ready" and "zero carbon life cycle" [4]. Several other plans such as the National *Definition for a Zero Emissions Building*, *The Long-Term Strategy (LTS) of the United States*, and *ACEEE Halfway There* present specific targets for decarbonization such as reducing GHG emissions by roughly 60–95% depending on the plan. In emission-related goals, the Department of Energy has also highlighted ways to "embed equity" and "energy justice" within communities through the energy transition. The plan considers economic, environmental, and social impacts that need to be achieved in the future.

To reach decarbonization goals, it will be necessary to become less dependent on fossil fuels and transition to renewable energy sources. With this transition, there is a need for energy storage technologies to balance intermittent energy supply with demand [12]. Grid-scale energy storage is primarily composed of electrical storage. Pumped hydroelectric makes up 96% of worldwide energy storage today, or about 162GW of deployed operating capacity [13]. Its scalability and long discharge time make pumped hydro one of the only suitable technologies for the largest storage applications (>100 MW). Several other electrical storage technologies have been deployed or are under development for large-scale deployment, including compressed air storage, electrochemical batteries, and thermal storage. Lithium-ion storage capacity has expanded rapidly in recent years, and their fast discharge time makes them suitable for peak shaving and load shifting [14]. Lithium-ion battery installations have a typical discharge time of about 4 hours, but Long Duration Energy Storage (LDES) with discharge times from 10 to 100 hours is required to enable a renewable power grid [15]. Ongoing efforts in high-temperature thermal storage, such as concentrating solar power, show promise for grid-scale applications, particularly in the context of LDES [16]. It is unlikely that a single technology will be able to satisfy grid-scale storage needs, rather a combination of technologies must be deployed. Thermal energy storage has been named one of the five thermal grand challenges that could enable full decarbonization of the grid [17].

The US Department of Energy supports ongoing research and development of energy storage technologies through the Energy Earthshots Initiative [18]. The Long Duration Storage Shot aims to reduce the cost of grid-scale energy storage by 90% for systems with discharge times greater than 10 hours. Additionally, the Industrial Heat Shot aims to reduce greenhouse gas emissions from processes that provide industrial heat by at least 85%. High-temperature thermal storage technologies can accelerate the decarbonization of industrial heat [19].

Thermal energy storage near room temperature can reduce energy consumption in buildings by shifting thermal loads near room temperature while still providing comfort to occupants [20]. The goal of thermal storage implemented in buildings today is to minimize operating costs due to energy consumption and peak demand. Load shifting can reduce energy consumption from air conditioners or heat pumps due to increased operation in more favorable conditions. Load-shifting strategies also take advantage of electricity rate structures that have large price differences between daily periods. An on-peak period is defined as the interval with the highest demand. These periods also have the highest electricity prices [21]. This incentivizes load shifting from on-peak periods to off-peak, as represented in Figure 12.1. In this case, the building with thermal energy storage (TES) has higher demand during the off-peak period while the TES is charged. During the on-peak period, the TES discharges, offsetting part of the building's thermal load. The IEA reported that with only a 5% reduction in peak demand, the wholesale price of electricity could be reduced by up to 50% in the United States [22]. For this reason, load-shifting strategies typically prioritize demand reduction over energy savings, and in some cases the total energy may increase [21, 23].

There are specific control schemes for load shifting that depend on the total storage capacity, such as partial or full storage. In a full storage scenario, the energy stored within the TES can offset the full cooling or heating demand during on-peak periods [21]. The full storage control strategy requires a higher capacity Heating, Ventilation, and Air Conditioning (HVAC) and TES system (and therefore capital expense) than

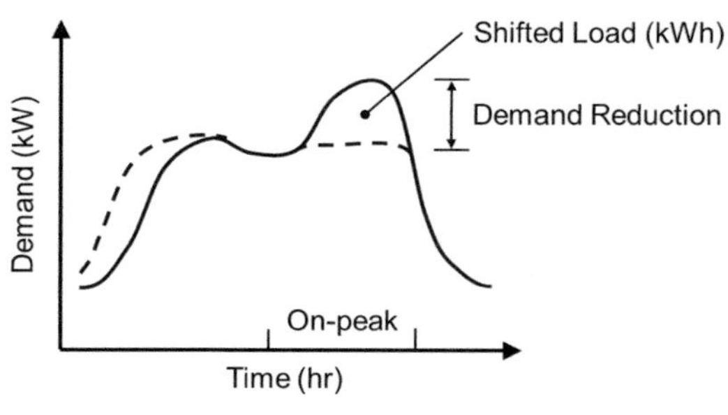

FIGURE 12.1 Qualitative thermal load shifting example (not to scale).

partial storage. Partial storage operates such that a fraction of the heating or cooling load (*e.g.*, 40%) is offset by the TES and the rest is achieved by the existing HVAC components. In the simplest case, the HVAC components operate under a constant load for the full day while TES shifts any excess supply or demand (load leveling). More complicated control systems can use TES to further limit demand, or even take into account the tradeoff between demand limiting and energy savings [21].

Building-integrated TES systems can contribute to decarbonization in more ways than just electrical demand response. For example, some thermal storage materials are suitable for long-duration storage. Short-term storage operates for a few hours (typically <10 hours), whereas long-term storage can be for days, weeks, or months [24]. Also, hot storage can enable the use of heat pumps in colder climates, reducing the need for on-site combustion. Finally, TES can be used to maintain comfort in buildings with respect to both temperature and humidity. Humidity loads make up about a third of annual global emissions from air conditioning, or about 1% of total emissions [25]. Some thermal storage materials (i.e., thermochemical or sorption-based) can directly uptake water vapor from building air and dehumidify it when implemented in an open system configuration, offsetting humidity and temperature loads [26, 27].

Cost and energy savings from thermal load shifting are sensitive to a range of parameters, including utility rate structure, occupancy, on-peak duration and onset, internal gains, building mass, temperature setpoint ranges, and weather [28]. Considering the variety of building types and climates where TES could be implemented, there is no one-size-fits-all approach. Different storage materials can be used depending on whether the application requires hot storage or cold storage, or both. These materials also have different requirements to build a thermal storage component, such as heat exchangers or chemical reactors. Finally, residential and commercial buildings will have different requirements for building integration and control schemes. In this chapter, we review various classes of thermal storage materials, how they are combined into thermal battery or storage components, and how they are integrated into buildings.

12.2 THERMAL ENERGY STORAGE MATERIALS

In this section the properties of different classes of thermal storage materials shown in Figure 12.2 are presented, categorized as sensible storage materials, phase change materials (PCMs), and thermochemical materials (TCMs).

12.2.1 Sensible

First, it is important to clarify some assumptions and highlight some differences. PCMs and TCMs make use of physical or chemical processes at defined states (specified temperature and pressure) to store heat. Their energy storage capacity can be defined on a gravimetric or volumetric basis (e.g., J g^{-1} or kWh m^{-3}). Sensible heat storage on the other hand occurs over a range of temperatures:

$$Q = \int mc_{\mathrm{p}}\mathrm{d}T$$

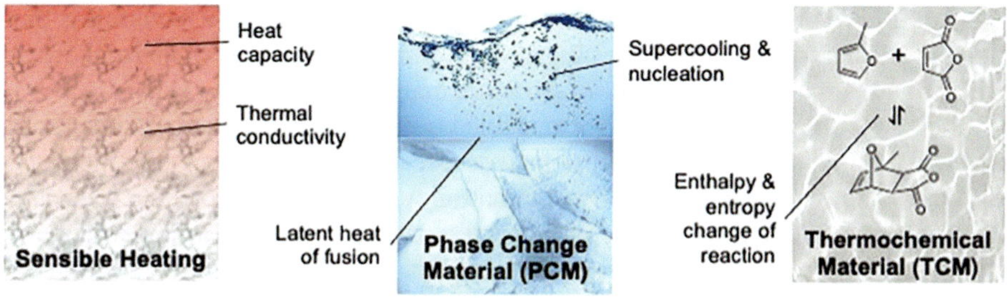

FIGURE 12.2 Visualization of the three classes of thermal energy storage materials and key properties that impact their storage performance.

(From Lilley et al. [29].)

where c_p is the specific heat at constant pressure in units of J g^{-1} K^{-1}. The storage capacity of sensible heating materials therefore depends on the temperature bounds of the storage application. The lower bound is the discharged state, which is defined as a shared reference point with the building. For example, a water heating application may use a lower bound of 10°C, referencing the temperature of mains water. In space heating the reference point is the return temperature, which may vary from 30 to 45°C depending on whether the building uses an air-based system or boilers and radiators. The upper bound also varies depending on the state of charge. Unlike latent heat processes, which occur at constant temperatures, sensible heat becomes less valuable as it is discharged, because smaller temperature differences between the storage medium and the load can limit the rate of heat transfer.

To compare storage materials in general, we discuss the storage capacity and other properties of sensible heating materials relative to a reference state of 25°C. Finally, since the temperature range considered for TES in buildings is relatively small, constant thermophysical properties based on the reference state can be assumed. The molar heat capacity of crystalline solids at room temperature falls within the range of 20–30 J mol^{-1} K^{-1}, following the law of Dulong and Petit [30]. So, a solid substance heated by 100°C will typically store no more than 3000 J mol^{-1}. Assuming a molecular weight of at least 50 g mol^{-1} (i.e., starting from common transition metals), that would correspond to a maximum storage capacity of 60 J g^{-1}. This is significantly lower than the specific energy densities that can be achieved using PCMs or TCMs, which can be on the order of 100s to 1000s J g^{-1}, respectively. This would indicate that sensible heat, while useful for high-temperature storage across large temperature gradients, is not viable for thermal storage in buildings. However, there are some exceptions that have been considered and are used in buildings today, due to their high specific heat, relevance to construction materials, and abundance. These include rocks, hydrocarbons, and liquid water, as discussed next.

12.2.1.1 Rocks

Rocks are considered for passive thermal storage since they are commonly used as building materials – this is referred to as building thermal mass. There are a variety of rocks and minerals to choose from, but their specific heat values fall within a narrow range of 0.7–0.9 J g^{-1} K^{-1} [31]. Another important criterion for these materials is their thermal effusivity, which describes the ability to exchange or absorb thermal energy [32]. Suitable materials include brick, granite, marble, and asphalt [31].

Rock beds are also considered for sensible heat storage in buildings. A heat transfer fluid such as air flows within the packed bed to charge or discharge the rocks. Various studies have explored their combination with solar collectors as a renewable heat source [33]. The thermal performance of a packed bed depends significantly on the heat transfer coefficient between the air and rocks, which is compounded by geometric factors such as particle size, porosity, and flow rate. A high air/solid heat transfer coefficient is desirable for fast rates of charge and discharge. Furthermore, the combination of high heat transfer coefficient, but poor heat conduction between the rocks, benefits packed bed storage by inducing thermal stratification [34].

Limestone is a sedimentary rock consisting primarily of calcium carbonate, and it can be used as a thermochemical storage material for high-temperature applications. The process of converting calcium carbonate into calcium oxide, the active ingredient in cement, is a thermochemical reaction that becomes spontaneous near 850°C. This process for manufacturing cement relies on fossil-fueled kilns at much higher temperatures, and it is responsible for ~8% of global CO_2 emissions [35]. When cement cures, it slowly absorbs CO_2 from the ambient and releases heat. In a controlled reactor, this reverse reaction can occur much faster and act as a high-temperature heat source. Calcium carbonate as a thermochemical storage material is currently being studied for grid-scale energy storage within the DEGREES Energy Earthshot Research Center [36].

12.2.1.2 Hydrocarbons

Hydrocarbons have been shown to have remarkably high molar heat capacity that exceeds the Dulong and Petit limit, but this largely results from their high molecular weight. Paraffin wax, for example, has a molar heat capacity of around 900 J mol^{-1} K^{-1}, which translates to a specific heat capacity on the order of 2 J g^{-1} K^{-1} [37]. When comparing materials for thermal storage applications, units should be on a gravimetric or volumetric basis. Considering paraffin wax, which typically melts around 50 to 65°C, an upper limit of 100 J g^{-1} in sensible heat storage can be estimated. Paraffin is usually considered a PCM due to its high latent heat of melting, which is on the order of 200 J g^{-1} [37, 38]. Also given that latent heat is accessible at the melting temperature, whereas the temperature of sensible heat degrades as it is discharged, this class of materials is better suited as PCMs.

12.2.1.3 Water

Liquid water is a practical substance for sensible heat storage in buildings. Its specific heat of 4.18 J g^{-1} K^{-1} is one of the highest among materials near room temperature. Heating water to its boiling point results in a maximum storage capacity of about 314 J g^{-1}. Additionally, water is advantageous because it is an effective heat transfer fluid, and it can be easily integrated with thermal loads. The availability of off-the-shelf components and established standards significantly lower the barrier to implementing hot-water storage [39]. Furthermore, buildings are already constructed with plumbing and storage tanks for hot water.

Similarly, cold-water storage is practical for large-scale applications due to its simplicity, based on existing chilled-water systems. The main challenge is that large storage volumes are required due to small temperature differences. For example, cold water stored at 5°C used to cool return air to 10°C would only have a 5°C temperature lift, resulting in a useful storage capacity of 40 J g^{-1}. Ice storage is more practical than chilled-water storage in applications with space constraints due to its high energy density [40].

12.2.2 Phase Change Materials (PCMs)

Phase change materials store energy in the form of latent heat associated with a physical process, typically a melting transition, which delivers a greater thermal energy density than sensible storage. Thermal properties such as energy density (latent heat), phase transition temperature, and thermal conductivity are the key parameters that determine the performance of PCMs as thermal storage materials [29]. The phase transition temperature must be compatible with the application (i.e., heating or cooling), and low thermal conductivity can inhibit the rate of thermal energy exchange in a demand response scenario.

In addition to thermal properties, material selection must also consider the following physical, kinetic, and chemical properties [38]. PCMs should have high physical density with minimal change in volume during phase change so they can be incorporated into building structures or heat exchangers. They should be resistant to phase segregation (incongruent melting) and supercooling, a phenomenon where a liquid can be lowered below its freezing point without undergoing phase transition (delay in crystallization or nucleation of the solid state) [41]. Finally, PCMs should be non-toxic and fire resistant, and they should not be corrosive to common building materials.

A comparison of the many available PCMs is provided in review articles in the literature [31, 38]. In this section instead, we focus on materials suitable for thermal storage in buildings, which can be categorized broadly into organic or inorganic materials. A third category dedicated to ice is also included due to its prevalence in real applications.

12.2.2.1 Organic PCMs

Organic PCMs are favored for being chemically stable, non-corrosive, and resistant to supercooling. Organic PCMs can be divided into two types: paraffins and non-paraffins. Commercially available paraffins are inexpensive, and they are available in a range of melting temperatures from 20 to 70°C, making them suitable for building applications [38]. Their thermal storage density ranges from 120 to 210 J g^{-1}, and they exhibit minimal degradation with thermal cycling [38].

Non-paraffins include other hydrocarbons such as fatty acids and glycols. These materials have comparable latent heat values to paraffins, ranging from 42 to 250 J g^{-1} [37]. Fatty acids have melting temperatures suitable for building heating applications, and their sharp melting points make them suitable for applications that require precise temperature control. However, their high cost (about 2–3 times more than paraffins) is a limiting factor [31, 37].

One of the primary advantages of organic PCMs is their chemical stability during the phase change process, which contributes to their long-term reliability. This is critical for building applications where the thermal storage component is expected to last from thousands to tens of thousands of cycles. Organic PCMs are also non-corrosive – having good compatibility with other materials reduces the need for encapsulation or protective coatings. This simplifies their integration with storage tanks and heat exchangers to make thermal storage components. Another advantage of organic PCMs is their ease of processing, which makes them tunable to specific applications. For example, the melting point in paraffin can be finely tuned by adjusting the chain length. This flexibility is critical for building applications, where the energy savings and demand response are sensitive to the TES temperature [42].

Despite these advantages, organic PCMs have significant drawbacks, including flammability and low thermal conductivity. Flammability is a major concern in building materials, especially in large-scale applications or densely populated locations. This problem is especially pertinent to hydrocarbons such as paraffins, which can limit their use. From a thermal point of view, the main limitation of organic PCMs such as paraffins is their low thermal conductivity (~ 0.2 W m^{-1} K^{-1}) [31]. Conductive fillers can be added to improve thermal conductivity, but large volume fractions may be required to achieve percolation paths, lowering the energy density. Furthermore, conductive fillers must be uniformly distributed in the material, but thermal cycling can lead to separation. Matrix structures can also be used to improve conductivity while providing shape stabilization. Physical properties also pose a challenge for paraffins, as they exhibit large volume changes during the phase transition. Various geometries of plastic containers can be used to contain paraffins, but it is a challenge to incorporate them within a building envelope [37].

12.2.2.2 Inorganic PCMs

Inorganic PCMs include materials such as salt hydrates and eutectic mixtures. Salt hydrates are a non-flammable alternative to organic PCMs with similarly high energy density. These salt hydrate PCMs are often the same compounds considered for TCMs, but the storage mechanisms and latent heat are different. Phase change in a salt hydrate is defined as the process in which increases in temperature cause the crystalline material in its highest hydrated state to release water as it enters a dehydrated state. If that dehydrated state is soluble in water, then it dissolves into an aqueous solution in a process called deliquescence. For example, calcium chloride hexahydrate ($CaCl_2 \cdot 6H_2O$) upon heating to ~30°C forms anhydrous $CaCl_2$ that is soluble in the released water, so the final product is a homogeneous solution. This process is referred to as congruent melting, even though it is a thermochemical process, and it is analogous to melting in other PCMs. This is not to be confused with the latent heat associated with transitioning between two crystalline hydration states, for example, $CaCl_2 \cdot 6H_2O$ to $CaCl_2 \cdot 2H_2O$. That process is a solid-gas reaction (sorption) and is classified as thermochemical storage.

A wide variety of salt hydrate PCMs are available with high energy densities and temperatures suitable for thermal storage in buildings, such as calcium chloride, magnesium chloride, and sodium sulfate. However, differences in physical properties at the phase transition significantly affect their performance. Heating the salt hydrate may result in a lower hydrate or anhydrous salt that is not soluble in water, resulting in a heterogeneous mixture of salt solution and salt crystals. This process is called incongruent melting. The salt crystals may settle out of the solution with repeated cycling, leading to phase segregation, which causes a loss in storage capacity. Phase segregation is one of the main limitations of salt hydrate PCMs. Researchers have explored solutions such as adding thickeners to prolong the lifetime of the storage material, and there is ongoing research on micro-encapsulation to limit phase segregation [43].

Eutectics are mixtures of two or more PCMs in a defined proportion so that their combined melting point is lower than that of the individual constituents. They often consist of mixtures of different salt hydrates or fatty acids. These are typically subdivided into three categories: inorganic-inorganic, organic-organic, and inorganic-organic eutectics. Organic eutectics tend to be better suited to thermal storage in buildings due to their lower melting temperatures and higher latent heat of fusion [31]. The materials must be uniformly mixed so that the phase transition occurs at the desired composition. Similar to salt hydrates, phase segregation can lead to variations in the melting temperature and a loss in storage capacity.

12.2.2.3 Ice

Ice is the most popular PCM for cold storage in buildings due to its high latent heat of fusion and accessibility [44]. The latent heat associated with melting ice is 334 J g^{-1}, making it one of the highest among PCMs, especially near room temperature. This makes ice particularly effective as it can store the same amount of heat as other PCMs using less material on a weight basis. This advantage extends to volumetric energy density (VED) as well, considering that the density of ice (1000 kg m^{-3}) is also higher than other PCMs such as paraffins (~900 kg m^{-3}). Ice has a reasonable thermal conductivity of 2.18 W m^{-1} K^{-1} at 0°C, or about 10 times higher than organic PCMs such as paraffins. This is a key advantage, as thermal conductivity influences the rate at which PCMs can be melted to meet cooling demands. Ice storage also has several practical advantages over other cold storage such as chilled water and eutectic salts. In addition to low storage volume requirements, its low temperature reduces sizing requirements for components such as pipes, ducts, and pumps [45]. It also has a greater potential for dehumidification than chilled-water systems [45].

Despite its advantages, there are several challenges related to its volume expansion during phase transition and melting temperature that limit the broad utilization of ice in building applications. When water freezes into ice its volume expands, causing mechanical stress on storage containers or heat exchanger components. Volumetric expansion must be addressed in the design of the TES component to prevent structural damage, which adds complexity. The phase transition temperature at 0°C also limits the use of ice to cold storage. Sub-zero transition temperatures that are relevant for refrigeration applications can be achieved with eutectic water-salt solutions [46]. Organic and inorganic PCMs have melting points that can be tuned to different temperature ranges, making them more flexible than ice/water-based storage for various heating and cooling demands.

12.2.2.4 Characterization

Characterization of the thermal properties of PCMs, such as melting point, latent heat of fusion, specific heat capacity, and thermal conductivity to a high degree of confidence, is critical to understand their performance and design a TES component. One of the most common methods to measure the relevant thermal properties is Differential Scanning Calorimetry (DSC), which measures the heat flow into or out of a sample. This can be used to measure specific heat when the temperature ramp rate is constant (typically 10–20 K min^{-1}) [47]. A phase transition can be observed as a peak in the heat flow, and the onset temperature is determined from the initial change in slope. The latent heat associated with the phase change is determined from the integral of the peak. This technique is widely used in lab-scale testing due to commercially available instrumentation and established measurement standards, but results may not be representative of properties in large-scale building components [48].

It is worth noting that many commercially available paraffins are blends that have broader melting peaks than high-purity materials, and properties can vary significantly across manufacturers [49]. Furthermore, complexities associated with phase transitions can add uncertainty. For example, non-uniform heating in PCMs can hide the onset of melting. PCMs can also exhibit supercooling, where the material remains in a liquid state below its freezing temperature, which can complicate the determination of the onset of solidification. These effects in turn lead to hysteresis in the observed phase transition temperature during cycling. Accurately measuring the enthalpy of fusion using DSC also poses a challenge, as small sample sizes may not be representative of the properties of bulk materials. Additionally, measured enthalpy is sensitive to heating rates – slow heating rates increase the overall measurement time and introduce errors due to drift in the instrumentation, but fast measurement times make the melting peak artificially wide and may result in incomplete phase transitions. Some PCMs such as salt hydrates are susceptible to phase separation (incongruent melting) and may not fully melt or crystallize during the measurement.

The T-history method has emerged as a measurement technique that can address these limitations of using DSC for characterizing PCMs [50]. This method involves larger sample sizes and a controlled temperature environment to observe the temperature within the PCM as a function of time. The latent heat, specific heat, and thermal conductivity can be estimated by comparing this function to that of a known reference material. It provides several advantages, as larger sample sizes and longer measurement duration result in more representative bulk properties. This method is also inherently less sensitive to supercooling and non-uniform melting, thus reducing hysteresis in the phase transition temperature. However, there are several challenges with the T-history method that limit its utility. The method does not provide the same level of precision as DSC, partly due to the large sample sizes, making it unsuitable for observing small thermal events that may be crucial to fully characterize a PCM. This is exacerbated by a lack of commercially available instrumentation or measurement standards – the method relies on custom equipment and several variations of the method have been used to publish PCM property data [51]. For these reasons, the T-history method should be considered as a complementary technique to DSC.

12.2.3 Thermochemical Materials

Thermochemical storage uses a reversible reaction to convert chemical energy into heat. TCMs offer significant advantages over other storage materials, including high storage density, a wide range of useful temperatures, and long-duration storage due to negligible self-discharge. However, they are more complex requiring mass exchange in addition to heat transfer. TCM-based storage involves gas-solid reactions, and the storage temperature is dependent on gas pressure as governed by the Clausius-Clapeyron relation [52]:

$$\ln\left(\frac{p}{p^{\circ}}\right) = \frac{\Delta_r S}{R} - \frac{\Delta_r H}{RT}$$

where $\Delta_r H$ is the enthalpy of the reaction (J mol^{-1}), $\Delta_r S$ is the entropy (J mol^{-1} K^{-1}), R is the universal gas constant, and p° is the atmospheric pressure. Therefore, the storage temperature can be increased by raising the pressure of the reaction gas. This is another significant advantage of TCMs, as the storage temperature is adjustable. However, building technologies incorporating TCMs are relatively new compared to those based on sensible storage or PCMs, and basic integration strategies are still being explored. For example, research in Europe has focused on long-term or seasonal storage, where the time scale can be up to days or even months [26, 53, 54]. Recent work in Europe and North America is exploring the use of TCMs in daily load shifting [27, 55]. This is because in the case of seasonal storage, a low number of cycles leads to high energy capacity cost or levelized cost of storage (LCOS), which we describe in Section 12.5. These approaches have different requirements for material performance, especially in terms of cycling stability. Nevertheless, the various approaches to thermochemical storage share some relevant material properties that can be used for an initial screening, such as energy density (enthalpy of the reaction), toxicity, and cost. The following section introduces some of the TCMs considered for thermal storage in buildings, the challenges associated with mass transfer and reaction kinetics, and some methods that are being explored to engineer improved materials.

Many reaction systems are available for thermochemical storage, including gases such as ammonia, carbon dioxide, hydrogen, and water vapor. Reaction pairs considered for buildings are centered around the use of water vapor due to its accessibility – it can be condensed and stored in a closed system without any safety concerns, and open systems can utilize water vapor available in humid air. Hundreds of potential TCMs are compatible with water vapor, and some classes considered for room temperature storage include salt hydrates, adsorbents such as silica and zeolites, and their composites. The storage material is charged by dehydration (endothermic reaction) and discharged by introducing water vapor for hydration (exothermic reaction). In both reactions, the amount of energy stored is directly proportional to the moles of water exchange (water uptake), as shown in the reversible chemical reaction below.

$$\text{Salt}\cdot(n+m)H_2O + \Delta_r H \leftrightarrow \text{Salt}\cdot nH_2O + mH_2O$$

Here it is important to note that unlike salt hydrates, adsorbents such as silica and zeolites do not store heat using thermochemical reactions. Instead, they interact with water vapor to store/release heat using a physical process and can be categorized as sorption materials [56]. The thermodynamics are also different – equilibrium in the salt is limited to specific hydration states, while adsorbents have a continuous uptake driven by the partial pressure of water vapor (or relative humidity in air). A class IV water sorption isotherm typical of mesoporous materials is drawn in Figure 12.3 [57]. Hydration state equilibrium in salts is described as monovariant and adsorption equilibrium in sorbents as bivariant [58]. However, physical sorption materials are often classified as TCMs as they are implemented in a similar way, and they can also be combined with salt hydrates within composite materials. This chapter focuses on salt hydrate TCMs given their higher energy density and lower dehydration/regeneration temperatures compared to sorption materials, which make them better suited for integration in buildings.

Screening studies of TCMs mainly focus on the energy density of thermal storage materials [55, 59–61]. Chemical databases of thermochemical reactions report their reaction enthalpies in units of kJ mol^{-1} [62]. These values are converted into gravimetric or VED using their molar mass and apparent density, although both of these properties change depending on the hydration state of the salt. Salt hydrates have been studied extensively as TCMs owing to energy densities that are 8-10 times higher than sensible heating materials and more than 2-5 times higher than PCMs [26].

Beyond energy density, there are other important material characteristics to consider that depend on the end-use application and reactor design. Some TCMs undergo multiple reactions to achieve their full energy storage capacity. In that case, there would be multiple Clausius-Clapeyron relations that govern the phase transition between each hydrated state. For example, strontium bromide has three crystalline phases: anhydrous, monohydrate, and hexahydrate. The phase transitions can be plotted as a function of temperature and water vapor pressure using the Clausius-Clapeyron relation, as shown in Figure 12.3. Building TES using $SrBr_2$ would likely

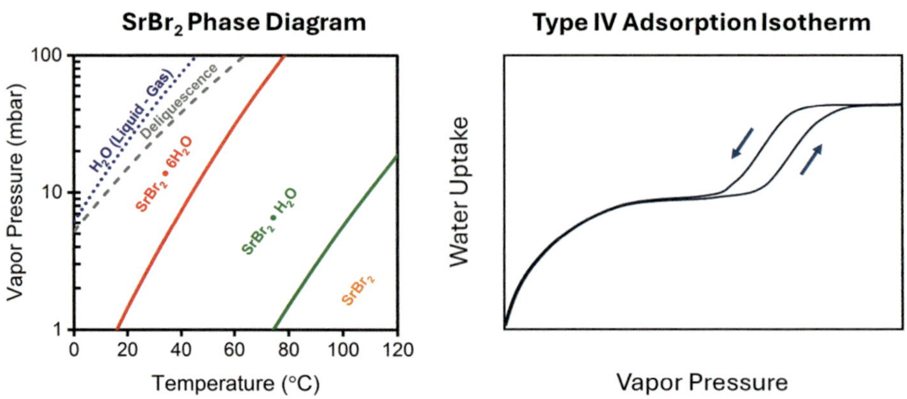

FIGURE 12.3 Phase diagram of a representative salt hydrate ($SrBr_2$) showing the distinct stable hydrates, and an adsorption isotherm typical of mesoporous adsorbents such as silica gel [57].

be limited to the transition between monohydrate and hexahydrate due to the high temperatures required to reach the anhydrous state. Other salt hydrates such as calcium chloride require multiple reaction steps to achieve high energy density. Salt hydrate mixtures have also been examined in an effort to optimize pressure-temperature reaction conditions, as well as improve reaction kinetics [63].

Reaction kinetics are another important aspect of TCMs. The charge/discharge rate that can be achieved in thermochemical storage is limited by a combination of reaction kinetics and mass diffusion. Single-step hydration and dehydration reactions may be described by a first-order rate model,

$$X(t) = 1 - \exp(-k \cdot t)$$

where $X(t)$ is the time-dependent reaction conversion and k is the rate coefficient, which is commonly modeled using the Arrhenius equation [64]. However, efforts to parametrize Arrhenius factors in hydration reactions from empirical data have obtained negative activation energies near equilibrium vapor pressures [65, 66]. Negative values for activation energy are physically nonsensical, but they imply additional intermediate reactions or physical processes limiting the reaction. For example, Sögütoglu et al. found that salt hydration involves the nucleation and growth of wetting layers, resulting in a metastable zone (prenucleation) near hydration state transitions. This causes the reaction rate to decelerate much faster close to equilibrium than predicted by rate models. This can have a significant impact on the power output versus temperature in thermochemical storage [67]. Birkelbach et al. developed a non-parametric model that captures the effect of the reverse reaction near equilibrium conditions [68]. Alternatively, efforts to characterize reaction kinetics for practical applications may neglect Arrhenius parameters and choose to develop empirical correlations [66]. However, studies involving densified, mm-scale samples have found the reaction rate is limited by water vapor diffusion rather than the chemical reaction [69].

The most significant limitation in TCMs is a loss in useful storage capacity after hygrothermal cycling. Since these are gas-solid reactions between different hydration states of salt in its crystalline form, the reaction rate is often limited by the available surface area for heat and mass transfer. Furthermore, these reactions are induced by water vapor or air, and the diffusion is impacted by the porosity at different length scales (within the salt and within a packed bed). We discussed previously that salt hydrates can be used as PCMs when the salt in its highest hydration state is heated, causing it to form a liquid solution. This is a physical process called deliquescence, and it is undesirable in salt hydrate-based TCMs because it leads to agglomeration of the storage material that impacts heat and mass transport.

Composite materials containing a sufficient volume fraction of active TCMs can still achieve relatively high energy densities, while the matrix prevents agglomeration of particles and maintains a large surface area for mass transfer. As early as 1996, Levitskij and Aristov presented two-phase systems consisting of a porous host matrix and hygroscopic salt crystals, which they called Chemical Heat Accumulators or Selective Water Sorbents [70, 71]. These are part of a series of papers from the Boreskov Institute of Catalysis that introduced impregnated TCM composites. They identified dehydration as one of the best reaction types for storing low-grade heat due to its high energy density and reaction reversibility, and highlighted calcium chloride in porous silica as a composite of particular interest. Encapsulation is another technique to fabricate composites by forming the matrix around the salt, or by containing the salt within a spherical shell. Research on composites is still ongoing as some challenges remain. In general, salt-matrix composites include inactive materials that do not contribute significantly to storage, lowering the overall energy density. Additionally, salt leakage from the composite leads to significant degradation of storage capacity with hygrothermal cycling. The fundamental understanding of degradation mechanisms and tools to engineer better materials is still being developed, driven by improvements in fabrication and characterization techniques.

Characterization of salt hydrate-based TCMs involves mapping the water vapor uptake and enthalpy of reaction between various states of temperature and vapor pressure. Measurement techniques used to characterize TCMs include Thermogravimetric Analysis (TGA) and DSC. TGA is used to measure the in situ mass change of a sample when the temperature or vapor pressure is changed. DSC is used simultaneously to measure the heat flow into or out of the sample. The time-dependent response of these outputs can be used to characterize the reaction kinetics.

Characterization of TCMs is more complex than PCMs, and there are several unique challenges associated with TGA-DSC measurements for these materials. It can be challenging to measure enthalpy using DSC because they are typically calibrated by the enthalpy of fusion of various transition metals, but the heat of reaction often has significantly greater energy density. Reaction completion and kinetics can be quantified by water mass uptake measured using TGA in parallel with DSC. These results are often normalized by the weight of the sample. When comparing properties in literature, it is essential to consider the mass basis being used (dry or hydrated mass). Another challenge with TGA-DSC characterization is that only the hydration kinetics can be studied, because dehydration is limited by the ramp rate of the instrument [66, 72]. Finally, these instruments are capable of characterizing reaction properties in humid air, but not in pure water vapor, which may be necessary to study reaction kinetics for closed-reactor systems. There is a need to develop new characterization techniques and standards to address these shortfalls.

12.2.4 Summary of Storage Materials

The energy density of various classes of storage materials is illustrated in Figure 12.4. Here the units have been converted to VED for a more practical comparison. Sensible heat storage is represented as a line with a constant slope depending on the specific heat of the storage material. The total capacity depends on the storage temperature. Energy density for water and rock-based materials is drawn with a load temperature of 25°C, but in practice this could vary from 10 to 40°C. PCMs generally have higher energy densities, and they are suitable for cold or hot storage. Efforts to improve PCM performance are generally focused on increasing physical density and thermal conductivity. A more fundamental limitation of PCMs is that the storage temperature is fixed, making them best suited to one application but not both. Recently, Lau et al. [74] developed a dynamically tunable PCM-based device that uses ion transport to

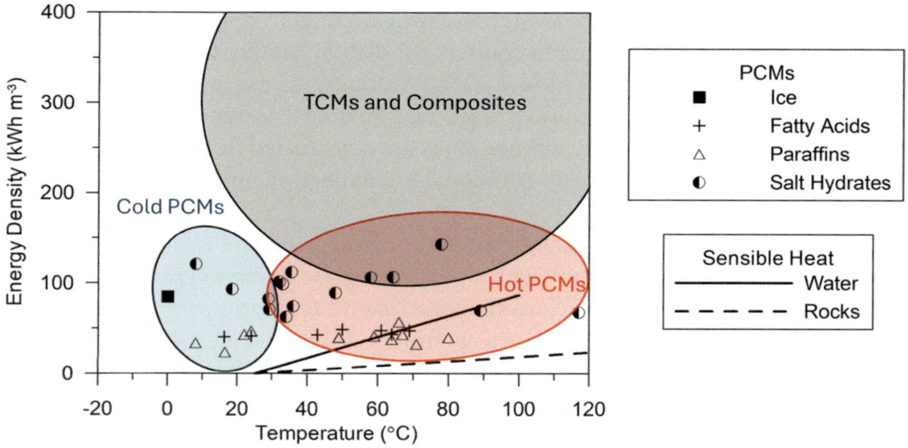

FIGURE 12.4 Energy density comparison for different storage materials relevant to buildings.

(Drawn using data from: [31, 73].)

change the phase change temperature. Pure TCMs have the highest energy density. Adsorbents and TCM composites are lower but still competitive with PCMs. Variable storage temperature is a significant advantage of TCMs.

12.3 THERMAL ENERGY STORAGE DEVICES

As discussed previously, thermal energy storage functions by exchanging the heat stored within a storage material and the building. This involves two layers of integration: (1) at the component level, where the storage material is paired with a heat exchanger, storage container, and insulation, and (2) at the system level, where heat is exchanged with the HVAC system. This section introduces various component-level designs, referred to as thermal storage devices. Containment of TES materials is not a trivial matter due to volume requirements, thermal expansion, and corrosion. Heat exchangers are also an important consideration, especially since many thermal storage materials have inherently low thermal conductivity. Devices based on TCMs introduce further complications because they must accommodate mass transfer in addition to heat transfer. Whether thermal storage materials are incorporated into storage tanks, building envelopes, or other heat exchangers, the design and integration vary significantly between sensible heat storage, PCMs, and TCMs.

12.3.1 Storage Tanks

Hot water is the most widely implemented form of thermal storage (sensible) for applications from room temperature up to 100°C. It has the highest specific heat of materials accessible to building applications, and it can be used as the heat transfer fluid, allowing the storage component to be compact and relatively simple.

Water-based storage also scales well, presenting a range of opportunities from hot-water tanks for domestic heating to aquifers for district heating [31, 75]. The design of the storage tank varies across these use cases, but it is governed by the same key criteria: total heat capacity, heat exchange rate, and heat losses during the charge/discharge cycle. To address this, storage tanks are constructed from water-tight materials to contain the water and any additional equipment, including heat exchangers and heating elements.

Domestic water tanks are made from a variety of parts and materials. They are commonly constructed from steel with a plastic or glass lining to prevent corrosion. Typically, an electric heating element heats the water during the charge cycle, although hot water may also be piped in from a solar heating source [75, 76] or charged with the condenser of a heat pump. During the discharge cycle, hot water is piped to the building while cool water returns to the tank. At this scale, water tanks must be insulated to reduce heat losses. Hot-water storage tanks can be thermally stratified, where buoyancy will keep hot water at the top and cold water at the bottom. This is beneficial because it allows hot and cool water to be accessed separately for charging or discharging. For example, the device can discharge water from the top of tank, where heat is concentrated. A thermodynamic analysis reveals that stratification improves the overall performance by increasing the exergy efficiency (or useful potential of the heat) [76, 77]. This can also be significant during the charge cycle, especially when driven by solar heating, as supplying cool water to the solar collector improves its performance [75]. Additional design factors involve hydrodynamics and thermodynamics of fluid flow, such as inlet velocity and momentum during charge or discharge, and natural convection currents within the tank. The design should aim to minimize the mixing of hot and cold water due to these effects and to minimize entropy generation. As such, storage devices often include baffle plates, stratification tubes, or multiple tanks to maintain the temperature difference for extended time periods when the device is idle. Inlet-outlet pairs must be positioned to make use of thermal stratification but also minimize dead water weight [76, 78].

At much larger scales, water storage can be used for district heating and cooling. Thermal storage devices can be divided into four basic design concepts: Tank (TTES), Pit (PTES), Borehole (BTES), and Aquifer Thermal Energy Storage (ATES) [79, 80]. Additionally, the storage container can be in the form of a solar pond [81]. Solar ponds share some similarities with PTES, but the mechanisms are sufficiently different to treat them as distinct thermal devices.

Aquifers are naturally occurring formations containing large quantities of groundwater, often mixed with sand, gravel, or fractured rocks. Groundwater is one of our most valuable resources, and large population centers tend to be located near aquifers to utilize it. ATES systems use groundwater for seasonally variable heating and cooling requirements. In summer, cool water is extracted from a well to offset building thermal loads, and the return water is rejected to a warm well. Over the course of the season, the cool well is depleted and the warm well is charged. In winter, the building draws water from the warm well and returns it to the cold well. Thus, there is no net consumption of groundwater. These systems are limited to sites where aquifers are able to deliver high flow rates [79]. Alternatively, water pits

(PTES) may be constructed to act as artificial aquifers, and boreholes (BTES) make use of vertical heat exchangers buried in the ground in combination with heat pumps to store heat in the earth. This has been an active area of development in Denmark, where several full-scale PTES systems have been implemented since the 2010s [82]. All these storage systems rely on seasonally variable heating and cooling requirements that are relatively balanced [79, 80].

In 1975, researchers began to investigate solar ponds as thermal storage for space heating [83]. Solar ponds are shallow, artificial bodies of water that utilize a salt concentration gradient to prevent the formation of natural convection currents. In essence, buoyancy forces that may arise beneath the surface due to solar heating are insufficient to overcome the concentration gradient. This effect can be used to develop thermal stratification for efficient heat storage. Solar ponds can be divided into three segments with increasing depth: an upper convective zone (UCZ) made up of low salinity brine, a non-convective zone (NCZ) where the concentration gradient is present, and a lower convective zone (LCZ) with high salinity brine. Unlike domestic water tanks, where heat is stored in the upper zone where it naturally concentrates, the UCZ is unsuitable for thermal storage due to evaporative losses. Instead, solar ponds are designed to store heat in the LCZ. Solar energy is absorbed throughout the depth of the pond, with different parts of the spectrum being absorbed as a function of depth. Heat conduction in liquid water is relatively low compared to convection, so the NCZ acts as an insulating layer and the bottom of the pond retains heat. Combined heat and mass transfer models based on experimental studies have been developed to confirm the understanding of this mechanism [83, 84].

Salt selection is a key parameter in solar pond design. Low cost and non-toxicity are among the most important characteristics, due to the large quantities of salt needed to construct a solar pond. Sea water (or natural brines derived from seawater) may be used if it is available locally, otherwise solid salts may be added. For a typical solar pond 1–2 m deep, about 1/3 ton of salt is required per m^2 of pond area [85]. Among individual salts, NaCl is the most likely candidate for solar ponds, although $MgCl_2$ is also commonly available as a byproduct in concentrated brines from seawater. Other factors in salt selection include transparency of the solution to solar radiation and a high degree of solubility that should have low variation with temperature [81]. In addition to salts, additives may be added to the LCZ to improve solar absorption within that layer, and additives may be added to the UCZ to prevent algae growth. Finally, the layers require ongoing maintenance to preserve the salt concentration gradient as water evaporates from the surface.

One of the major limitations of solar ponds is the low efficiency of solar collection. The upper limit for collection efficiency is expected to be around 50%, but in practice it is closer to 20% [85]. Therefore, solar ponds require large use of land area to store thermal energy. Physical design constraints of solar ponds are primarily their size and surrounding geology – the location should be flat to minimize construction costs, and it should not be near underground aquifers, especially freshwater sources. Even if the pond is lined and water-tight, nearby aquifers can present structural issues, thermal losses due to underground convection, and contamination of fresh water [85].

12.3.2 Building Envelope

12.3.2.1 Passive Storage

Building envelopes present another opportunity to implement thermal storage. Sensible heating materials or PCMs can be incorporated into building components such as walls, ceilings, and floors. These are referred to as thermal mass, and they are able to regulate indoor temperature passively, leading to a delay or reduction in peak loads [86]. Thermal mass ultimately affects the HVAC system, but it acts indirectly and is strongly dependent on regional climate. Heat transfer between the conditioned space and the building envelope is governed by ambient boundary conditions, so these components have less flexibility than a storage device integrated directly with the HVAC loop. A review of passive PCM applications found that paraffins are by far the most common materials [87]. However, the majority of studies were located in temperate climates, and there was a lack of full-scale building data. A more recent review found that cold PCM storage applications in hot climates are challenging [88]. If ambient night temperatures do not fall sufficiently below the phase transition temperature, then it will not fully recharge.

12.3.2.2 Active Storage

Building thermal mass can also be actively charged by the HVAC system. For example, air conditioning can precool a building at night when temperatures are relatively low. This has been an active area of research in commercial buildings due to their large thermal mass and electricity pricing based on a combination of energy consumption and peak demand [21]. This has led to the development of a number of control schemes to minimize demand costs. Historically, there has been limited effort in precooling residential buildings due to their relatively low thermal mass and generally uniform electricity rates. Recently, there has been interest in implementing PCM walls due to the adoption of renewable energy sources and time-of-use electricity rates [89].

Some studies have investigated the effects of incorporating active thermal components in building envelopes to improve their performance. For example, thermal switches are active elements that use a non-thermal input such as an electric field to amplify or attenuate the flow of heat, even when the temperature difference is fixed [90]. Thermal switches integrated with the building envelope can change their effective resistance, reducing the heat exchanged between the PCM and the environment at disadvantageous times. For example, Miao et al. [91] developed a switching mechanism that can isolate or couple the inside and outside surfaces of a building wall. When combined with a PCM layer, this device provides more control to utilize ambient temperatures to charge/discharge the PCM while lowering the overall thermal load on the building.

Finally, PCMs can exchange heat with the building envelope remotely. The Thermally Anisotropic Building Envelope (TABE) is a recent technology that uses a sealed storage container to store the PCM [92]. Not only is the storage container well suited to handle the volumetric expansion of the PCM, but it is able to make use of the change in pressure to determine the state of charge of the thermal battery. A heat transfer fluid exchanges heat directly between the storage container and a heat exchanger integrated into the building envelope. The thermal performance of the test system at Oak Ridge National Lab is promising, but the cost of heat exchangers integrated into the building could be a limiting factor.

12.3.3 PCM Heat Exchangers

PCMs are promising materials for thermal storage when coupled to HVAC systems directly. However, they require a heat exchanger that can effectively melt the PCM by adding heat or freeze the PCM by removing heat. As discussed in Section 12.2.2, PCMs have low thermal conductivity, less than 1 W/m-K. By incorporating thermally conductive fillers such as graphite, thermal conductivity can be increased up to 25 W/m-K [93], making it easier to design this heat exchanger.

Figure 12.5 shows an example of a PCM heat exchanger, in this case being melted by an incoming hot heat transfer fluid. Between the heat exchanger tubes, the PCM melts – first near the surface of the hot tube, and then toward the centerline between the tubes. Reducing the spacing between the heat exchange tubes will speed up the melting process but with additional weight and cost.

The design process for PCM heat exchangers can be summarized by two desired quantities: the stored thermal energy (kWh) and the thermal power (kW). Like many other engineering systems, there is a tradeoff between energy and power. Woods et al. described this tradeoff, and how to quantify it, by drawing an analogy between electrochemical batteries and PCM thermal storage devices [94]. Their framework describes thermal rate capability plots, with the outlet temperature from the device plotted against the cumulative discharged energy. Higher powers will lead to discharge temperatures further from the ideal temperature – the PCM transition temperature. If the outlet temperature exceeds (for cooling) or falls below (for heating) a cutoff temperature, then the remainder of the battery is deemed unusable for this power, which will significantly decrease the available stored thermal energy. This is similar to cutoff voltages used for electrochemical batteries.

After evaluating a storage device at multiple discharge power levels, Ragone plots can be generated that are often used to quantify the power-energy tradeoff for energy

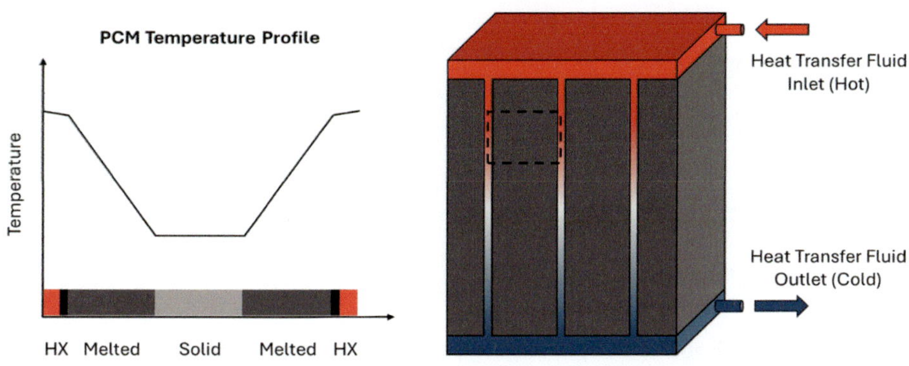

FIGURE 12.5 PCM heat exchanger schematic with a PCM temperature profile representing cold storage (discharge).

storage systems. A more recent paper showed how this Ragone framework can be used to optimize a PCM storage device for energy density, initial cost, or life-cycle cost [95].

12.3.4 Thermochemical Reactors

Thermochemical energy storage (TCES) reactors have unique design complexities due to a requirement for mass transfer between the storage material and heat transfer fluid. Considering that the materials for building applications are salt hydrates and adsorbents, these can be categorized as two types of designs: (1) open reactors, where the material reacts with water vapor present in ambient air, and (2) closed reactors, where the material reacts with pure water vapor in a controlled $p-T$ chamber, as illustrated in Figure 12.6. Each type of reactor has distinct mechanisms to control the reaction, with advantages and challenges.

Open systems utilize water vapor available in moist air to hydrate the storage material. This configuration requires fewer additional components to interface with the HVAC system, as the salt hydrate can heat or dehumidify the building air directly, making it a cost-effective option. The charge/discharge cycle is governed by the ambient air temperature and vapor pressure, as well as the phase diagram of the storage material. Therefore, salt hydrates must be selected according to the local climate and seasonal variability [96, 97]. Open reactors face regional limitations if the goal is load shifting via diurnal charge/discharge, but they are naturally well-suited to seasonal storage. Researches have established thermodynamic analyses and conducted experimental studies to guide the design of open reactors and predict their performance specifically for these applications [50, 98].

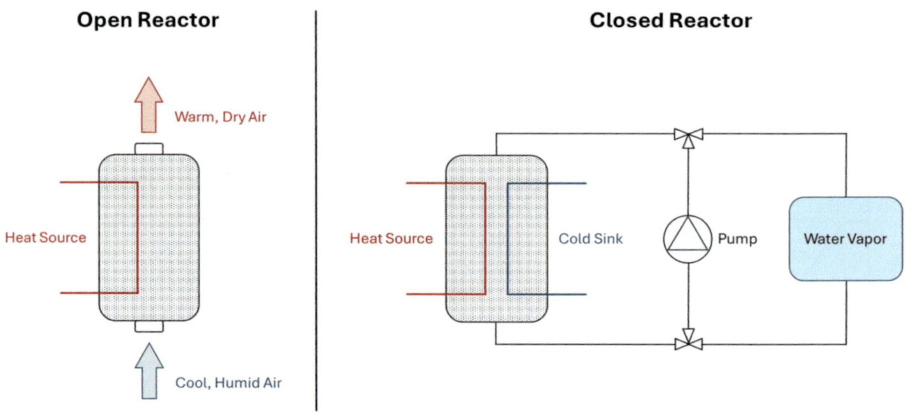

FIGURE 12.6 Schematics of open and closed thermochemical reactors that utilize packed beds of TCMs for energy storage. The open reactor exchanges heat and mass directly with ambient air (moist air is the reacting gas). The closed reactor operates under a vacuum to control reaction conditions (pure water vapor is the reacting gas), and it exchanges only heat with the surroundings.

Closed systems interface with the HVAC system using additional components such as heat exchangers to charge/discharge the storage material. The additional complexity and heat exchangers are likely to drive up the cost of storage, but closed systems can control operating conditions to achieve ideal reaction conditions. Water vapor is condensed and stored in a water tank, and HVAC components including the evaporator and condenser can control the temperature and vapor pressure. Additionally, the absence of inert gases allows for greater temperature lift and improved reaction kinetics [54]. Experimental demonstrations of closed reactors are an active area of research. In a system based on K_2CO_3, it was shown that powder beds are unsuitable for large-scale applications, and mm-scale particles should be investigated [99]. Additionally, impurities in storage materials such as $KHCO_3$ can lead to the formation of noncondensable gases, which significantly affect reaction rates. Purifying the storage material and degassing the reactor water may be necessary to achieve consistent performance over multiple charge/discharge cycles.

Whether the reactive gas is moist air or pure water vapor, thermochemical reactors are promising for long-duration energy storage. Abedin et al. [27] applied an energy- and exergy-based analysis of both types of reactors to evaluate the efficiency of thermochemical storage and associated losses. The results indicated that thermochemical storage is generally capable of similar efficiencies as other forms of thermal storage with a more compact form, but differences in thermodynamic performance between open and closed reactors are insufficient to choose one or the other. Rather, reactor selection must consider other factors such as differences in environment, economics, degradation, and maintenance requirements [27]. Detailed reactor models developed by Michel et al. [100] reach a similar conclusion with performance quantified as specific reaction power (W kg^{-1}) for a given set of operating parameters. The results indicated that heat transfer is the main limitation to the reaction rate in a closed reactor, whereas mass transfer is the limitation in an open reactor. These findings provide guidance for materials development for thermochemical storage, such as fillers to increase thermal conductivity or permeability.

12.4 HVAC INTEGRATED THERMAL STORAGE

There is significant research on thermal storage materials and devices, but to help decarbonize buildings and the electric grid, these materials and devices must ultimately be combined with HVAC systems. HVAC systems use electricity or combust fossil fuels to heat or cool the building, and therefore the storage must support these HVAC systems to improve their efficiency, enable the installation of electrified heating equipment, or shift HVAC system operation from times when the grid carbon intensity is high to times when it is low. This is true even for envelope-integrated thermal storage, except that the only interaction between the envelope and the HVAC system is related to controls, since the HVAC system is not connected physically to the TES materials in the envelope.

The options for integrating thermal storage with HVAC systems are numerous. In this section, we first describe these diverse HVAC systems and the types of buildings where they are used. We then describe options for integrating thermal storage with HVAC, using both existing, commercially available systems, and other approaches described in the literature.

12.4.1 HVAC System Types

Buildings are diverse, and the HVAC systems used in buildings are varied, which also necessitates different types of TES. There are other uses of energy storage with diverse applications – the most ubiquitous is electrochemical battery storage, which is used in watches, laptops, automobiles, and trucks. However, the basic design of these batteries is similar, even if their applications are numerous. Some electric vehicles were designed around packs of laptop batteries, for example.

Figure 12.7 shows the types of equipment used for space heating and cooling of buildings. The remainder of this section describes these HVAC system types in more detail, and some high-level considerations for adding TES to these systems.

For commercial buildings in the United States, central plants account for 20–25% of heating and cooling systems [101]. This is typically achieved using gas-driven boilers for heating and large electric chillers for cooling [39]. Both systems are coupled with hydronic loops, pumping either water or an antifreeze solution around the building for either radiant heating or for water-based air-handling units that use this hot or cold water to heat or cool air in the building [39]. Integrating storage with these central plants is a common TES solution used today. Figure 12.8 shows an example of a small central chiller system, which connects to four ice storage tanks. The four 160-tonh tanks can be charged at night by the chiller, and then these tanks can supplement the chiller during the day, lowering the building's utility costs that normally come from high daytime energy and demand charges. The glycol is pumped between the chiller, ice tank, and the building load, depending on the operating mode.

Nearly half of the commercial building floor area in the United States is served by packaged HVAC systems [101]. Many of these are rooftop units (RTU), which pull in return air from the building, mix it with outdoor ventilation air, and then return it to the building after heating or cooling the air. Rooftop units sit on a "curb," which adapts the inlet and outlet ports of the RTU to the return and supply ducts in the building.

One challenge with adding TES to rooftop units is the additional load this gives to the roof. For example, if ice storage is added to a 6-ton rooftop unit (21 kW_{th}), and it enables compressor-less operation for 6 hours, this requires 36 tonh (127 kWh_{th}) of storage. Ice has an enthalpy of fusion of 333 kJ kg^{-1}, so this application needs 1370 kg, or 3000 lbs of water (not including any pumps, valves, or heat exchangers). Considering that a high-efficiency 6-ton rooftop air conditioner weighs less than 1000 lbs [103], this ice storage would increase the weight by 300%. Depending on the building, this can require additional structural support for the roof, which can drive up installation costs significantly. Partial load shifting with rooftop units could allow them to still use TES, by operating the compressor at reduced capacity instead of turning it off entirely for 6 hours (e.g., see Goyal et al. [104]).

For residential buildings, most air conditioners are split systems – the outdoor unit includes the outdoor condenser heat exchanger, compressor, and outdoor fan, and the indoor unit includes the indoor fan, expansion valve, and indoor fan. Most of these systems are installed and then *charged* with refrigerant in the field. Charging is the process of first evacuating air from the pipes, compressor, and heat exchangers, and then filling them with refrigerant. The indoor unit is often in a tight space inside the

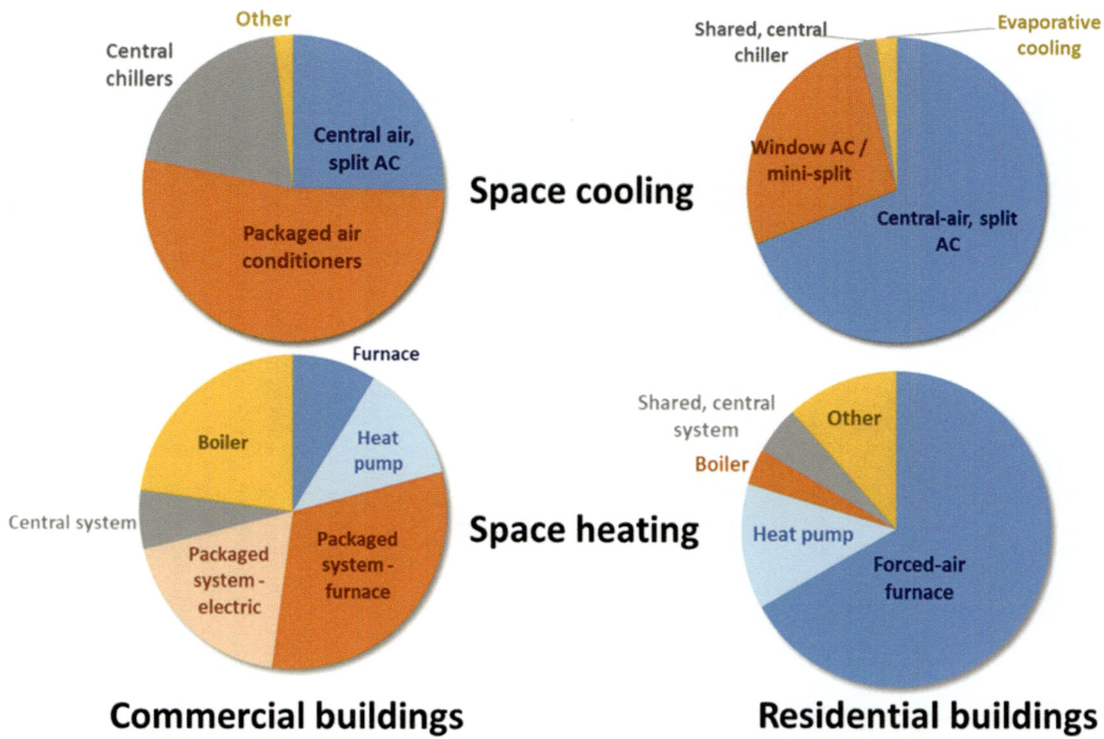

FIGURE 12.7 Type of space heating and cooling equipment for commercial and residential buildings. The data for commercial buildings is shown as the fraction of conditioned floor area served by that type of HVAC system, while residential is shown as the fraction of households served by that type of HVAC system.

(Source: [101, 102].)

FIGURE 12.8 Central chiller plant with four 160-tonh ice storage tanks (right). The chiller (top right) is used to charge the tanks at night. The balance of plant (left) includes pumps and valves to direct the fluid between the ice tanks, chiller, and the building cooling system.

(Photo credit: Trane Technologies, used with permission.)

building, and it is difficult to add TES in this space. Adding it outdoors is easier, but this means it is difficult to have the TES interact with the indoor load except through the existing refrigerant lines going into the building.

The above paragraphs focus on adding TES to cooling systems, but electrified heating systems can also benefit from TES. For commercial buildings with central plants, eliminating on-site fossil fuel combustion for heating requires replacing gas-driven boilers with hydronic heat pumps, such as air-to-water heat pumps. These are essentially chillers but acting in reverse. It is possible to use the same storage system that is used for cooling. A system that uses a storage-source heat pump was recently analyzed, which showed that the same ice storage tanks that are used for cooling could also be used for heating [105]. In this system, the ice acts as the heat source for the heat pump when the ambient temperature is low.

For packaged commercial HVAC systems, sensible storage can be used if an application can provide a large change in temperature. This could be done for heating the cold ventilation air in cold climates [106]. Sensible storage in domestic hot-water tanks can also be used to provide load shifting for the electric grid [107].

For residential buildings, PCM storage can be integrated into the heat pumps in different ways. A recent paper analyzed how it can be used as the heat source (during heating) or heat sink (during cooling), allowing the same PCM to be used for both heating and cooling. The study showed how the choice of transition temperature will impact both the reduction in peak demand and the total energy use [42].

TCMs can also be integrated into buildings. For open systems, it is important to consider the importance of having sufficient humidity available to drive the reaction. A recent paper showed how integrating these thermochemical reactors to heat incoming ventilation air can help overcome this issue [97].

12.4.2 Additional Considerations

There are two questions to consider when determining the best way to integrate TES into a certain HVAC system:

1. What is the most effective way to move heat into and out of the thermal storage?
2. How do we minimize additional engineering, manufacturing, and installation costs?

These questions are general, so we consider some examples. For question (1), the TES needs to interact with a heat transfer fluid, which then interacts with the building (discharge the storage) or the HVAC system (charge the storage). Fluids common in buildings are refrigerants (used in heat pumps), water or antifreeze solutions, and air. Figure 12.9 shows possible locations for TES in a typical HVAC system, in this case a system for space heating (a similar diagram, not shown, could be made for space cooling). There are seven numbered locations in Figure 12.9, indicating possible locations for TES. The first (TES1) is when thermal storage is added to the heat-source heat exchanger. This could be the ground in a borehole energy storage [79, 80], for example. TES2 is located in a secondary fluid loop, such as water/glycol, and could be used as the heat source instead of the ambient heat exchanger

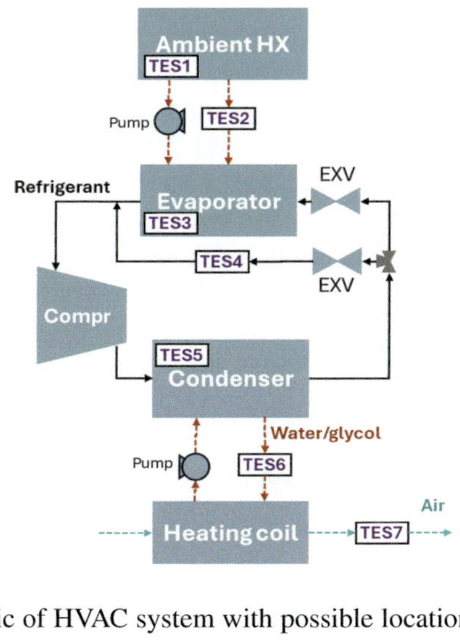

FIGURE 12.9 Schematic of HVAC system with possible locations for thermal energy storage (TES).

(HX). This would be advantageous if the TES2 is charged with the heat pump at some earlier time, and it is now much warmer than the ambient. TES3 would require a combined refrigerant-water-TES heat exchanger [104]. TES4 is similar but is only a refrigerant-to-TES heat exchanger. It can be charged with the heat pump, assuming a reversing valve, and then used as the heat source as an alternative to the conventional evaporator that gets its heat from ambient (e.g., Moreno et al.) [108]. TES5 is integrated into the condenser, and like TES3 includes refrigerant, water, and a TES material. The condensing refrigerant would charge the TES, and then the TES could be used to meet the load when the compressor is off. TES6 is similar to TES2 but is kept at a high enough temperature so it can directly heat the building while the heat pump is off. Finally, TES7 is integrated into the air duct. It can be charged with hot air from the heat pump, and then later used to heat the air entering the building when the heating coil is inactive.

Each of the options shown in Figure 12.9 will work to provide load shifting, or potential reduce energy use. Some examples of these can be found in more in-depth review articles [108–110]. However, the system must also include other valves and dampers (not shown) to be able to charge and discharge the TES during different operating modes. There are pros and cons to each location, and many of these are related to which fluid they interact with.

Refrigerants, which change phase while transferring heat, are very effective heat transfer fluids, and could effectively move heat into and out of a TES. This would also require minimal additional components because refrigerants are already used in heat pumps, air conditioners, and chillers. However, one challenge with using refrigerant is that it requires many tubes inside the TES, to transfer heat between the TES and the refrigerant. This adds a significant amount of refrigerant charge to the system, and managing this in each of the operating modes can be challenging [111]. This approach is used in some commercially available systems, as well as research prototypes.

Water and water/glycol mixtures are commonly used in commercial buildings [39]. This could be from a chiller that creates cold water that is then pumped throughout a building to individual air-handling units. These air-handling units exchange heat between the indoor air and the water [39]. Integrating storage with these systems is relatively straightforward because the working fluid (water) can be a storage medium. For example, campuses can use large hot-water and/or cold-water storage tanks [112]. Adding phase-change material heat exchangers (e.g., an ice tank) that exchange heat with water as the heat transfer fluid is commercially available.

Air is not a good heat transfer fluid due to its low density, low specific heat, and low thermal conductivity. However, it is the most common heat transfer fluid in buildings since it is ultimately where the heating or cooling ends up – building occupants are in an air-filled room. Many of the packaged and split HVAC systems described thus far directly heat or cool air, which is then delivered to the building. Designing a TES that directly interacts with air can be challenging, because air is a poor heat transfer fluid. But it likely has one of the lowest incremental integration costs because the HVAC system, and its controls, can largely remain unchanged – the TES is simply added in a return or supply-air duct. Dampers may also be used to control when it is charged or discharged or to bypass the TES when it is not in use, to minimize pressure drop. One example of integrating TES on the airside is using open-system thermochemical reactors, which require humidity from the air to drive the hydration reaction [97]. Another example is using PCMs to provide free cooling during the day, after charging the PCM at night [113, 114].

12.4.3 Commercially Available HVAC Systems with TES

There are few TES systems available for residential or light commercial systems, including both split systems and packaged rooftop units. The rooftop units integrated with ice storage have seen little market penetration, primarily because of the weight constraints mentioned previously, and due to high cost. There continues to be research in this space, including developing novel heat exchangers [115, 116], and novel systems with PCM storage [117–119].

This section lists commercially available TES systems that integrate with HVAC systems:

- **Hot-water storage – water coupled**: Water tanks are already used in hot-water heaters because the thermal load for domestic water heating is very large but also intermittent, especially for uses like showers. The tank allows for the heating capacity to be much smaller than the load – a 15 kW water heater may be suitable for a building that has periodic 25 kW heating loads, if the tank is large enough. These tanks are also used by electric utilities to increase or decrease demand to balance generation and load on the grid.
- **PCM thermal batteries – water/glycol coupled**: A heat exchanger inside a PCM tank can add or remove heat from the tank when the water/glycol is above or below the PCM transition temperature, respectively. These thermal batteries are commercially available and have been explored in the literature [93, 120], but no HVAC system on the market uses this TES currently.
- **Chilled-water tanks or ice storage with chiller plants – water/glycol coupled**: There are many examples of sensible chilled-water tanks, or ice-based storage systems, for building cooling in central chiller plants. Some of these systems use off-the-shelf ice tanks, while others use custom-built ice or chilled-water storage tanks [121].
- **Electric thermal storage bricks – air coupled**: Sensible storage can be energy dense if the change in temperature is very large. Heating ceramic bricks with embedded electric resistance heaters is one of achieving this. In some commercially available systems, the bricks are heated to 650°C, and then later used to directly heat air that passes across these bricks [122].
- **Air-to-PCM heat exchangers – air coupled**: There are commercially available air-to-PCM heat exchangers that can be directly integrated into air ducts in buildings [93].
- **Ice storage for packaged air conditioners – refrigerant coupled**: Packaged rooftop units can be coupled with an ice maker that uses cold refrigerant coils to directly make ice. The ice is then used during the day to provide cooling, by pumping the refrigerant into the air conditioner's heat exchanger coil [123].

12.4.4 Other TES Applications in Buildings

In addition to the heating and cooling applications discussed thus far, there are two other potential uses for TES in buildings. This corresponds to two thermal loads that have high electricity consumption: refrigeration, and dehumidification.

Unlike comfort heating and cooling, refrigeration of products is a year-round load in all climates. Refrigeration systems remove heat from these products, and the surrounding air, and reject that heat either into the building or outside. Integration of TES into refrigeration systems mirrors many of the same integration decisions as above, including multiple locations in the heat pump system where the TES can interact with refrigerant, a secondary fluid, and/or air. Several reviews have covered this topic [124, 125].

Dehumidification is the removal of water from the air. Keeping buildings comfortable requires the removal of water vapor from the air, in addition to lowering the air temperature. The most common method for air dehumidification is to condense moisture from the air using a cold surface, and large amounts of energy are required to maintain this cold surface while removing the heat released from condensation. Recent research has shown that the energy use and carbon footprint of dehumidification are equal to that of controlling temperature [25]. An alternative to the cold-surface condensation method is to use liquid desiccants to remove water from the air [126]. There are potential energy savings benefits of this technology, but more relevant to this chapter is the ability to store the desiccant [127].

12.5 COST CONSIDERATIONS

The previous sections outlined the different types of thermal energy storage (materials and devices) and the associated integration strategies for heating and cooling in buildings. Key advantages include abundant materials, decoupling storage capacity and power, storing energy over a range of timescales from hourly to seasonal, and integration with HVAC to boost overall system efficiency. TES adoption however is driven primarily by cost considerations and incentives, as discussed in this section.

12.5.1 Levelized Cost of Storage

Like any storage, thermal storage systems have an initial capital cost and an operating cost. To compare systems fairly, a levelized cost is often used that considers both these aspects and includes the time value of money. The LCOS framework for TES was developed by Odukomaiya et al. [24], which is calculated per unit of electric energy (US$/kWh$_e$). The LCOS includes the initial cost, electrical efficiency of charging the TES, electrical efficiency of the equipment that the TES is displacing, TES lifetime, thermal roundtrip efficiency, discount rate, and the number of equivalent full cycles the thermal storage is used per year. The number of cycles is an important parameter – a storage used only once per year, such as for seasonal storage, will have a much higher LCOS. This means seasonal storage materials and systems must have very low upfront costs to be cost-effective. TES can be less expensive than electrochemical batteries, primarily because of the lower cost materials used in thermal storage (e.g., water, salts, oils) compared with batteries (e.g., lithium, cobalt, other metals). However, additional components are needed to move heat in/out of TES, as well as to integrate it with HVAC in buildings that can contribute to the overall cost – this is discussed in the following section.

12.5.2 Cost Analysis

12.5.2.1 Cost Scaling in PCMs

Kocher et al. [128] developed a cost scaling analysis for PCM-based thermal batteries represented in Figure 12.5. The contributions to cost include the storage material, heat exchanger, and insulation. The storage material is a volumetric cost, and it is proportional to the PCM density and cost per mass. Heat exchanger costs scale with PCM surface area, and they are proportional to the heat transfer coefficient [129]. The insulation cost scales with the external surface area of the component and the R-value of the insulation material. Therefore, a general expression for total cost of a PCM battery can be developed in terms of simple material properties and geometric inputs.

Like electrochemical batteries, thermal batteries do not fully deplete their storage capacity during discharge. They must discharge at the desired heat transfer rate, and there is a cutoff point in the state of charge where the battery can no longer meet this requirement, as discussed in Section 12.3.2 [94]. Deriving the cost per unit of energy storage (US$/kWh) relies on solving for that cutoff in the state of charge [128]. This approach is used to generate a cost regime map and figures of merit to minimize cost based on which contribution is dominant. Heat exchanger costs often dominate in PCM thermal batteries. In that case, lowering the storage material cost would have minimal effect on the overall cost of the component. Instead, the figure of merit indicates that thermal effusivity should be maximized in the storage material. Thus, efforts to fabricate PCMs with thermally conductive fillers will be important to drive down the cost of thermal storage.

12.5.2.2 Cost Scaling in Sensible Storage

Cost scaling for large-scale, water-based TES is straightforward compared to other thermal storage devices. Normally, the capital cost of a thermal battery consists of a balance of contributions from the storage material, heat exchangers, and insulation as discussed above. However, aquifers make use of freely available water as the storage material. Additionally, heat losses to the ambient environment become less significant at this scale due to an increase in the ratio of volume to surface area. Schmidt et al. compared the investment cost per volume for completed projects across all four design concepts of large-scale TES [79]. Specifically, this is defined as a water equivalent storage volume, which takes into account a decrease in heat capacity from sand or gravel being mixed into the water. While there are significant cost differences between these designs at intermediate scales, these do not take into account maintenance costs or thermal performance at the system level. The overwhelming trend is that cost decreases as the storage volume increases. Ultimately, storage costs below 50 €/m^3 were achieved, and a minimum appropriate size for underground thermal storage was defined as 2000 m^3 [79].

12.5.3 Incentives

The capital cost of thermal energy storage for buildings is heavily influenced by incentives. The United States recently introduced the first federal incentives for thermal storage. Previously, energy storage projects could only qualify for the Investment Tax Credit (ITC) if they were tied to solar facilities. The Inflation Reduction Act of 2022 expanded the ITC to include standalone energy storage [130]. It specifically applies to TES, which is directly connected to the HVAC system in residential or commercial buildings. There are also three defined exclusions: swimming pools, combined heat and power systems, and buildings or their structural components. The ITC includes a base credit of 6% of the basis of the energy property, and the bonus rate structure makes most TES eligible for up to 40% [131].

Incentives in the United States are lacking at the state level, even in states at the forefront of renewable sources and energy storage. Notably, California is a leader in regulations and incentives surrounding electrochemical storage but lacks any guidance on thermal storage. Instead, incentives are applied at the utility level, especially for cold storage in commercial buildings. These are often in the form of rebates for offsetting demand (US$/kW) during peak hours in the summer [132]. In addition to demand response, new programs can be implemented to encourage peak shifting. Xcel Energy offers a Dynamic Thermal Storage (DTS) program for commercial buildings in Minnesota [133]. This is a daily, 4-hour shifting program where customers use cold storage to lower refrigeration or air conditioning demands during the afternoon. Thermal storage devices such as PCM pallets or large ice tanks are eligible to receive 10% of the project cost after installation, and 5% of the project cost annually while the operation remains in compliance. This sort of incentive model can be beneficial to both parties by incentivizing proper maintenance and load shifting over the lifetime of the thermal storage device.

Incentive programs are also being tested for residential thermal storage. Arizona Public Service recently piloted the Reserve Rewards program in 2018, which offered customers an instant rebate for the full installed cost of a smart water heater that generates domestic hot water when excess solar generation is available [134]. This type of incentive program can address the problem with small-scale TES in that energy savings may be small at the individual level, but they become significant when aggregated. A significant challenge is to develop standards for the controls and communications between the utility and the water heaters. A similar study was conducted between Pacific Northwest National Lab (PNNL), Bonneville Power Administration, Portland General Electric, and 10 utilities in the northwest United States. In this test, about 250 water heaters were installed using the CTA-2045 communications protocol to shift peak loads and take advantage of renewable energy sources [135]. The CTA-2045 protocol was developed specifically for demand response programs, providing a standard module (US$150 per unit) with two-way communications between the electricity provider and the end use [136].

12.6 CONCLUSIONS AND OUTLOOK

Natural gas and other fossil fuels dominate building heating systems, contributing significantly to greenhouse gas emissions. The need for decarbonization of the buildings sector is driving a shift toward more sustainable heating and cooling solutions.

Recently, there has been an emphasis on electrification, such as the widespread adoption of heat pumps, which decrease on-site combustion sources and offer a more efficient alternative for heating. Recent market trends underscore this transition – Americans bought over 4 million heat pumps in 2022, outpacing sales of gas furnaces [137]. Notably, the trend precedes federal incentives from the Inflation Reduction Act. Annual sales of heat pumps more than doubled over the prior 10 years, which may be attributed to decreasing costs and technology improvements for cold weather [137]. The heat pump market is also expanding globally – while North America has the largest adoption, the European Union has the fastest-growing market [6]. The IEA estimates that heat pumps have the potential to reduce CO_2 emissions globally by over 500 million tons by 2030 [6].

While beneficial for decarbonization, the widespread adoption of heat pumps presents a challenge for energy infrastructure. Peak loads are a limiting factor, especially in cold climates, where fossil fuel heat would still be required even with advanced heat pumps [138]. Projections indicate that the United States faces an increased risk of electricity shortfalls by the 2030s due to increased demand from data centers and the electrification of buildings and transportation [139]. In this scenario, energy storage can improve the reliability of the electric grid and contribute to decarbonization efforts by shifting peak loads. A variety of storage technologies will be necessary to offset different types of loads. This chapter discussed the role of thermal energy storage technologies based on sensible, latent, and thermochemical heat to manage thermal loads in buildings.

Water is the most common medium in sensible heat storage due to its favorable thermophysical properties, such as high density and specific heat, which lead to relatively high volumetric storage capacity. Its low cost and environmental impact also present significant advantages in large-scale applications. Additionally, water acts as the heat transfer fluid linking the storage component to the load. This allows for practical design considerations such as thermal stratification in hot-water tanks or combining water-based storage with rock beds. Unlike other forms of thermal storage, where there is ongoing fundamental research and materials development, advances in water-based storage are focused on practical challenges. For example, implementing communication modules and standards would enable control of hot-water tanks in aggregate at the utility scale for efficient load shifting.

PCMs are commercially available for building thermal management – ice is the most common substance for cold storage and paraffin for hot storage. Material characteristics such as energy density and phase transition temperature are generally well established, but characterization techniques on small samples may differ from real properties in a PCM thermal battery. Ongoing research efforts in PCMs include a mix of fundamental materials development and practical design considerations. One of the key limitations of PCMs is low thermal conductivity, which significantly increases the cost of storage. Integrating conductive fillers and PCMs into stable form factors can both increase thermal conductivity and eliminate challenges like incongruent melting.

Thermochemical storage is promising due to its high energy density, but it requires more research and development to be considered for building applications. Material degradation is a limitation to the long-term performance of TCMs. Heat and mass transfer impacts on reaction kinetics are not fully understood, so fundamental studies on reaction mechanisms in storage materials are necessary to develop new strategies to mitigate degradation. This will also require efforts to standardize material characterization techniques. Experimental demonstrations have shown that thermochemical reactors can match the performance of other storage systems, and they have also highlighted some of the practical challenges in reactors. Optimal reactor design appears to be heavily dependent on storage material properties, so these efforts must continue to evolve as improved materials are developed.

The link between building thermal loads and electricity use is the HVAC system. This thus makes the cost-effective and efficient integration of thermal storage with HVAC systems an important research topic. Most commercially available systems include water or ice-based storage integrated with hydronic systems, such as chiller plants. These systems are relatively mature, but integrating thermal storage with smaller packaged systems, which are more common in buildings, is challenging. Research is underway to develop thermal storage that minimizes the cost of this integration and maximizes the impact each kWh of thermal storage can have on the electric load profile of HVAC systems.

REFERENCES

1. Jackson, R.B., et al., Global energy growth is outpacing decarbonization. *Environmental Research Letters*, 2018. **13**(12).
2. Intergovernmental Panel on Climate, C, *Global Warming of 1.5°C: IPCC Special Report on Impacts of Global Warming of 1.5°C above Pre-industrial Levels in Context of Strengthening Response to Climate Change, Sustainable Development, and Efforts to Eradicate Poverty*. 2022, Cambridge: Cambridge University Press.
3. IEA, *World Energy Outlook 2021*. 2021, Paris: International Energy Agency (IEA).
4. Langevin, J., et al., *Decarbonizing the U.S. Economy by 2050: A National Blueprint for the Buildings Sector*. 2024: United States.
5. Levesque, A., et al., Deep decarbonisation of buildings energy services through demand and supply transformations in a 1.5°C scenario. *Environmental Research Letters*, 2021. **16**(5).
6. International Energy Agency (IEA), *The Future of Heat Pumps*. 2022.
7. Camarasa, C., et al., A global comparison of building decarbonization scenarios by 2050 towards 1.5-2 degrees C targets. *Nature Communications*, 2022. **13**(1): p. 3077.
8. Ang, Y.Q., et al., Carbon reduction technology pathways for existing buildings in eight cities. *Nature Communications*, 2023. **14**(1): p. 1689.
9. Commission, E, *Energy Performance of Buildings Directive*. 2024; Available from: https://energy.ec.europa.eu/topics/energy.efficiency/energy.efficient-buildings/energy-performance-buildings-directive_en#documents
10. EEA, *Annual European Union greenhouse gas inventory 1990-2021 and inventory report 2023*. 2023.
11. European Commission, *Renovation Wave: Doubling the Renovation Rate to Cut Emissions, Boost Recovery and Reduce Energy Poverty*. 2020: Brussels.
12. Lizana, J., et al., Advances in thermal energy storage materials and their applications towards zero energy buildings: A critical review. *Applied Energy*, 2017. **203**: pp. 219–239.
13. Gür, T.M., Review of electrical energy storage technologies, materials and systems: Challenges and prospects for large-scale grid storage. *Energy & Environmental Science*, 2018. **11**(10): pp. 2696–2767.

14. Dunn, B., H. Kamath, and J.-M. Tarascon, Electrical energy storage for the grid: A battery of choices. *Science*, 2011. **334**(6058): pp. 928–935.
15. Albertus, P., J.S. Manser, and S. Litzelman, Long-duration electricity storage applications, economics, and technologies. *Joule*, 2020. **4**(1): pp. 21–32.
16. Ma, Z., et al., Thermal energy storage for energy decarbonization, in *Annual Review of Heat Transfer*, Z.M. Zhang, V. Prasad, and Y. Jaluria, Editors. 2022, Begell House Inc. pp. 51–115.
17. Henry, A., R. Prasher, and A. Majumdar, Five thermal energy grand challenges for decarbonization. *Nature Energy*, 2020. **5**(9): pp. 635–637.
18. U.S. Department of Energy. *Energy Earthshots Initiative*. [cited 2025].
19. Gilbert, T., et al., Heat source and application-dependent levelized cost of decarbonized heat. *Joule*, 2023. **7**(1): pp. 128–149.
20. Booten, C., et al., Theoretical minimum thermal load in buildings. *Joule*, 2021. **5**(1): pp. 24–46.
21. Sun, Y., et al., Peak load shifting control using different cold thermal energy storage facilities in commercial buildings: A review. *Energy Conversion and Management*, 2013. **71**: pp. 101–114.
22. IEA, *The Power to Choose: Demand Response in Liberalised Electricity Markets*. 2003, Paris: International Energy Agency (IEA).
23. Braun, J.E., *Load control using building thermal mass. Journal of Solar Energy Engineering*, 2003. **125**: pp. 292–301.
24. Odukomaiya, A., et al., Addressing energy storage needs at lower cost via on-site thermal energy storage in buildings. *Energy & Environmental Science*, 2021. **14**(10): pp. 5315–5329.
25. Woods, J., et al., Humidity's impact on greenhouse gas emissions from air conditioning. *Joule*, 2022. **6**(4): pp. 726–741.
26. Aydin, D., S.P. Casey, and S. Riffat, The latest advancements on thermochemical heat storage systems. *Renewable and Sustainable Energy Reviews*, 2015. **41**: pp. 356–367.
27. Abedin, A.H. and M.A. Rosen, *Closed and open thermochemical energy storage: Energy- and exergy-based comparisons*. Energy, 2012. **41**(1): pp. 83–92.
28. Henze, G.P., T.H. Le, and A.R. Florita. Sensitivity analysis of optimal building thermal mass control. in *International Solar Energy Conference*. 2005. Orlando, Florida: ASME.
29. Lilley, D., et al., Phase change materials for thermal energy storage: A perspective on linking phonon physics to performance. *Journal of Applied Physics*, 2021. **130**(22).
30. Ashcroft, N.W. and N.D. Mermin, *Solid State Physics*. 1976: Cengage Learning.
31. Tatsidjodoung, P., N. Le Pierrès, and L. Luo, A review of potential materials for thermal energy storage in building applications. *Renewable and Sustainable Energy Reviews*, 2013. **18**: pp. 327–349.
32. Verbeke, S. and A. Audenaert, Thermal inertia in buildings: A review of impacts across climate and building use. *Renewable and Sustainable Energy Reviews*, 2018. **82**: pp. 2300–2318.
33. Singh, H., R.P. Saini, and J.S. Saini, A review on packed bed solar energy storage systems. *Renewable and Sustainable Energy Reviews*, 2010. **14**(3): pp. 1059–1069.
34. Alkilani, M.M., et al., Review of solar air collectors with thermal storage units. *Renewable and Sustainable Energy Reviews*, 2011. **15**(3): pp. 1476–1490.
35. Shah, I.H., et al., Cement substitution with secondary materials can reduce annual global CO(2) emissions by up to 1.3 gigatons. *Nature Communications*, 2022. **13**(1): pp. 5758.
36. DEGREES: Degradation Reactions in Electrothermal Energy Storage. Available from: www.nrel.gov/degrees/

37. Hasnain, S.M., Review on sustainable thermal energy storage technologies, part I: Heat storage materials and techniques. *Energy Conversion and Management*, 1998. **39**(11): pp. 1127–1138.
38. Baetens, R., B.P. Jelle, and A. Gustavsen, Phase change materials for building applications: A state-of-the-art review. *Energy and Buildings*, 2010. **42**(9): pp. 1361–1368.
39. American Society of Heating, R and I. Air-Conditioning Engineers, 2020 *ASHRAE Handbook - HVAC Systems and Equipment (I-P Edition)*. American Society of Heating, Refrigerating and Air-Conditioning Engineers, Inc. (ASHRAE).
40. Heier, J., C. Bales, and V. Martin, Combining thermal energy storage with buildings – A review. *Renewable and Sustainable Energy Reviews*, 2015. **42**: pp. 1305–1325.
41. Safari, A., et al., A review on supercooling of Phase Change Materials in thermal energy storage systems. *Renewable and Sustainable Energy Reviews*, 2017. **70**: pp. 905–919.
42. Hirschey, J., et al., Demand reduction and energy saving potential of thermal energy storage integrated heat pumps. *International Journal of Refrigeration*, 2023. **148**: pp. 179–192.
43. Kumar, N., et al., Review of stability and thermal conductivity enhancements for salt hydrates. *Journal of Energy Storage*, 2019. **24**.
44. Yau, Y.H. and B. Rismanchi, A review on cool thermal storage technologies and operating strategies. *Renewable and Sustainable Energy Reviews*, 2012. **16**(1): pp. 787–797.
45. Hasnain, S.M., Review on sustainable thermal energy storage technologies, part II: Cool thermal storage. *Energy Conversion and Management*, 1998. **39**(11): pp. 1139–1153.
46. Li, G., et al., Review of cold storage materials for subzero applications. *Energy*, 2013. **51**: pp. 1–17.
47. ASTM, *Standard Test Method for Determining Specific Heat Capacity by Differential Scanning Calorimetry*. 2024.
48. Shukla, N., et al., Lab-scale dynamic thermal testing of PCM-enhanced building materials, in *Next-Generation Thermal Insulation Challenges and Opportunities*, T.K. Stovall and T. Whitaker, Editors., West Conshohocken, PA: ASTM International, pp. 1–13.
49. Hasnain, S.M. and B.M. Gibbs, Determination of thermophysical properties of heat storage materials for solar heating applications, in *Clean and Safe Energy Forever*. 1990. pp. 601–605.
50. Zhang, Y.N., R.Z. Wang, and T.X. Li, Experimental investigation on an open sorption thermal storage system for space heating. *Energy*, 2017. **141**: pp. 2421–2433.
51. Solé, A., et al., Review of the T -history method to determine thermophysical properties of phase change materials (PCM). *Renewable and Sustainable Energy Reviews*, 2013. **26**: pp. 425–436.
52. DeHoff, R., *Thermodynamics in Materials Science*. 2nd Edition ed. 2006: CRC Press.
53. Rommel, M., A. Hauer, and W. van Helden, IEA SHC task 42/ECES annex 29 compact thermal energy storage. *Energy Procedia*, 2016. **91**: pp. 226–230.
54. Fumey, B., R. Weber, and L. Baldini, Sorption based long-term thermal energy storage – Process classification and analysis of performance limitations: A review. *Renewable and Sustainable Energy Reviews*, 2019. **111**: pp. 57–74.
55. N'Tsoukpoe, K.E., et al., A systematic multi-step screening of numerous salt hydrates for low temperature thermochemical energy storage. *Applied Energy*, 2014. **124**: pp. 1–16.
56. Zhang, Y. and R. Wang, Sorption thermal energy storage: Concept, process, applications and perspectives. *Energy Storage Materials*, 2020. **27**: pp. 352–369.
57. Thommes, M., et al., Physisorption of gases, with special reference to the evaluation of surface area and pore size distribution (IUPAC Technical Report). 2015. **87**(9–10): pp. 1051–1069.

58. Meunier, F., et al., A comparative thermodynamic study of sorption systems: second law analysis. *International Journal of Refrigeration*, 1996. **19**(6): pp. 414–421.
59. Kiyabu, S., et al., Computational screening of hydration reactions for thermal energy storage: New materials and design rules. *Chemistry of Materials*, 2018. **30**(6): pp. 2006–2017.
60. Donkers, P.A.J., et al., A review of salt hydrates for seasonal heat storage in domestic applications. *Applied Energy*, 2017. **199**: pp. 45–68.
61. Richter, M., et al., A systematic screening of salt hydrates as materials for a thermochemical heat transformer. *Thermochimica Acta*, 2018. **659**: pp. 136–150.
62. Glasser, L., Thermodynamics of inorganic hydration and of humidity control, with an extensive database of salt hydrate pairs. *Journal of Chemical & Engineering Data*, 2014. **59**(2): pp. 526–530.
63. Barbosa, E. and A.K. Menon, Thermochemical energy storage using salt mixtures with improved hydration kinetics and cycling stability. *Journal of Energy Storage*, 2024. **90**.
64. Chapter 7 Thermal dehydration of hydrated salts, in *Studies in Physical and Theoretical Chemistry*, A.K. Galwey and M.E. Brown, Editors. 1999, Elsevier. pp. 217–268.
65. Schaube, F., et al., A thermodynamic and kinetic study of the de- and rehydration of $Ca(OH)_2$ at high H_2O partial pressures for thermo-chemical heat storage. *Thermochimica Acta*, 2012. **538**: pp. 9–20.
66. Stengler, J., I. Bürger, and M. Linder, Thermodynamic and kinetic investigations of the $SrBr_2$ hydration and dehydration reactions for thermochemical energy storage and heat transformation. *Applied Energy*, 2020. **277**.
67. Sögütoglu, L.-C., et al., Understanding the hydration process of salts: The impact of a nucleation barrier. *Crystal Growth & Design*, 2019. **19**(4): pp. 2279–2288.
68. Birkelbach, F., M. Deutsch, and A. Werner, The effect of the reaction equilibrium on the kinetics of gas-solid reactions — A non-parametric modeling study. *Renewable Energy*, 2020. **152**: pp. 300–307.
69. Aarts, J., et al., Diffusion limited hydration kinetics of millimeter sized salt hydrate particles for thermochemical heat storage. *Journal of Energy Storage*, 2022. **47**.
70. Levitskij, E.A., et al., Chemical heat accumulators: A new approach to accumulating low potential heat. *Solar Energy Materials and Solar Cells*, 1996. **44**: pp. 219–235.
71. Aristov, Y.I., et al., Selective water sorbents for multiple applications, 1. $CaCl_2$ confined in mesopores of silica gel: Sorption properties. *Reaction Kinetics and Catalysis Letters*, 1996. **59**: pp. 325–333.
72. Clark, R.-J., et al., Experimental screening of salt hydrates for thermochemical energy storage for building heating application. *Journal of Energy Storage*, 2022. **51**.
73. Zalba, B., et al., Review on thermal energy storage with phase change: Materials, heat transfer analysis and applications. *Applied Thermal Engineering*, 2003. **23**: pp. 251–283.
74. Lau, J., et al., Dynamic tunability of phase-change material transition temperatures using ions for thermal energy storage. *Cell Reports Physical Science*, 2021. **2**(10): p. 100613.
75. Cabeza, L.F., Advances in thermal energy storage systems: methods and applications, in *Advances in Thermal Energy Storage Systems*. 2021. pp. 37–54.
76. Chandra, Y.P. and T. Matuska, Stratification analysis of domestic hot water storage tanks: A comprehensive review. *Energy and Buildings*, 2019. **187**: pp. 110–131.
77. Badar, M.A., S.M. Zubair, and A.A. Al-Farayedhi, Second-law-based thermoeconomic optimization of a sensible heat thermal energy storage system. *Energy*, 1993. **18**(6): pp. 641–649.
78. Furbo, S. and S. Knudsen, Improved design of mantle tanks for small low flow SDHW systems. *International Journal of Energy Research*, 2006. **30**(12): pp. 955–965.

79. Schmidt, T., et al. Design aspects for large-scale pit and aquifer thermal energy storage for district heating and cooling. in *16th International Symposium on District Heating and Cooling*. 2018. Hamburg, Germany.
80. Novo, A.V., et al., Review of seasonal heat storage in large basins: Water tanks and gravel–water pits. *Applied Energy*, 2010. **87**(2): pp. 390–397.
81. Akbarzadeh, A., J. Andrews, and P. Golding, Solar pond technologies: A review and future directions, in *Advances in Solar Energy*, Y. Goswami, Editor. 2005, Routledge. pp. 233–294.
82. Schmidt, T. and P.A. Sørensen. Monitoring results from large scale heat storages for district heating in Denmark. in *14th International Conference on Energy Storage*. 2018. Adana, Turkey.
83. Rabl, A. and C.E. Nielsen, Solar ponds for space heating. *Solar Energy*, 1975. **17**: pp. 1–12.
84. Kurt, H., F. Halici, and A.K. Binark, Solar pond conception - Experimental and theoretical studies. *Energy Conversion and Management*, 2000. **41**(9): pp. 939–951.
85. Tabor, H., Solar ponds. *Solar Energy*, 1981. **27**(3): pp. 181–194.
86. Sadineni, S.B., S. Madala, and R.F. Boehm, Passive building energy savings: A review of building envelope components. *Renewable and Sustainable Energy Reviews*, 2011. **15**(8): pp. 3617–3631.
87. Cui, Y., et al., A review on phase change material application in building. *Advances in Mechanical Engineering*, 2017. **9**(6).
88. Zahir, M.H., et al., Challenges of the application of PCMs to achieve zero energy buildings under hot weather conditions: A review. *Journal of Energy Storage*, 2023. **64**.
89. Kishore, R.A., et al., Modulating thermal load through lightweight residential building walls using thermal energy storage and controlled precooling strategy. *Applied Thermal Engineering*, 2020. **180**.
90. Wehmeyer, G., et al., Thermal diodes, regulators, and switches: Physical mechanisms and potential applications. *Applied Physics Reviews*, 2017. **4**(4).
91. Miao, R., et al., A non-volatile thermal switch for building energy savings. *Cell Reports Physical Science*, 2022. **3**(7).
92. Biswas, K., et al., Thermally anisotropic composites for improving the energy efficiency of building envelopes. *Energies*, 2019. **12**(19).
93. Mehling, H., M. Brütting, and T. Haussmann, PCM products and their fields of application - An overview of the state in 2020/2021. *Journal of Energy Storage*, 2022. **51**.
94. Woods, J., et al., Rate capability and Ragone plots for phase change thermal energy storage. *Nature Energy*, 2021. **6**(3): pp. 295–302.
95. James, N., A. Mahvi, and J. Woods, Optimizing phase change composite thermal energy storage using the thermal Ragone framework. *Journal of Energy Storage*, 2022. **56**.
96. Zaleski, N., E. Barbosa, and A.K. Menon, Evaluating the potential of thermochemical energy storage to decarbonize buildings: A case study in the United States. *MRS Energy & Sustainability*, 2024.
97. Zeng, Y., et al., Open-cycle thermochemical energy storage for building space heating: Practical system configurations and effective energy density. *Applied Energy*, 2024. **376**.
98. Marias, F., et al., Thermodynamic analysis and experimental study of solid/gas reactor operating in open mode. *Energy*, 2014. **66**: pp. 757–765.
99. Houben, J., et al., K_2CO_3 in closed heat storage systems. *Renewable Energy*, 2020. **166**: pp. 35–44.
100. Michel, B., P. Neveu, and N. Mazet, Comparison of closed and open thermochemical processes, for long-term thermal energy storage applications. *Energy*, 2014. **72**: pp. 702–716.

101. U.S. Energy Information Agency, *Commercial Buildings Energy Consumption Survey (CBECS)*. 2018.
102. U.S. Energy Information Agency, *Residential Energy Consumption Survey (RECS)*. 2020.
103. TRANE, *Product Catalog: PrecedentTM Packaged Rooftop Units*. 2024.
104. Goyal, A., E. Kozubal, and J. Woods. Integrated thermal energy storage system for air-conditioners with phase-change composites. in *International Refrigeration and Air Conditioning Conference*. 2021.
105. MacCracken, M.M., Electrification, heat pumps and thermal energy storage. *ASHRAE Journal*, 2020 (July 2020 Issue): pp. 32–39.
106. Huang, R., et al. Numerical investigation of high delta-T sensible storage integrated CO_2 heat pump. in *International Refrigeration and Air Conditioning Conference*. 2024. West Lafayette, IN.
107. Clift, D.H., et al., Assessment of advanced demand response value streams for water heaters in renewable-rich electricity markets. *Energy*, 2023. **267**.
108. Moreno, P., et al., The use of phase change materials in domestic heat pump and air-conditioning systems for short term storage: A review. *Renewable and Sustainable Energy Reviews*, 2014. **39**: pp. 1–13.
109. Hussein, N.A., A.M. Al Jubori, and A.S. Abdul-Zahra, Enhancing the performance of air conditioning systems by integrating phase change materials: A comprehensive review. *Journal of Energy Storage*, 2024. **101**.
110. Takudzwa Muzhanje, A., M.A. Hassan, and H. Hassan, Phase change material based thermal energy storage applications for air conditioning: Review. *Applied Thermal Engineering*, 2022. **214**.
111. Wan, T.S., *Direct Refrigerant Ice Thermal Storage Systems*. 1995.
112. Guelpa, E. and V. Verda, Thermal energy storage in district heating and cooling systems: A review. *Applied Energy*, 2019. **252**.
113. Zeinelabdein, R., S. Omer, and G. Gan, Experimental performance of latent thermal energy storage for sustainable cooling of buildings in hot-arid regions. *Energy and Buildings*, 2019. **186**: pp. 169–185.
114. Dardir, M., et al., Opportunities and challenges of PCM-to-air heat exchangers (PAHXs) for building free cooling applications—A comprehensive review. *Journal of Energy Storage*, 2019. **22**: pp. 157–175.
115. Choure, B.K., T. Alam, and R. Kumar, A review on heat transfer enhancement techniques for PCM based thermal energy storage system. *Journal of Energy Storage*, 2023. **72**: p. 108161.
116. Freeman, T.B., et al., Advanced materials and additive manufacturing for phase change thermal energy storage and management: A review. *Advanced Energy Materials*, 2023. **13**(24): p. 2204208.
117. Huang, R., et al., Evaluation of phase change thermal storage in a cascade heat pump. *Applied Energy*, 2024. **359**: p. 122654.
118. Lin, Y., et al., Performance investigation on an air source heat pump system with latent heat thermal energy storage. *Energy*, 2022. **239**: p. 121898.
119. Emhofer, J., et al., Experimental demonstration of an air-source heat pump application using an integrated phase change material storage as a desuperheater for domestic hot water generation. *Applied Energy*, 2022. **305**: p. 117890.
120. Waser, R., et al., Fast and experimentally validated model of a latent thermal energy storage device for system level simulations. *Applied Energy*, 2018. **231**: pp. 116–126.
121. Lindsay, B.B. and J.S. Andrepont, Evolution of thermal energy storage for cooling applications. *ASHRAE Journal*, 2019 (October 2019 Issue): pp. 42–59.

122. Xu, G., et al., Experimental and numerical study of an electrical thermal storage device for space heating. *Energies*, 2018. **11**(9).
123. Heine, K., P.C. Tabares-Velasco, and M. Deru, Design and dispatch optimization of packaged ice storage systems within a connected community. *Applied Energy*, 2021. **298**.
124. Selvnes, H., et al., Review on cold thermal energy storage applied to refrigeration systems using phase change materials. *Thermal Science and Engineering Progress*, 2021. **22**.
125. Bista, S., et al., Performance improvement and energy consumption reduction in refrigeration systems using phase change material (PCM). *Applied Thermal Engineering*, 2018. **142**: pp. 723–735.
126. Qi, R., C. Dong, and L.-Z. Zhang, A review of liquid desiccant air dehumidification: From system to material manipulations. *Energy and Buildings*, 2020. **215**.
127. Kozubal, E. *Energy Storing Efficient HVAC*. Available from: https://www.energy.gov/sites/default/files/2023-06/bto-peer-2023-39160-energy-storing-eff-hvac-nrel-kozubal.pdf
128. Kocher, J.D., et al., Thermal battery cost scaling analysis: Minimizing the cost per kW h. *Energy & Environmental Science*, 2024. **17**(6): pp. 2206–2218.
129. Shah, R.K. and D.P. Sekulić, *Fundamentals of Heat Exchanger Design*. John Wiley & Sons.
130. *Inflation Reduction Act of 2022*. 2022. pp. 1818.
131. The White House *Inflation Reduction Act Guidebook*. 2023.
132. Florida Power & Light Company. *Business Heating Ventilation & Air Conditioning ("HVAC") Program Standards*. 2020 [cited 2025]; Available from: https://partner.fpl.com/content/dam/partner/us/en/images/contractor/pdfs/business-hvac-standards.pdf
133. Xcel Energy. *Dynamic Thermal Storage (DTS) Information Sheet*. 2023 [cited 2025]; Available from: https://www.xcelenergy.com/staticfiles/xe-responsive/Energy%20Solutions/Business%20Solutions/23-10-508_MN-DynamicThermoStorage_IS_P04.pdf
134. Arizona Public Service. *APS offers three new programs to help customers use more solar energy*. 2018; Available from: https://www.aps.com/en/About/Our-Company/Newsroom/Articles/APS-offers-three-new-programs-to-help-customers-use-more-solar-energy
135. Office of Energy Efficiency and Renewable Energy. *Heat Pump Water Heaters Achieve Significant Peak Reduction and Energy Savings*. 2019; Available from: https://www.energy.gov/eere/buildings/articles/heat-pump-water-heaters-achieve-significant-peak-reduction-and-energy
136. Dayem, K. and Xergy Consulting, Standardized Communications for Demand Response, in *Business and Technology Report*. 2018, National Rural Electric Cooperative Association.
137. Olano, M.V. Chart: Americans bought more heat pumps than gas furnaces last year. *Home of the Future* 2023 [cited 2025]; Available from: https://www.canarymedia.com/articles/heat-pumps/chart-americans-bought-more-heat-pumps-than-gas-furnaces-last-year
138. Waite, M. and V. Modi, Electricity load implications of space heating decarbonization pathways. *Joule*, 2020. **4**(2): pp. 376–394.
139. North American Electric Reliability Corporation (NERC), *2024 Long-Term Reliability Assessment*. 2024.

Part V

Strategies and Innovations for a Decarbonized Future

13 Circular Economy Perspectives and Innovations for Decarbonizing the Construction Industry

Brandon S. Byers, Qingxu (Bill) Jin, Zain Karsan, and Luke Lombardi

13.1 INTRODUCTION

Decarbonization is a dangerous term. Originally, decarbonization referred to removing the build-up of carbon deposits inside a combustion engine from organic fuel sources. Appropriating the word away from its original meaning to imply anything that mitigates climate change grossly simplifies the complex problem space of environmental management. The risk of using decarbonization as a flagship strategy is myopic and subject to tunnel vision. Early work on the dangers of tunnel vision for environmental management was from Alan Miller, who commented that its dangers contain selective attention that inhibits problem-solving competence for more complex problems (1982). He continues to comment that there is an over-emphasis on applying a scientific decision-making process, which is moot without sufficient attention to the proper problem formulation.

What, then, is the problem that decarbonization aims to address? If this solution is concerned with CO_2 emissions, then the US and Europe should already be lauded because they have been consistently dropping their CO_2 emissions for about the past two decades (Figure 13.1) (EIA, 2024). Russia and Japan, too, have decreased their rate of CO_2 emissions since 1990 as shown in Figure 13.2 (EDGAR/JRC, 2024). Yet, when comparing greenhouse gases (GHGs), one unit of Methane has an equivalent impact to about 28 units of CO_2 in the atmosphere (EPA, 2024), but there is no rallying behind a *demethanization*. In the most recent report of planetary boundaries (of which six of nine have been crossed) biogeochemical flows, biosphere integrity, novel entities, and radiative forcing are all in higher risk zones than CO_2 concentration (Richardson et al., 2023). Global environmental management is a multi-objective optimization problem beyond just minimizing CO_2- and CO_2-equivalent (eCO_2) emissions and must consider the nonlinear interactions between each issue.

DOI: 10.1201/9781003655442-18

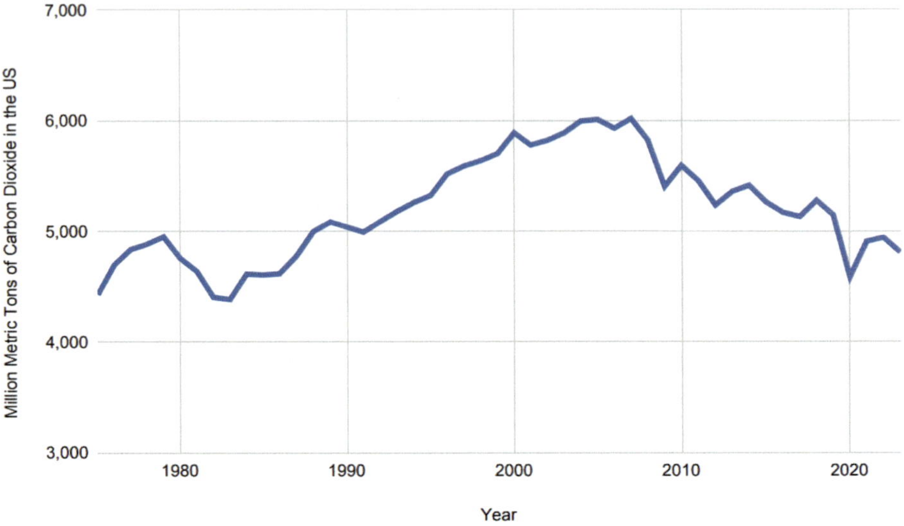

FIGURE 13.1 Carbon dioxide emissions in the US, 1975–2023.

(Data from EIA (2024), figure by author.)

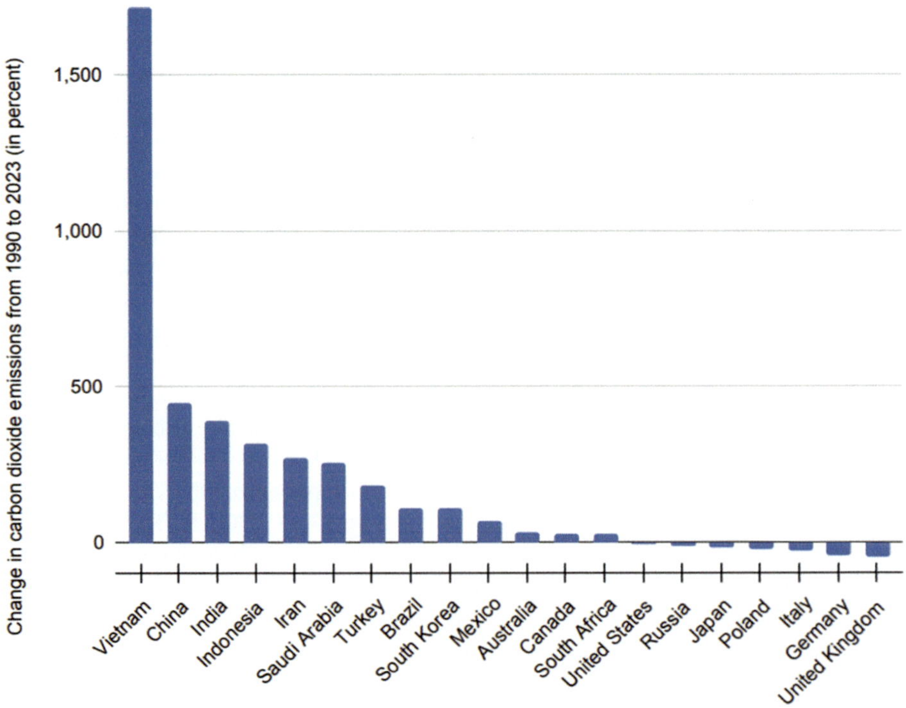

FIGURE 13.2 Global CO_2 emissions change 1990–2023, by country.

(Data from EDGAR/JRC (2024), figure by author.)

In addition, a lot of the discussion around decarbonization in the built environment is limited to operational energy or operational carbon, particularly around Scope 1 and Scope 2 emissions. Yet, if the concerns of decarbonization are concentrated on renewable energy, it should be noted that the production of renewables has nearly doubled in the last 10 years (IRENA, 2024). Beyond the operational impacts are the embodied impacts from upstream processes and the production of materials, the embodied carbon or embodied energy is more often captured in Scope 3 emissions. These dual sources of carbon are analogous to companies' expenditures, such that capital expenditure (CapEx) and operational expenditure (OpEx) are parallel to capital or embodied carbon (CapCarb or EmCarb) and operational carbon (OpCarb) (Enzer et al., 2013). The conversation around decarbonization must not be limited to OpCarb and acknowledge the impact of EmCarb, particularly as some industries and countries are already greening their energy sourcing.

Adopting the principles of a circular economy, where resources are kept in use and circulation for as long as possible before disposal, is a growing approach to minimizing embodied carbon, increasing responsible resource management, and increasing climate change adaptation (Wang et al., 2022). The climate is already changing (Lenssen et al., 2019). Thus, proper focus needs to be given not just to the resilience of the earth's biosphere but the resilience of human civilization through proper resource management and responsible living for the future. Therefore, the objective of this chapter is to position circular economy practices as pivotal for resource efficiency and EmCarb decarbonization in the construction industry.

13.1.1 Impact of the Construction Industry

In the global effort to mitigate the effects of climate change, the construction industry is unfortunately responsible for a significant amount of the impact. Construction contributes 37% of global eCO_2 emissions, which is split between operational and embodied carbon (Pasanen et al., 2021). Embodied carbon emissions from material sourcing and construction amounted to 5.4 Gt eCO_2 in 2020 (Shell and Deloitte Consulting, 2023). These numbers are from both the real estate and infrastructure parts of the built environment where real estate is responsible for most of the operational emissions and infrastructure splits the embodied emissions with real estate. This shows that researchers and practitioners in the built environment have great leverage for making change.

However, the global distribution of eCO_2 emissions is not proportional and is increasing and decreasing at different rates per country. With limited changes to business-as-usual, the eCO_2 emissions from the built environment could increase by 10–15% (Boland et al., 2023), driven by places with rapid urbanization. The choice of construction materials is an important factor in the net impact of construction itself. Cement and metals contribute significantly to GHG emissions, where cement is responsible for 2.6% of emissions and metals for 7.8% of the global total emissions (Wang et al., 2022). The inefficient over-design of these materials is a known contributor to waste and superfluous expenditure in construction. Material efficiency strategies are estimated to reduce embodied eCO2 emissions in residential buildings by 80–100% in G7 countries and China (Wang et al., 2022). Informed material choices, better design, and cleaner production may help mitigate the waste in the industry.

What's more is that Europe and the US have already seen a decrease in eCO_2 emissions within recent years implying the success of recent efforts (EIA, 2024). Nonetheless, other parts of the world are developing rapidly, subject to new construction, and need large amounts of raw construction materials to meet demands. Figure 13.2 shows that China, India, countries in South-East Asia, and the Middle East are showing increased rates of CO_2 emissions. China, the US, and India are still responsible for over 50% of global CO_2 emissions (Figure 13.3) (Statista, 2024), and are expected to be the top three contributors to global construction growth between 2020 and 2030 (followed closely by Indonesia) (Marsh et al., 2021). Thus, how these countries build will influence how the environment is affected.

Beyond eCO_2 emissions, the construction industry suffers from significant waste and inefficiency. Construction and demolition waste (CDW) accounts for about one-third of global solid waste generation (Purchase et al., 2022). The significant amount of waste indicates inefficiencies in the processes of construction and the low incentives for improvement. Targeting existing inefficiencies in construction is an important approach to decarbonizing the industry.

When considering innovative construction products and processes, significant trade-offs between economic and environmental factors often slow the adoption of new methods. Interest is growing in analyzing the environmental and economic costs

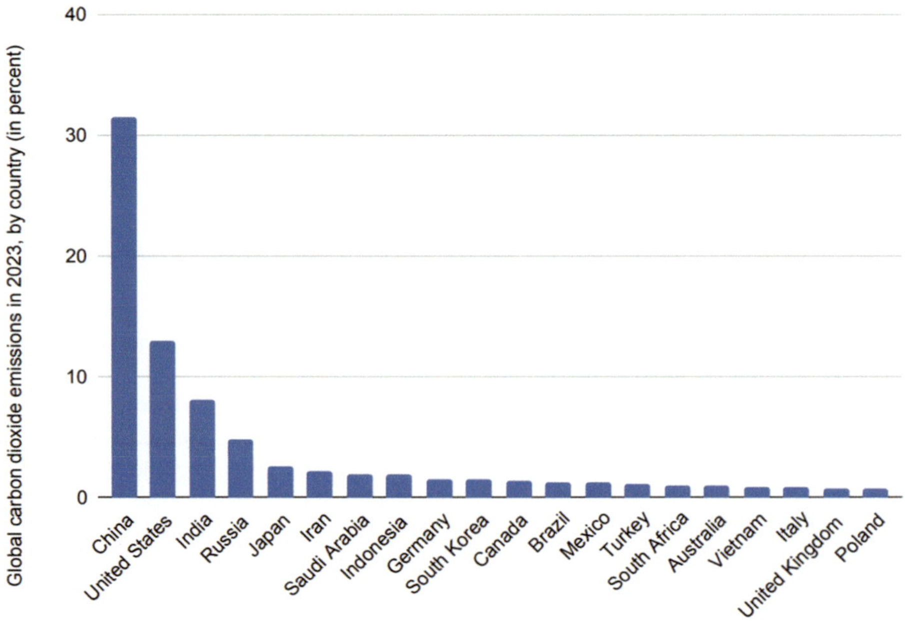

FIGURE 13.3 Largest global emitters of carbon dioxide, 2023, in percent by country.

(Data from Statista (2024), figure by author.)

associated with demolishing old buildings for new construction versus refurbishing and reusing existing structures ("Adaptis," 2024). Some researchers have found for construction material reuse the optimum strategy varies wildly based on existing policies, skilled labor, and cultural preferences (Mollaei et al., 2023b). In more developed regions of the world, new green technologies still require significant capital investment. For example, implementing carbon capture for cement production may increase total material costs by up to 15% by 2030 (Boland et al., 2023). To decarbonize the construction industry, the existing costs for new products and processes must be recognized, but not discouraged, for change.

13.2 FOUNDATIONS OF THE CIRCULAR ECONOMY

The aim of a circular economy is not new and is aligned with early and pre-industrial human civilizations where resource use was more conservative. Industrialization has been a result of access to denser forms of energy, primarily fossil fuels, which saw a boom in innovation at the cost of environmental damage, GHG emissions, and waste generation. Modern circular economy concepts are proposed as a mental model to understand human's relationship with resource use in the present and the future as an alternative to traditional industrial economies. Walter Stahel was an early pioneer of the concept opposing typical linear economy processes of material production and consumption (a take-make-waste model) and proposed a framework for a circular economy as well as a performance economy, which focuses on the services available from materials rather than their consumption (Stahel, 2016).

Later evolutions of the concept of a circular economy further illustrate the repeated use of raw materials over several life cycles. This includes work from Michael Braungart and William McDonough in their cradle-to-cradle (C2C) conceptualization. In this model, C2C refers to the life cycle terminology indicating that the materials can repeat life cycles, particularly when split into their constituent biological and technical nutrients (Braungart and McDonough, 2009). Later developments emerged from the Ellen MacArthur Foundation, which focused on eliminating waste and pollution, circulating products and materials, and regenerating nature (Ellen MacArthur Foundation, 2013). The Ellen MacArthur Foundation developed the popularized butterfly diagram, based on Braungart and McDonough's work, which distinguishes the flows between renewables and finite materials (Ellen MacArthur Foundation, 2013).

Throughout its evolution, the conceptualization of a circular economy has changed several times. Circular economy principles emerged out of the late 1980s and had initial conceptions of reduce, reuse, and recycle as defined as early as 2008 in China (Standing Committee of the National People's Congress, 2008) and have since evolved to include many more R-strategies (Kirchherr et al., 2023). The 10 R-strategies have been popularized by Dutch researchers and implemented in Dutch policies related to consumer products (Potting et al., 2017). These strategies act as operational tools and address different aspects of a circular economy in production, consumption, and circulation. There is some semantic overlap and occasional confusion on the strategy definitions, and the scale of the impact is not necessarily linear through the progression. Table 13.1 aggregates the R-strategies based on their proposal in the literature.

TABLE 13.1
List of R-Strategies Used to Enact a Circular Economy

R - Strategy	Definition	Reference
Re-mine	Retrieval of materials after the landfilling phase	Reike et al. (2018)
Recover	Incineration of material with energy recovery	Potting et al. (2017)
Recycle	Process materials to obtain the same or lower quality	Potting et al. (2017)
Repurpose	Use discarded product or its parts in a new product with a different function	Potting et al. (2017)
Remanufacture	Use parts of discarded products in a new product with the same function	Potting et al. (2017)
Refurbish	Restore an old product and bring it up to date	Potting et al. (2017)
Repair	Repair and maintenance of defective product so it can be used with its original function	Potting et al. (2017)
Reuse	Reuse discarded product that is still in good condition and fulfills its original function	Potting et al. (2017)
Renew	Redesign product in view of circularity	Potting et al. (2017)
Reduce	Increase efficiency in product manufacture or use by consuming fewer natural resources	Potting et al. (2017)
Rethink	Make product use more intensive for the same material	Potting et al. (2017)
Refuse	Prevent the use of resources by abandoning their function or by offering the same function with a radically different product	Cramer (2014) and Potting et al. (2017)
Regenerate	Use non-toxic material, use renewable energy, and regenerate natural ecosystems	Konietzko et al. (2020)

An additional understanding of a circular economy is through the *decoupling* of economic growth from resource use and emissions intensity (Scheel et al., 2020). This statistical relationship was first explored in a working paper by the National Bureau of Economic Research that found an N-shaped cubic curve explained sulfur dioxide emissions against GDP per capita (Grossman and Krueger, 1991). This concept was soon further popularized in a report by the World Bank that claimed increased incomes would cause demands for improvements in environmental quality (World Bank, 1992). The report showed a figure similar to Figure 13.4 with data from 1970 to 1988 decoupling the rate of GDP growth from particulates, nitrogen oxides, sulfur oxides, and lead (World Bank, 1992). The inverse U-shaped curve describing how as GDP per capita initially increases emissions rise but will eventually fall again as the country continues to increase its proportional GDP is called the environmental Kuznets curve hypothesis. This relationship has been shown to not hold true for all emissions nor all countries at variable income levels (Freire-González et al., 2024; Stern, 2017).

Although this view of a circular economy (CE) is perhaps the simplest to understand at a conceptual level, and one that directly acknowledges the economic aspect of a CE, it may not capture the nuance of what is meant by circularity. Figure 13.4 shows an example of decoupling in the US by comparing GHG emissions per capita and emissions per USD GDP extracted from EPA data (EPA, 2024). Although this is

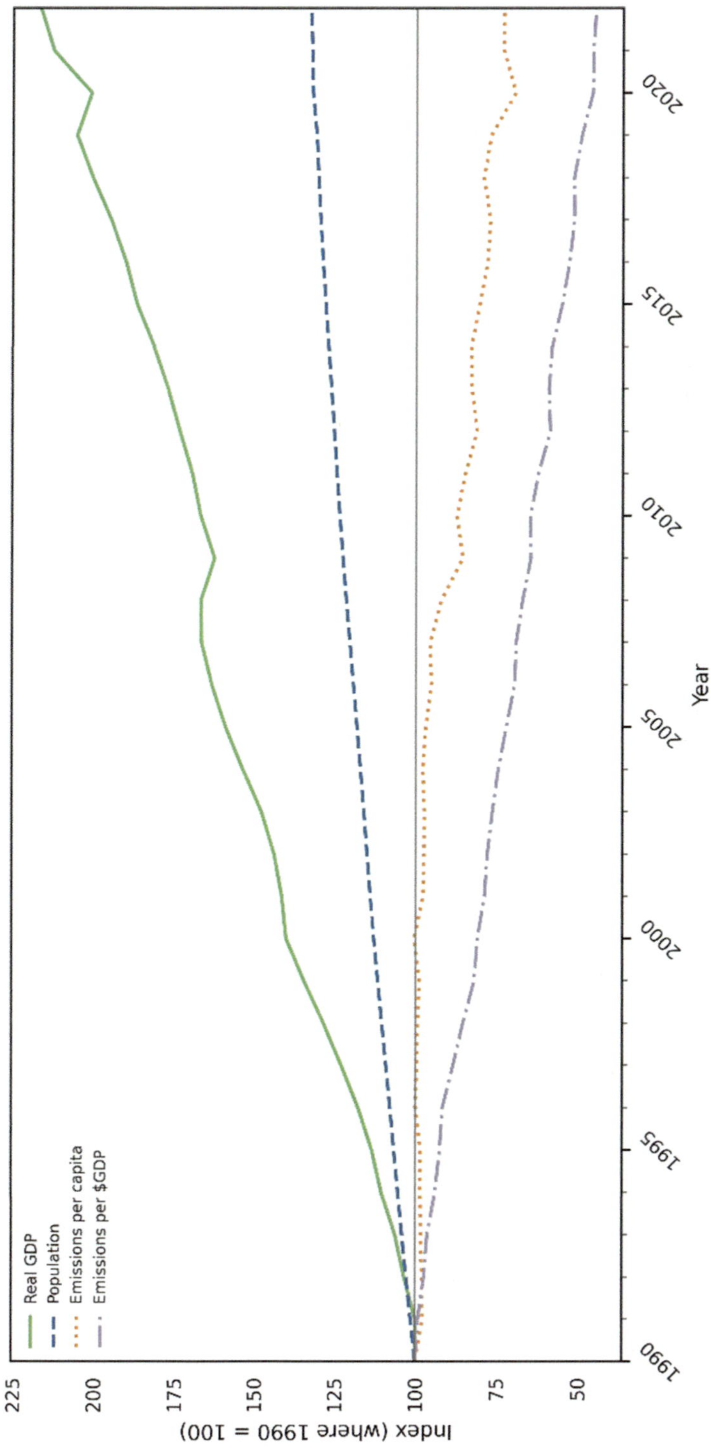

FIGURE 13.4 US Greenhouse Gas Emissions per Capita and per Dollar of GDP, 1990–2022. (Data from EPA (2024), figure from author.)

partially believed to be enabled in economies that rely on services and shift manufacturing abroad, the means to calculate decoupling are also still contested and often claimed to not be statistically robust, universally true, nor sustainable over the long term (Stern, 2017).

The International Organization for Standardization (ISO) provides a definition of a circular economy in ISO 59004:2024, which is an "economic system that uses a systemic approach to maintain a circular flow of resources, by recovering, retaining or adding to their value, while contributing to sustainable development," where "the inflow of virgin resources is kept as low as possible, and the circular flow of resources is kept as closed as possible to minimize waste, losses and releases from the economic system" (ISO, 2024). An economic system is a "system by which a society organizes and allocates resources" (ISO, 2024). In essence, a circular economy can be viewed as a system that manages resources to minimize new resource consumption and maintain or add value to existing resources in the system. The ISO defines sustainable development as "development that meets the environmental, social and economic needs of the present without compromising the ability of future generations to meet their own needs" as originally proposed by the Brundtland Report (Brundtland, 1987). This definitional landscape should be used to understand the context and use of the phrase *circular economy* in this chapter.

13.3 CIRCULAR ECONOMY STRATEGIES AND CONSTRUCTION CASE STUDIES

For each of the R-strategies for a circular economy, there are different levels of application, such as in material production, construction processes, or end-of-life management. An action like reuse can be done at the material level, component level, assembly level, or the building and site level. The optimum level of application will vary based on the technical and economic feasibility of each strategy. Below is a collection of case study examples from industry and research applied in the construction industry for each of the circular R-strategies. Strategies are introduced in increasing order of impact, that is, reuse has greater resource savings than recycling.

13.3.1 Re-mine

Researchers explored the informal reuse economy for construction materials in India (Raghu and De Wolf, 2024). They found several different types of actors emerge in an informal ecosystem, which includes experts in deconstruction but also waste pickers that collect components from landfills and dump sites. These waste pickers can collect between 60 and 90 kg of disposed materials per day (Raghu and De Wolf, 2024). Additional actors in this ecosystem are the waste segregators and processors before the material re-enters the construction process. Re-mining recontextualizes waste as an underexplored mine for materials.

13.3.2 Recover

Recovery in the context of materials management is often conceptualized as waste-to-energy (WtE) technology. China leads the world with over 300 WtE plants processing over 130 million metric tons of waste (Martinez-Marquez et al., 2022). Yet, broadly, recovery for the built environment can expand to many other applications. A novel approach to waste heat recovery in London captures wasted energy from the London Underground transportation network to supply thermal energy to the Bunhill Heat Network through heat recovery coils and a two-stage heat pump (Lagoeiro et al., 2022a). The London Underground consumed over 1,700 GWh of electricity in 2016–2017, of which 500 GWh was lost as waste heat (Lagoeiro et al., 2022b). The system is estimated to reduce the levelized cost of heating by about 9% and save 18% in carbon abatement costs (Lagoeiro et al., 2022a). Though there are several approaches to saving wasted energy in the built environment, decarbonizing can be pursued by recovering wasted energy from our materials and processes.

13.3.3 Recycle

An increasingly common approach in recycling construction materials is to crush recovered concrete and use it as aggregate in new concrete. Zirkulit AG is a European company that does this concrete recycling in combination with carbonizing the demolished materials with the aim of net-zero concrete (Zhang, 2024). Using recycled aggregate in new concrete could replace up to 50% of new aggregate without considerable impact on strength and durability (Lotfi et al., 2015).

Structural steel is another common construction material that is frequently recycled and requires up to 10 times less energy than the production of virgin steel (Harvey, 2021). Steel can be infinitely recycled, but new products globally only contain about 30% recycled steel on average (World Steel Association, 2022). This is critical considering one analysis that showed with steady-state global steel stocks with present-day collection efficiencies the mining requirements for iron ore would still be about 300 MT per year higher (Harvey, 2021). Another report estimated that switching to an electric arc furnace has the potential to abate up to 50% of emissions relative to a traditional blast furnace (Boland et al., 2023). An additional approach would be using renewable primary energy and steel scrap for 45% emissions reduction as a share of the baseline (Boland et al., 2023).

13.3.4 Repurpose

Repurposing is a strategy of using discarded entities in a new application with a new purpose than its original design. This happens at multiple scales as it can occur at the component level or building level. Repurposing construction products and materials is also a creative process that imagines new purposes for the used materials. Case studies have shown successful implementation of the design and construction of structural elements from repurposed bike frames (Nian et al., 2022) or repurposed snow skis (Colabella et al., 2017). In a review of reused concrete elements, cases showed their use for low-cost housing projects, as pavers, or as pure compression elements if the steel detailing is unknown (Küpfer et al., 2023). Another example of repurposing is the use of agricultural waste as new building insulation (Bakatovich et al., 2018; Gaspar et al., 2020) or ceramic waste as material for new brick production (Fiala et al., 2020).

Repurposing at the building level is also called adaptive reuse and refers to the continued use of the structure with updating programmatic needs, which often includes upgrading materials and modified adherence to code requirements (Assefa and Ambler, 2017). One analysis in Canada showed that if only 10% of all commercial and institutional buildings in the country were repurposed instead of demolished, it would save up to 165 megatons of eCO_2, around 23% of total Canadian GHG emissions in 2013 (Assefa and Ambler, 2017). The design practice Perkins & Will reported on the increase of adaptive reuse and repurposing of buildings and stated that it can be faster and more cost-effective than traditional construction and gave examples from their practice (Hines, 2024). Though this strategy is harder to scale and repeat because of its custom nature, it facilitates new mental models for decarbonizing and optimizing material flows in the built environment.

13.3.5 Remanufacture

Remanufacturing is a strategy that uses discarded products and restores them to use in their original function. In Chile, Neptuno Pumps, originally established in 1972, evolved its business model to use Industry 4.0 technologies to aid in the remanufacturing of its products (Cerda-Suarez et al., 2023). One of their case studies showed that in 2016 the reuse and remanufacturing of pumps allowed one client to save €120,000, 70% of carbon emissions, 60% of energy consumption, and 75% of the potential waste (Cerda-Suarez et al., 2023). Another case of remanufacturing products in the built environment is with Siemens Wind Power. The company uses an on-site remanufacturing approach that modifies the wind turbine blades to increase the power output without needing to replace the wind turbines (Jensen et al., 2019). This approach to decarbonization decreases the environmental impact per energy unit produced and increases revenue for power suppliers between 1.5 and 3% (Jensen et al., 2019).

Takeback programs offer a mechanism for remanufacturing, such as Davies Office, Inc. Suppliers offer users the service of returning their used goods to be restored back to their original purpose. In the case of Davies Office, their remanufacturing of office furnishings has proven to be environmentally preferable, economically viable, and keeps products in circulation (Krystofik et al., 2018). Similar programs have been developed for building elements such as carpets or lighting.

13.3.6 Refurbish

Refurbishing materials focuses on restoring a used product or materials. Urban Machine is a startup based out of San Francisco, US, which uses computer vision, deep learning, and purpose-built robots to refurbish wood from construction and demolition waste for resale ("Urban Machine," 2024). Their mobile assembly line begins with the preliminary cleaning of the collected lumber with heavy-duty wire brushes. Next, the lumber is induction heated so that the lodged metal fasteners will scorch the surrounding wood, making their location more conspicuous to object detection algorithms in the subsequent nail and staple picking step. Finally, the lumber is stamped with a material passport to contextualize the history of the refurbished material ("Urban Machine," 2024). This approach helps divert CDW and brings value back to a common construction resource, all through a system where the energy demand is met through mobile solar panels. This approach is also seen with other construction materials. For example, FRACO has developed a machine called Brick Recyc that cleans old bricks on-site to revitalize and reuse them ("FRACO | Brick Recyc," 2024). Refurbishment can also be applied beyond the product level to the building level. Combining seismic retrofits and energy refurbishments is a common approach to increase the longevity of a structure (Marini et al., 2022).

13.3.7 Repair

Traditional examples across humanity, like the Parthenon in Greece or the Anji Bridge in China, illustrate that buildings and infrastructure can exist for millennia if society focuses on repairing and maintaining them. For over 2,500 years, the acropolis of Athens has been subject to wars, fires, poor repairs, and earthquakes, but the building remains, and active restoration works continue (Figure 13.5). Repairing infrastructure has been associated with cultural and archaeological purposes, but in modern contexts it is associated with maintaining performance for civil society while minimizing costs or drops in levels of service (Cagle, 2003). Decisions on repair are often politically and economically driven, and sometimes made based on estimated loss of life, but only recently has the framing of decarbonization, circularity, and sustainability been used to encourage repair.

FIGURE 13.5 Repair work of the Parthenon on the Athenian Acropolis.

In order to facilitate repair, building systems must be designed so that the structure and services are accessible. For example, repair of the building structure will not occur on the same time horizon as building electrical. If these systems are compositely linked, damage or replacement of the structure may have to occur far before the end of its useful life to repair the lighting. One example of this is an energy-dissipating fuse that absorbs seismic energy from a rocking structural frame and is easily replaced and repaired (Eatherton et al., 2008).

13.3.8 Reuse

Reuse is becoming an increasingly valuable approach to circularity in the built environment. In Northwest Europe, only about 1% of materials at a building's end of life are salvaged for reuse (Deweerdt and Mertens, 2020; Menny et al., 2024). One of the pioneering companies leading these efforts is Rotor, a design practice that focuses on material salvaging, reuse, reselling, and educating ("Rotor – Brussels | Rotor," n.d.). Although in practice a lot of reused materials are architectural elements, Küpfer et al. reviewed 77 case studies of piecewise reused concrete components in new construction across the US and Europe (2023). Their analysis showed that many limitations are around industry standards, norms, confusion on connection details, and excess costs, but the reuse of structural elements is possible. When examining the impacts of reusing structural steel (thereby removing the energy-intensive smelting process for recycling), the GWP from stages A1–A3 is estimated at 46.6 kg/CO_2/t, about 5% of the impact from new hot-rolled steel sections ("Reusable Steel Environmental Product Declaration," 2022).

One study from Singapore states reusing building components has precedent in informal settlements and rural communities, but new strategies and norms need to be developed for upscaling in formal economies in addition to a more discretionary use of the term "waste" when referring to products at a building's end of life (Arora et al., 2020). Their analysis showed Singapore could support the construction of 2,000–6,000 low-cost houses each year with reused architectural elements (i.e., kitchen, lighting, toilet, doors, windows, and tiles) (Arora et al., 2020). Mollaei et al. analyzed the optimal balance of landfilling, recycling, and reusing building components for a case study building in Switzerland, Canada, China, Nigeria, and Brazil (2023b). The results showed that reuse is particularly favorable in countries that have deconstruction culture and policies to encourage the practice and that the cost of labor is not always the main contributor (Mollaei et al., 2023b).

13.3.9 Reduce

Reduce as a circularity strategy is also conceptualized as the *narrowing* of resource use through design optimization of structures, reducing harmful materials, and efficient planning (Konietzko et al., 2020). The reduction of structural material through design optimization in particular holds promise in decarbonizing the construction industry, considering the majority of embodied carbon is typically attributed to structural components (Chau et al., 2015). This can be seen in waffle slabs popularized by Pierre Luigi Nervi. Additional precedent of expressive yet structurally efficient constructions can be traced from Gaudi to Guastavino (Baroš et al., 2022; Ochsendorf and Freeman, 2010). A contemporary example that comes out of this tradition is the company VAULTED AG, which produces prefabricated funicular floor systems that are optimized for the forces transferred through floor slabs ("VAULTED AG," n.d.). The strategy of optimizing slab design allows for up to 80% reduction in eCO_2 due to the reduction in concrete and steel and allows for disassembly and reuse. The technology is a spin-off out of ETH Zurich and has been approved for its first commercial use in a 10-story office building in Zug, Switzerland, after finding success in the experimental and small-scale HiLO unit of Empa's Nest building (Liew et al., 2017; "VAULTED to build first large commercial project," 2024). In one UK study of 30 projects, there was a 40% under-utilization of structural members by mass, indicating ample opportunities for the reduction of materials (Gholam, 2020).

13.3.10 Rethink

There are many exciting initiatives on rethinking business models, products, and services for the built environment that contribute to a circular economy. A large trend in Europe is the servitization of buildings, particularly in the Netherlands and Scandinavia. Parmaco is a company that offers rental services, not for the space, but for the building itself ("Parmaco/Smart buildings – with a sustainable solution," 2024). A customer with land can have a contract for a certain building type and when the contract ends the building can be sold or returned. This is related to trends in industrialized construction, also being led by BoKlok, which prefabricates modular homes in Sweden and the UK, streamlining the housing process ("Welcome to the BoKlok world! | BoKlok," n.d.).

There is also a rise in the marketization and servitization of building products (product-service systems). Work in the Netherlands out of TU Delft focused on the feasibility, management, and financing of a facades-as-a-service business model (Azcárate-Aguerre et al., 2023). Skellet, a steel fabrication company out of Belgium, designed a light, symmetrical steel profile that allows for easy assembly and disassembly ("Skellet," n.d.). The company also has a buy-back guarantee that refunds up to 130% of the original purchase value after 15 years of use, which is supported by the backcasted and forecasted value of steel products. Another example, CircoTrade, is a company out of France that has built a market for future options on steel elements within a building that enable the reuse and retention of steel products before a building reaches its end of life ("Circotrade," n.d.). By not limiting construction processes to the status quo, new opportunities for decarbonization emerge.

13.3.11 Refuse

One of the most straightforward circular economy strategies to remove embodied emissions from construction is to refuse new construction. A landmark decision by the Secretary of State for Levelling Up, Housing and Communities in London denied Marks & Spencer's permit for demolishing and redeveloping one of its buildings based on the environmental impact (Murphy, 2023). The final decision was made after an open letter from SAVE Britain's Heritage appealed the initial decision based on the waste of nearly 40,000 tons of CO_2 from the demolition and new construction (SAVE Britain's Heritage, 2022). These sorts of decisions in the early stages of planning are pivotal strategies for decarbonizing the built environment.

13.4 DECARBONIZATION OPPORTUNITIES AND INNOVATIONS IN CONSTRUCTION

The role of new technologies and digital transformation in decarbonizing the construction industry should not be understated. As supported by many of the identified case studies, using innovative processes, materials, and technologies can enable resource savings and positive additionalities, which contribute to circular strategies like reducing, rethinking, or potentially regenerating.

13.4.1 Innovative Processes

Construction is a project-based, not product-based, industry, which is what contributes to some of the inefficiencies as the processes between each project cannot be perfectly replicated or scaled. By adopting some of the processes of and learning from the manufacturing industry, industrialized construction can achieve potential cost and environmental savings in addition to decreasing the time of construction by up to 20% (Boland et al., 2023). Similarly, design for manufacturing and assembly could reduce up to 40% of transportation and equipment emissions, savings of up to 30% on construction material, and a reduction of up to 5% of total embodied emissions (the latter found for multifamily residential construction) (Boland et al., 2023). Case study examination of modular design found that it might be able to reduce material waste by up to 80% by streamlining production and assembly offsite (Boafo et al., 2016; Loizou et al., 2021). Similarly, the use of automation has already been shown to increase material circularity and reuse of construction lumber and bricks ("FRACO I Brick Recyc," 2024; "Urban Machine," 2024). Implementing new processes of design, manufacturing, and construction shows promise in decarbonizing the industry.

With increased adoption of life cycle assessment (LCA) (in combination with life cycle costing and social-LCA) and whole-life carbon assessments (WLCA) for built assets, a greater understanding of trade-offs between costs, resource consumption, and emissions can improve the state-of-the-art (De Wolf et al., 2023). Projects that utilize LCAs during the design phase have increased potential for savings, but this becomes more challenging later in the construction stage when an LCA is considered (Goldstein and Rasmussen, 2018). New digital tools and platforms are emerging to support these decision-making processes (e.g., Adaptis, OneClickLCA, or tallyLCA) that aim to help calculate the impact of a building. An important part of this accounting comes from the Environmental Product Declaration (called EPD), a third-party-verified document that openly reports on the environmental impact of a product. Tools and platforms like 2050 Materials, Pathways, or Embodied Carbon in Construction Calculator are on the market to help professionals collect and use this EmCarb data. Emerging tools and platforms will aid in providing better data for better decisions in the early-stage and procurement processes in construction.

New market mechanisms to represent, trade, and gamify environmental impact will continue to be critical in the holistic approach toward decarbonization. The use of tradeable put options in carbon markets has shown promise in aiding carbon dioxide removal (Lockley and Coffman, 2018). In the context of circularity, CircoTrade out of the Circular Buildings Coalition is a company offering futures trading to encourage financing reusing building materials when they are disassembled in the future ("Circotrade," n.d.). A market allows for the compression of information on a material for optimizing its use against other similar assets. Financial incentives like tax credits and grants have helped scale the adoption of low-carbon technologies. Liang (2025) reviewed green project financing for infrastructure and building construction and found that the main instrument is through debt financing. Findings also revealed that green financing is often not transparent in its reporting within the private sector and is thus hard to defend against claims of greenwashing (Liang, 2025). The move toward a circular and net-zero built environment needs to be funded and financially sustainable and new initiatives must help support the transition.

13.4.2 Innovative Technologies

The industrialization of construction may be regarded as a subset of a broader shift in manufacturing termed Industry 4.0, whereby processes become increasingly automated or autonomous, and the description of digital and physical objects becomes entwined. These technologies have been reviewed in a circular construction context by Byers et al. (2024b), and include technologies like the Internet of Things (Byers and De Wolf, 2023; Tang et al., 2019), AI and big data (Raghu and De Wolf, 2022), extended reality (Soman et al., 2024), cloud and blockchain (Byers et al., 2024a; Hunhevicz and De Wolf, 2023), and robotics.

Robotics have been introduced in efforts to reduce potential harm to humans, enable custom fabrication, and increase efficiency. Robotics can be seen for uses of automating steel cutting and disassembly (Karsan et al., 2024), human-robot collaboration in construction (Reisach et al., 2024), and CDW picking and recycling through robotics and computer vision (Sarc et al., 2019; "Urban Machine," 2024). Process automation with collaborative, semi, or fully autonomous robots is quickly becoming a reality on and off construction sites. This trend is tightly coupled with advances in machine learning and the availability of large-scale training data, enabling a growing set of construction tasks.

The promise of robotics in construction is as a pathway to material efficiency, waste reduction, and enhanced safety, by leveraging more intelligent machines to perform challenging, dangerous, and repetitive tasks (Petersen et al., 2019). Robotic automation spans surveying, excavation, material deposition and handling, finishing, inspection, maintenance, and deconstruction. Site surveying and construction layout work have been automated by mobile robots such as HP's SitePrint, the CIVDOT by CIV robotics, or Fieldprinter by Dusty Robotics. Autonomous excavation has been commercialized by Gravis Robotics using deep reinforcement learning to enable robotized backhoes to perform complex tasks on-site (Egli et al., 2022). Reinforcement learning can enable robots to learn many things, as demonstrated by Anymal, the quadrupedal robot traversing uncertain terrain (Rudin et al., 2022). A similar approach is taken by Built Robotics, that outfit construction equipment with multi-modal sensors to develop autonomous excavation and pile-driving robots.

Robotics in material deposition range from welding arc additive manufacturing of steel assemblies, by MX3D, or bricklaying robots like HadrianX by FBR. Finishing of drywall application has been commercialized by CANVAS Robotics with a mobile robot that sprays and sands drywall compounds. Quadrupedal robots like Spot of Boston Dynamics support on-site inspection and monitoring. Gekko is a wall-scaling robot that supports facade cleaning by Serbot. Finally, Husqvarna and Brokk have a family of demolition equipment, including chain-track mobile robots with articulated arms and heavy-duty attachments for breaking, crushing, sawing, and shearing.

Disassembly is an area of increasing interest in enabling a circular economy, considering the potential for the reuse of recovered components. At the scale of consumer products and electronics, robotic automation of disassembly tasks has had substantial research since the 1990s (Dario et al., 1994; Jorgensen et al., 1996). A key challenge in this field is developing robust control with the uncertainty of the context or workpiece (Foo et al., 2022). Hence, contemporary commercial applications of disassembly robotics heavily rely on accurate information regarding the product to take apart, such as Apple's Daisy Robot, which can dismantle and recover 200 phones per hour (Foo et al., 2022). Nevertheless, general-purpose datasets, for example in identifying screw locations in electronics, could enable automation. Practical experiments in robotic steel cutting demonstrate how vision-based workflows, schematically shown in Figure 13.6, can support on-site disassembly automation (Karsan et al., 2024).

3D printing in construction is typically associated with greater design flexibility, high performance, waste reduction, and lower labor costs. Experimental processes in 3D printing show promising pathways to making construction more intelligent, either by upcycling waste like in binder jetted aggregates from quarrying (Odaglia

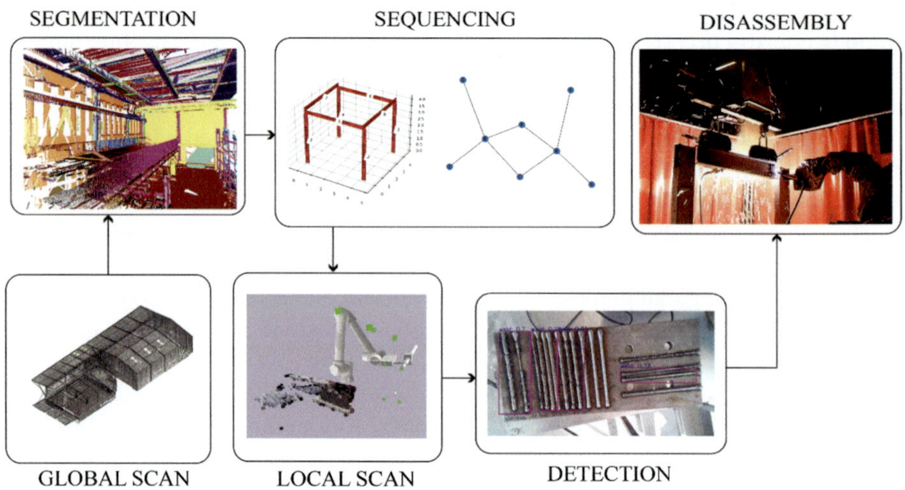

FIGURE 13.6 Workflow for automating disassembly tasks with computer vision, task and motion planning, and purpose-built robotic hardware.

et al., 2023), producing lightweight and materially efficient wall assemblies with nonstandard cross sections (Leschok et al., 2024), or reducing waste in construction by printing thermally insulating formwork (Bedarf et al., 2024). The research in additive manufacturing for sustainable construction spans across materials, computational geometry, and process design.

3D printing of materials is one approach that promises increased material optimization and rapid construction speed, although there are still challenges with end-of-life management and repair. Brenner et al. (2023) report on the use of advanced techniques in digital reconstruction to repair a "high-tech" façade element using welding additive manufacturing. The use of wire arc additive manufacturing for repair follows a broad history of structural steel repair using arc methods (Manai, 2020). There are examples of 3D printing in concrete materials (Buswell et al., 2018), customized steel connections (Ariza, 2018), and even bio-based materials (Curth et al., 2024).

Beyond industrializing construction, other technologies hold promise for decarbonizing. The uptake of building information modeling (BIM) is useful for potential materials savings in optimizing design, procurement, and limiting rework (Shell and Deloitte Consulting, 2023). BIM is already a commonly used platform for collaborative planning, promising savings through clash detection as well as enhanced communication among consultants and project team members (van der Heijden, 2023). Additional layers of information can support analysis in end-of-life management (Akanbi et al., 2019) and adaptive reuse scenarios (Sanchez and Haas, 2018) that extend beyond a building's current life cycle. AI can be used for new creative designs (Bucher et al., 2023) or rapid information acquisition on the existing building stock (De Wolf et al., 2024; Raghu et al., 2023). Using reality capture technologies, including photogrammetry and LiDAR scanning, the existing building stock can be reconstructed to better understand the existing embodied carbon of the built environment (Byers et al., 2023; Heisel et al., 2022; Pătrăucean et al., 2015; Xiong et al., 2022). The adoption of new technologies can enable the construction industry to meet future goals of decarbonizing.

13.4.3 Innovative Materials

The structural materials used in construction account for most of a project's lifetime emissions, creating clear opportunities for improvement. In the construction supply chain, cement contributes about 43% of embodied emissions, steel accounts for 25%, and other materials collectively make up 24% (Shell and Deloitte Consulting, 2023). Finding ways to adapt these common building materials to reduce or even reverse their environmental impact has become a critical research priority.

In efforts to create low-carbon construction materials, the steel industry is showing interest in on-site clean energy generation to make greener products. Some companies across the US are considering building on-site nuclear or hydrogen energy plants so that the primary energy source for steel recycling and fabrication becomes significantly cleaner (Devlin et al., 2023). This is supplementary to existing steel recycling processes, where 40% of global steel production is secondary steel (Harvey, 2021) and 1.5 tons of CO_2 are saved from recycling one ton of steel (World Steel

TABLE 13.2
Comparison of Steel Production Processes by Select Regions in 2023

Country	Million tons	BOF %	EAF %	Other %
Asia	1395.2	81.2	18.8	–
European Union (27)	126.3	55.2	44.8	–
North America	110.2	30.8	69.2	–
Russia & Other CIS + Ukraine	90.5	62.2	32.4	5.3
Middle East	54.5	4.5	95.5	–

Source: Data from World Steel Association (2024), table from author.
Note: EAF is Electric Arc Furnace and BOF is Basic Oxygen Furnace.

Association, 2022). The adoption of the electric arc furnace over basic oxygen furnace allows for 100% recycled steel content and is an increasingly prevalent technology (Table 13.2). These innovations in steel production can significantly help to drive down the negative impact of the material.

To reduce the embodied carbon and energy of cement and concrete materials, there are several strategies, including increasing the use of supplementary cementitious materials (SCMs) to reduce Ordinary Portland cement (OPC) in concrete, developing sustainable alternative cement, improving the production and utilization efficiency of cement, and carbon capture and utilization (CCU) (K. L. Scrivener et al., 2018b). For SCMs, such as coal fly ash, blast-furnace slag, and natural pozzolans, research has shown that concretes with high replacement rates (around 50%) of these materials can exhibit better durability and mechanical properties. However, the use of SCMs in concrete is limited by their availability. The availability of traditional SCMs in the US has tightened as steel production has shifted internationally. Additionally, the use of coal-fired power plants is being replaced by more environmentally friendly technologies. Therefore, it has become important to expand the use of off-specification fly ash, including re-mined and reclaimed coal combustion products from landfills and ash ponds as alternative SCMs (TRB, 2018).

As put forward by the American Concrete Institute (ACI), the most effective and scalable approach so far (and in the foreseeable future) is to expand the applications of SCMs in concrete, such as using indigenously abundant and non-traditional SCMs (Zhang et al., 2022). To maximize the substitution level for OPC, a combination of non-traditional SCMs leveraging their synergistic interactions is desirable (Jin et al., 2024). An effective example of such strategies is to create limestone calcined clay cement (LC^3) using indigenously abundant limestone and clay and strategically blend it with SCMs for high substitution levels (30–50%) (K. Scrivener et al., 2018a). Additionally, many waste clay materials (e.g., river/channel sludge) can be used to produce calcined clay in LC^3, and the high reactivity of calcined clay and its synergy with limestone, after optimization, may enable approximately 50% substitution levels. Compared to OPC, studies have shown LC^3-based cementitious materials exhibit higher compressive strength, better chloride and Alkali-Silica reaction resistance, lower autogenous shrinkage and creep, and lower embodied carbon and energy (Avet and Scrivener, 2018; Sánchez Berriel et al., 2016; Scrivener et al., 2019).

An alternative to the mix design of concrete is its utilization. Researchers have developed an engineered cementitious composite (ECC), known as "flexible/bendable concrete" that exhibits tensile ductility over 300 times that of regular concrete (Li, 2019). While retaining a similar appearance and mechanical properties to conventional concrete, bendable concrete shows unique ductile behavior, producing cracks finer than human hair (<100 μm) (Leon-Miquel et al., 2023). These properties enable a self-healing mechanism, like minor paper cuts healing on human skin. The exceptional ductility of bendable concrete reduces the need for repair, maintenance, and replacement, cutting carbon emissions over time. This composite has been successfully applied to various surface infrastructures, including earthquake-resistant high-rise buildings, concrete towers for wind turbines, pavements, permanent formwork, and tunnel linings (Jin et al., 2013; Jin and Leung, 2011; Jin and Li, 2019a, 2019b). Due to the high resilience of the materials, it could potentially be used for reuse in modular and circular construction practices (Bao and Li, 2020).

For CCU in cement and concrete, widely studied and commercialized technologies include in-mixing carbonation (e.g., CarbonCure) and post-mixing carbonation (e.g., CarbonBuilt and Solidia). However, in-mixing carbonation has been limited by an overall carbon uptake of 0.2% of the weight of cement, beyond which the concrete properties could be adversely affected (Michael Thomas, 2019). Post-mixing carbonation is limited by macroscopic diffusion of CO_2 (Zhang et al., 2017), such that they are only effectively applicable to thin-wall precast concrete elements with large surface area to volume ratios, like hollow concrete masonry units. Apart from these technological limitations, the near-future large-scale deployment of CCU technologies can also be limited by the supply chain of CO_2, as carbon capture has not yet been realized on an industrial scale.

Beyond the typical construction materials, some of the most traditional and reliable materials are bio-based and continue to show promise in achieving functionality with minimal impact. For example, one LCA of a university residence building in Arkansas found that a mass timber structure saved 19% carbon emissions in the A1-A4 phases relative to a steel structure with the same purpose (Hemmati et al., 2024). Although some critiques of engineered timber include the impact from glue often used between laminations, dowel-based connectors, or lumber-based construction with joinery-based connections can reduce negative impacts (Fang and Mueller, 2023). Wood is often considered a regenerative material because of the rate of replenishment and sustainable forestry practices. Bamboo is particularly interesting because of its high rate of growth and availability in many rapidly developing areas of the world. One study performed an LCA of bamboo construction materials relative to concrete beams and columns and found the environmental impact values of the bamboo were less than 1% of the concrete across the whole lifetime (Xu et al., 2022). Looking for alternative bio-based materials is a critical strategy for decarbonizing the built environment, and wood is one of the most reliable materials for construction humans have ever used.

13.5 FUTURE PATHWAYS FOR GROWTH

There is much effort needed to change the way construction happens, and the services conceptualized for the built environment. Lessons can be drawn from successful policy implementation in the past, but this must be balanced with the growing challenges of the future.

13.5.1 Regulatory Drivers

Europe has focused a lot on developing regulatory drivers for a circular economy and reducing the negative environmental externalities from construction. One example is from the European Union (EU) Sustainable Finance Taxonomy, which requires whole-life carbon emissions accounting for buildings over 5,000 m^2, but no thresholds will be set until 2025 (Pasanen et al., 2021). Similarly, Level(s) is a European-developed framework for measuring and reporting the environmental performance of buildings throughout their life cycle to align with the European Green Deal, Circular Economy Action Plan, and Renovation Wave Strategy (Honic and De Wolf, 2023). Europe continues to be a global leader in proposing unified strategies for circularity and sustainability across countries.

Many initiatives for carbon accounting exist at the national level in hopes of unifying the approach across the industry. Several countries have already started with LCA and carbon requirements for federal buildings, like Germany, Belgium, the US, and Canada (Pasanen et al., 2021). Sweden has enforced national carbon reporting for all new buildings since 2022 and then will develop and set limits by 2027 (Pasanen et al., 2021). Denmark has implemented mandatory limitations on embodied carbon emissions for new buildings that are normalized per square meter (Shell and Deloitte Consulting, 2023). Since 2013, the Netherlands has set regulatory requirements to use an LCA for new buildings (Pasanen et al., 2018). Similarly, London's building proposals and new Dutch highway constructions are required to perform an LCA before construction (Shell and Deloitte Consulting, 2023). From a 2018 review of 156 sustainability certifications and regulations, only 105 address embodied carbon directly, and only 26 countries had national systems (Pasanen et al., 2018). Increasing the requirement for carbon accounting can help increase accountability and baselining for new projects to understand the extent of the impact.

Outside of carbon accounting and a focus on buildings is an emphasis on general construction and demolition waste reduction. Some estimates in the US find CDW to account for a third of all solid waste (Rios et al., 2015). The US has initiated the Federal Buy Clean Initiative that prioritizes the procurement of local and low-carbon materials in federal construction projects (The White House, 2023). In addition to cleaner infrastructure is a strategy to tax the disposal of construction and demolition waste both during the construction process and end-of-life management. This approach aims to minimize waste production and incentivize innovative ways to incorporate waste as material input for construction (The White House, 2023). Additional efforts on deconstruction ordinances can be seen at the municipal level across the US, such as in San Antonio (City of San Antonio, TX, 2022), Palo Alto (City of Palo Alto, CA, 2020), Milwaukee (City of Milwaukee, WI, 2018), and Portland (City of Portland, OR, 2016).

New York City Economic Development Corporation released circular design and construction guidelines to decarbonize the City's development (NYCEDC, 2024). This report was shortly followed by a white paper discussing the environmental, resource, and economic potential of circular construction across all of New York State (Circular Construction Lab et al., 2024). It is important to think beyond sustainable buildings and how policy can encourage effective resource management across all construction types and the asset's life cycle.

Additional consideration should be given to rapidly growing and urbanizing countries. China is responsible for about 57% of global cement production capacity and 53% of steel production capacity (Shell and Deloitte Consulting, 2023). However, China already initiated circular economy policies as early as 2005, with the primary policy focus being on waste recycling (Gu et al., 2017). Being responsible for about one-third of carbon emissions, China has made commitments to net-zero by 2060 (Shell and Deloitte Consulting, 2023). In addition, India has about 7% of the global capacity for steel and cement production and is responsible for about 7% of the carbon emissions, though committed to net-zero only by 2070 (Shell and Deloitte Consulting, 2023).

Shifting from linear to circular material processes and moving toward a decarbonized built environment is a collaborative effort that requires collective action. Economic and cultural drivers have been found to be significant variables for justifying material reuse and can be incorporated into policy (Mollaei et al., 2023a). Combining the incentives and resources of private industry and public goods and funding through public-private partnerships is also a promising pathway for scaling research into low-carbon construction materials (Shell and Deloitte Consulting, 2023). To standardize and scale reporting and broader circular practices industry-wide coalitions and working groups should continue to coordinate efforts.

13.5.2 Emerging Synergy of Circular Economy and Climate Resilience

Natural disasters (specifically hydrological, meteorological, and climatic events) have been reported with increasing frequency and severity, and, subsequently, climate resilience and climate adaptation are increasingly important topics (Hoeppe, 2016). Although this phenomenon of increased disasters is partially because of higher resolution data and increased accounting for minor disasters, the resulting impacts remain and can be seen in Figure 13.7. (Jones et al., 2022; Ritchie and Rosado, 2024). The current fires in Southern California at the time of writing in January 2025 have destroyed over 16,000 structures and burned over 50,000 acres at the time of writing (CAL FIRE, 2025). This is after the 27 weather and climate disasters causing over $1 billion in damages in the US in 2024 (Smith, 2025). Post-disaster waste significantly contributes to the levels of embodied carbon. One estimate showed that in Japan concrete rubble constitutes over 42% of waste generated from disasters (MOE (Japan), 2023). Building back quickly is essential for restoring livelihoods and a level of service to society, and also reducing the necessary processing of additional construction materials to aid in decarbonization.

The academic discussion of "resilience" is rich and has evolved greatly from different fields of mechanics, psychology, managerial science, systems theory, ecology,

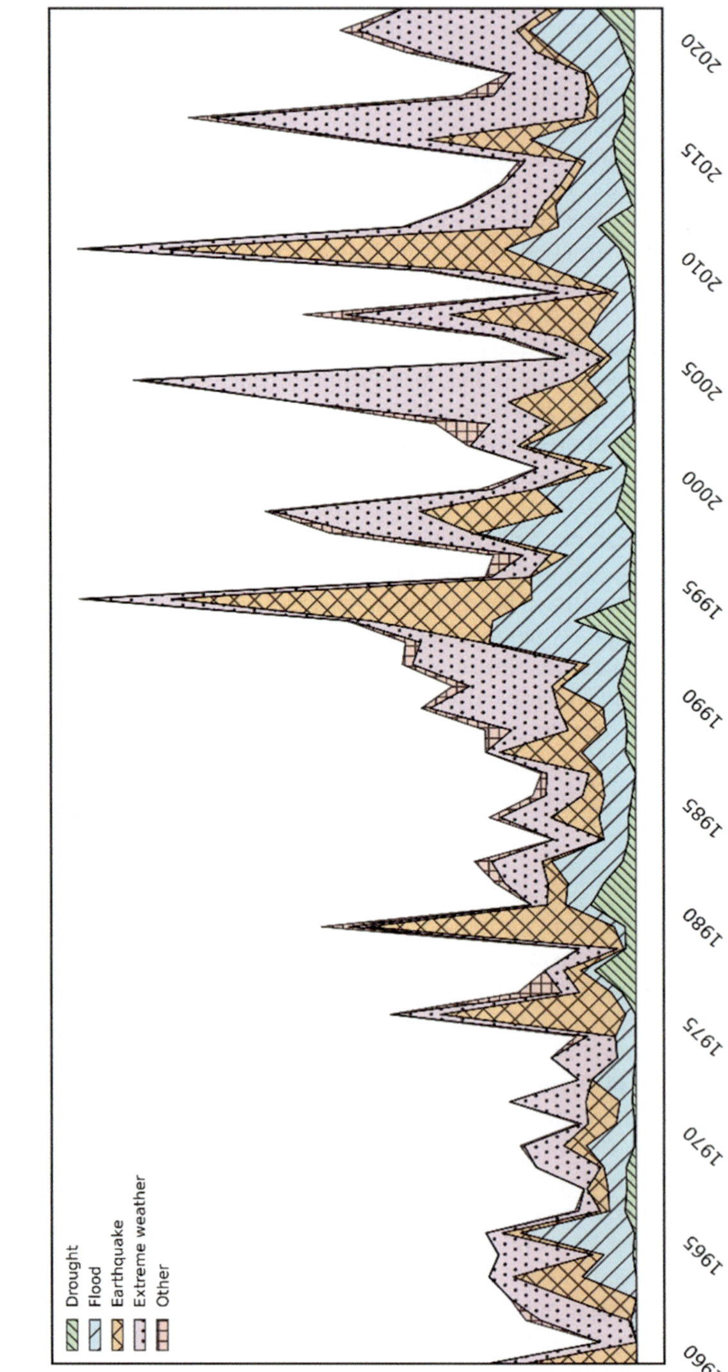

FIGURE 13.7 Economic damages from disasters as a share of GDP, World, 1960 to 2022.

(Data from EM-DAT et al. (2024), figure from authors.)

and, only more recently, climate change adaptation (Alexander, 2013). In the context of infrastructure, one of the earliest papers examining resilience is from 2003 and defines community seismic resilience as "the ability of social units to mitigate hazards, contain the effects of disasters when they occur, and carry out recovery activities in ways that minimize social disruption and mitigate the effects of future earthquakes" (Bruneau et al., 2003). The authors continue to discuss the criticality of infrastructure lifelines and emergency response management for community resilience in responding and recovering from disasters.

The application of sustainability to resilience frameworks in the built environment is a bit more recent. Bocchini et al. (2014) examine how sustainability and resilience complement each other in civil infrastructure. They note a key distinction: sustainability is assessed continuously over time, while resilience focuses on specific moments of failure and recovery. However, both fields are increasingly adopting similar approaches, including risk analysis, probability-based methods, and comprehensive assessments of economic, social, and ecological impacts (Bocchini et al., 2014). Important features of resilience include resourcefulness and faster recovery or regeneration of infrastructure systems (Bocchini et al., 2014; Bruneau et al., 2003). Regeneration, as a circular economy strategy, is an overarching description of a system that focuses on narrowing, slowing, and closing energy and material loops (Geissdoerfer et al., 2017; Konietzko et al., 2020). Regeneration is applicable to natural resources and ecosystems and indicates societal well-being as well.

The conceptual and practical unification of circular economy and resilience principles is more nascent. Kennedy and Linnenluecke (2022) published a paper proposing an agenda on advancing combined research in these fields by overlaying narrowing, slowing, and closing resource loops with socio-ecological, industry, and firm-level resilience. At a high level, the research themes include reducing reliance on virgin resources, reducing waste, increasing manufacturing capabilities and adaptabilities, increasing diversity of material supply and interconnectivity across industries (Kennedy and Linnenluecke, 2022). Some intermittent case studies of combined action exist, such as using recycled plastic for floating structures in flood-prone communities in Vietnam (Rijke et al., 2021) or the recycling of decommissioned solar panels for construction materials (Goh et al., 2024). In recent work in response to the 2023 earthquakes in Türkiye, Çetin and Kirchherr (2025) developed a Build Back Circular framework from interviews and workshops. The framework proposes 10 strategies, including policy-, technology-, design-, and community-based actions, to enhance resilience and disaster recovery (Çetin and Kirchherr, 2025). Nevertheless, the academic and professional body of work is still underdeveloped.

There are several potential technological advancements that could combine the aims of a circular economy and resilience in the built environment. In cases of damages from fire, the residual ash may be able to be used as SCM in new concrete, which has been previously studied through wood fly ash by-products from timber processing plants (Cheah and Ramli, 2011; Rumman et al., 2023). Demolition and disaster waste might be able to be processed to be used as material input for new construction with 3D printing, as demonstrated with some types of industrial waste (Dey et al., 2022). The deployment of robots and digital fabrication may facilitate a rapid reconstruction of built infrastructures supplementing skilled labor. Some research has been conducted on using reality capture technologies, Artificial Intelligence of Things, and digital twins to speed up reconstruction in post-war environments (Chashyn et al., 2024). Unmanned aerial vehicle photography data can be combined with laser point cloud data for 3D reconstruction of CDW piles, useful for auditing materials both damaged and at hand (Shi et al., 2024). Similarly, Ruggiero et al. (2024) are developing a digital workflow to upcycle wood waste from post-disaster areas using reality capture, BIM technologies, and cloud databases of the quality and quantity of recovered wood products. These technologies can be utilized after natural disasters for resilient and circular reconstruction with indigenous materials.

In a recent report funded by the US National Renewable Energy Laboratory, circular economy strategies were purported as important measures for decarbonization and climate change adaptation (Wang et al., 2022). Primary suggested strategies include reducing demand for virgin materials, regenerative agriculture, better flood resilience through waste management, circular water solutions, and circular value chains (Wang et al., 2022). A strong circular economy will be like a strong local economy that enables resource sharing and reuse. Both resilience and circular economy for the built environment will require strong circular construction supply chains. Ultimately, as called by many researchers (Bocchini et al., 2014; Çetin and Kirchherr, 2025; Kennedy and Linnenluecke, 2022; Wang et al., 2022; Byers et al., 2025), though a circular economy is often touted as a strategy for climate change mitigation through decarbonization and minimizing emissions, more research needs to be conducted on the synergies between circular economy, climate change adaptation, and infrastructure resilience.

13.6 CONCLUSIONS ON CIRCULARITY FOR THE CONSTRUCTION INDUSTRY

The complexity of global environmental challenges requires solutions that extend beyond the reductive narrative of "decarbonization." While the reduction of carbon dioxide emissions is important, a narrow focus on this metric risks obscuring broader environmental imperatives, including EmCarb, resource efficiency, and waste management, particularly for the construction industry. This chapter demonstrated how circular economy principles offer a more comprehensive approach to addressing these challenges, supported by case studies spanning material reuse, innovative design, and emerging technologies.

The integration of circular strategies with climate resilience presents a particularly important pathway forward. As natural disasters increase in frequency and severity, the ability to rapidly reconstruct using low-carbon materials becomes increasingly crucial. From robotics and AI to bio-based materials and advanced concrete, emerging technologies are creating new possibilities for material optimization, waste reduction, and building reuse. However, successful implementation requires careful consideration of local contexts, including regulatory frameworks, cultural preferences, and economic conditions.

Looking forward, meaningful progress will require collective action across industry stakeholders, policymakers, and researchers. While some regions have made significant strides in implementing circular economy policies and practices, broader adoption across developing economies will be crucial, given their projected construction growth. As noted by a report from the US National Renewable Energy Laboratory, "applying circular economy strategies for the decarbonisation technologies themselves is necessary for their scaling" (Wang et al., 2022). Success will require not just technological innovation but also supportive regulatory frameworks, new business models, and a fundamental shift in how the built environment resources are valued and managed. Through this holistic approach, the construction industry can not only reduce its environmental impact but also contribute to the regeneration of natural systems and the long-term resilience of human civilization.

REFERENCES

Adaptis [WWW Document], 2024. *Adaptis.ai.* URL https://www.adaptis.ai (accessed 27 January 2025).

Akanbi, L.A., Oyedele, L.O., Omoteso, K., Bilal, M., Akinade, O.O., Ajayi, A.O., Davila Delgado, J.M., Owolabi, H.A., 2019. Disassembly and deconstruction analytics system (D-DAS) for construction in a circular economy. *Journal of Cleaner Production* 223, 386–396. https://doi.org/10.1016/j.jclepro.2019.03.172

Alexander, D.E., 2013. Resilience and disaster risk reduction: An etymological journey. *Natural Hazards and Earth System Sciences* 13, 2707–2716. https://doi.org/10.5194/nhess-13-2707-2013

Ariza, I.M., 2018. In Place Detailing. Combining 3D printing and robotic assembly, in: *ACADIA//2018: Recalibration. On Imprecision and Infidelity. [Proceedings of the 38th Annual Conference of the Association for Computer Aided Design in Architecture (ACADIA) ISBN 978-0-692-17729-7]* Mexico City, Mexico, 18–20 October, 2018, pp. 312–321. CUMINCAD.

Arora, M., Raspall, F., Cheah, L., Silva, A., 2020. Buildings and the circular economy: Estimating urban mining, recovery and reuse potential of building components. *Resources, Conservation and Recycling* 154, 104581. https://doi.org/10.1016/j.resconrec.2019.104581

Assefa, G., Ambler, C., 2017. To demolish or not to demolish: Life cycle consideration of repurposing buildings. *Sustainable Cities and Society* 28, 146–153. https://doi.org/10.1016/j.scs.2016.09.011

Avet, F., Scrivener, K., 2018. Investigation of the calcined kaolinite content on the hydration of limestone calcined clay cement (LC3). *Cement and Concrete Research* 107, 124–135. https://doi.org/10.1016/j.cemconres.2018.02.016

Azcárate-Aguerre, J.F., den Heijer, A.C., Arkesteijn, M.H., Vergara d'Alençon, L.M., Klein, T., 2023. Facades-as-a-service: Systemic managerial, financial, and governance innovation to enable a circular economy for buildings. Lessons learnt from a full-scale pilot project in the Netherlands. Frontiers in Built Environment 9. https://doi.org/10.3389/fbuil.2023.1084078

Bakatovich, A., Davydenko, N., Gaspar, F., 2018. Thermal insulating plates produced on the basis of vegetable agricultural waste. *Energy and Buildings* 180, 72–82. https://doi.org/10.1016/j.enbuild.2018.09.032

Bao, Y., Li, V.C., 2020. Feasibility study of lego-inspired construction with bendable concrete. *Automation in Construction* 113, 103161. https://doi.org/10.1016/j.autcon.2020.103161

Baroš, T., Kabošová, L., Baroš, M., Katunský, D., 2022. Experimental form-finding: A review. *IOP Conference Series: Materials Science and Engineering* 1252, 012002. https://doi.org/10.1088/1757-899X/1252/1/012002

Bedarf, P., Szabo, A., Zanini, M., Dillenburger, B., 2024. Robotic 3D printing of geopolymer foam for lightweight and insulating building elements. *3D Printing and Additive Manufacturing* 11, 1–9. https://doi.org/10.1089/3dp.2023.0183

Boafo, F.E., Kim, J.-H., Kim, J.-T., 2016. Performance of modular prefabricated architecture: Case study-based review and future pathways. *Sustainability* 8, 558. https://doi.org/10.3390/su8060558

Bocchini, P., Frangopol, D.M., Ummenhofer, T., Zinke, T., 2014. Resilience and sustainability of civil infrastructure: Toward a unified approach. *Journal of Infrastructure Systems* 20, 04014004. https://doi.org/10.1061/(ASCE)IS.1943-555X.0000177

Boland, B., Lekhwani, S., Reiter, S., Sjödin, E., 2023. *Building Value by Decarbonizing the Built Environment*. McKinsey & Company.

Braungart, M., McDonough, W., 2009. *Cradle to Cradle*. Random House.

Brenner, M., Vandenabeele, L., Langenberg, S., 2023. The potential of reverse engineering and digital fabrication for the repair of high-tech architecture. *The International Archives of the Photogrammetry, Remote Sensing and Spatial Information Sciences* XLVIII-M-2–2023, 303–309. https://doi.org/10.5194/isprs-archives-XLVIII-M-2-2023-303-2023

Brundtland, G.H., 1987. Our common future—Call for action. *Environmental Conservation* 14, 291–294. https://doi.org/10.1017/S0376892900016805

Bruneau, M., Chang, S.E., Eguchi, R.T., Lee, G.C., O'Rourke, T.D., Reinhorn, A.M., Shinozuka, M., Tierney, K., Wallace, W.A., von Winterfeldt, D., 2003. A framework to quantitatively assess and enhance the seismic resilience of communities. *Earthquake Spectra* 19, 733–752. https://doi.org/10.1193/1.1623497

Bucher, M.J.J., Kraus, M.A., Rust, R., Tang, S., 2023. Performance-based generative design for parametric modeling of engineering structures using deep conditional generative models. *Automation in Construction* 156, 105128. https://doi.org/10.1016/j.autcon.2023.105128

Buswell, R.A., Leal de Silva, W.R., Jones, S.Z., Dirrenberger, J., 2018. 3D printing using concrete extrusion: A roadmap for research. *Cement and Concrete Research, SI: Digital Concrete* 2018 112, 37–49. https://doi.org/10.1016/j.cemconres.2018.05.006

Byers, B.S., De Wolf, C., 2023. QR code-based material passports for component reuse across life cycle stages in small-scale construction. *Journal of Circular Economy* 1, 1–16. https://doi.org/10.55845/IWEB6031

Byers, B.S., Gordon, M., Iuorio, O., De Wolf, C., 2023. Calculating embodied carbon for reused structural components with laser scanning, in: *Life-Cycle of Structures and Infrastructure Systems. Presented at the 8th International Symposium on Life-Cycle Civil Engineering (IALCCE 2023)*, CRC Press, pp. 149–156. https://doi.org/10.1201/9781003323020-15

Byers, B.S., Hunhevicz, J.J., Honic, M., Wolf, C.D., 2024a. Exploring Tokenized Product Passports for Circular Construction Supply Chains, in: *Computing in Construction. Presented at the EC3 Conference 2024, European Council on Computing in Construction*, https://doi.org/10.35490/EC3.2024.265

Byers, B.S., Raghu, D., Olumo, A., De Wolf, C., Haas, C., 2024b. From research to practice: A review on technologies for addressing the information gap for building material reuse in circular construction. *Sustainable Production and Consumption* 45, 177–191. https://doi.org/10.1016/j.spc.2023.12.017

Byers, B., Bompa, D., Jin, Q., & De Wolf, C., 2025. Circular economy for post-hazard infrastructure restoration: Strategies for resilience and resource efficiency (A. Bendiek-Laranjo, Trans.). *Journal of Sustainability*, 1(1). https://doi.org/10.55845/jos-2025-117

Cagle, R.F., 2003. *Infrastructure Asset Management: An Emerging Direction*. AACE International Transactions PM21–PM26.

CAL FIRE, 2025. Current Emergency Incidents [WWW Document]. *CAL FIRE*. URL https://www.fire.ca.gov/incidents (accessed 24 January 2025).

Cerda-Suarez, L.M., Paz-Gil, I., Sutil-Martin, D.L., 2023. Providing sustainable solutions in Chile: Neptuno pumps—Innovation that flows, in: Bianchi, E.C., Vazquez Burguete, J.L., Galan-Ladero, M.M., Lanero Carrizo, A. (Eds.), *Responsible Consumption and Sustainability: Case Studies from Corporate Social Responsibility, Social Marketing, and Behavioral Economics*. Springer International Publishing, Cham, pp. 165–181. https://doi.org/10.1007/978-3-031-30742-3_10

Çetin, S., Kirchherr, J., 2025. The build back circular framework: Circular economy strategies for post-disaster reconstruction and recovery. *Circular Economy and Sustainability*. https://doi.org/10.1007/s43615-024-00495-y

Chashyn, D., Khurudzhi, Y., Daukšys, M., 2024. Integration of building information modeling and artificial intelligence of things in the post-war reconstruction and renovation of buildings. *Journal of Sustainable Architecture and Civil Engineering* 35, 117–132. https://doi.org/10.5755/j01.sace.35.2.35160

Chau, C.K., Leung, T.M., Ng, W.Y., 2015. A review on life cycle assessment, life cycle energy assessment and life cycle carbon emissions assessment on buildings. *Applied Energy* 143, 395–413. https://doi.org/10.1016/j.apenergy.2015.01.023

Cheah, C.B., Ramli, M., 2011. The implementation of wood waste ash as a partial cement replacement material in the production of structural grade concrete and mortar: An overview. *Resources, Conservation and Recycling* 55, 669–685. https://doi.org/10.1016/j.resconrec.2011.02.002

Circotrade [WWW Document], n.d. URL https://www.circotrade.com/ (accessed 18 December 2024).

Circular Construction Lab, Just Places Lab, Finger Lakes Reuse, Susan Christopherson Center for Community Planning, 2024. Constructing a Circular Economy in New York State: Deconstruction and Building Material Reuse. New York.

City of Milwaukee, WI, 2018. Deconstruction Ordinance.

City of Palo Alto, CA, 2020. Deconstruction Ordinance.

City of Portland, OR, 2016. Deconstruction Ordinance.

City of San Antonio, TX, 2022. Vacant Buildings and Deconstruction Ordinance.

Colabella, S., D'amico, B., Hoxha, E., Fivet, C., 2017. Structural Design with Reclaimed Materials: an Elastic Gridshell out of Skis. *Proceedings of IASS Annual Symposia 2017*, 1–10.

Cramer, J., 2014. Moving towards a circular economy in the Netherlands: Challenges and directions. Utrecht University 1–9.

Curth, A., Pearl, N., Castro-Salazar, A., Mueller, C., Sass, L., 2024. 3D printing earth: Local, circular material processing, fabrication methods, and life cycle assessment. *Construction and Building Materials* 421, 135714. https://doi.org/10.1016/j.conbuildmat.2024.135714

Dario, P., Guadagnini, C., Laschi, C., Rucci, M., 1994. An experimental multisensorial robotic system for disassembly automation, in: *Proceedings of IEEE/RSJ International Conference on Intelligent Robots and Systems (IROS'94). Presented at the Proceedings of IEEE/RSJ International Conference on Intelligent Robots and Systems (IROS'94)*, pp. 2103–2110 vol. 3. https://doi.org/10.1109/IROS.1994.407575

De Wolf, C., Byers, B.S., Raghu, D., Gordon, M., Schwarzkopf, V., Triantafyllidis, E., 2024. A five-step D5 workflow of digital circular construction towards matchmaking for material reuse. *NPJ Materials Sustainability*, in press.

De Wolf, C., Cordella, M., Dodd, N., Byers, B., Donatello, S., 2023. Whole life cycle environmental impact assessment of buildings: Developing software tool and database support for the EU framework Level(s). *Resources, Conservation and Recycling* 188, 106642. https://doi.org/10.1016/j.resconrec.2022.106642

Devlin, A., Kossen, J., Goldie-Jones, H., Yang, A., 2023. Global green hydrogen-based steel opportunities surrounding high quality renewable energy and iron ore deposits. *Nature Communications* 14, 2578. https://doi.org/10.1038/s41467-023-38123-2

Deweerdt, M., Mertens, M., 2020. *A Guide for Identifying the Reuse Potential of Construction Products*. Working Draft Version.

Dey, D., Srinivas, D., Panda, B., Suraneni, P., Sitharam, T.G., 2022. Use of industrial waste materials for 3D printing of sustainable concrete: A review. *Journal of Cleaner Production* 340, 130749. https://doi.org/10.1016/j.jclepro.2022.130749

Eatherton, M., Deierlein, G., Krawinkler, H., Billington, S., Ma, X., 2008. Controlled rocking of steel-framed buildings with replaceable energy-dissipating fuses. *Presented at the World Conference on Earthquake Engineering, China.*

EDGAR/JRC, 2024. Change in carbon dioxide emissions in selected countries worldwide from 1990 to 2023 (in percent).

Egli, P., Gaschen, D., Kerscher, S., Jud, D., Hutter, M., 2022. Soil-adaptive excavation using reinforcement learning. *IEEE Robotics and Automation Letters* 7, 9778–9785. https://doi.org/10.1109/LRA.2022.3189834

EIA, 2024. Carbon dioxide emissions from energy consumption in the United States from 1975 to 2023 (in million metric tons of carbon dioxide).

Ellen MacArthur Foundation, 2013. Towards the circular economy Vol. 1: an economic and business rationale for an accelerated transition.

EM-DAT, CRED/UCLouvain, World Bank, 2024. Annual economic damages as a share of GDP from droughts – EM-DAT.

Enzer, M., Manidaki, M., Radford, J., Ellis, T., 2013. *Infrastructure Carbon Review*. His Majesty's Treasury.

EPA, 2024. Inventory of U.S. greenhouse gas emissions and sinks: 1990–2022.

Fang, D., Mueller, C., 2023. Mortise-and-tenon joinery for modern timber construction: Quantifying the embodied carbon of an alternative structural connection. *Architecture, Structures and Construction* 3, 11–24. https://doi.org/10.1007/s44150-021-00018-5

Fiala, L., Konrád, P., Fořt, J., Keppert, M., Černý, R., 2020. Application of ceramic waste in brick blocks with enhanced acoustic properties. *Journal of Cleaner Production* 261, 121185. https://doi.org/10.1016/j.jclepro.2020.121185

Foo, G., Kara, S., Pagnucco, M., 2022. Challenges of robotic disassembly in practice. *Procedia CIRP, The 29th CIRP Conference on Life Cycle Engineering*, April 4–6, 2022, Leuven, Belgium. 105, pp. 513–518. https://doi.org/10.1016/j.procir.2022.02.085

FRACO|Brick Recyc [WWW Document], 2024. URL https://www.fraco.com/en/products/BrickRecyc (accessed 29 January 2025).

Freire-González, J., Padilla Rosa, E., Raymond, J.L., 2024. World economies' progress in decoupling from CO_2 emissions. *Science Reports* 14, 20480. https://doi.org/10.1038/s41598-024-71101-2

Gaspar, F., Bakatovich, A., Davydenko, N., Joshi, A., 2020. 8 - Building insulation materials based on agricultural wastes, in: Pacheco-Torgal, F., Ivanov, V., Tsang, D.C.W. (Eds.), *Bio-Based Materials and Biotechnologies for Eco-Efficient Construction*, Woodhead Publishing Series in Civil and Structural Engineering. Woodhead Publishing, pp. 149–170. https://doi.org/10.1016/B978-0-12-819481-2.00008-8

Geissdoerfer, M., Savaget, P., Bocken, N.M.P., Hultink, E.J., 2017. The circular economy – A new sustainability paradigm? *Journal of Cleaner Production* 143, 757–768. https://doi.org/10.1016/j.jclepro.2016.12.048

Gholam, B., 2020. What do we mean by efficiency? A holistic approach to reducing embodied carbon. *The Structural Engineer: Journal of the Institution of Structural Engineer* 98, 14–17.

Goh, K.C., Kurniawan, T.A., Goh, H.H., Zhang, D., Jiang, M., Dai, W., Khan, M.I., Othman, M.H.D., Aziz, F., Anouzla, A., Meidiana, C., 2024. Harvesting valuable elements from solar panels as alternative construction materials: A new approach of waste valorization and recycling in circular economy for building climate resilience. *Sustainable Materials and Technologies* 41, e01030. https://doi.org/10.1016/j.susmat.2024.e01030

Goldstein, B., Rasmussen, F.N., 2018. LCA of buildings and the built environment, in: Hauschild, M.Z., Rosenbaum, R.K., Olsen, S.I. (Eds.), *Life Cycle Assessment: Theory and Practice*. Springer International Publishing, Cham, pp. 695–722. https://doi.org/10.1007/978-3-319-56475-3_28

Grossman, G.M., Krueger, A.B., 1991. *Environmental impacts of a North American Free Trade Agreement* (No. 3914). National Bureau of Economics Research, Cambridge, MA.

Gu, Y., Wu, Y., Xu, M., Wang, H., Zuo, T., 2017. To realize better extended producer responsibility: Redesign of WEEE fund mode in China. *Journal of Cleaner Production* 164, 347–356. https://doi.org/10.1016/j.jclepro.2017.06.168

Harvey, L.D.D., 2021. Iron and steel recycling: Review, conceptual model, irreducible mining requirements, and energy implications. *Renewable and Sustainable Energy Reviews* 138, 110553. https://doi.org/10.1016/j.rser.2020.110553

Heisel, F., McGranahan, J., Boghossian, A., 2022. ScanR: A composite building scanning and survey method for the evaluation of materials and reuse potentials prior to demolition and deconstruction. *IOP Conference Series: Earth and Environmental Science* 1078, 012012. https://doi.org/10.1088/1755-1315/1078/1/012012

Hemmati, M., Messadi, T., Gu, H., Seddelmeyer, J., Hemmati, M., 2024. Comparison of embodied carbon footprint of a mass timber building structure with a steel equivalent. *Buildings* 14, 1276. https://doi.org/10.3390/buildings14051276

Hines, M., 2024. 5 ways an evolving market is compelling owners and developers to rethink their existing real estate [WWW Document]. Perkins&Will. URL https://perkinswill.com/insights/repurposing-adaptive-reuse/ (accessed 23 December 2024).

Hoeppe, P., 2016. Trends in weather related disasters – Consequences for insurers and society. Weather and climate extremes, observed and projected (longer-term) changes in weather and climate. *Extremes* 11, 70–79. https://doi.org/10.1016/j.wace.2015.10.002

Honic, M., De Wolf, C., 2023. Applying EU Level(s) framework indicators to improve circularity: A case study. *IOP Conference Series: Earth and Environmental Science* 1196, 012041. https://doi.org/10.1088/1755-1315/1196/1/012041

Hunhevicz, J., De Wolf, C., 2023. Blockchain for a circular digital built environment. *Presented at the Construction Blockchain Conference 2023*. https://doi.org/10.47330/CBC.2023.MYTX8932

IRENA, 2024. Production of renewable energy worldwide from 2009 to 2022 (in terawatt hours).

ISO, 2024. ISO 59004:2024(en), Circular economy—Vocabulary, principles and guidance for implementation.

Jensen, J.P., Prendeville, S.M., Bocken, N.M.P., Peck, D., 2019. Creating sustainable value through remanufacturing: Three industry cases. *Journal of Cleaner Production* 218, 304–314. https://doi.org/10.1016/j.jclepro.2019.01.301

Jin, Q., Leung, C.K.Y., 2011. Fiber-reinforced-cementitious-composites plate for anchoring FRP sheet on concrete member. *Journal of Composites for Construction* 15, 790–798. https://doi.org/10.1061/(ASCE)CC.1943-5614.0000211

Jin, Q., Leung, C.K.Y., Yu, C., 2013. Effective joining method for pseudo-ductile permanent formwork. *Material Structures* 46, 345–360. https://doi.org/10.1617/s11527-012-9905-3

Jin, Q., Li, V.C., 2019a. Development of lightweight engineered cementitious composite for durability enhancement of tall concrete wind towers. *Cement and Concrete Composites* 96, 87–94. https://doi.org/10.1016/j.cemconcomp.2018.11.016

Jin, Q., Li, V.C., 2019b. Structural and durability assessment of ECC/concrete dual-layer system for tall wind turbine towers. *Engineering Structures* 196, 109338. https://doi.org/10.1016/j.engstruct.2019.109338

Jin, Q., Liao, W., Ni, X., Ma, H., 2024. Low-grade fly ash in Portland cement blends: A decoupling approach to evaluate reactivity and hydration effects. Cement 18, 100119. https://doi.org/10.1016/j.cement.2024.100119

Jones, R.L., Guha-Sapir, D., Tubeuf, S., 2022. Human and economic impacts of natural disasters: can we trust the global data? *Science Data* 9, 572. https://doi.org/10.1038/s41597-022-01667-x

Jorgensen, T.M., Andersen, A.W., Christensen, S.S., 1996. Shape recognition system for automatic disassembly of TV-sets, in: *Proceedings of 3rd IEEE International Conference on Image Processing. Presented at the Proceedings of 3rd IEEE International Conference on Image Processing*, pp. 653–656, vol. 2. https://doi.org/10.1109/ICIP.1996.560962

Karsan, Z., De Wolf, C., Dillenburger, B., 2024. Vision assisted robotic plasma cutting setup for on-site disassembly of welded steel structures. https://doi.org/10.2139/ssrn.4942936

Kennedy, S., Linnenluecke, M.K., 2022. Circular economy and resilience: A research agenda. *Business Strategy and the Environment* 31, 2754–2765. https://doi.org/10.1002/bse.3004

Kirchherr, J., Yang, N.-H.N., Schulze-Spüntrup, F., Heerink, M.J., Hartley, K., 2023. Conceptualizing the circular economy (revisited): An analysis of 221 definitions. *Resources, Conservation and Recycling* 194, 107001. https://doi.org/10.1016/j.resconrec.2023.107001

Konietzko, J., Bocken, N., Hultink, E.J., 2020. Circular ecosystem innovation: An initial set of principles. *Journal of Cleaner Production* 253, 119942. https://doi.org/10.1016/j.jclepro.2019.119942

Krystofik, M., Luccitti, A., Parnell, K. and Thurston, M., 2018. Adaptive remanufacturing for multiple lifecycles: A case study in office furniture. *Resources, Conservation and Recycling* 135, 14–23. https://doi.org/10.1016/j.resconrec.2017.07.028

Küpfer, C., Bastien-Masse, M., Fivet, C., 2023. Reuse of concrete components in new construction projects: Critical review of 77 circular precedents. *Journal of Cleaner Production* 383, 135235. https://doi.org/10.1016/j.jclepro.2022.135235

Lagoeiro, H., Revesz, A., Davies, G., Curry, D., Vivian, J., Faulks, G., Murphy, D., Maidment, G., 2022a. Recovering Waste Heat from the London Underground: Sizing the Opportunity. Presented at the *World Sustainable Energy Days 2022*.

Lagoeiro, H., Revesz, A., Davies, G., Gysin, K., Curry, D., Faulks, G., Murphy, D., Vivian, J., Maidment, G., 2022b. Investigating the opportunity for cooling the London underground through waste heat recovery. *Building Services Engineering Research and Technology* 43, 347–359. https://doi.org/10.1177/01436244221084913

Lenssen, N.J.L., Schmidt, G.A., Hansen, J.E., Menne, M.J., Persin, A., Ruedy, R., Zyss, D., 2019. Improvements in the GISTEMP Uncertainty Model. *Journal of Geophysical Research: Atmospheres* 124, 6307–6326. https://doi.org/10.1029/2018JD029522

Leon-Miquel, M., Silva-Retamal, J., Aparicio, D., Rangelov, M., Jin, Q., Paul, A., 2023. Novel application of Chilean natural pozzolan for sustainable strain-hardening cementitious composite. *Resources, Conservation and Recycling* 197, 107098. https://doi.org/10.1016/j.resconrec.2023.107098

Leschok, M., Piccioni, V., Lydon, G., Seshadri, B., Schlueter, A., Gramazio, F., Kohler, M., Dillenburger, B., 2024. Thermal and manufacturing properties of hollow-core 3D-printed elements for lightweight facades. *Developments in the Built Environment* 19, 100485. https://doi.org/10.1016/j.dibe.2024.100485

Li, V.C., 2019. *Engineered Cementitious Composites (ECC): Bendable Concrete for Sustainable and Resilient Infrastructure*. Springer, Berlin, Heidelberg. https://doi.org/10.1007/978-3-662-58438-5

Liang, Y., 2025. Green project finance for infrastructure and building construction: instruments and benchmarks. *Journal of Infrastructure Systems* 31, 06025001. https://doi.org/10.1061/JITSE4.ISENG-2573

Liew, A., López, D.L., Van Mele, T., Block, P., 2017. Design, fabrication and testing of a prototype, thin-vaulted, unreinforced concrete floor. *Engineering Structures* 137, 323–335. https://doi.org/10.1016/j.engstruct.2017.01.075

Lockley, A., Coffman, D., 2018. Carbon dioxide removal and tradeable put options at scale. *Environmental Research Letters* 13, 054034. https://doi.org/10.1088/1748-9326/aabe96

Loizou, L., Barati, K., Shen, X., Li, B., 2021. Quantifying advantages of modular construction: Waste generation. *Buildings* 11, 622. https://doi.org/10.3390/buildings11120622

Lotfi, S., Eggimann, M., Wagner, E., Mróz, R., Deja, J., 2015. Performance of recycled aggregate concrete based on a new concrete recycling technology. *Construction and Building Materials* 95, 243–256. https://doi.org/10.1016/j.conbuildmat.2015.07.021

Manai, A., 2020. A framework to assess and repair pre-fatigued welded steel structures by TIG dressing. *Engineering Failure Analysis* 118, 104923. https://doi.org/10.1016/j.engfailanal.2020.104923

Marini, A., Passoni, C., Belleri, A., Feroldi, F., Preti, M., Metelli, G., Riva, P., Giuriani, E., Plizzari, G., 2022. Combining seismic retrofit with energy refurbishment for the sustainable renovation of RC buildings: A proof of concept. *European Journal of Environmental and Civil Engineering* 26, 2475–2495. https://doi.org/10.1080/19648189.2017.1363665

Marsh, Guy Carpenter, Oxford Economics, 2021. Future of Construction: A Global Forecast for Construction to 2030. London.

Martinez-Marquez, D., Florin, N., Hall, W., Majewski, P., Wang, H., Stewart, R.A., 2022. State-of-the-art review of product stewardship strategies for large composite wind turbine blades. *Resources, Conservation & Recycling Advances* 15, 200109. https://doi.org/10.1016/j.rcradv.2022.200109

Menny, T., Le Guirriec, S., De Wolf, C., 2024. The butterfly matchmaking model for circular construction: Towards a digital matchmaking platform tailored to French policy. *Sustainable Production and Consumption* 49, 130–143. https://doi.org/10.1016/j.spc.2024.06.011

Thomas, Michael, 2019. *Impact of CO2 Utilization in Fresh Concrete on Corrosion of Steel Reinforcement*. CarbonCure Technologies Inc., Dartmouth, NS, Canada.

Miller, A., 1982. Tunnel vision in environmental management. *The Environmentalist* 2, 223–231. https://doi.org/10.1016/S0251-1088(82)93407-6

MOE (Japan), 2023. Share of waste generated by disasters in Japan in 2021, by type.

Mollaei, A., Bachmann, C., Haas, C., 2023a. Assessing the impact of policy tools on building material recovery. *Resources, Conservation and Recycling* 198, 107188. https://doi.org/10.1016/j.resconrec.2023.107188

Mollaei, A., Byers, B., Christovan, C., Olumo, A., De Wolf, C., Bachmann, C., Haas, C., 2023b. "A global perspective on building material recovery incorporating the impact of regional factors." *Journal of Cleaner Production* 429, 139525. https://doi.org/10.1016/j.jclepro.2023.139525

Murphy, W., 2023. Deep dive: Retrofit, embodied carbon and the M&S Decision - refusal and recrimination on Oxford Street [WWW Document]. *Cornerstone Barristers*. URL https://cornerstonebarristers.com/deep-dive-retrofit-embodied-carbon-and-the-ms-decision-refusal-and-recrimination-on-oxford-street/ (accessed 17 December 2024).

Nian, S., Pham, T., Haas, C., Ibrahim, N., Yoon, D., Bregman, H., 2022. A functional demonstration of adaptive reuse of waste into modular assemblies for structural applications: The case of bicycle frames. *Journal of Cleaner Production* 348, 131162. https://doi.org/10.1016/j.jclepro.2022.131162

NYCEDC, 2024. *Clean and Circular: Design & Construction Guidelines (No. 1.0)*. New York City Economic Development Corporation, New York City.

Ochsendorf, J., Freeman, M., 2010. *Guastavino Vaulting: The Art of Structural Tile*. Princeton Architectural Press.

Odaglia, P., Posani, M., Voney, V., Habert, G., Dillenburger, B., 2023. *Geopolymer Binder Jetting with Recycled Aggregates: Testing and Validation for Construction Applications*. https://doi.org/10.2139/ssrn.4594048

Parmaco/Smart buildings - with a sustainable solution [WWW Document], 2024. Parmaco. URL https://www.parmaco.fi/en/ (accessed 18 December 2024).

Pasanen, P., Sipari, A., Terranova, E., Castro, R., Bruce-Hyrkas, T., 2018. *The Embodied Carbon Review—Embodied Carbon Reduction 100+ Regulations and Rating Systems Globally*. One Click LCA Ltd, Bionova Ltd.: Hertfordshire, UK.

Pasanen, P., Tikka, S., Le Gouvello, L., Koukoulopoulos, K., Kalfountzos, V., Bounds, L., 2021. *Decarbonizing Construction: Guidance for Investors and Developers to Reduce Embodied Carbon*. World Business Council for Sustainable Development (WBCSD), Geneva, Switzerland.

Pătrăucean, V., Armeni, I., Nahangi, M., Yeung, J., Brilakis, I., Haas, C., 2015. State of research in automatic as-built modelling. *Advanced Engineering Informatics, Infrastructure Computer Vision* 29, 162–171. https://doi.org/10.1016/j.aei.2015.01.001

Petersen, K.H., Napp, N., Stuart-Smith, R., Rus, D., Kovac, M., 2019. A review of collective robotic construction. Science *Robotics* 4, eaau8479. https://doi.org/10.1126/scirobotics.aau8479

Potting, J., Hekkert, M.P., Worrell, E., Hanemaaijer, A., others, 2017. *Circular Economy: Measuring Innovation in the Product Chain*. PBL Netherlands Environmental Assessment Agency.

Purchase, C.K., Al Zulayq, D.M., O'Brien, B.T., Kowalewski, M.J., Berenjian, A., Tarighaleslami, A.H., Seifan, M., 2022. Circular economy of construction and demolition waste: A literature review on lessons, challenges, and benefits. *Materials* 15, 76. https://doi.org/10.3390/ma15010076

Raghu, D., Bucher, M.J.J., De Wolf, C., 2023. Towards a 'resource cadastre' for a circular economy – Urban-scale building material detection using street view imagery and computer vision. *Resources, Conservation and Recycling* 198, 107140. https://doi.org/10.1016/j.resconrec.2023.107140

Raghu, D., De Wolf, C., 2024. India's informal reuse ecosystem towards circular construction, in: Thomsen, M.R., Ratti, C., Tamke, M. (Eds.), *Design for Rethinking Resources, Sustainable Development Goals Series*. Springer International Publishing, Cham, pp. 127–137. https://doi.org/10.1007/978-3-031-36554-6_10

Raghu, D., De Wolf, C., 2022. Digitalizing existing buildings: Moving towards circularity using big data and machine learning. *Presented at the Artificial Intelligence in Architecture, Engineering and Construction Conference (AI in AEC 2022), AI in AEC 2022.*

Reike, D., Vermeulen, W.J.V., Witjes, S., 2018. The circular economy: New or refurbished as CE 3.0?—Exploring controversies in the conceptualization of the circular economy through a focus on history and resource value retention options. *Resources, Conservation and Recycling, Sustainable Resource Management and the Circular Economy* 135, 246–264. https://doi.org/10.1016/j.resconrec.2017.08.027

Reisach, D., Schütz, S., Willmann, J., Schneider, S., 2024. Digital fabrication for circular timber construction: A case study. *Journal of Circular Economy* 1, 1–18. https://doi.org/10.55845/vwgd7873

Reusable Steel Environmental Product Declaration, 2022.

Richardson, K., Steffen, W., Lucht, W., Bendtsen, J., Cornell, S.E., Donges, J.F., Drüke, M., Fetzer, I., Bala, G., von Bloh, W., Feulner, G., Fiedler, S., Gerten, D., Gleeson, T., Hofmann, M., Huiskamp, W., Kummu, M., Mohan, C., Nogués-Bravo, D., Petri, S., Porkka, M., Rahmstorf, S., Schaphoff, S., Thonicke, K., Tobian, A., Virkki, V., Wang-Erlandsson, L., Weber, L., Rockström, J., 2023. Earth beyond six of nine planetary boundaries. *Science Advances* 9, eadh2458. https://doi.org/10.1126/sciadv.adh2458

Rijke, J., Geerling, L., Quan, N.H., Trung, N.H., 2021. Removing challenges for building resilience with support of the circular economy, in: de Graaf-van Dinther, R. (Ed.), *Climate Resilient Urban Areas: Governance, Design and Development in Coastal Delta Cities*. Springer International Publishing, Cham, pp. 109–127. https://doi.org/10.1007/978-3-030-57537-3_6

Rios, F.C., Chong, W.K., Grau, D., 2015. Design for disassembly and deconstruction - Challenges and opportunities. *Procedia Engineering, Defining the Future of Sustainability and Resilience in Design, Engineering and Construction* 118, 1296–1304. https://doi.org/10.1016/j.proeng.2015.08.485

Ritchie, H., Rosado, P., 2024. Is the number of natural disasters increasing? Our World in Data.

Rotor – Brussels | Rotor [WWW Document], n.d. URL https://rotordb.org/en (accessed 23 December 2024).

Rudin, N., Hoeller, D., Reist, P., Hutter, M., 2022. Learning to Walk in Minutes Using Massively Parallel Deep Reinforcement Learning, in: *Proceedings of the 5th Conference on Robot Learning. Presented at the Conference on Robot Learning, PMLR*, pp. 91–100.

Ruggiero, R., Cognoli, R., Cocco, P.L., 2024. From debris to the data Set (DEDA) a digital application for the upcycling of waste wood material in post disaster areas, in: Barberio, M., Colella, M., Figliola, A., Battisti, A. (Eds.), *Architecture and Design for Industry 4.0: Theory and Practice*. Springer International Publishing, Cham, pp. 807–835. https://doi.org/10.1007/978-3-031-36922-3_41

Rumman, R., Kamal, M.R., Bediwy, A., Alam, M.S., 2023. Partially burnt wood fly ash characterization and its application in low-carbon mortar and concrete. *Construction and Building Materials* 402, 132946. https://doi.org/10.1016/j.conbuildmat.2023.132946

Sanchez, B., Haas, C., 2018. A novel selective disassembly sequence planning method for adaptive reuse of buildings. *Journal of Cleaner Production* 183, 998–1010. https://doi.org/10.1016/j.jclepro.2018.02.201

Sánchez Berriel, S., Favier, A., Rosa Domínguez, E., Sánchez Machado, I.R., Heierli, U., Scrivener, K., Martirena Hernández, F., Habert, G., 2016. Assessing the environmental and economic potential of Limestone Calcined Clay Cement in Cuba. *Journal of Cleaner Production* 124, 361–369. https://doi.org/10.1016/j.jclepro.2016.02.125

Sarc, R., Curtis, A., Kandlbauer, L., Khodier, K., Lorber, K.E., Pomberger, R., 2019. Digitalisation and intelligent robotics in value chain of circular economy oriented waste management – A review. *Waste Management* 95, 476–492. https://doi.org/10.1016/j.wasman.2019.06.035

SAVE Britain's Heritage, 2022. SAVE Open Letter on the decision to suspend the demolition of Marks & Spencer's flagship store.

Scheel, C., Aguiñaga, E., Bello, B., 2020. Decoupling economic development from the consumption of finite resources using circular economy. *A Model for Developing Countries. Sustainability* 12, 1291. https://doi.org/10.3390/su12041291

Scrivener, K., Avet, F., Maraghechi, H., Zunino, F., Ston, J., Hanpongpun, W., Favier, A., 2019. Impacting factors and properties of limestone calcined clay cements (LC3). *Green Materials* 7, 3–14. https://doi.org/10.1680/jgrma.18.00029

Scrivener, K., Martirena, F., Bishnoi, S., Maity, S., 2018a. Calcined clay limestone cements (LC3). Cement and Concrete Research, Report of UNEP SBCI Working Group on Low-CO2 Eco-efficient Cement-Based Materials 114, 49–56. https://doi.org/10.1016/j.cemconres.2017.08.017

Scrivener, K.L., John, V.M., Gartner, E.M., 2018b. Eco-efficient cements: Potential economically viable solutions for a low-CO2 cement-based materials industry. Cement and Concrete Research, Report of UNEP SBCI Working Group on Low-CO2 Eco-efficient Cement-Based materials 114, 2–26. https://doi.org/10.1016/j.cemconres.2018.03.015

Shell and Deloitte Consulting, 2023. Decarbonising Construction: Building a Low-Carbon Future.

Shi, L., Zhao, W., Wang, Y., Yan, X., 2024. High-precision 3D Modeling of Construction Waste Pile Bodies by Integrating Multi-source Data. *The International Archives of the Photogrammetry, Remote Sensing and Spatial Information Sciences* XLVIII-4–2024, 423–429. https://doi.org/10.5194/isprs-archives-XLVIII-4-2024-423-2024

Skellet [WWW Document], n.d. Skellet. URL https://www.skellet.com (accessed 18 December 2024).

Smith, A., 2025. 2024: An active year of U.S. billion-dollar weather and climate disasters | NOAA Climate.gov. Beyond the Data. URL http://www.climate.gov/news-features/blogs/beyond-data/2024-active-year-us-billion-dollar-weather-and-climate-disasters (accessed 24 January 2025).

Soman, R.K., Nikolić, D., Sanchez, B., 2024. Extended reality as a catalyst for circular economy transition in the built environment, in: De Wolf, C., Çetin, S., Bocken, N.M.P. (Eds.), *A circular built environment in the digital age*. Springer International Publishing, Cham, pp. 171–193. https://doi.org/10.1007/978-3-031-39675-5_10

Stahel, W.R., 2016. The circular economy. *Nature* 531, 435–438. https://doi.org/10.1038/531435a

Standing Committee of the National People's Congress, 2008. Circular Economy Promotion Law of the People's Republic of China.

Statista, 2024. World's biggest CO_2 emitters 2023.

Stern, D.I., 2017. The environmental Kuznets curve after 25 years. *Journal of Bioeconomics* 19, 7–28. https://doi.org/10.1007/s10818-017-9243-1

Tang, S., Shelden, D.R., Eastman, C.M., Pishdad-Bozorgi, P., Gao, X., 2019. A review of building information modeling (BIM) and the internet of things (IoT) devices integration: Present status and future trends. *Automation in Construction* 101, 127–139. https://doi.org/10.1016/j.autcon.2019.01.020

The White House, 2023. Federal-State Buy Clean Partnership Principles. The White House, Washington.

TRB, 2018. NCHRP 10-104 - Recommendations for Revision of AASHTO M 295 Standard Specification to Include Marginal and Unconventional Source Coal Fly Ashes. [WWW Document]. URL https://apps.trb.org/cmsfeed/TRBNetProjectDisplay.asp?ProjectID=4565 (accessed 2 December 2019).

Urban Machine [WWW Document], 2024. Urban Machine. URL https://urbanmachine.build/ (accessed 17 December 2024).

van der Heijden, J., 2023. Construction 4.0 in a narrow and broad sense: A systematic and comprehensive literature review. *Building and Environment* 244, 110788. https://doi.org/10.1016/j.buildenv.2023.110788

VAULTED AG [WWW Document], n.d. URL https://www.vaulted-floors.com/ (accessed 23 December 2024).

VAULTED to build first large commercial project [WWW Document], 2024. Startup ticker. URL https://www.startupticker.ch/en/news/vaulted-to-build-first-large-commercial-project (accessed 23 December 2024).

Wang, K., Costanza-van den Belt, M., Heath, G., Walzberg, J., Curtis, T., Berrie, J., Schröder, P., Lazer, L., Altamirano, J.C., 2022. Circular economy as a climate strategy: current knowledge and calls-to-action (No. NREL/TP-6A20-84141). National Renewable Energy Laboratory, Golden, CO. https://doi.org/10.2172/1897625

Welcome to the BoKlok world! | BoKlok [WWW Document], n.d.. BoKlok. URL https://www.boklok.se/global/ (accessed 18 December 2024).

World Bank, 1992. World Development Report 1992: Development and the Environment, 2. print. ed. Oxford University Press.

World Steel Association, 2024. World Steel Data.

World Steel Association, 2022. Steel Facts. World Steel Association.

Xiong, Z., Gordon, M., Byers, B., De Wolf, C., 2022. Reality capture and site-scanning techniques for material reuse planning, in: *Proceedings of IASS Annual Symposia. International Association for Shell and Spatial Structures (IASS)*, pp. 1–11.

Xu, P., Zhu, J., Li, H., Wei, Y., Xiong, Z., Xu, X., 2022. Are bamboo construction materials environmentally friendly? A life cycle environmental impact analysis. *Environmental Impact Assessment Review* 96, 106853. https://doi.org/10.1016/j.eiar.2022.106853

Zhang, D., 2024. CO_2 utilization for concrete production: Commercial deployment and pathways to net-zero emissions. *Science of the Total Environment* 931, 172753. https://doi.org/10.1016/j.scitotenv.2024.172753

Zhang, D., Ghouleh, Z., Shao, Y., 2017. Review on carbonation curing of cement-based materials. *Journal of CO_2 Utilization* 21, 119–131. https://doi.org/10.1016/j.jcou.2017.07.003

Zhang, W., Gallagher, B., Jin, Q., Huang, C.-H., 2022. Mineralogy optimization for immobilization of potentially toxic elements and halides in Co-disposed flue gas desulfurization brines and bituminous coal fly ash. *Journal of Hazardous Materials Advances* 8, 100154. https://doi.org/10.1016/j.hazadv.2022.100154

14 Separation Technologies for Decarbonization

Junqiang Liu and Xue (Ida) Chen

14.1 CO_2 CAPTURE BY SECTORS

The International Energy Agency (IEA) has compiled an extensive list of both existing and planned carbon capture, utilization, and storage (CCUS) projects. These projects are systematically categorized into nine distinct sectors, providing an overview of the current landscape and future developments in this critical field.[1] Table 14.1 offers a comparative analysis between the current operational projects and those that are scheduled for development by 2030 within each sector. This comparison not only highlights the significant strides made in the implementation of CCUS technologies but also underscores the dynamic nature of the sector. As the industry evolves, it continuously adapts to emerging needs and incorporates innovative solutions to enhance efficiency and effectiveness. This comprehensive analysis by the IEA serves as a testament to the ongoing efforts and advancements in the field of CCUS. It emphasizes the importance of these technologies in addressing climate change and showcases the proactive measures being taken to meet future environmental challenges.

The power and heat sector, recognized as the largest contributor to global CO_2 emissions, is increasingly focusing on carbon capture technologies to mitigate its environmental impact. Due to the sector's vast scale of operations, the implementation of new technologies is inherently complex and time-consuming. A prominent example of this effort is the Petro Nova carbon capture plant in Texas. This facility stands as a testament to the sector's commitment to reducing emissions. Despite its potential, the Petro Nova plant has faced significant challenges. The substantial capital investment required and operational difficulties led to a temporary shutdown of the plants. However, it has recently resumed operations, showcasing the resilience and determination within the sector to overcome these hurdles. This scenario underscores both the immense potential and the significant challenges associated with integrating carbon capture technologies into large-scale energy production. It highlights the sector's ongoing efforts to innovate and adapt, striving to balance environmental responsibilities with the practical realities of implementation. The experience of the Petro Nova plant serves as a valuable lesson in the journey toward a more sustainable future, illustrating the complexities and the critical importance of continued investment and innovation in carbon capture technologies.

Natural gas sweetening has emerged as the most significant sector for CO_2 capture, driven by stringent requirements for CO_2 concentration in the final natural gas product. This process is crucial for ensuring the quality and safety of natural gas.

DOI: 10.1201/9781003655442-19

TABLE 14.1
Operational and Planned Capture Capacity (million tons per year) by IEA[1]

Sector	Current Mitigation in 2023 (MM ton/yr)	Projected Mitigation in 2030 (MM ton/yr)	Sector Example Project	Country	Partners	Project Status	Announced Capacity (Mt CO_2/yr)	Fate of CO_2	Current Separation Method
Power and heat	3.2	89.1	Petra Nova Carbon Capture (TX)	United States	Equity: NRG and JX Nippon, Debt: JBIC & Mizhou bank (backed by NEXI)	Operational	1.4	EOR	Absorption
Natural gas processing/ LNG	33	76.1	Petrobras Santos Basin pre-salt oilfield CCS	Brazil	Petrobras	Operational	3–10.6	EOR	Membrane
Hydrogen or ammonia	None	75.7	Eastern Louisiana Clean Hydrogen Complex (LA)	United States	Air Products	Under construction	5	Dedicated storage	Absorption
DAC	0.01	63.7	STRATOS (Ector County) (TX)	United States	1PointFive, Oxy low carbon ventures (Occidental), Blackrock	Under construction	0.5	Mixed	KOH reaction

Other fuel transformation	8.2	56.9	Great Plains Synfuel Plant (ND) Weyburn-Midale (SK)	United States	Dakota Gasification Company, Cenovus, Whitecap, Cardinal	Operational	3	EOR	Absorption
Biofuels	1.3	29.9	Bonanza BioEnergy CCUS (KS)	United States	Conestoga Energy, Gary Climate Solutions Gary Climate Solutions	Operational	0.1–0.16	EOR	Compression
Cement	None	27.4	Heidelberg Materials Brevik cement	Norway	Heidelberg Materials (Norcem), Aker CC (tech)	Under construction	0.4	Dedicated storage	Absorption
Chemicals	3.8	16.5	Enid Fertilizer (OK)	United States	Koch Nitrogen Company, Daylight Petroleum	Operational	0.68	EOR	Absorption
Iron and steel	0.9	5	Emirates Steel Industries (Abu Dhabi CCS Phase 1)	United Arab Emirates	ADNOC, Masdar	Operational	0.8	EOR	Absorption

Additionally, the feasibility of sequestering captured CO_2 in nearby wells for enhanced oil recovery (EOR) has further bolstered its importance. Since the 1970s, CO_2-EOR has been specifically applied to increase the recovery factor of oil fields. This technique not only enhances oil production but also provides a practical solution for CO_2 sequestration. A notable example is the Petrobras Santos Basin Pre-Salt Oil Field CCUS project, which is the world's largest CO_2 sequestration initiative. It boasts an impressive injection rate of over 10 million metric tons (MMT) of CO_2 per year, demonstrating the scale and potential of such projects.[2] In the past decade, membrane separation has gained prominence for biomethane separation due to its promising economic viability. This technology stands out with investment costs from €3500 to €7500 per cubic meter per hour (m^3/h) and operational costs of between €7.5 and €12.5 per cubic meter per hour (m^3/h).[3] These cost-effective solutions make membrane separation an attractive option for the industry.

Overall, the advancements in natural gas sweetening and CO_2-EOR, along with the economic feasibility of membrane separation for biomethane, highlight the sector's commitment to reducing emissions and enhancing efficiency. These technologies play a pivotal role in the ongoing efforts to address environmental challenges and promote sustainable energy practices.

Blue hydrogen and ammonia are set to become the third-largest sector utilizing CO_2 capture technologies, driven by their dual roles as both energy sources and chemical feedstocks. The U.S. National Clean Hydrogen Strategy and Roadmap has outlined strategic pathways for clean hydrogen production, aiming for 10 MMT annually by 2030, 20 MMT by 2040, and 50 MMT per year by 2050 in the United States.[4] Blue hydrogen, produced from fossil fuels with CO_2 capture, is currently viewed as a transitional solution between high-emission grey hydrogen and the limited-scale zero-emission green hydrogen.[5] The levelized cost of hydrogen (LCOH) for blue hydrogen ranges from \$1.5 to 2.3 per kilogram, whereas green hydrogen costs approximately twice as much, with prices ranging from \$2.7 to \$6.8 per kilogram.[6] Hydrogen's versatility makes it a crucial enabler for future CO_2 utilization, advancing sustainable energy solutions. Several commercial technologies exist for blue hydrogen separation, including cryogenic separation, pressure swing adsorption (PSA), and amine absorption. An exemplary project is the East Louisiana Clean Hydrogen Complex, currently under construction, which will use amine absorption technology to capture approximately 5 million tons of CO_2 annually.

The hydrogen production of 10 MMT per year by steam methane reforming (SMR) with CO_2 sequestration would annually require 1.9 billion GJ of natural gas resources, withdraw 0.30 cubic kilometers (km^3) of water resources, and occupy 25.1 square kilometers (km^2) of land resources. Although the future cost of blue hydrogen is expected to decrease due to cumulative experience and tax incentives, it may not be economically competitive with grey hydrogen in the near term. Therefore, to jumpstart a clean hydrogen economy appropriate government policy needs to be established.[7] This strategic approach underscores the importance of hydrogen in the transition to a low-carbon economy, highlighting both the potential and the challenges of integrating CO_2 capture technologies in large-scale hydrogen production.

DAC is projected to experience the most explosive growth, becoming the fourth-largest sector for CO_2 capture by 2030. DAC is a revolutionary tool to reverse the historical CO_2 emission by directly removing CO_2 from the atmosphere. According to the Intergovernmental Panel on Climate Change (IPCC), to limit global warming to under 2°C, globally we will need to remove one billion tons of CO_2 from the atmosphere by 2025, 10 billion tons by 2050, and 100 billion tons by the end of the century.[8] Despite criticisms regarding the high energy consumption and land usage associated with DAC, the sector remains optimistic due to continuous technological advancements. Currently, there are about 19 DAC plants in operation worldwide, collectively removing around 0.01 MMT CO_2 per year. However, even larger plants are either under-constructed or in advanced planning stages. For instance, Occidental Petroleum has strategically shifted its focus from flue gas capture to DAC. Through its subsidiary 1PointFive, Occidental has announced an ambitious plan to build the world's largest DAC plant, STRATOS (in Ector, Texas, United States), which aims to capture 0.5 MMT CO_2 annually by 2025.[9]

A recent study projects the cost of achieving a cumulative DAC capacity of 1000 MMT of CO_2 per year, including transport and storage, at \$341 per ton of CO_2 (\$226–\$544 at 90% confidence) for caustic solution (Carbon Engineering), \$374 per ton of CO_2 (\$281–\$579) for solid sorbent (Climeworks), and \$371 per ton of CO_2 (\$230–\$835) for CaO ambient weathering (Heirloom).[10] The technology is still evolving, with ongoing efforts to reduce cost and scale-up operations before widespread commercial deployment can be achieved.[11] The rapid growth and development in the DAC sector highlight its potential to play a crucial role in global efforts to combat climate change. As the technology continues to advance, it promises to become a key component in the portfolio of solutions needed to achieve significant CO_2 reductions and move toward a more sustainable future.

Other fuel transformation, particularly in clean coal research, is set to become the fifth-largest sector for CO_2 capture. A notable example is the \$2.1 billion Great Plains Synfuel Plant in North Dakota. Since 1984, this facility has been converting 18,000 tons of lignite coal per day into synthetic natural gas and 13 other products, while capturing and sequestering a significant portion of the CO_2 emissions. The plant has successfully sequestered more than 44 MMT of CO_2 since 2000. By demonstrating effective CO_2 capture and storage technologies, the Great Plains Synfuel Plant extends the viable timeframe of fossil fuel usage.[12]

Four other significant sectors employing CO_2 capture technologies are biofuels, cement, chemicals, and iron and steel production. Each of these sectors presents distinct challenges and opportunities for CO_2 capture due to their varying processes. For instance, CO_2 captured from biofuel fermentation typically involves high concentrations of CO_2, which simplifies the separation process compared to other sectors. In cement production, a major source of CO_2 emissions release is the calcination of calcium carbonate ($CaCO_3$). The emissions from iron and steel production also have unique compositions in their exhaust gases. Consequently, CO_2 mitigation strategies in these industries must be tailored to their specific emission profiles.[13]

Overall, the advancements in CO_2 capture technologies across these diverse sectors highlight the ongoing efforts to reduce emissions and promote sustainability. By addressing the unique challenges of each sector, these technologies play a crucial role in the global effort to combat climate change and transition to a low-carbon economy.

14.2 CO_2 SEPARATION TECHNOLOGIES

Table 14.2 shows the TRL progression of different technologies across application areas. The values of 2005 and 2015 were derived from previous reviews, providing a historical perspective on the development of these technologies.[14] The 2024 values are newly added, reflecting the latest advancements and innovations in each field. This table serves as a comprehensive overview of technological evolution, highlighting the significant strides made over the years. Each technology's TRL progression is discussed in detail in the sections below, offering insights into their status, recent developments, and future potential.

14.2.1 Amine

14.2.1.1 Amine for Pre-Combustion

Amine absorption for CO_2 capture has been a well-established technology for several decades. This process utilizes aqueous amine solutions to absorb CO_2 from flue gases or other gas streams. Over the years, technology suppliers have continuously improved the formulations and processes. Key development trends are improving efficiency, reducing costs, and enhancing environmental performance. Research into alternative absorbents, such as ionic liquids and solvents with novel chemical properties, offers promising opportunities for further performance enhancements. Additionally, combining amine absorption with other CO_2 capture technologies, such as membrane separation or adsorption, can create hybrid systems that leverage the strengths of multiple technologies.[30] The chemicals of some commonly used amines in CO_2 capture include monoethanolamine (MEA), diethanolamine (DEA), methyldiethanolamine (MDEA), piperazine (PZ), diisopropanolamine (DIPA), and triethanolamine (TEA).

These amines are selected for their ability to effectively absorb and release CO_2 through reversible chemical reactions. The structural versatility of these compounds, characterized by the presence of hydroxyl groups, facilitates aqueous dissolution while offering kinetic and thermodynamic benefits.

14.2.1.2 Amine for Post-Combustion

Amine absorption for post-combustion CO_2 capture is the best-proven technology, but it faces several significant challenges, including long-term stability, tolerance to impurities, and gas-liquid mass transfer limitation. Degradation of amine solutions can occur due to thermal, oxidative, and chemical factors, leading to a significant loss in absorption capacity. Addressing these challenges requires a combination of advanced amine formulations, effective feed gas pre-treatment, and optimized equipment design. Improving the mass transfer rate in amine absorption systems is essential to enhance CO_2 capture efficiency and reduce the size of the contactor. Membrane contactors offer an improved interfacial area (up to 10,000 m^2/m^3) significantly greater than most traditional absorber columns (80–800 m^2/m^3), resulting in high overall mass transfer rates. The weight and dimensions of the membrane contactor can be reduced by 66% and 72%, respectively.[31]

TABLE 14.2
Advancement of CO_2 Capture Technologies[14]

Separation Process	Application	TRL			Comment
		2005	2015	2024	
Absorption					
Physical solvent	Industry	9	9	9	Commercial for syngas H_2/CO_2
	Pre-combustion	8	8	8	Viable but not as competitive to chemical solvents for high recovery and purity of CO_2[15,16]
Chemical solvent	Industry	9	9	9	Commercial for refineries and natural gas sweetening
	Post-combustion	7	8	9	Boundary Dam CCS has been operational since 2014. Petro Nova project restart, Century project termination
Cryogenics					
Air Separation	Industry	9	9	9	<4000 ton/day commercial production[14]
	Oxy-combustion	5	7	7	8 Rivers plan to start commercial 300MW process in 2027, conservative 2032.[17,18]
CO_2 anti-sublimation	Post-combustion	3	3	3	Lack of major pilot demo[19]
Chemical Looping					
Oxygen Carrier	Oxy-combustion	3	6	6	brief demo at 1MW[20]
Calcium looping	Post-combustion	2	6	7	continuous demo >1MW by multiple plants[21]
	DAC	2	3	6	0.5MMT CO_2/year plant by 1PointFive in 2025[22]
Solid Sorbents					
	Industry	9	9	9	PSA for industrial H_2 production
	Pre-combustion	8	8	9	Exxon Baytown project using UOP technology[23]
	Post-combustion	2	5	7	Svante Rotating Bed, Kawasaki Moving Bed[24]
	Oxy-combustion	6	6	6	Not competitive compared to cryogenic at large scale[25]
	DAC	2	3	5	Small pilot scale plant 0.01MMT CO_2/year[26]
Membranes					
	Industrial	9	9	9	Commercial technology for natural gas sweetening and EOR
	Post-combustion	3	5	7	150 ton/day largest pilot in 2024[27]
	Pre-combustion	3	3	5	Pd membrane small pilot scale[28]
	Oxy-combustion	4	5	5	Ion Transport Membrane's largest demonstration is ~5 t/day, lack further scale up[29]

The Petra Nova carbon capture project is a notable example of post-combustion CO_2 capture technology in operation. The plant started in 2016 but was temporarily shut down in 2020 due to the plunging oil price. During the initial three years of operation, the plant operated at less than 85% of its target capacity. In 2023, the plant was restarted by JX Nippon Oil & Gas Exploration Corp.[32] The viability of such projects is closely tied to stable economic conditions and revenue streams, such as EOR. Incentives, carbon pricing, and regulatory frameworks can help mitigate financial risks and encourage investment in carbon capture technologies. The Petra Nova project offers valuable insights into both the potential and challenges of implementing such technologies at scale.

14.2.1.3 Physical Solvent

Physical solvents use physical absorption rather than chemical reactions to capture CO_2. These solvents absorb CO_2 based on its solubility in the solvents. One of the key advantages of physical solvents is that they typically require less energy for regeneration compared to amine solutions. This is because the regeneration process relies on pressure changes rather than high temperatures, making it more energy efficient. Physical solvents generally exhibit better stability and lower degradation rates compared to amines, leading to longer operational lifetimes. Additionally, they tend to be less corrosive than amine solutions, reducing maintenance issues and extending the lifespan of the equipment used in the CO_2 capture process. However, physical solvents do have some limitations. They generally have a lower capacity and selectivity for CO_2 absorption compared to amines. This means that while they are effective, they may not capture as much CO_2 as amine-based systems. Physical solvents also require low-temperature operation and are sensitive to temperature changes, which can affect their performance.[15]

Common physical solvents used for CO_2 capture include:

- Selexol: A mixture of dimethyl ethers of polyethylene glycol, effective at absorbing CO_2 at high pressures.
- Rectisol: Uses cold methanol to absorb CO_2 and other acid gases, suitable for high-pressure applications.
- Propylene Carbonate: Known for its high CO_2 solubility and low volatility, making it effective in certain industrial applications.
- N-Methyl-2-pyrrolidone (NMP): A solvent with good CO_2 absorption capacity and stability.
- Polyethylene Glycol (PEG): Used in various formulations for its ability to dissolve CO_2 efficiently.
- Sulfinol: A mixture of physical and chemical solvents, combining the benefits of both types for enhanced CO_2 capture.

Despite the challenges, physical solvents remain a valuable option for CO_2 capture, particularly in scenarios where energy efficiency and equipment longevity are prioritized. The ongoing research and development in this field aim to enhance the performance of physical solvents, making them more competitive with amine-based systems and expanding their applicability in various industrial settings.

14.2.2 Cryogenic

14.2.2.1 Cryogenic for Oxy-Combustion

Cryogenic separation is currently the most effective method for producing high-purity oxygen at large scales, particularly for applications requiring more than 500 tons per day. This process involves cooling air to extremely low temperatures to liquefy its components, allowing for the separation of oxygen from other gases like nitrogen and argon. One of the primary advantages of cryogenic separation is its ability to produce oxygen with very high purity levels, which is essential for efficient oxy-combustion. However, the main barrier to the widespread commercial deployment of the technology is the high energy consumption associated with oxygen production. The process typically requires approximately 200–300 kWh per ton of O_2, which significantly impacts the overall thermal-to-electric efficiency of the plant. This energy penalty results in an efficiency reduction of about 8.5–12% compared to traditional air combustion plants.[33] Despite these challenges, cryogenic separation remains a critical technology for applications that demand high-purity oxygen. Ongoing research and development efforts aim to reduce energy consumption and improve the efficiency of the process. Innovations in cryogenic technology, such as advanced heat exchangers and more efficient refrigeration cycles, hold promise for making cryogenic separation more economically viable.

In summary, while cryogenic separation is highly effective for large-scale oxygen production, its commercial deployment is hindered by high energy consumption. Addressing this challenge through technological advancements is essential for enhancing the efficiency and feasibility of oxy-combustion and other applications requiring high-purity oxygen.

14.2.2.2 Cryogenic for Pre-Combustion

Cryogenic CO_2 capture is a sophisticated and effective method for separating CO_2, particularly when high purity is required. This process involves cooling the compressed, chilled CO_2 stream to very low temperatures, typically below –50°C, to transform CO_2 into a liquid or solid state. This step may involve additional refrigeration cycles or expansion processes. At these temperatures, CO_2 can be more easily separated from other gases that remain in the gaseous phase. The captured liquid or solid CO_2 is then collected, stored, and transported for further use or sequestration. It can be stored in tanks or pipelines designed to handle cryogenic temperatures. This method has the advantage of scalability and the ability to produce high-purity CO_2. Therefore, it is sometimes used as a second polishing step after amine, membrane, or PSA processes for high-purity CO_2 production. Air Liquide's Cryocap® process is a notable example of this approach.

However, the cryogenic method also has some disadvantages. It requires special equipment for low-temperature handling, and the operational costs are high due to the extensive compression and cooling required. Managing these high energy requirements and associated costs is crucial to making cryogenic CO_2 capture a viable option in broader CO_2 reduction strategies.

Despite these challenges, cryogenic CO_2 capture remains a valuable technology for applications where high-purity CO_2 is essential. Ongoing research and development efforts aim to improve the efficiency and cost-effectiveness of this process, making it a more attractive option for large-scale CO_2 capture.[19]

14.2.3 Cryogenic for Post-Combustion, Anti-Sublimation

The cryogenic method for post-combustion CO_2 capture, also known as anti-sublimation or solidification, is a promising technology due to its potential high energy efficiency and the ability to produce high-purity CO_2. This process involves cooling the CO_2-laden gas stream to very low temperatures, causing the CO_2 to solidify. The solid CO_2 can then be separated from the remaining gas components. Despite its potential, the cryogenic method faces several technical and operational challenges that need to be addressed for it to become a viable large-scale solution. One of the primary challenges is the handling of solid CO_2. Managing solid CO_2 requires specialized equipment and processes to ensure efficient separation and transportation. Another significant challenge is related to mass and heat transfer. Efficiently transferring heat in and out of the system is crucial for maintaining the low temperatures required for CO_2 solidification. This requires advanced heat exchanger designs and effective insulation to minimize energy losses.

Additionally, the lack of large-scale pilot demonstrations poses a barrier to the widespread adoption of this technology. While laboratory and small-scale studies have shown promising results, scaling up the process to industrial levels requires extensive testing and validation to ensure reliability and cost-effectiveness.[19]

Addressing these challenges will require continued research and development, as well as investment in pilot projects to demonstrate the feasibility of cryogenic CO_2 capture at scale. With further advancements, this technology has the potential to play a significant role in reducing CO_2 emissions.

14.2.4 Adsorption

Adsorption is a process that separates molecules (adsorbates) based on their affinity to a solid adsorbent surface. Adsorbents are categorized into two main types: physical and chemical, depending on the strength of the interaction between the adsorbate and the adsorbent. Physical adsorption involves weaker interactions, resulting in lower capacity and generally less selectivity. However, it is easier to desorb the adsorbate during the regeneration step, making PSA typically sufficient. This method is often used in pre-combustion CO_2 capture with adsorbents such as activated carbon or zeolite. In contrast, chemical adsorption involves stronger interactions, offering higher capacity and greater selectivity. Amine functionalized adsorbent belongs to this group.[34] Although chemical adsorption can achieve high-purity CO_2, it requires more energy for regeneration, typically employing temperature swing adsorption (TSA). This method is more suitable for post-combustion CO_2 capture or direct air capture (DAC) to achieve effective capacity even at low CO_2 concentrations.[24]

In summary, while physical adsorption is advantageous for its ease of regeneration and lower energy requirements, chemical adsorption provides higher capacity and selectivity, making it ideal for applications requiring high-purity CO_2 capture.

Each method has its specific applications and benefits, contributing to the overall effectiveness of CO_2 capture technologies.

14.2.4.1 Adsorption for Pre-Combustion CO_2 Capture

The ExxonMobil Baytown CO_2 capture plant employs an innovative hybrid approach that combines PSA and cryogenic processes. The hybrid approach leverages the strengths of both PSA and cryogenic techniques to optimize CO_2 capture efficiency and purity. PSA is utilized for the bulk separation of CO_2. PSA is effective in handling large volumes of gas, efficiently separating CO_2 from other components in the gas stream. This process operates by cycling the pressure within the adsorption columns, allowing CO_2 to be adsorbed at high pressures and desorbed at low pressures. PSA is particularly advantageous due to its lower energy requirements for regeneration compared to other methods. Cryogenic processes are then employed to achieve high-purity CO_2. By cooling the CO_2-laden gas stream to very low temperatures, CO_2 is transformed into a liquid or solid state, which can be easily separated from other gases. This step ensures that the CO_2 captured is of high purity, suitable for various industrial applications or sequestration. The hybrid approach at the Baytown plant combines these two methods to maximize efficiency and minimize energy consumption. PSA handles the initial bulk separation, reducing the load on the cryogenic process. This synergy allows the cryogenic stage to operate more effectively, achieving high-purity CO_2 with potentially lower energy input compared to cryogenic methods alone.[23]

This innovative combination of PSA and cryogenic processes exemplifies the potential for hybrid systems in CO_2 capture technology. By leveraging the strengths of multiple methods, the Baytown plant demonstrates a practical and efficient solution for pre-combustion CO_2 capture, contributing to the broader efforts to reduce greenhouse gas emissions and combat climate change.

14.2.4.2 Adsorption for Post-Combustion

Using traditionally packed bed adsorption for post-combustion CO_2 capture presents several challenges due to factors such as low concentration of CO_2, low pressure and large volume of flue gas, and high level of impurities, particularly moisture. These issues make traditionally packed bed systems infeasible from a footprint perspective.[35] However, advancements in non-conventional adsorption technologies, such as moving bed and rotating bed systems, offer promising alternatives. These technologies improve efficiency by allowing separate control of temperature and environmental conditions in the adsorption and desorption zones. As a result, the adsorption/desorption cycles can be achieved at rates 10 to 100 times faster, significantly reducing the system's footprint.[36] The adsorption process for post-combustion CO_2 capture must also consider competitive adsorption by moisture. In most studies, CO_2 and H_2O are assumed to follow single-component isotherms without interactions.[37] However, water vapor in flue gas can reduce the specific adsorption capacity of CO_2 due to competing adsorption. Additionally, the presence of water increases the heat penalty of molecular sieve regeneration due to water adsorption.

There are several non-conventional adsorption processes employing fluidized, moving, or rotating beds of adsorbent, developed to the TRL level of about 7. The following are a few examples.

The pilot Hadong dry-sorbent CO_2 capture system led by the Korea Electric Power Research Institute (KEPRI) is the largest adsorption project registered, with a capture capacity of 200 t-CO_2/day, equivalent to a 10 MW coal-fired SCPC Supercritical Pulverized Coal (SCPC) power plant. This pilot unit is located at the Hadong power plant in South Korea and operated from April 2014 to 2017 using a specially developed adsorbent: KEPCO2P2. The material consisted of 35 wt% K_2CO_3 and 65 wt% of support material. It consists of a fluidized-bed carbonator and regenerator, sorbent cooler, and several cyclones. The pilot-scale plant is housed in a 34 m (L) × 15 m (W) × 59 m (H) building. The adsorption was carried out at 80°C and regeneration of the sorbent at 200°C. Over 1000-hr continuous operation of the pilot plant has been performed and the results showed that the dry-sorbent gave a stable performance of 80% CO_2 capture rate and 95% CO_2 purity in the regenerator throughout the test campaign.[38]

Kawasaki Heavy Industries, Ltd. (KHI) has researched and developed a CO_2 capture system, which is called Kawasaki Carbon Capture (KCC), utilizing solid sorbent. The amine-impregnated solid sorbent development stemmed from DAC for submarines and can also be applied to direct capture from ambient air. The greatest feature of the KCC system is that it can desorb CO_2 using low-temperature steam below 100°C (e.g., 60°C), and this steam can be generated from unused exhaust heat in various plants. It comprises three towers: adsorption tower, desorption tower, and drying tower. A long-term continuous operation has been demonstrated in a pilot-scale moving bed plant with recovery of 40 t-CO_2/day at Kansai Electric Power Company's Maizuru Power Station. Next, maximum recovery tests, parameter change tests, and continuous operation tests will be conducted using this pilot plant by 2024. In the future, a thousand tons of CO_2/day commercial capture plant is planned.[24] In 2023, a Japan-U.S. collaboration was started to construct a demonstration test plant that will be built at the Integrated Test Center (ITC) adjacent to Dry Fork Station (a pulverized coal-fired power plant) in Wyoming. Tests will be conducted using actual flue gas from the power plant.[39]

Svante (formerly Inventys) has developed a rapid cycle TSA (VeloxoTherm™) process using structured adsorbents and steam-assisted regeneration with fast kinetics (< 1.5 minutes cycle time). It developed a new Metal Organic Framework (MOF), CALF-20, sorbent, and achieved semi-commercial scale production. This MOF is robust with regard to steam, O_2, and acidic contaminant gases (such as NO_X/SO_X), which make it an ideal candidate for the flue gas CO_2 capture application. CALF-20 performance after 2300 hours of VeloxoTherm™ capture process has been studied.[40] The inventive rotating bed adsorption technology enables faster adsorption cycle and reduces the adsorbent inventory and equipment footprint. The scale-up from the current low t-CO_2/day to a few hundred t-CO_2/day scale is ongoing and should be closely monitored.[41]

14.2.4.3 Adsorption for DAC

DAC is expected to become increasingly important in the future. The U.S. Infrastructure Investment and Jobs Act introduced significant tax credit incentives for carbon-free energy and clean technologies. CCUS is eligible for large tax credits of \$85/t-CO_2 for permanent storage and \$60/t-CO_2 for utilization. DAC is eligible for tax credits of \$180/t-CO_2 for permanent storage and \$130/t-CO_2 for utilization. These tax credits have an enormous impact and make DAC a realistic investment prospect. Depending on the CO_2 adsorption affinity and regeneration temperature, DAC can be separated into high-temperature (HT) and low-temperature (LT) processes. A recent study reported a 111–133 €/t_{CO2} cost for HT DAC, which is higher than the projection from Carbon Engineering at 85–87 €/t_{CO2}. LT DAC costs have been calculated to be 38–69 €/t_{CO2}, which is in line with the projected costs of Climeworks (75 €/t_{CO2}) and Global Thermostat (11–38 €/t_{CO2}) for long-term or large-scale deployment of DAC technologies.[26]

A recent report by the National Academies of Science, Engineering, and Medicine estimated that the sorbent will account for 80% of all DAC costs.[42] There are three key approaches to reduce the DAC cost from adsorbent innovation: (1) reduce the cost of manufacturing, (2) increase the cyclic stability (reduce degradation) of materials, and (3) increase the capacity of materials. The long-term goal is to improve the lifetime capacity from the current 130 to 1000 kg CO_2/kg sorbent.[43]

14.2.5 Chemical Looping

14.2.5.1 Oxygen Carrier Chemical Looping

A chemical looping concept has been pursued in different processes since the 1900s: H_2 from coal and steam using iron oxide as an oxygen carrier, Atlantic Richfield Company (ARCO) developed a gas-to-gasoline (GTG) process, which uses two fluidized-bed reactors to convert CH_4 to an ethylene-rich intermediate via oxidative coupling of methane (OCM). DuPont developed a process to produce maleic anhydride through selective butane oxidation. A vanadium phosphorus oxide was used as the chemical intermediate. Producing syngas from CH_4 through a fixed bed reactor system using CeO_2 as an oxygen carrier was also attempted in the 1990s. Dow is scaling up an enhanced ethane cracking technology using oxygen carrier for selective burning of the H_2 product.[44] A large-scale oxygen carrier chemical looping process has not been achieved due to inefficient process, reactivity and recyclability of oxygen carriers, and mechanical strength.[45]

Oxygen carrier chemical looping is a promising CO_2 capture technology, which is expected to achieve 100% capture efficiency with a low energy penalty. The principle utilizes metal particles as a medium to transport oxygen between two reactors, thereby avoiding nitrogen in the combustion exhaust and achieving high-purity CO_2. There are about 40 small-scale pilot plant demonstration units worldwide. The largest nominal thermal power is about 4 MWth.[20]

14.2.5.2 Calcium Looping CO_2 Capture

Calcium looping is a leading technology for CO_2 capture that utilizes calcium-based materials to absorb and release CO_2. This process employs calcium oxide (CaO) as the active material, relying on the carbonation and calcination reactions of calcium carbonate ($CaCO_3$). Calcium looping can achieve high CO_2 removal efficiencies from flue gases through carbonation reaction. Calcium-based materials are abundant and relatively inexpensive. The spent CaO can potentially be recycled as cement feedstock, which helps reduce operational costs. However, the calcium oxide sorbent can experience deactivation over time due to sintering and attrition, impacting its performance and lifespan. The calcination process requires high temperatures, leading to substantial energy consumption. The initial capital investment for calcium looping systems can be high due to the need for circulating fluidized beds. The technology has been demonstrated at a large pilot scale for CO_2 capture with a TRL of 7. A few small-scale pilot plants are shown below.

The Ohio State University's (OSU) calcium looping process, using an inventive hydration step to prevent CaO sintering, was licensed to, commercialized, and utilized by the Industrial Technology Research Institute (ITRI) in Taiwan with a 1.9 MW_{th} nominal thermal power. The calcium looping unit is operated in a synergistic scheme with a cement plant, while CO_2 capture is also used in the production of high-value-added commercial products. The operation captures the CO_2 in the flue gas stream from a coal power plant with spent calcium sorbent used as feedstock for cement production. Apart from ITRI, the calcium looping process has been demonstrated by a few other institutions and organizations, such as a 1 MW_{th} unit by Darmstadt University in Germany, a 1.7 MW_{th} unit by Consejo Superior de Investigaciones Cientificas (CSIC) in Spain, and a 3 MW_{th} unit by Alstom power in the United States.[21]

14.2.5.3 DAC

The carbon engineering DAC process employs two closed chemical loops. The first brings air into contact with an aqueous basic KOH solution to capture CO_2 and form potassium carbonate. Then, calcium hydroxide solution is added to regenerate the contactor KOH solution and precipitate the calcium carbonate using a pellet reactor, which is borrowed from the wastewater treatment industry. The $CaCO_3$ solids collected from the centrifuge are then heated in a calciner at 300°C to 900°C to release CO_2. This step also leaves behind calcium oxide, which is then mixed with water in the slaker for rehydration. The rehydrated $Ca(OH)_2$ is then fed back into the pellet reactor, beginning the cycle again.[22] Occidental acquired Carbon Engineering in 2023 and announced an ambitious plan to build the world's largest DAC plant (STRATOS, Ector, TX) to capture 0.5 MMT CO_2 year by 2025.[9]

14.2.6 Membrane

Membrane separation technology, driven by the pressure difference, is gaining popularity due to its chemical-free operation, lack of moving parts, and modular scalability. Membranes have been extensively used in natural gas sweetening since the 1980s, especially at high CO_2 feed concentrations (e.g., >10%). Their modular nature makes them ideal for the emerging biomethane applications. Traditional polymeric membranes, such as those made from cellulose acetate, polysulfone, and polyimide, offer effective separation for these established industrial applications. It is well-suited for the above bulk separation where CO_2 concentrations in the feed are high. However, it is generally less effective at achieving high CO_2 recovery or purity (e.g., >95%) due to its limited separation factor.[46]

14.2.6.1 Pre-Combustion

Pre-combustion processes typically require high purity for both hydrogen (H_2, >99% for low carbon emission fuel) and CO_2 (>95% for sequestration), imposing stringent selectivity requirements (e.g., >100) that traditional polymeric membranes cannot meet. Separating H_2 from CO_2 is particularly challenging because both molecules permeate similarly through conventional membrane materials, making the separation difficult. Two promising technologies under development are palladium (Pd) membranes and amine-facilitated transport membranes, which may offer sufficient separation factors.

Pd membranes selectively permeate H_2 by dissociating H_2 molecules into absorbed atoms, which then recombine to form H_2 as they permeate through the membrane. However, Pd membranes are approximately 100 times more expensive than polymeric membranes and operate at high temperatures. Several small-scale pilot-scale demonstration units have been reported. Reinertsen AS has developed a 19-tube hydrogen separation module for the demonstration activity, providing 2.7 m^2 of active membrane surface area. Additionally, a membrane skid capable of handling up to 2.7 tons per day (200 Nm^3 h^{-1}) of synthesis gas slipstream was designed, constructed, and installed at the Statoil 2500 MTPD methanol synthesis plant at Tjeldbergodden, Norway, for a 12-month field test starting in October 2016.[47] Tokyo Gas Co., Ltd scaled up Pd-based membrane reformer system with a nominal hydrogen production capacity of 40 $Nm^3 \cdot h^{-1}$. The system demonstrates the potential advantages of the membrane reformer: the simplicity of the system configuration. High purity (99.999% level) hydrogen is produced in a single step from natural gas. It's compact and high efficiency. Other benefits of the membrane reformer include the fact that the CO_2 in the offgas can easily be liquefied and captured, due to the high CO_2 concentration of 90%.[28] Many other explorations at a small pilot scale are also being tested under relevant industrial conditions.[48]

Microporous membrane tight-size sieving pores, such as zeolite, MOF, and carbon molecular sieves, are also proposed. Once a perfect and defect-free membrane is obtained, almost all of the resulting emerging microporous material-based membranes will show promising performance for CO_2 separations. However, translating this promising performance into a real industrial application for CO_2 separation is still a major challenge. From a membrane fabrication point of view, this includes membrane fabrication into efficient packing forms (i.e., hollow fiber). Mechanical strength is another issue for these thin inorganic membranes. It is also imperative to test the successful membranes in the mixed gas scenario or use the real feed gas to elucidate the membrane robustness. Future development of these membranes will need to address moisture tolerance, scalability, cost, and separation efficiency.[49] Finally, membrane aging should also be thoroughly investigated to evaluate its long-term performance.[50]

14.2.6.2 Post-Combustion

Post-combustion CO_2 capture using membranes presents several challenges, including the low partial pressure of CO_2, the large volume of flue gas, and the presence of oxidative environments and impurities. To address these challenges, two operational modes for membrane systems have been proposed. The first mode involves compressing the flue gas feed to create the necessary pressure driving force, while the second mode utilizes a vacuum at the permeate side to enhance CO_2 permeation. The vacuum permeate method has been demonstrated to be the most efficient process. However, it will need large-scale industrial vacuum equipment.[51]

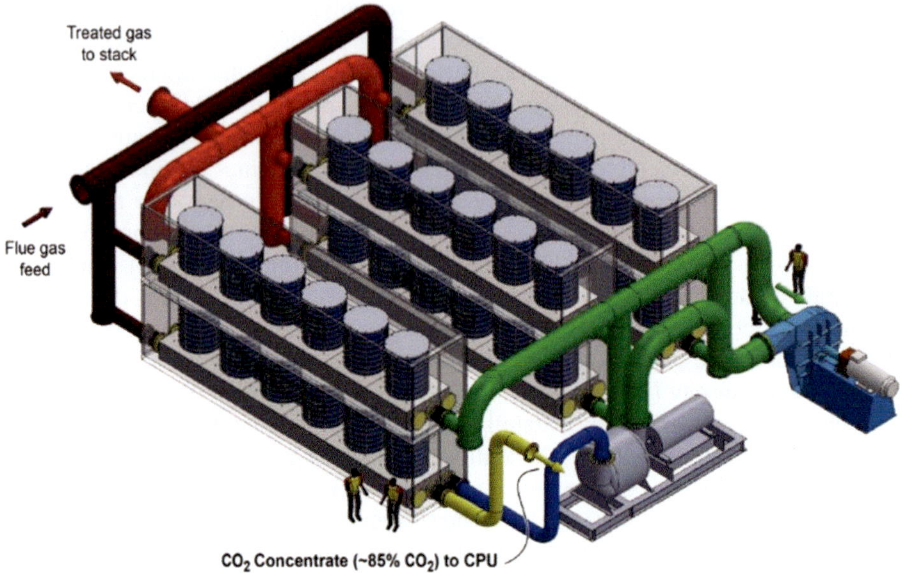

FIGURE 14.1 MTR Polaris membrane process, courtesy of Membrane Research and Technology.

The MTR Polaris membrane is one of the most advanced technologies, operating at a TRL of approximately 7. It is commissioning the world's largest membrane process, 150 t-CO_2/day for post-combustion by 2024, as shown in Figure 14.1. It uses a robust crosslinked polyethylene glycol material as the selective layer. Pilot tests indicate that it is resistant to contaminants and provides a sufficient separation factor. However, due to the low pressure of the flue gas feed, which only utilizes a blower for pressure boost, a plate-and-frame membrane packing is required. This packing has about 10 times less efficiency compared to state-of-the-art hollow fiber membranes used in large-scale gas separations, resulting in an exceptionally large footprint for the process. MTR and other developers are working on next-generation membranes with higher permeance to reduce the footprint.[27]

MTR invented a two-stage membrane process for flue gas CO_2 capture; the first stage recovers a portion of the CO_2 to produce a high-purity permeate. The second stage employs an air purge to recover the remaining CO_2, which is then used as furnace air feed. This configuration improves CO_2 recovery and increases the CO_2 concentration in the flue gas, making subsequent membrane separation more efficient.[27]

Ohio State University is also developing amine-facilitated transport (FTM) membranes for post-combustion CO_2 capture. Amine FTM uses mobile or fixed amine sites to selectively bond with and release CO_2. Polymeric FTMs typically possess excellent scalability. This feature allows for the facile fabrication of prototype membranes and membrane modules even in the early stage of material development.[52] OSU University is pioneering this technology mainly under DOE funding. Pilot scale modules have been fabricated and continuous operation demonstrated with high CO_2

purity.[52] However, FTM membranes require proper hydration during operation, which adds operational complexity and cost.

14.2.6.3 Oxygen Ion Transport Membrane

Oxygen ions are transported through a mixed-metal oxide lattice by jumping between oxygen vacancies in the lattice structure. The vacancies are in effect fixed carrier sites facilitating oxygen transport. Electrons are liberated at the permeate side as the ions recombine into oxygen molecules that are conducted back through the matrix to maintain electrical neutrality. A high electrical conductivity is therefore a desired characteristic for such a membrane.

The limited oxygen permeability and stability of ITM pose significant challenges to their use in carbon capture technology, membrane reactors, industrial oxygen production, and solid oxide fuel cells. To address these issues, it is essential to modify ITM to enhance their oxygen permeability performance and stability. Modification strategies have been used, including membrane surface modification, membrane composition modification, and membrane configuration design.[53]

ITM membrane scale-up has been pursued since 1999. Partly funded by the U.S. Department of Energy National Energy Technology Laboratory, Air Products and partners started a five tO_2/day prototype unit in 2005. The perovskite-based metal oxide ITM modules have a wafer stack configuration. Multiple modules are housed within the high-pressure membrane vessel. The plant delivers a >99% pure oxygen permeate stream and successfully demonstrated continuous operation. This technology reduces the energy cost of oxygen production by a third compared to cryogenic methods. An intermediate scale pilot capable of producing 100 tO_2/day has been constructed and tested but future development work, which aimed to scale capacity to more than 1000 tO_2/day, was terminated in 2015.[54] Future research is needed to reduce the cost of ITM membranes. Additionally, ITM technology is integral to solid oxide fuel cells and is one of three technologies for producing green hydrogen through electrolysis. The development of ITM materials is also closely related to chemical looping processes, making it important to closely monitor advancements in this technology.

14.3 TRENDS AND FORECAST

Three sectors are poised for explosive growth in the next few years for CCUS: energy and heat, blue hydrogen and ammonia, and DAC. The energy and heat sector is expected to become the largest for CCUS due to its vast market size. As the largest contributor to global CO_2 emissions, the energy and heat sector is increasingly focusing on carbon capture technologies to mitigate its environmental impact.

The blue hydrogen and ammonia sector is gaining significant momentum as technologies for pre-combustion CO_2 capture are steadily being proven at large scales. Blue hydrogen, produced from fossil fuels with CO_2 capture, serves as a bridge between high-emission grey hydrogen and zero-emission green hydrogen.

DAC is set to experience the most explosive growth even though none of the technologies have yet been fully proven at large scales. Governmental policy plays a crucial role in driving this sector forward, providing the necessary support and incentives for development.

In addition to these sectors, three technologies show significant theoretical potential but lack substantial advancement: ITM membrane for oxy-combustion, oxygen carrier chemical looping, and anti-sublimation for CO_2 solid separation. Further scale-up and long-term operation presents the major engineering challenges and the primary bottleneck.

There are three potential breakthrough technologies: membranes for post-combustion capture, non-conventional adsorption for post-combustion, and solid sorbents for DAC. The first two are likely to steadily advance toward TRL 7 and beyond in the coming years. Continuous development of adsorbents and processes will enhance the economic viability of DAC, attracting numerous stakeholders to the field. Overall, the advancements in these sectors and technologies highlight the dynamic nature of the CCUS field and its critical role in addressing climate change.

REFERENCES

1. IEA CCUS Projects Explorer. https://www.iea.org/data-and-statistics/data-tools/ccus-projects-explorer
2. Junior, A. N. et al. Assessing EOR strategies for application in Brazilian pre-salt reservoirs. *Geoenergy Science and Engineering* **223**, 211508 (2023).
3. Baena-Moreno, F. M., Saché, E. Le, Pastor-Pérez, L. & Reina, T. R. Membrane-based technologies for biogas upgrading: A review. *Environmental Chemistry Letters* **18**, 1649–1658 (2020).
4. U.S. Department of Energy. U.S. national clean hydrogen strategy and roadmap. https://www.hydrogen.energy.gov/docs/hydrogenprogramlibraries/pdfs/us-national-clean-hydrogen-strategy-roadmap.pdf
5. Ueckerdt, F. et al. On the cost competitiveness of blue and green hydrogen. *Joule* **8**, 104–128 (2024).
6. Tetteh, D. A. & Salehi, S. The blue hydrogen economy: A promising option for the near-to-mid-term energy transition. *Journal of Energy Resources Technology* **145**, (2022).
7. Wu, W., Zhai, H. & Holubnyak, E. Technological evolution of large-scale blue hydrogen production toward the U.S. Hydrogen Energy Earthshot. *Nature Communications* **15**, 5684 (2024).
8. Masson-Delmotte, V. et al. Climate change 2021: the physical science basis. *Working Group I Contribution to the Sixth Assessment Report of the Intergovernmental Panel on Climate Change* **2**, 2391 (2021).
9. 1PointFive Stratos project. https://www.1pointfive.com/projects/ector-county-tx
10. Sievert, K., Schmidt, T. S. & Steffen, B. Considering technology characteristics to project future costs of direct air capture. *Joule* **8**, 979–999 (2024).
11. Chowdhury, S., Kumar, Y., Shrivastava, S., Patel, S. K. & Sangwai, J. S. A review on the recent scientific and commercial progress on the direct air capture technology to manage atmospheric CO_2 concentrations and future perspectives. *Energy & Fuels* **37**, 10733–10757 (2023).
12. Dakota Gasification Company. https://www.dakotagas.com/about-us/index
13. Bahman, N., Al-Khalifa, M., Baharna, S. A., Abdulmohsen, Z. & Khan, E. Review of carbon capture and storage technologies in selected industries: Potentials and challenges. *Reviews in Environmental Science and Bio/Technology* **22**, 451–470 (2023).
14. Abanades, J. C. et al. Emerging CO_2 capture systems. *International Journal of Greenhouse Gas Control* **40**, 126–166 (2015).

15. Smith, K. H., Ashkanani, H. E., Morsi, B. I. & Siefert, N. S. Physical solvents and techno-economic analysis for pre-combustion CO_2 capture: A review. *International Journal of Greenhouse Gas Control* **118**, 103694 (2022).
16. Lummus Blue H2 Technologies. https://www.lummustechnology.com/sustainability/hydrogen-synthesis-gas-production/blue-hydrogen
17. 8 Rivers Oxy-combustion technology. https://8rivers.com/portfolio/the-allam-fetvedt-cycle/
18. Maxson, A., Jukkola, G., Myhre, R., Weitzel, P. D. & Marasigan, J. Oxy-combustion Roadmap to Commercialization. in *Proceedings of the 15th Greenhouse Gas Control Technologies Conference* 15–18 (2021).
19. Song, C., Liu, Q., Deng, S., Li, H. & Kitamura, Y. Cryogenic-based CO_2 capture technologies: State-of-the-art developments and current challenges. *Renewable and Sustainable Energy Reviews* **101**, 265–278 (2019).
20. Abuelgasim, S., Wang, W. & Abdalazeez, A. A brief review for chemical looping combustion as a promising CO_2 capture technology: Fundamentals and progress. *Science of The Total Environment* **764**, 142892 (2021).
21. Chen, J., Duan, L. & Sun, Z. Review on the Development of Sorbents for Calcium Looping. *Energy & Fuels* **34**, 7806–7836 (2020).
22. Carbon Engineering DAC. https://carbonengineering.com/our-technology/
23. Exxon Mobil Baytown CO2 Capture Project. https://www.honeywell.com/us/en/press/2023/02/exxonmobil-to-deploy-honeywell-carbon-capture-technology
24. Kawasaki Moving Bed Amine-impregnated Solid Sorbent. https://global.kawasaki.com/en/corp/rd/magazine/185/pdf/n185en04.pdf
25. Šulc, R. & Ditl, P. A technical and economic evaluation of two different oxygen sources for a small oxy-combustion unit. *Journal of Cleaner Production* **309**, 127427 (2021).
26. Fasihi, M., Efimova, O. & Breyer, C. Techno-economic assessment of CO2 direct air capture plants. *Journal of Cleaner Production* **224**, 957–980 (2019).
27. MTR Large Pilot Membrane Plant. https://mtrccs.com/
28. Shirasaki, Y. & Yasuda, I. Membrane reactor for hydrogen production from natural gas at the Tokyo Gas Company: A case study. in *Woodhead Publishing Series in Energy* vol. 2 487–507 (Woodhead Publishing, 2013).
29. Anderson, L. L. et al. Advances in ion transport membrane technology for oxygen and syngas production. *Proceedings of the 20th International Conference on Solid State Ionics SSI-20* **288**, 331–337 (2016).
30. Vega, F. et al. Current status of CO_2 chemical absorption research applied to CCS: Towards full deployment at industrial scale. *Applied Energy* **260**, 114313 (2020).
31. Shiravi, A., Maleh, M. S., Raisi, A. & Sillanpää, M. Hollow fiber membrane contactor for CO_2 capture: A review of recent progress on membrane materials, operational challenges, scale-up and economics. *Carbon Capture Science & Technology* **10**, 100160 (2024).
32. Petro Nova CCS Project. https://www.nex.jx-group.co.jp/english/newsrelease/upload_files/20230913EN.pdf
33. García-Luna, S., Ortiz, C., Carro, A., Chacartegui, R. & Pérez-Maqueda, L. A. Oxygen production routes assessment for oxy-fuel combustion. *Energy* **254**, 124303 (2022).
34. Gelles, T., Lawson, S., Rownaghi, A. A. & Rezaei, F. Recent advances in development of amine functionalized adsorbents for CO_2 capture. *Adsorption* **26**, 5–50 (2020).
35. Rodrigues, G., Dubettier, R., TRAN, M. & Cruz, B. Capturing the Benefits of Increasing CO_2 Concentration in a Flue Gas through Cryocap^{TM} FG. in vol. *Proceedings of the 16th Greenhouse Gas Control Technologies Conference 23–24* (2022).
36. Wang, S. et al. Adsorption behaviour of molecular sieve and activated carbon for CO_2 adsorption at cold temperatures. *Carbon Neutrality* **1**, 16 (2022).

37. Okumura, T., Yamaguchi, T., Nishibe, S., Numaguchi, R. & Kawajiri, Y. Modeling of moving bed CO2 capture process using amine solid adsorbent and its validation using pilot plant test data. *International Journal of Greenhouse Gas Control* **133**, 104094 (2024).
38. Park, Y. C. et al. Test operation results of the 10 MWe-scale dry-sorbent CO_2 capture process integrated with a real coal-fired power plant in Korea. *Energy Procedia* **63**, 2261–2265 (2014).
39. Kawasaki demo plant in Wyoming. https://global.kawasaki.com/news_231011-1e.pdf
40. Hovington, P. Rapid Cycle Temperature Swing Adsorption Process Using Solid Structured Sorbent for CO_2 capture from Cement Flue Gas. *Proceedings of the 16th Greenhouse Gas Control Technologies Conference 22–23* (2022).
41. Rezaei, S., Liu, A. & Hovington, P. Emerging technologies in post-combustion carbon dioxide capture & removal. *Catalysis Today* **423**, 114286 (2023).
42. Sciences, N. A. et al. Negative emissions technologies and reliable sequestration: A research agenda. *National Academies of Sciences* (2019).
43. Husk, J. C. & Wenz, G. B. Inside-out: Driving down direct air capture costs with high-efficiency adsorbents. *Frontiers in Climate* **3**, (2022).
44. ICIS: Dow's strategy on EDH, e-cracking, and blue H2, 2021. https://www.icis.com/explore/resources/news/2021/05/20/10642416/dow-advances-edh-e-cracking-to-lower-carbon-emissions-execs/
45. Wang, D., Joshi, A. & Fan, L.-S. Chemical looping technology – A manifestation of a novel fluidization and fluid-particle system for CO_2 capture and clean energy conversions. *Powder Technology* **409**, 117814 (2022).
46. Iulianelli, A. & Drioli, E. Membrane engineering: Latest advancements in gas separation and pre-treatment processes, petrochemical industry and refinery, and future perspectives in emerging applications. *Fuel Processing Technology* **206**, 106464 (2020).
47. Peters, T. et al. Palladium (Pd) membranes as key enabling technology for pre-combustion CO_2 capture and hydrogen production. *Energy Procedia* **114**, 37–45 (2017).
48. Jokar, S. M. et al. The recent areas of applicability of palladium based membrane technologies for hydrogen production from methane and natural gas: A review. *International Journal of Hydrogen Energy* **48**, 6451–6476 (2023).
49. Sun, S., Li, S., Wang, S. & Chen, Y. Design and development of highly selective and permeable membranes for H_2/CO_2 separation—A review. *Chemical Engineering Journal* **494**, 152972 (2024).
50. Prasetya, N., Himma, N. F., Sutrisna, P. D., Wenten, I. G. & Ladewig, B. P. A review on emerging organic-containing microporous material membranes for carbon capture and separation. *Chemical Engineering Journal* **391**, 123575 (2020).
51. Kárászová, M. et al. Post-combustion carbon capture by membrane separation, Review. *Separation and Purification Technology* **238**, 116448 (2020).
52. Han, Y. & Ho, W. S. W. Recent advances in polymeric facilitated transport membranes for carbon dioxide separation and hydrogen purification. *Journal of Polymer Science* **58**, 2435–2449 (2020).
53. Zhao, J., Pang, Y., Su, C., Jiang, S. & Ge, L. Toward high performance mixed ionic and electronic conducting perovskite-based oxygen permeable membranes: An overview of strategies and rationales. *Energy & Fuels* **37**, 7042–7061 (2023).
54. Rackley, S. Chapter 9 - Membrane CO_2 separation. in (eds. Rackley, S. et al.) 163–190 (Elsevier, 2023). doi:10.1016/b978-0-12-819663-2.00007-1

Index

Pages in *italics* refer to figures, pages in **bold** refer to tables, and pages followed by "n" refer to notes.

F

I

J

K

L

N

O

P

Q

R

S

U

V

W

X

Z

For Product Safety Concerns and Information please contact our
EU representative GPSR@taylorandfrancis.com Taylor & Francis
Verlag GmbH, Kaufingerstraße 24, 80331 München, Germany